Ultrafast Physical Processes in Semiconductors

SEMICONDUCTORS
AND SEMIMETALS
Volume 67

Semiconductors and Semimetals

A Treatise

Edited by R. K. Willardson
CONSULTING PHYSICIST
12722 EAST 23RD AVENUE
SPOKANE, WA 99216-0327

Eicke R. Weber
DEPARTMENT OF MATERIALS SCIENCE
AND MINERAL ENGINEERING
UNIVERSITY OF CALIFORNIA
AT BERKELEY
BERKELEY, CA 94720

Ultrafast Physical Processes in Semiconductors

SEMICONDUCTORS
AND SEMIMETALS

Volume 67

Volume Editor

K. T. TSEN

DEPARTMENT OF PHYSICS AND ASTRONOMY
ARIZONA STATE UNIVERSITY
TEMPE, ARIZONA

ACADEMIC PRESS
San Diego   San Francisco   New York   Boston
London   Sydney   Tokyo

This book is printed on acid-free paper.

COPYRIGHT © 2001 BY ACADEMIC PRESS

ALL RIGHTS RESERVED.

NO PART OF THIS PUBLICATION MAY BE REPRODUCED OR TRANSMITTED IN ANY FORM OR BY ANY MEANS, ELECTRONIC OR MECHANICAL, INCLUDING PHOTOCOPY, RECORDING, OR ANY INFORMATION STORAGE AND RETRIEVAL SYSTEM, WITHOUT PERMISSION IN WRITING FROM THE PUBLISHER.

Requests for permission to make copies of any part of the work should be mailed to: Permissions Department, Harcourt, Inc., 6277 Sea Harbor Drive, Orlando, Florida 32887-6777

The appearance of code at the bottom of the first page of a chapter in this book indicates the Publisher's consent that copies of the chapter may be made for personal or internal use of specific clients. This consent is given on the condition, however, that the copier pay the stated per-copy fee through the Copyright Clearance Center, Inc. (222 Rosewood Drive, Danvers, Massachusetts 01923), for copying beyond that permitted by Sections 107 or 108 of the U.S. Copyright Law. This consent does not extend to other kinds of copying, such as copying for general distribution, for advertising or promotional purposes, for creating new collective works, or for resale. Copy fees for pre-1999 chapters are as shown on the title pages; if no fee code appears on the title page, the copy fee is the same as for current chapters. 0080-8784/01 $35.00

ACADEMIC PRESS
A Harcourt Science & Technology Company
525 B Street, Suite 1900, San Diego, CA 92101-4495, USA
http://www.academicpress.com

ACADEMIC PRESS
Harcourt Place, 32 Jamestown Road, London NW1 7BY, UK
http://www.hbuk.co.uk/ap/

International Standard Book Number: 0-12-752176-3
International Standard Serial Number: 0080-8784

PRINTED IN THE UNITED STATES OF AMERICA
00 01 02 03 04 05 MB 9 8 7 6 5 4 3 2 1

Contents

PREFACE xi
LIST OF CONTRIBUTORS xiii

Chapter 1   Ultrafast Electron–Phonon Interactions in Semiconductors: Quantum Kinetic Memory Effects   1

Alfred Leitenstorfer and Alfred Laubereau

I. INTRODUCTION   1
II. GENERAL CONSIDERATIONS   4
   1. *Choice of Interaction Process and Material Systems*   4
   2. *Ultrafast Generation of Nonequilibrium Charge Carriers* 11
III. EXPERIMENTAL TECHNIQUE 13
   1. *The Two-Color Femtosecond Ti:Sapphire Laser Oscillator* 13
   2. *Ultrasensitive Pump-Probe Spectroscopy at the Shot-Noise Limit* . . 14
   3. *Optimum Conditions to Observe Electron–Phonon Dynamics* 18
IV. RESULTS AND DISCUSSION 25
   1. *The Weakly Coupled Case: GaAs* 25
   2. *Intermediate Coupling Regime: CdTe* 30
V. CONCLUSION 33
   REFERENCES 35

Chapter 2   Spatially and Temporally Resolved Near-Field Scanning Optical Microscopy Studies of Semiconductor Quantum Wires 39

Christoph Lienau and Thomas Elsaesser

I. INTRODUCTION 39
II. SEMICONDUCTOR NANOSTRUCTURES 41
   1. *General Aspects* 41
   2. *Semiconductor Quantum Wires* 43

III. NEAR-FIELD SCANNING OPTICAL MICROSCOPY 46
  1. Introduction 46
  2. Near-Field Optics 49
  3. Theoretical Description of Near-Field Optics of Semiconductor
     Nanostructures 53
  4. Near-Field Probes 56
  5. Probe-to-Sample Distance Control 58
  6. Low-Temperature Near-Field Microscopy 59
  7. Temporally and Spatially Resolved Near-Field Spectroscopy 61
IV. STATIONARY NEAR-FIELD SPECTROSCOPY OF SEMICONDUCTOR
    NANOSTRUCTURES 65
  1. Low-Temperature Near-Field Spectroscopy 65
  2. Near-Field Spectroscopy of Quantum Wires on Patterned (311)A GaAs
     Surfaces 66
V. TIME-RESOLVED NEAR-FIELD SPECTROSCOPY 79
  1. Lateral Carrier Transport Studied by Picosecond Near-Field Luminescence
     Spectroscopy 79
  2. Femtosecond Near-Field Pump-and-Probe Spectroscopy 89
VI. OUTLOOK AND CONCLUSIONS 98
REFERENCES 98

Chapter 3   Ultrafast Dynamics in Wide Bandgap Wurtzite GaN . . . 109

K. T. Tsen

I. INTRODUCTION 109
II. RAMAN SPECTROSCOPY IN SEMICONDUCTORS 110
  1. Theory of Raman Scattering from Carriers in Semiconductors 110
  2. Theory of Raman Scattering by Lattice Vibrations in Semiconductors . 122
III. SAMPLES, EXPERIMENTAL SETUP, AND APPROACH 125
IV. PHONON MODES IN THE WURTZITE STRUCTURE GaN 128
V. EXPERIMENTAL RESULTS 128
  1. Studies of Electron–Phonon Interactions in Wurtzite GaN by Nonlinear
     Laser Excitation Processes 128
  2. Nonequilibrium Electron Distributions and Electron–Longitudinal Optical
     Phonon Scattering Rates in Wurtzite GaN Studied by an Ultrashort,
     Ultraviolet Laser 132
  3. Anharmonic Decay of the Longitudinal Optical Phonons in Wurtzite GaN
     Studied by Subpicosecond Time-Resolved Raman Spectroscopy 139
VI. CONCLUSIONS AND FUTURE EXPERIMENTS 147
REFERENCES 148

Chapter 4   Ultrafast Dynamics and Phase Changes in Highly Excited GaAs 151

J. Paul Callan, Albert M.-T. Kim, Christopher A. D. Roeser, and Eric Mazur

I. INTRODUCTION 152
II. MEASURING ULTRAFAST PHENOMENA WITH LIGHT 154

1. Ultrafast Measurements: The Pump-Probe Technique 154
  2. Linear Optical Properties and the Dielectric Function 154
  3. Microscopic Theory of the Dielectric Function 157
  4. The Dielectric Function of Crystalline Solids: Interband Contributions 158
  5. The Dielectric Function of Crystalline Solids: Intraband Contributions 160
  6. Lattice Structure Effects on the Dielectric Function 162
  7. Electronic Configuration Effects on the Dielectric Function 163
  8. Why We Measure the Dielectric Function 165
III. DYNAMICS OF ELECTRONS AND ATOMS FROM FEMTOSECONDS
    TO MICROSECONDS 166
  1. Mechanisms of Carrier Excitation 167
  2. Carrier Redistribution, Thermalization, and Cooling 169
  3. Carrier–Lattice Thermalization 172
  4. Carrier Recombination 172
  5. Carrier Diffusion 174
  6. Structural Effects 175
  7. Summary 177
IV. MEASURING THE TIME-RESOLVED DIELECTRIC FUNCTION 179
  1. White-Light Probe Generation 180
  2. Measuring Transient Reflectivity Spectra 180
  3. Extracting the Transient Dielectric Function 181
V. TIME-RESOLVED DIELECTRIC FUNCTION OF HIGHLY EXCITED GaAs 182
  1. Low Fluence Regime ($F < 0.5$ kJ/m^2) 183
  2. Medium Fluence Regime ($F = 0.5$–0.8 kJ/m^2) 185
  3. High Fluence Regime ($F > 0.8$ kJ/m^2) 189
VI. CARRIER AND LATTICE DYNAMICS 191
  1. Carrier Dynamics: Excitation, Scattering, and Relaxation 192
  2. Structural Dynamics: Lattice Heating, Disordering, and Phase Transitions . . 197
VII. CONCLUSIONS 200
    REFERENCES 202

Chapter 5 Quantum Kinetics for Femtosecond Spectroscopy in Semiconductors. 205

Hartmut Haug

  I. INTRODUCTION 205
 II. SEMICONDUCTOR BLOCH EQUATIONS FOR PULSE EXCITATION WITH QUANTUM
     KINETIC SCATTERING INTEGRALS 209
III. LOW-EXCITATION FEMTOSECOND SPECTROSCOPY 215
  1. Femtosecond FWM with LO-Phonon Quantum Kinetic Scattering 217
  2. Femtosecond DTS with LO-Phonon Quantum Kinetic Scattering 219
 IV. FEMTOSECOND DTS FOR SCREENED COULOMB AND LO-PHONON QUANTUM
     KINETIC SCATTERING 220
  V. RESONANT FWM WITH SCREENED COULOMB AND LO-PHONON QUANTUM
     KINETIC SCATTERING 223
     REFERENCES 228

Chapter 6 Coulomb Correlation Signatures in the Excitonic Optical Nonlinearities of Semiconductors ... 231

T. Meier and S. W. Koch

   I. Introduction ... 231
  II. Theoretical Approach and Model ... 242
     1. The Coherent $\chi^{(3)}$ Limit ... 242
     2. Nonlinear Optical Signals ... 248
     3. One-Dimensional Model System ... 250
 III. Applications to Pump-Probe Spectroscopy ... 254
     1. Resonant Absorption Changes for Low Intensities ... 254
     2. Off-Resonant Absorption Changes for Low Intensities ... 266
     3. Higher Intensities up to the Coherent $\chi^{(5)}$ Limit ... 274
 IV. Absorption Changes Induced by Incoherent Occupations ... 280
  V. Applications to Four-Wave-Mixing Spectroscopy ... 290
     1. Biexcitonic Beats ... 290
     2. Disorder-Induced Dephasing ... 297
 VI. Summary ... 304
VII. Outlook ... 307
     References ... 309

Chapter 7 Electronic and Structural Response of Materials to Fast, Intense Laser Pulses ... 315

Roland E. Allen, Traian Dumitrică, and Ben Torralva

   I. Introduction ... 315
  II. Tight-Binding Electron–Ion Dynamics ... 317
 III. $\epsilon(\omega)$ and $\chi^{(2)}$ as Signatures of a Nonthermal Phase Transition ... 327
 IV. Detailed Information from Microscopic Simulations ... 336
     1. A Simple Picture ... 336
     2. Excited-State Tight-Binding Molecular Dynamics ... 339
     3. Detailed Model ... 343
     4. Electronic Excitation and Time-Dependent Band Structure ... 344
     5. Dielectric Function as a Signature of the Change in Bonding ... 346
     6. Second-Order Susceptibility as a Signature of the Change in Symmetry ... 348
     7. Summary of Results for GaAs ... 354
  V. Formula for the Second-Order Susceptibility $\chi^{(2)}$ ... 356
     1. Tight-Binding Hamiltonian in an Electromagnetic Field ... 357
     2. Second-Order Susceptibility in a Tight-Binding Representation ... 358
     3. Formula for the Dielectric Function ... 360
     4. Calculation of $\chi^{(2)}(\omega)$ for GaAs ... 360
 VI. Response of Si to Fast, Intense Pulses ... 363
VII. Density Functional Simulation for Si ... 370
VIII. Response of C_{60} to Ultrafast Pulses of Low, High, and Very High Intensity ... 371
 IX. The Simplest System, H_2^+ ... 377
  X. Biological Molecules ... 382
 XI. Conclusion ... 384
     References ... 386

Chapter 8 Coherent THz Emission in Semiconductors 389
E. Gornik and R. Kersting

- I. INTRODUCTION 390
- II. PRINCIPLES OF PULSED THz EMISSION AND DETECTION 392
  - 1. Dipole Emission 392
  - 2. Temporal and Spatial Coherence 394
  - 3. Detection of THz Pulses 395
- III. MACROSCOPIC CURRENTS AND POLARIZATIONS 400
  - 1. Drude–Lorentz Model of Carrier Transport 400
  - 2. The Instantaneous Polarization 403
  - 3. Phases of Transport and Polarization 404
  - 4. Emission from Field-Induced Transport 405
  - 5. Bandwidth Increase by Ultrafast Currents 409
  - 6. Emission from an Instantaneous Polarization 410
- IV. COHERENT CHARGE OSCILLATIONS 410
  - 1. Plasma Oscillations of Photoexcited Carriers 411
  - 2. Plasma Oscillations of Extrinsic Carriers 414
  - 3. Plasma Oscillations in 2D Layers 417
  - 4. Coherent Phonons and Plasmon–Phonon Coupling 418
- V. THz EMISSION FROM QUANTUM STRUCTURES 420
  - 1. Quantum Beats 420
  - 2. Bloch Oscillations 422
  - 3. Intersubband Transitions 426
- VI. APPLICATIONS IN SEMICONDUCTOR SPECTROSCOPY 427
  - 1. Optical Pump–THz Probe Spectroscopy 428
  - 2. THz Time-Domain Spectroscopy 429
- VII. CONCLUSIONS 433
  - REFERENCES 434

INDEX 441
CONTENTS OF VOLUMES IN THIS SERIES 449

Preface

This book consists of eight chapters covering new developments in the field of ultrafast dynamics in semiconductors. Chapter 1 reviews ultrafast electron–phonon interactions in semiconductors, with special emphasis on quantum kinetic memory effects. Chapter 2 deals with spatially and temporally resolved near-field scanning optical microscopy studies of semiconductor quantum wires. Chapter 3 is devoted to ultrafast dynamics in wide bandgap wurtzite GaN. Chapter 4 discusses ultrafast dynamics and phase changes in highly excited GaAs. Chapter 5 reviews quantum kinetics for femtosecond spectroscopy in semiconductors. Chapter 6 demonstrates Coulomb correlation signatures in the excitonic optical nonlinearities of semiconductors. Chapter 7 reviews the electronic and structural response of materials to intense laser pulses of ultrashort duration. Chapter 8 presents coherent THz emission in semiconductors.

There are many books in the market that are devoted to the review of certain fields. This volume is different from those in that the authors not only provide reviews of their field, but also present their own important contributions in a tutorial way. As a result, researchers who are already in the field of ultrafast dynamics in semiconductors and its device applications — as well as researchers and graduate students just entering the field — will benefit from it.

The editing of a book with eight different chapters written by authors in several countries is not an easy task. I would like to thank all authors for their patience and cooperation. I would also like to thank Dr. Zvi Ruder of Academic Press for help in many different aspects of this book. Last but not least, I wish to express my appreciation to my wife and children for their encouragement, understanding, and support.

K. T. TSEN

List of Contributors

Numbers in parentheses indicate the pages on which the authors' contribution begins.

ROLAND E. ALLEN (315), *Department of Physics, Texas A&M University, College Station, Texas*

J. PAUL CALLAN (151), *Department of Physics and Division of Engineering and Applied Sciences, Gordon McKay Laboratory, Harvard University, Cambridge, Massachusetts*

TRAIAN DUMITRICĂ (315), *Department of Physics, Texas A&M University, College Station, Texas*

THOMAS ELSAESSER (39), *Max-Born-Institut für Nichtlineare Optik und Kurzzeitspektroskopie, Berlin, Germany*

E. GORNIK (389), *Institut für Festkörperelektronik, Technische Universität Wien, Wien, Austria*

HARTMUT HAUG (205), *Institut für Theoretische Physik, J. W. Goethe Universität Frankfurt, Frankfurt, Germany*

R. KERSTING (389), *Department of Physics, Applied Physics, and Astronomy, Rensselaer Polytechnic Institute, Troy, New York*

ALBERT M.-T. KIM (151), *Department of Physics and Division of Engineering and Applied Sciences, Gordon McKay Laboratory, Harvard University, Cambridge, Massachusetts*

S. W. KOCH (231), *Department of Physics and Material Sciences Center, Philipps University, Marburg, Germany*

ALFRED LAUBEREAU (1), *Physik-Department E11, Technische Universität Munchen, Garching, Germany*

ALFRED LEITENSTORFER (1), *Physik-Department E11, Technische Universität Munchen, Garching, Germany*

CHRISTOPH LIENAU (39), *Max-Born-Institut für Nichtlineare Optik und Kurzzeitspektroskopie, Berlin, Germany*

ERIC MAZUR (151), *Department of Physics and Division of Engineering and Applied Sciences, Gordon McKay Laboratory, Harvard University, Cambridge, Massachusetts*

T. MEIER (231), *Department of Physics and Material Sciences Center, Philipps University, Marburg, Germany*

CHRISTOPHER A. D. ROESER (151), *Department of Physics and Division of Engineering and Applied Sciences, Gordon McKay Laboratory, Harvard University, Cambridge, Massachusetts*

BEN TORRALVA (315), *Department of Physics, Texas A&M University, College Station, Texas*

K. T. TSEN (109), *Department of Physics and Astronomy, Arizona State University, Tempe, Arizona*

CHAPTER 1

Ultrafast Electron–Phonon Interactions in Semiconductors: Quantum Kinetic Memory Effects

Alfred Leitenstorfer and Alfred Laubereau

PHYSIK-DEPARTMENT E11
TECHNISCHE UNIVERSITÄT MÜNCHEN
GARCHING, GERMANY

I. INTRODUCTION	1
II. GENERAL CONSIDERATIONS	4
1. *Choice of Interaction Process and Material Systems*	4
2. *Ultrafast Generation of Nonequilibrium Charge Carriers*	11
III. EXPERIMENTAL TECHNIQUE	13
1. *The Two-Color Femtosecond Ti:Sapphire Laser Oscillator*	13
2. *Ultrasensitive Pump-Probe Spectroscopy at the Shot-Noise Limit*	14
3. *Optimum Conditions to Observe Electron–Phonon Dynamics*	18
IV. RESULTS AND DISCUSSION	25
1. *The Weakly Coupled Case: GaAs*	25
2. *Intermediate Coupling Regime: CdTe*	30
V. CONCLUSION	33
REFERENCES	35

I. Introduction

Since the advent of quantum theory more than seven decades ago, the intermediate region between the macroscopic world, where the laws of classical physics are valid, and the microscopic regime, which has to be described by wave mechanics, has always received special attention. In many cases, the question about a transition from conditions dominated by quantum phenomena to a scenario described by classical equations of motion is asked in the context of spatial variables. As an example of fundamental importance for future solid-state electronics, the nanometer length scale of semiconductor devices is mentioned where the operation becomes influenced by the wave nature of the elementary excitations. In analogy, an ultrashort time scale does exist, where the charge carriers in a semiconductor show a dynamics that deviates strongly from the classical picture of interacting particles in a box. This so-called quantum kinetic regime is governed by wave phenomena such as interference and diffraction. The scope of the investigations presented in the present chapter is to give experimental proof of these phenomena and to map out the conditions under which they occur in the ultrafast dynamics of electronically excited semiconductors.

Within the past 15 years, investigations of the femtosecond dynamics of photogenerated nonthermal carriers in semiconductors have become an important branch of solid-state physics (Shah, 1999). The excitement of research is fueled both by the fundamental questions arising in this field and by its technological relevance. Until very recently, most of the experiments have been explained theoretically on the basis of the semiclassical Boltzmann equation (Elsaesser *et al.*, 1991; Young *et al.*, 1994; Langot *et al.*, 1996). In this context, the term "semiclassical" has the following meaning: The various mechanisms leading to the ultrafast carrier dynamics are calculated assuming that every single interaction event between elementary excitations strictly conserves energy and momentum. Collisions between quasiparticles are regarded as instantaneous, that is, they are pointlike in time without any internal dynamics. The scattering rates are determined via quantum mechanics, but within the framework of Fermi's Golden Rule, representing the quasistationary case. In addition, Pauli's exclusion principle for fermions has to be obeyed. The dynamics, however, is fully described by a distribution of classical particles without taking into account effects connected to the phase of the quantum mechanical wave functions.

In high-purity semiconductors carrier–carrier interactions via the Coulomb potential and carrier–phonon scattering provide the most relevant ultrafast relaxation processes. Influences of the interband polarization created, for example, by a coherent laser pulse enter in an extension of the Boltzmann model, the semiconductor Bloch equations (Leitenstorfer *et al.*, 1994).

In any scattering process, the assumptions of the semiclassical picture hold if the particle distributions vary slowly on the time scale of the collision itself. Under these conditions the individual interaction events do not interfere and become decoupled. The oscillation period of the energy quantum exchanged gives an estimate for the duration of a collision. In experiments with ultrahigh temporal resolution, this time interval may no longer be regarded as infinitely short. Consequently, typical wave mechanical features such as energy non-conserving transitions and memory effects might show up. In contrast to the semiclassical case, where the dynamics is determined by purely statistical, dissipative processes, this regime is dominated by quantum kinetics with the quantum mechanical phase playing a central role (Haug and Jauho, 1996).

In theory, these phenomena are studied using models for the carrier dynamics derived within either the technique of the Keldysh nonequilibrium

Green's functions (Lipavsky et al., 1986) or the density matrix formalism (Zimmermann, 1990). These non-Markovian equations contain memory kernels that allow for typical quantum effects on ultrafast time scales but drive the distributions toward the semiclassical limit as time proceeds.

It turns out that the emission of longitudinal optical (LO) phonons by energetic electrons in the Γ-valley of bulk GaAs is a good candidate for the investigation of quantum kinetic effects: This system has been studied in a large number of theoretical publications (Brunetti et al., 1989; Kuznetsov, 1991; Bányai et al., 1992; Zimmermann, 1992; Hartmann and Schäfer, 1992; Schilp et al., 1994; Heiner, 1996; Kenrow, 1997; Schönhammer and Wöhler, 1997; Schmenkel et al., 1998; Zimmermann et al., 1998). However, an experimental proof of the influences of quantum kinetics on the dynamics of free carrier distributions was not available before publication of the data presented here. It had even been doubted whether these effects could be accessible by measurements because of fundamental limits given by the uncertainty relation.

As a first phenomenon that has to be described on a level beyond the semiclassical Boltzmann–Bloch equations, ultrafast LO phonon quantum beats have been found in four wave mixing experiments investigating the decay of the interband polarization with ultrahigh time resolution (Bányai et al., 1995; Steinbach et al., 1996). The coherent control of these LO phonon quantum beats has been demonstrated with phase locked pulses (Wehner et al., 1998), yielding further insight into non-Markovian dynamics of excitons. In the regime of high carrier excitation densities, where Coulomb scattering among the electrons dominates the dynamics, the experimental results were shown to be well described by simulations based on quantum kinetic models (Camescasse et al., 1996; Vu et al., 1999). Even the buildup of Coulomb screening has been included in quantum kinetic calculations for the electron–electron interaction (Bányai et al., 1998b). As a first application of this model, strongly damped Rabi oscillations of free carrier transitions in highly excited semiconductors were simulated that had been found experimentally (Fürst et al., 1997b). Much progress has also been achieved by measurements investigating a field closely connected to quantum kinetics: studies on higher-order excitonic correlations (Kner et al., 1998; Bartels et al., 1998).

In this chapter, the first experiments demonstrating quantum kinetic effects in the relaxation of hot carrier distributions are reviewed (Fürst et al., 1997a). In comparison to other studies on related subjects, these measurements allow a relatively direct demonstration of non-Markovian dynamics and an intuitive access to memory effects in semiconductors. The paper is organized as follows: After the introduction in Section I, some general considerations concerning the observability of quantum kinetic effects in ultrafast experiments monitoring carrier distributions are introduced in Section II. A detailed insight into the highly sensitive experimental technique allowing the measurements is given in Section III. The principle of two-color femtosecond pump-probe spectroscopy with time and energy resolution at the uncertainty limit is outlined. We describe the experimental conditions necessary to study the LO phonon emission of highly energetic electrons as unobscured by other processes as currently possible. The main sets of experimental data are presented in Section IV and the underlying physics is discussed. As a first example, the case of weak electron–phonon scattering is studied in bulk GaAs. References to model calculations simulating the experiments are given. A detailed analysis of such simulations is found in the chapter contributed to this book by H. Haug (Chapter 5). In order to map out the influence of increasing electron–phonon coupling strength on the quantum kinetics, we have investigated the II–VI semiconductor CdTe. This material allows study of the intermediate regime where the interaction can no longer be regarded as a weak perturbation. On the other hand, the ionicity of CdTe is still small enough to prevent the formation of strongly localized polarons, as would be the case in typical insulators. Concluding remarks are given in Section V.

II. General Considerations

1. Choice of Interaction Process and Material Systems

Many ultrafast processes governing the dynamics in physics, chemistry, and molecular biology actually occur under conditions where non-Markovian effects are expected to play a significant role. Nevertheless, it turns out that in most cases, unambiguous experimental evidence of quantum kinetic phenomena is readily obtained only when selected systems with strongly limited degrees of freedom are investigated. Generally speaking, this fact originates from the uncertainty principle inherent in quantum phenomena. As a consequence, the experimentalist has to choose a well-defined model system where the non-Markovian effects are not masked, for example, by intrinsic broadening effects that would also occur in the case of semiclassical dynamics.

For several reasons, the polar-optical Fröhlich interaction of highly energetic electrons with LO phonons offers ideal conditions to study quantum kinetic phenomena (see Fig. 1): The electrons can scatter into a continuum of states with different wave vectors $\mathbf{k}$ and energies E_e. At the same time, their interaction partners, the LO phonons, exhibit a well-defined energy of $\hbar\omega_{LO}$ with negligible dispersion in the center of the Brillouin zone. Within the Γ-valley of the conduction band (cb) of direct-gap semiconductors with zincblende symmetry, polar-optical scattering is the dominant interaction mechanism with the crystal lattice. Collisions of electrons with optical phonons via the deformation potential are not allowed to first order

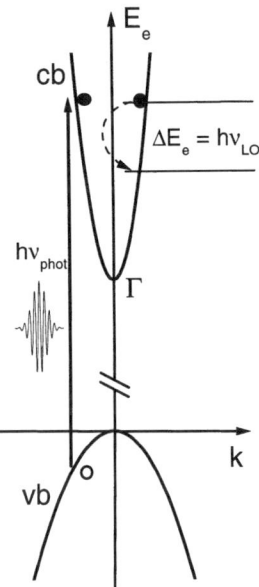

FIG. 1. Band structure excitation and relaxation scheme in GaAs and CdTe for the ultrafast pump-probe experiments on quantum kinetic electron–phonon scattering: Electrons (filled dots) are generated in the conduction band (cb) by an ultrashort laser pulse with central photon energy $h\nu_{phot} > E_{gap}$, leaving holes (open circles) in the valence band (vb). The photogenerated electrons relax via emission of longitudinal optical (LO) phonons of an energy $h\nu_{LO}$.

by the selection rules (Ridley, 1993) and scattering times with acoustic phonons are quite long, typically in the tens of picoseconds regime. As discussed in detail in Subsection 3b of Section III, Coulomb scattering among the electrons may be eliminated by working at low carrier concentrations.

As a result, if an electron distribution is generated in the Γ-valley, for example, by an ultrashort laser pulse with a central photon energy $\hbar\omega_{phot}$ larger than the fundamental gap of the semiconductor (see Fig. 1), the electrons will relax toward the band minimum almost exclusively via LO phonon emission. Phonon absorption and stimulated emission is suppressed by operating at low lattice temperatures T_L with $k_B T_L \ll \hbar\omega_{LO}$. Assuming a parabolic conduction band with effective mass m_e^*, one obtains the following relationship for the spontaneous phonon emission rate Γ_{e-LO} of electrons of an excess energy E_e above the band minimum (Ridley, 1993):

$$\Gamma_{e-LO} = \sqrt{\frac{m_e^*}{2E_e}} \frac{e^2 \omega_{LO}}{2\pi\hbar\varepsilon_0} \left(\frac{1}{\varepsilon_\infty} - \frac{1}{\varepsilon_s}\right) \times \text{Arsinh}\left(\sqrt{\frac{E_e}{\hbar\omega_{LO}} - 1}\right), \quad (1)$$

TABLE I

MATERIAL PARAMETERS USED FOR THE CALCULATION OF THE
SPONTANEOUS POLAR-OPTICAL SCATTERING RATE OF
ELECTRONS IN GaAs AND CdTe (Eq. (1) and Fig. 2)
(see also Madelung, 1996)

	GaAs	CdTe
m_e^*/m_0	0.067	0.090
$\hbar\omega_{LO}$ (meV)	36	21
ε_∞	10.9	7.1
ε_s	12.5	10.2

with the high-frequency and static limits for the dielectic function ε_∞ and ε_s. Equation (1) has been evaluated for electrons close to the minimum of the conduction band in GaAs and CdTe, adopting the material parameters given in Table I. The results are depicted in Fig. 2: At low lattice temperatures T_L, the polar-optical scattering rate Γ_{e-LO} is zero for electron energies below $\hbar\omega_{LO}$. The LO phonon energy in the zone center is 36 meV in GaAs and 21 meV in CdTe. For higher excess energies, the emission rate increases rapidly and saturates at a value that remains nearly constant. In the III–V semiconductor GaAs, this constant LO emission rate amounts to approximately 4 ps^{-1}, corresponding to a scattering time of 250 fs. For the II–VI compound CdTe, we calculate a significantly larger phonon emission rate on the order of 14 ps^{-1}, equivalent to a scattering time of 70 fs. For both materials, these values have been found to agree excellently with

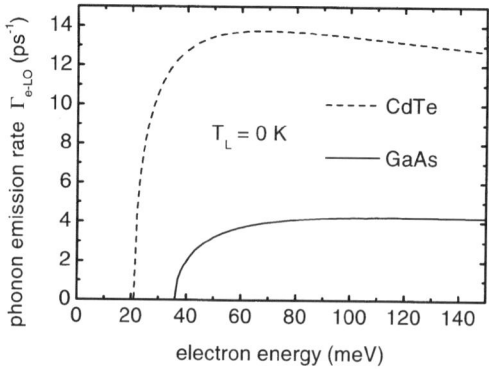

FIG. 2. Polar-optical LO phonon emission rate Γ_{e-LO} in the conduction band of GaAs and CdTe versus electron excess energy above the Γ-minimum, as given by Eq. (1). The data are calculated via Fermi's Golden Rule, assuming zero phonon occupation and a parabolic band structure. The material parameters used are collected in Table I.

electron–phonon collision rates determined via ultrafast nondegenerate four-wave-mixing measurements (Betz *et al.*, 1999). At large excess energies, Eq. (1) is expected to underestimate the scattering rates since the effective mass m_e^* increases because of the nonparabolicity of the conduction band. At the highest electron energies investigated in this chapter (135 meV in GaAs) the emission rate is enhanced by 25% with respect to the value calculated in a parabolic band. This deviation is not relevant for the discussion and simulation of the results presented here.

The band structure region close to the fundamental gap is very similar in GaAs and CdTe. The large discrepancy in polar-optical scattering rates is mainly a result of the different amount of ionicity in both materials. Conventionally, the strength of the Fröhlich interaction is characterized by the dimensionless polaron coupling constant α with

$$\alpha = \sqrt{\frac{m_e^*}{2\hbar\omega_{LO}}} \frac{e^2}{4\pi\hbar\varepsilon_0} \left(\frac{1}{\varepsilon_\infty} - \frac{1}{\varepsilon_s}\right). \tag{2}$$

Adopting the material parameters in Table I, one obtains $\alpha = 0.06$ for GaAs. This result places the electron–phonon interaction in the III–V compound in the category of weak coupling with $\alpha \ll 1$ where the scattering can be handled as a small perturbation. In contrast, $\alpha = 0.33$ is calculated for the more polar II–VI semiconductor CdTe. Consequently, this material represents an example for intermediate electron–phonon coupling where $\alpha \approx 1$. Under these conditions, higher-order electron–phonon correlations are expected to play a dominant role while a separation between electronic and lattice degrees of freedom is still meaningful. The strong coupling regime with $\alpha \gg 1$ will not be addressed in this contribution. In the latter case, the Born–Oppenheimer approximation breaks down. Localized polarons are formed consisting of electrons dressed with a strongly polarized phonon cloud. This small polaron picture is valid for many insulators such as SiO_2 and Al_2O_3.

As we shall see, the different regimes of electron–phonon coupling strength leave a characteristic fingerprint in the dynamics of the distribution function of nonequilibrium carriers, respectively. Therefore, we turn back to the model problem of a single electron excited with a well-defined excess energy E_e above the Γ-minimum (Fig. 1): In the semiclassical picture, the scattering rate between an initial and a final state is given by Fermi's Golden Rule enforcing strict energy conservation. As a result, an energetically sharp electron distribution will relax via LO emission, forming an exact replica of the initial distribution, which is energetically narrow at any time. However, the uncertainty principle tells us that strict energy conservation cannot be a good concept on short time scales that are determined by the typical energies involved in the scattering process. For the case of electron–phonon scattering, the phonon energy is the typical energy exchanged. These energies are linked to the temporal duration of the phonon oscillation cycle via Planck's constant. The LO phonon energies of 36 meV in GaAs and 21 meV in CdTe correspond to LO oscillation periods of 115 and 200 fs, respectively. By comparing these values to the inverse phonon emission rates, we recognize immediately that in both materials significant dynamics occurs on a time scale where a semiclassical description is expected to fail and quantum kinetic effects might show up.

There is, however, a qualitative difference between the two materials: Whereas the LO oscillation cycle of 115 fs in GaAs is shorter than the scattering time of approximately 250 fs, the opposite case is realized in CdTe where $v_{LO}^{-1} = 200$ fs and $\Gamma_{e-LO}^{-1} \approx 70$ fs. In order to achieve a feeling for the dynamics of the electron distributions under these quantum kinetic conditions, we will look at the results of calculations employing a simplified model that can be solved analytically. Based on the Tomonaga–Luttinger model, an expression for the electron–phonon quantum kinetics in a one-dimensional band structure with linear dispersion may be obtained (Meden et al., 1995). Although far from realistic, the analytic solution has the major advantage of being exact for arbitrary strength of electron–phonon coupling. Since the model does not possess a band edge, it is expected to give qualitatively reasonable results only as long as a limited interval of electron energies far from the band minimum is taken into account. Under the assumption that electrons with zero energy are injected into the empty system at time $t = 0$, the solution for the time evolution of the electron distribution consists of a single integral that can be computed easily (Meden et al., 1995). Calculated electron distributions for the material parameters of GaAs and CdTe at various times t after injection of the electron are depicted in Fig. 3. In each graph, a sharp δ spike at zero electron energy represents the fraction of electrons that have not yet undergone scattering. The relative height of these lines has been chosen to represent the exponential decay of the initial state due to phonon emission. Note that the absolute value of the excitation peak is reduced arbitrarily in order to match the ordinate scale for the dynamics of the scattered distribution. In GaAs (left column in Fig. 3), a broadened maximum appears after 50 fs centered approximately 36 meV below the excitation energy at 0 meV. This feature consists of electrons having emitted one LO phonon. Although we have prepared a narrow initial distribution and the LO phonons possess no dispersion, the energetic width of the so-called phonon replica is comparable to the phonon energy itself. We recall that in a semiclassical description, a sharp phonon satellite would be expected at any time. The initial broadening is a direct consequence of the uncertainty principle: On a very early time scale, scattering events are allowed that actually violate strict energy conservation. Metaphorically speaking, the system does not yet know the phonon energy. The first phonon replica has narrowed significantly when a time of 100 fs, comparable to the LO oscillation cycle of 115 fs, has elapsed after injection. This phenomenon is a pure quantum kinetic feature: As time proceeds, destructive quantum

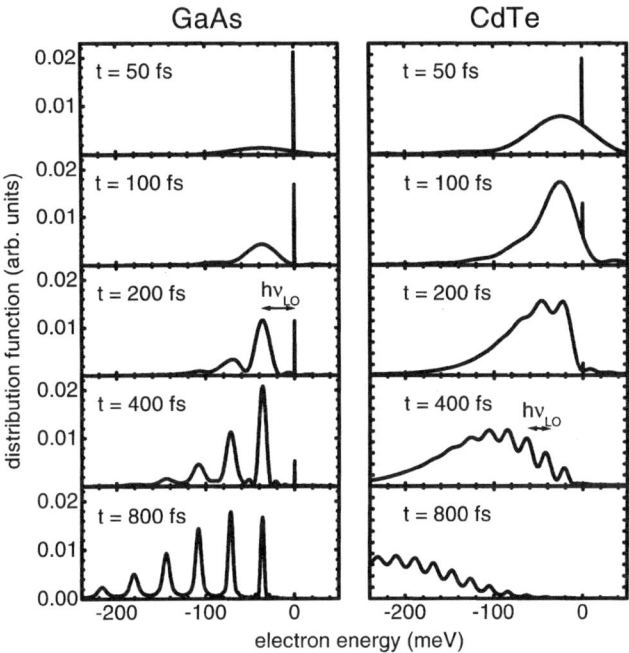

FIG. 3. Temporal evolution of the distribution function of electrons generated with zero energy at time $t = 0$ fs, emitting LO phonons. The results are based on an analytical solution for the quantum kinetic electron dynamics in a one-dimensional band structure with linear dispersion (Meden et al., 1995). For GaAs (left) and CdTe (right), phonon emission rates of 4 and 14 ps^{-1} are assumed, respectively. The decay of the initial state is indicated by the height of the sharp peak at $E_e = 0$ meV, which is arbitrarily scaled relative to the scattered distribution.

interference is established in the energetic wings of the replica. In this region, the wave functions of the electrons scattered off resonance run out of phase with components transferred at later times, leading to a reduction of the total probability amplitude. In contrast, constructive interference leads to an increasing value of the distribution function in the center of the phonon satellite. The de Broglie waves of the electrons and their driving forces, the zero-point oscillations of the phonon bath, are at resonance and remain in phase over a longer period of time. Interestingly, analogous phenomena also occur at the buildup of the second phonon replica consisting of electrons that have emitted two LO phonons. At a time delay of 100 fs, the second satellite is still energetically very broad. After 200 fs, there is a minimum between the first and the second LO phonon replica. A nice cascade of sharpened LO phonon peaks, as expected for the long time limit, is calculated in GaAs for $t = 400$ fs. This finding is the most direct manifestation of quantum kinetic memory effects, mediated by quantum interference. Despite the transient energetic broadening in the distributions, the system has "remembered" its history beginning as an energetically sharp feature. After 800 fs, the peak value of the first satellite starts to decrease and the second maximum becomes the most pronounced peak as energy relaxation proceeds.

A substantially different picture is encountered if we adopt the CdTe material parameters (right column in Fig. 3). A very broad maximum made up of scattered electrons is seen after 50 fs. Since the scattering time is as short as 70 fs, a substantial portion of the electrons have already interacted with a phonon at this point. As in GaAs, the distribution narrows with time, but the peak starts to develop some substructure only as late as 200 fs, that is, after approximately three phonons per electron have been emitted on average. Some weak maxima corresponding to the phonon satellites become visible after 400 fs. However, the distribution still consists more or less of a coherent feature that is completely smeared out on its low-energy tail. The maximum of the distribution relaxes to lower energies with a constant loss of average energy per time interval. Two points are fundamentally different from the GaAs case described previously: (1) Because of the fast scattering time (70 fs) as compared to the phonon cycle (200 fs), the system does not evolve toward a semiclassical limit, where energetically sharp phonon replicas are obtained. Quantum kinetics completely dominates the relaxation. (2) At early delay times below 100 fs, a substantial portion of electrons are even scattered to positive energies with respect to the excitation energy, although the phonon modes are not occupied at the low temperature assumed in the simulation. This phenomenon corresponds to a *virtual absorption* of LO phonons that is possible only on an extremely fast time scale. It also leads to a nonexponential relaxation of the electron occupation at the excitation energy ($E_e = 0$ meV): Even if the unscattered distribution decays exponentially, it cannot be distinguished from a contribution of electrons having scattered virtually with negligible energy exchange (compare Bányai et al., 1998a). In principle, this phenomenon does also occur in GaAs, but because of the weaker coupling only few electrons will undergo an interaction event within the relevant time window.

As a conclusion, we have seen that GaAs is an almost ideal model system to unambiguously identify quantum kinetic effects in nonequilibrium carrier relaxation. In less strongly coupled materials, no significant dynamics beyond the semiclassical description is indicated. In more polar materials with emission time faster than the phonon cycle, the distribution functions are expected to remain featureless even in the long time limit. As a representative of this intermediate coupling regime, we study the situation in CdTe. As discussed in the next subsection, even for favorable conditions such as in GaAs, the non-Markovian memory effects may be observable only after finding out the correct balance between time resolution and spectral selectivity for the carrier generation.

2. ULTRAFAST GENERATION OF NONEQUILIBRIUM CHARGE CARRIERS

In the calculations leading to the results in Fig. 3, we assumed an initial state that may hardly be prepared in a realistic experiment. The electron distribution is generated instantaneously at $t = 0$ and exhibits an energetic width that is arbitrarily small. It is clear that such a situation may not be achieved in nature because of time–energy uncertainty. This fact is best understood discussing the details of the carrier excitation process as it occurs in the measurements to be presented in Section IV. Presently, the best experimental clue to nonequilibrium carrier dynamics in semiconductors is provided by spectroscopy with ultrafast laser pulses. In the studies discussed in this chapter, electrons are excited inducing an interband optical transition with a femtosecond laser pulse. This process is indicated schematically by the vertical arrow in Fig. 1. In order to achieve the time resolution needed to monitor the ultrafast dynamics, one has to work with a pulse width for excitation that is comparable or even shorter than the inverse scattering rates involved. A typical example for a second-order field autocorrelation trace used to characterize the duration of femtosecond pulses is depicted in Fig. 4a. Assuming a Gaussian profile of the temporal envelope of the pulse, we deduce a full width at half maximum (FWHM) of the intensity of 107 fs.

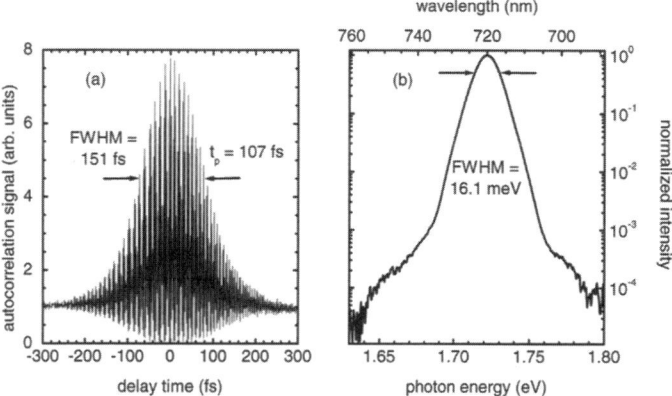

FIG. 4. Second-order electric field autocorrelation function (a) and intensity spectrum (b) of typical femtosecond light pulses used for the excitation of nonequilibrium photocarriers in semiconductors. Assuming a Gaussian intensity profile, the deduced pulse duration is 107 fs. The spectral bandwidth is determined to be 16.1 meV. Both values correspond to the full width at half maximum (FWHM) of the intensity. From these measurements, a time–bandwidth product of $\Delta v \cdot \Delta t = 0.42$ is obtained, in close agreement with the theoretical value of $\Delta v \cdot \Delta t = 0.441$ expected for a perfectly coherent Gaussian wave packet. Such transform-limited pulses are essential for the preparation of a nonequilibrium electron distribution with a minimum uncertainty product.

In the case of a fully coherent wave packet, characterized by a flat phase relationship, Fourier transform into the frequency domain predicts a spectral linewidth (FWHM) of these pulses of 17 meV (Sala *et al.*, 1980). As seen in Fig. 4b, the measured pulse spectrum exhibits a full width at half maximum of 16.1 meV. The experimental situation is close to the ideal case of a perfectly coherent Gaussian allowing measurements at the uncertainty limit.

The laser pulses characterized in Fig. 4 were carefully optimized: The pulse duration of 107 fs allows a temporal resolution adequate to the phonon emission times in both GaAs ($\Gamma_{e-LO} \approx 250$ fs) and CdTe ($\Gamma_{e-LO} \approx 70$ fs). Equally important, the energetic uncertainty brought in by the excitation bandwidth of 16 meV is still smaller than the LO phonon energy in both compounds of 36 meV and 21 meV, respectively. This balancing of temporal and energetic selectivity in the carrier generation process has turned out to be crucial for the observation of quantum kinetic electron–phonon relaxation. If the pulses are too short, the excitation bandwidth is too large and the spectral features due to memory effects are masked. In the case of long pulses allowing energetically sharp carrier generation, the time resolution may be insufficient to produce the temporal information desired.

Similar to the quantum kinetic effects during electron–phonon scattering, the carrier generation process by an ultrashort coherent light pulse also possesses an internal dynamic (Leitenstorfer *et al.*, 1994). During the initial part of the excitation pulse, interband polarization is excited in a frequency interval that is broader than the bandwidth of the pump light. As time proceeds, the polarization components created off resonance run out of phase with the driving light field and recombine via stimulated emission. In the presence of interband dephasing, some electron–hole pairs in the wings of the distribution lose their phase relationship with the pump field. They cannot be pumped down any more, leading to a broadened generation as compared to the case without phase relaxation. Toward the end of the pump pulse, carrier generation remains operative preferentially for the transitions resonant with the pump frequency. Finally, these phenomena lead to a photogenerated distribution that is expected from the energy spectrum of the optical pump convoluted with the homogeneous linewidth of the interband transitions of the semiconductor.

Although conceptually related to the non-Markovian phenomena predicted for electron–phonon interaction, the coherent dynamics due to the excitation pulse is not a quantum kinetic effect: Whereas the carrier generation can be described taking into account a classical light field as a driving force, the subsequent phonon emission processes occur via interaction with an incoherent bath of quantized lattice vibrations, and zero-point oscillations are playing a major role. Nevertheless, the initial energy broadening during femtosecond excitation has to be kept in mind when interpreting experimental data. We wish to mention that these effects have been fully included in the model calculations presented in Chapter 5 of this volume. It turns out that the coherent dynamics due to the pump plays a role only at very early delay times in the experiment (usually only a fraction of the phonon emission time). Especially the features connected to electrons that have emitted more than one LO phonon are dominated by quantum kinetics alone with negligible influence of the coherent excitation.

III. Experimental Technique

1. THE TWO-COLOR FEMTOSECOND Ti:SAPPHIRE LASER OSCILLATOR

As discussed in the previous section, bandwidth-limited pump pulses of a duration of typically 100 fs are well suited for the excitation of the nonthermal electron distributions to be studied in this chapter. In order to monitor the temporal relaxation of the energy distributions via transient transmission spectroscopy, a second pulse train of broadband ($\Delta E_{phot} > 100$ meV) and extremely short ($t_p < 25$ fs) pulses is required that is exactly synchronized to the pump. In addition, the stability and repetition rate of the laser pulses has to allow for measurements at low carrier concentrations where an extremely high sensitivity is needed.

In order to meet the requirements, a special laser source was developed, the two-color femtosecond Ti:sapphire laser (Leitenstorfer et al., 1995). A schematic view of the laser setup may be seen in Fig. 5. In principle, the laser system consists of two mode-locked Ti:sapphire oscillators that are both pumped by a Nd:vanadate laser with intracavity frequency doubling. A beam splitter (BS) divides the output of this laser to pump the two branches

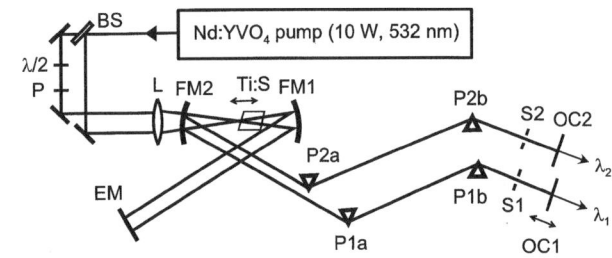

FIG. 5. Setup of the two-color femtosecond Ti:sapphire laser. BS, beam splitter; $\lambda/2$, half-wave plate; P, polarizer; L, pump lens; FM1, FM2, folding mirrors ($R = 10$ cm); Ti:S, gain crystal; EM, end mirror (flat); P1a–P2b, Brewster prisms (fused silica); S1, S2, tuning slits; OC1, OC2, output couplers.

of the femtosecond oscillator separately. A combination of a half-wave plate ($\lambda/2$) and a polarizer (P) allows independent adjustment of the pump power. The pump beams are focused by an incoupling lens ($f = 10$ cm), pass through the folding mirror FM2, and cross inside the gain crystal (Ti:S). The two branches of the Ti:sapphire oscillator are parallel at the flat end mirror (EM) with a distance of approximately 4 mm. They are both focused by folding mirror FM1, become amplified in the gain rod Ti:S, and are recollimated by FM2. After the second folding mirror, the two branches are again nearly parallel. Two separate prism sequences (P1a–P2b) compensate for the positive group velocity dispersion of the gain crystal and are used to adjust the pulse duration in each branch. Mechanical slits (S1 and S2) are employed to tune the central frequency of the femtosecond pulses independently. Partially reflecting flat mirrors serve as output couplers (OC1 and OC2).

When the cavity roundtrip times of the lasers are matched by precise positioning of output coupler OC1, which is mounted on a translation stage, the two pulse trains become synchronized with an accuracy better than 2 fs (Leitenstorfer *et al.*, 1995). This phenomenon has been explained by an interplay between the cross-phase modulation of the pulses when they copropagate in the gain crystal and the negative group velocity dispersion in both cavities (Fürst *et al.*, 1996). For a detailed discussion of the coupling mechanism and the operational principle of the two-color femtosecond laser, the reader is referred to this publication. A summary of the performance data of the laser system may be found in Table II.

2. ULTRASENSITIVE PUMP-PROBE SPECTROSCOPY AT THE SHOT-NOISE LIMIT

a. Optical Setup

Transient transmission spectroscopy is an ideal tool for ultrafast measurements close to the uncertainty limit: A test pulse can be chosen substantially shorter than the pump pulse, and spectral selection giving the energy information on the excited electron distribution is performed after the sample. In this way, the additional time–energy uncertainty inferred by the probing process is kept small and the time–bandwidth product of the measurement is dominated by the excitation process. It must be stressed that this property is a major advantage of the pump-probe scheme applied here in comparison to other ultrafast techniques such as luminescence upconversion or two-photon photoemission.

The experimental setup for the transient transmission experiments is depicted schematically in Fig. 6. The source of the ultrashort light pulses for excitation and probing is the two-color femtosecond Ti:sapphire laser

TABLE II

OPERATING PARAMETERS OF THE TWO-COLOR FEMTOSECOND Ti:SAPPHIRE LASER

Pulse duration: Independently adjustable between 14 fs and 120 fs
Average output power per branch: 100–500 mW
Maximum pulse energy: 7 nJ
Peak power: 400 kW
Maximum pulse bandwidth: 35 THz; 145 meV
Tuning range central frequency: 705–860 nm; 1.76–1.44 eV
Pulse repetition rate: 76 MHz
Synchronization: Better than 2 fs

oscillator discussed previously. Free electron–hole pairs far from equilibrium are generated in an optically thin semiconductor sample (S) with transform-limited light pulses of branch 2 of the laser (λ_2). The carrier dynamics is probed with broadband pulses of a duration of 25 fs, corresponding to a spectral width of 70 meV from branch 1 of the two-color oscillator (λ_1). The probe pulse generates a pair density that is approximately a factor of 5 smaller than that of the pump. The spectra of the light pulses employed for pumping and probing are depicted in Fig. 7 for the case of GaAs. Temporal resolution is obtained by passing the test pulses over a variable delay line (VD). The beam diameters of pump and probe at the sample are 1 mm and 0.5 mm, respectively.

MBE-grown films of high-purity GaAs (thickness $d = 500$ nm) and CdTe ($d = 300$ nm) are studied as samples (S). The epilayers are antireflection

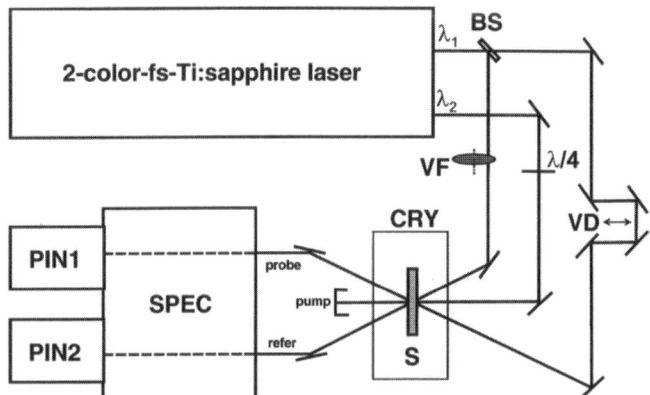

FIG. 6. Scheme of the time and energy resolved transient transmission experiment. BS, beam splitter; VF, variable filter; $\lambda/4$, quarter-wave plate; VD, variable optical delay; CRY, closed-cycle He-cryostat; S, semiconductor sample; SPEC, double monochromator; PIN1, PIN2, Si photodiodes.

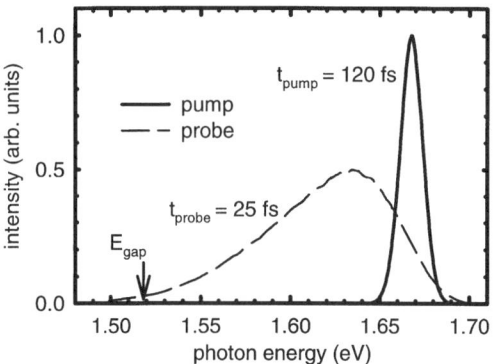

FIG. 7. Spectra of the synchronized pump (solid line) and probe (dashed line) pulses employed in the femtosecond experiments on quantum kinetic electron–phonon scattering in GaAs. The bandgap energy of GaAs is marked with an arrow.

coated on both sides and glued to a transparent fused silica substrate. The GaAs substrate required for the crystal growth of both compounds is removed by selective etching for the transmission experiments. The sample is held at a lattice temperature of $T_L = 15$ K inside a closed-cycle helium cryostat (CRY).

The probe pulses are analyzed in a double monochromator (SPEC) after having passed through the sample. In the experiments presented here, the spectral resolution is set to 6 meV. Working with only a single probe beam, the sensitivity to detect pump-induced transmission changes is limited by the excess noise of the laser source. Therefore, a reference signal is derived from the probe beam by a beam splitter BS; see Fig. 6. The reference pulses pass the sample at a negative time delay, approximately 0.5 ns before the excitation, and are sent through the spectrometer on a slightly different path. At the exit slit of the monochromator, probe and reference are detected with separate Si photodiodes (PIN1 and PIN2).

b. *Electronic Amplification Scheme*

In order to measure the pump-induced transmission changes in the semiconductor samples, we work with the differential detection system depicted schematically in Fig. 8. After detection of the probe and reference photons by the photodiodes PIN1 and PIN2, the photocurrent of the two channels is measured as a voltage drop over a load resistor ($R = 200$ kΩ), amplified and the background is separated from the signal amplitude by active high-pass filters (HP,PA). The prefiltered signals from both probe and reference beams are subtracted from each other in a differential amplifier (DA). In this way, the amplitude fluctuations of the laser system are

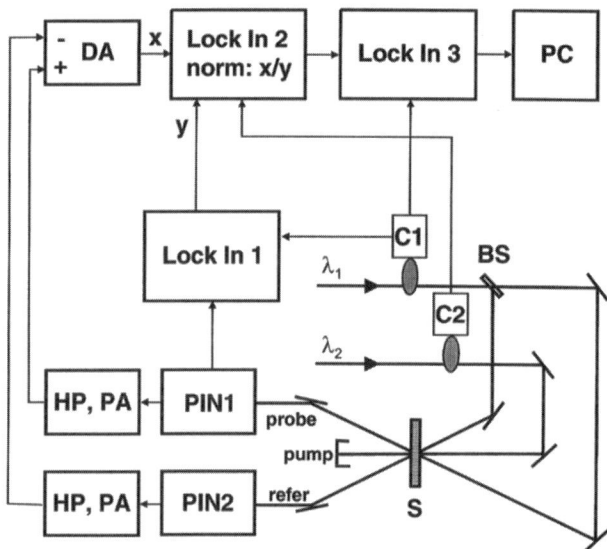

FIG. 8. Electronic detection scheme of the pump-induced transmission changes. C1, optical chopper ($f = 12$ Hz); C2, optical chopper ($f = 10$ kHz); BS, beam splitter; S, semiconductor sample (500 nm GaAs or 300 nm CdTe on fused silica substrate, ar-coated); PIN1, PIN2, Si photodiodes; HP, PA, active high-pass filter and amplifier; DA, differential amplifier; Lock In 1 and 3, analog lock-in amplifiers; Lock In 2, digital lock-in amplifier and normalization; PC, personal computer.

eliminated. Only the small pump-induced probe signal survives, since it is superimposed exclusively on the probe beam, not on the reference. If the probe and reference intensities are properly equalized by a variable filter (VF; see Fig. 6), the sensitivity of our setup is limited only by the combined shot noise of the photon current of the probe and reference pulses. At a typical probing power of $10 \mu W$, the relative noise level is as small as 4×10^{-7} $Hz^{-1/2}$. The pump-induced transmission changes $\Delta T/T$ are measured with a digital lock-in amplifier system (Lock In 2), which is synchronized to the chopper blade C2 in the pump beam operating at 10 kHz. The probe and reference beams are modulated at a frequency of 12 Hz by optical chopper C1. The digital lock-in amplifier 2 normalizes the detected signal to the momentary probing intensity detected with lock-in 1. In a final step, the scattered pump light from the sample surface is subtracted from the output of lock-in 2 by a third lock-in amplifier (Lock In 3) that is synchronized to chopper C1. The resulting data are directly proportional to the absolute transmission change induced by the pump pulse measured at a certain time delay and energy position. The values are read out by a personal computer (PC).

3. OPTIMUM CONDITIONS TO OBSERVE ELECTRON–PHONON DYNAMICS

This subsection presents important issues that have to be realized in the experiments in order to obtain signals that can be interpreted as directly as possible. The discussion is made up along the results in GaAs. Because of the identical crystal symmetry and similar band structures, the conditions in CdTe are analogous, as has been checked experimentally.

a. Reduction of Light-Hole Contributions via Interband Selection Rules

In all bulk semiconductors with zincblende symmetry, the heavy and light hole bands are energetically degenerate in the center of the Brillouin zone. Because of the spin–orbit splitting, the maximum of the split-off band is separated from the top of the valence band by 0.34 eV in GaAs and by 0.8 eV in CdTe (Madelung, 1996). For the typical photon energies for carrier excitation employed in the present experiments, transitions from both heavy and light hole bands into the conduction band are possible. The presence of the split-off band is negligible for the interpretation of the data. To simplify the situation even further, we have developed special polarization schemes for pumping and probing in order to suppress influences of the light-hole band, as discussed in the following.

The energy resolved transmission changes for a pump-probe delay of $t_D = 40$ fs are depicted for various polarization geometries in Fig. 9. The delay time is chosen such that the probe pulse passes the sample approximately at the end of the pump pulse. If the pump and probe beams are linearly polarized with the electric field vectors aligned perpendicular to each other, the transient transmission spectra reflect the entire electron distribution excited from both the light and heavy hole bands (Fig. 9a). The most pronounced positive maximum, corresponding to an increased transmission, is found close to the pump photon energy (hh/hh). It is connected to the electrons and heavy holes created in the strong transition between the hh and cb bands, bleaching the same transition because of phase space filling effects. The electrons generated out of the light-hole band but blocking absorption out of the heavy-hole band lead to a smaller maximum at lower photon energies (lh/hh). A third feature (hh/lh) is observed blue-shifted with respect to the pump spectrum: It originates from the electrons created by the hh transition that are probed via the lh band. This maximum is less pronounced and more strongly broadened than the lh/hh peak, since the lh band exhibits a smaller density of states and a higher dispersion than the hh band. Surprisingly, the hh/hh maximum is red-shifted from the excitation spectrum by approximately 10 meV and an induced absorption appears at the high-energy side. This phenomenon is connected

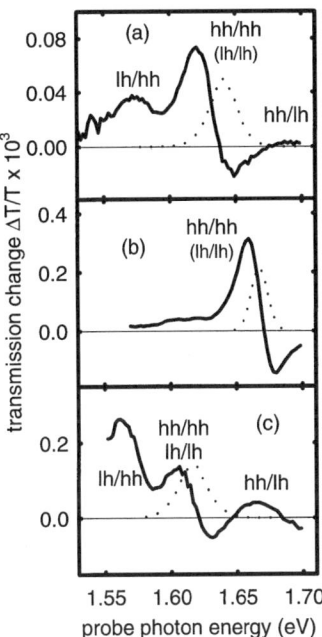

FIG. 9. Spectrally resolved transmission changes in GaAs at a pump-probe delay of $t_D = 40$ fs and at a lattice temperature of 15 K demonstrating the influences of interband selection rules depending on different polarization geometries: (a) cross-linearly, (b) cocircularly, and (c) countercircularly polarized pump and probe pulses. The spectrum of each excitation pulse is indicated by a dotted line.

to the fact that optical absorption in a direct semiconductor is strongly influenced by the Coulomb attraction between the photogenerated electrons and holes. In addition to the transmission changes due to the Pauli blocking of occupied electron and hole states, the carriers also modify the Coulomb enhancement factor of the continuum absorption, resulting in the dispersive-looking feature around the excitation energy (Foing *et al.*, 1992; El Sayed and Stanton, 1997). The Coulomb effects in the transmission changes due to photoexcited electron distributions have been readily included in simulations of the pump-probe measurements presented in this chapter (Fürst *et al.*, 1997a; Schmenkel *et al.*, 1998; Zimmermann *et al.*, 1998), and quantitative agreement with the experiment is found.

Spin selectivity may be exploited to suppress disturbing influences of the light-hole (lh) band: If both pump and probe pulses are circularly polarized with a photon spin of $+1$ (σ^+), we measure the transient spectrum depicted in Fig. 9b. The satellite peaks lh/hh and hh/lh seen in Fig. 9a have now vanished almost completely. In GaAs and CdTe, the hh band belongs to

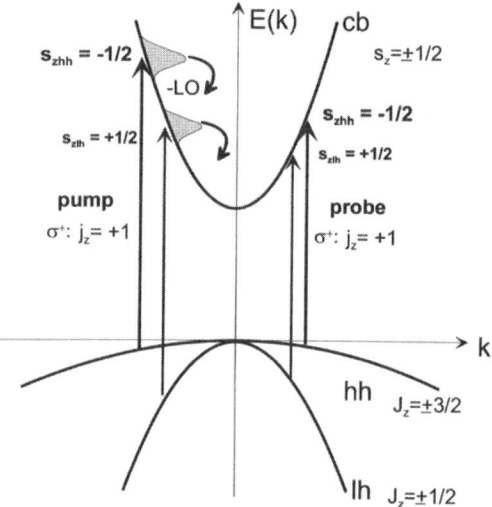

FIG. 10. Band structure scheme close to the fundamental gap in GaAs and CdTe, including conduction (cb), heavy-hole (hh), and light-hole (lh) bands. Direct dipole transitions relevant for the experiments are marked with vertical arrows. The spin alignment is indicated for the situation of cocircularly polarized pump and probe pulses.

projections of the total angular momentum of $j_z = \pm 3/2$, the lh band to $j_z = \pm 1/2$, and the conduction band to $s_z = \pm 1/2$ (see Meier and Zakarchenya, 1984). Exciting with σ^+, we generate spin-polarized electrons with opposite projections of s_z via the transitions out of the two valence bands (see Fig. 10). The spin alignment of the electrons persists on time scales substantially longer than the range of the present study (Heberle et al., 1994). Figure 10 demonstrates that if cocircular excitation is used for pumping and probing, the spin systems become decoupled and the electrons generated involving the hh (lh) band will bleach only the transition out of the hh (lh) band, respectively. In bulk semiconductors, this picture is strictly valid only for carriers generated with a wave vector parallel to the beam axis (Meier and Zakarchenya, 1984). However, the strong anisotropy of the photogenerated distributions further reduces the influences of the cross bleaching peaks involving different transitions for pumping and probing. Because of the small density of states and strong dispersion of the lh band, the absorption changes seen in cocircular geometry essentially originate from the carriers generated involving the hh band. Therefore, this situation is adapted in our measurements in order to keep the degree of complexity in the spectra as low as possible.

To demonstrate the opposite effect of countercircularly polarized excitation and test pulses, we show the transient spectrum obtained in this geometry in Fig. 9c. The cross bleaching peaks lh/hh and hh/lh are now strongly enhanced with respect to the direct bleaching maximum hh/hh, as expected. However, we find that hh/hh can never be eliminated completely because of the modified spin selection rules for off-axis carrier generation and possibly also because of the fast spin relaxation of the holes.

b. Suppression of Carrier–Carrier Scattering

Besides the reduction of disturbing influences of the lh-band, a key issue for observing electron–phonon quantum kinetics is the elimination of carrier–carrier scattering as a competing process. Monitoring of carrier dynamics in a density regime where carrier–LO phonon scattering predominates compared to carrier–carrier interaction was demonstrated previously (Leitenstorfer et al., 1996). To investigate the dependence of carrier relaxation on particle concentration, we have measured the temporal decay of the most prominent bleaching maximum in Fig. 9a by keeping the monochromator position at a photon energy of 1.62 eV and taking time traces in a wide interval of electron–hole excitation densities N ranging from 7×10^{14} cm^{-3} to 1.4×10^{17} cm^{-3} (Fig. 11). In all experiments presented in this chapter, the excitation densities are calculated from the sheet density of photons per light pulse in the central area of the excitation spot on the sample and from the optical absorption coefficient α of the material studied ($\alpha = 13{,}000$ cm^{-1} in GaAs and $\alpha = 28{,}000$ cm^{-1} in CdTe at the excitation wavelength, respectively). An initial time constant τ_1 is connected to the fast energy relaxation of electrons in the conduction band (see Figs. 11a–11f). For carrier concentrations below approximately 2×10^{15} cm^{-3}, this decay time is constant with a value of 240 ± 20 fs. This result is in agreement with a calculation of the LO phonon emission time of electrons with an excess energy in the order of 100 meV, assuming a pure Fröhlich-type interaction mechanism and GaAs material parameters (see Fig. 2). Consequently, this scattering process can be expected to dominate the relaxation in the low density regime. At higher concentrations N, the electron relaxation times τ_1 become substantially faster because of the increasing influence of carrier–carrier scattering via the Coulomb interaction.

c. The Role of the Hole Distributions and Hot Phonon Effects

As a last point, we want to discuss the influence of the photoexcited holes on the transmission changes measured in the experiments. Although until now, only the electrons in the conduction band have been mentioned, it is clear that the holes created in the valence band may also lead to significant nonlinear optical effects.

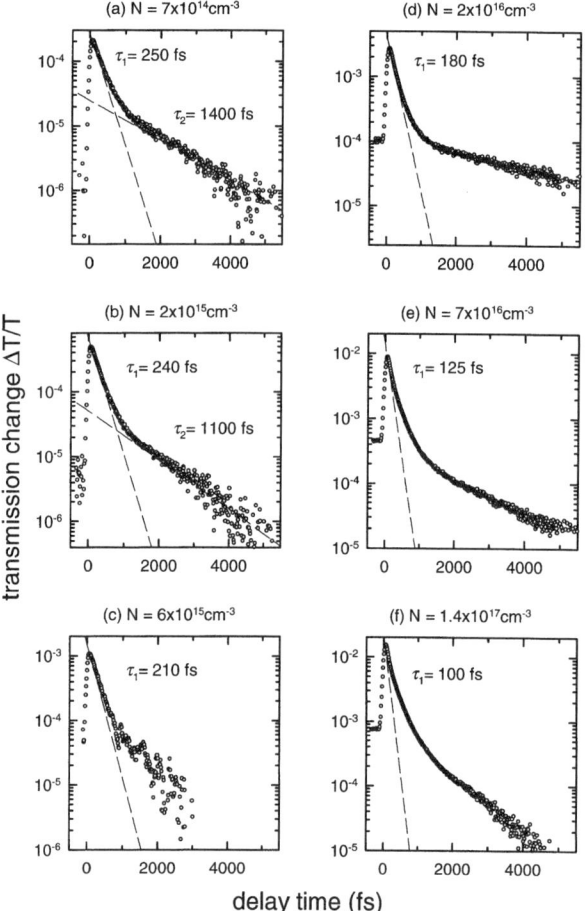

FIG. 11. (a–f) Temporal dynamics of the maximum of the initial bleaching peak hh/hh in GaAs (see Fig. 9a) at different excitation densities N. A fast decay at early delay times is connected to the energy relaxation of photogenerated electrons. The following picosecond dynamics is attributed to nonthermal heavy holes.

As an example, we point to the slower dynamics seen in the time-resolved data in Fig. 11 at later delay times that is connected to the relaxation of the photogenerated heavy holes. Because of the high effective mass in the heavy-hole band, this type of carrier is generated with an excess energy of approximately 15 meV, that is, below the LO phonon threshold. Consequently, the fast emission of LO phonons is not possible for these particles and they are able to bleach the transition from the heavy-hole to the conduction band near the excitation energy on a longer time scale given by acoustic phonon and hole–hole scattering.

Since the electrons and heavy holes generated with light pulses of an excess energy on the order of 100 meV experience different relaxation channels, the distributions may be clearly distinguished in a pump-probe experiment at advanced delay times. As an example, transient transmission changes in GaAs recorded 700 fs after pumping at 1.64 eV with different excitation densities N are shown in Fig. 12. At a photoexcited electron–hole concentration of 3×10^{14} cm^{-3} (Fig. 12a), a rectangular bleaching peak is seen close to the band edge, between probe photon energies of 1.52 and 1.55 eV. This feature is connected to the photoexcited electrons having almost completely relaxed to excess energies below the threshold for LO phonon emission after 700 fs. The rectangular shape of the bleaching feature indicates that the electron distribution is still nonthermal, because electron–electron and electron–acoustic phonon scattering are slow at low densities and lattice temperatures. A second, less pronounced bleaching maximum is observed around 1.62 eV, slightly red-shifted to the excitation spectrum (dashed line). This feature corresponds to the photoexcited heavy holes that cannot emit LO phonons because of their low excess energy. As a result,

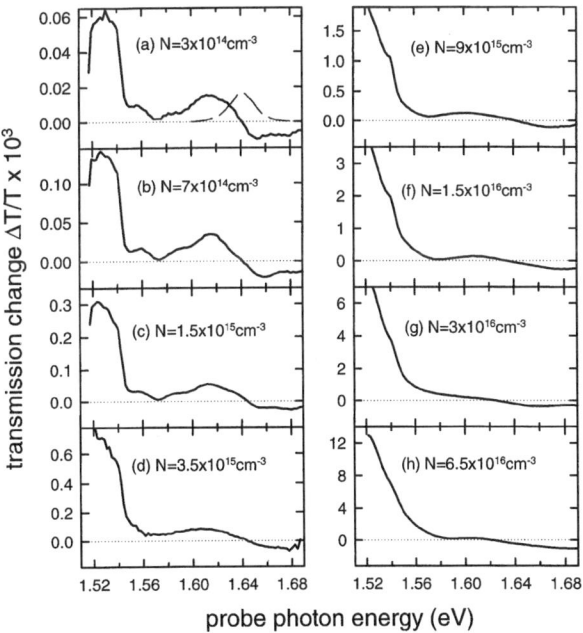

FIG. 12. (a–h) Transmission change versus probe photon energy 700 fs after excitation of different electron–hole densities N in GaAs. The lattice temperature is 15 K. Note that the ordinate scales are normalized to the excitation density N. The spectra show two clearly separated bleaching features due to (i) electrons relaxed to the band edge region between 1.52 and 1.55 eV and (ii) nonthermal heavy holes around 1.62 eV.

these carriers are bleaching the transitions close to the excitation energy on a longer time scale, and the distribution remains clearly nonthermal. As in the case of the electrons at early times, the red shift of the bleaching peak and the induced absorption at probing energies higher than the excitation position is a result of changes in the Coulomb-enhancement factor of the absorption due to the nonequilibrium carrier distribution. The maximum connected to the heavy holes appears broadened in comparison to the energetic width of the pump spectrum as a result of scattering with acoustic phonons becoming relevant on a picosecond time scale. This interpretation is supported by the observation that the transient absorption spectrum at $t_D = 700$ fs hardly changes when the excitation density is increased to 7×10^{14} cm^{-3} (Fig. 12b), indicating that carrier–carrier interaction is still negligible. An additional broadening of the heavy-hole peak is found at $N = 1.5 \times 10^{15}$ cm^{-3} (Fig. 12c), indicating the onset of hole–hole scattering. When the density is increased even further, the bleaching maximum becomes less pronounced and is hardly discernible 700 fs after excitation at $N = 6.5 \times 10^{16}$ cm^{-3} (Fig. 12h).

We wish to point out that in the case of the heavy holes, a change in the shape of the bleaching features need not be a result of energy relaxation, as in the case of electrons. Because of the strong anisotropy ("warping") of the heavy-hole effective mass, even pure momentum relaxation leads to a large change of the energy position for the induced transmission change since the interband transitions are probed.

We do not observe any pronounced features connected to the photoexcited light holes in our experiments. The reason may be found in the following arguments: First, the number of generated light holes is approximately a factor of three lower than that of the heavy holes because of the small density of states in the lh-band. Second, the light holes are expected to scatter into the hh-band via LO phonon emission on a time scale of 100 fs (Brudevoll *et al.*, 1990; Scholz, 1994).

Also, with increasing carrier concentration, the feature at the band edge connected to the electrons in Fig. 12 changes from a rectangular pattern to a smooth feature resembling an exponential tail. This finding is connected to the faster thermalization of the electrons into a Maxwell–Boltzmann distribution as carrier–carrier scattering becomes more efficient. Interestingly, the distribution at a time delay of 700 fs appears to be hotter at $N = 6.5 \times 10^{16}$ cm^{-3} (Fig. 12h), as compared to the densities between 9×10^{15} cm^{-3} (Fig. 12e) and 3×10^{16} cm^{-3} (Fig. 12g), where the electron temperature seems to be fairly similar. This fact indicates that at carrier concentrations close to 10^{17} cm^{-3}, the occupation numbers of LO phonons that have been emitted by the hot electrons are no longer small compared to unity. As a consequence of these hot phonons, phonon reabsorption may occur, slowing down the carrier cooling rate. On the other hand, it becomes clear that influences of nonequilibrium phonon distributions are negligible at excitation densities in the order of $N = 10^{15}$ cm^{-3} where the experiments on electron–phonon quantum kinetics have been performed.

IV. Results and Discussion

1. THE WEAKLY COUPLED CASE: GaAs

In this subsection, we present results on ultrafast electron–phonon relaxation in the III–V semiconductor GaAs, having a bandgap energy of $E_{gap} = 1.52$ eV at low temperatures. We recall that the LO phonon energy in this compound is 36 meV, corresponding to an oscillation period of 115 fs. This time interval is shorter but in the same order of magnitude as the spontaneous LO phonon emission time of electrons of approximately 250 fs. These conditions are a typical example for weak electron–phonon coupling, as expressed by the polaron constant of $\alpha = 0.06 \ll 1$ in GaAs.

a. Time- and Energy-Resolved Transmission Changes

Having discussed the conditions necessary to investigate the dynamics of nonthermal electron distributions in a regime dominated by LO phonon emission, we now point out the experimental findings concerning the influences of quantum kinetic effects. In Fig. 13 we present energy-resolved transmission changes $\Delta T/T$ measured at delay times t_D ranging from 0 to 500 fs. Electron–hole pairs are generated using ultrashort Gaussian light pulses of a duration of 120 fs and an energetic width of 15 meV. As explained in detail in Subsection 2 of Section II, these pulses represent an optimized compromise allowing both the time resolution to monitor the LO phonon emission of a time constant of 250 fs and the spectral selectivity to excite an initial distribution narrower than the LO phonon energy of 36 meV. The excitation density of 8×10^{14} electron–hole pairs per cubic centimeter is held extremely low to keep carrier–carrier scattering inefficient (compare Subsection 3b of Section III. The central photon energy of the pump pulses of 1.67 eV is substantially higher than the bandgap energy of GaAs of 1.52 eV (see Fig. 7). As a consequence, the carriers initially possess large kinetic energies. Excitation out of the heavy-hole (hh) band yields an electron distribution centered at an excess energy of 135 meV and a hh distribution around 15 meV. For the transition from the light-hole (lh) to the conduction band (cb), more similar effective masses lead to initial energies of 90 and 60 meV, respectively. Cocircular excitation and probe beams are applied in order to minimize transmission changes connected to the transitions involving the light hole band, as discussed in Subsection 3a of Section III.

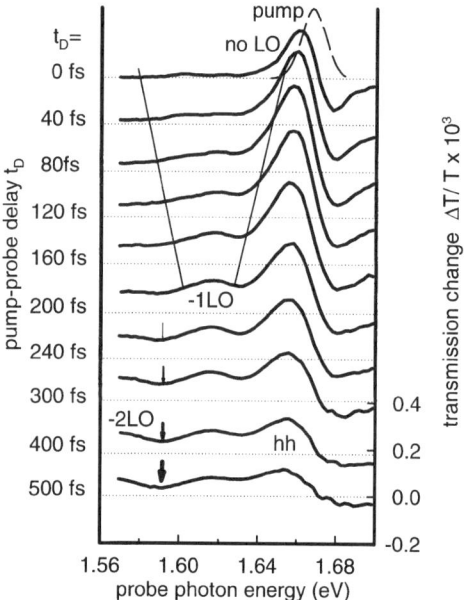

FIG. 13. Spectrally resolved transmission changes $\Delta T/T$ in GaAs measured for different delay times t_D after photogeneration of 8×10^{14} electron–hole pairs per cubic centimeter. The lattice temperature is 15 K. The spectrum of the pump pulse is shown as a dashed line. To guide the eye, the narrowing of the first phonon replica (-1LO) between $t_D = 0$ and 200 fs is marked by the thin converging lines. The delayed buildup of a minimum separating the first and the second phonon replica (-2LO) is indicated by vertical arrows. Both phenomena are a manifestation of quantum kinetic memory effects at relatively weak electron–phonon coupling ($\alpha = 0.06$ in GaAs) and would not be expected in Markovian dynamics.

At $t_D = 0$ fs, a well-pronounced peak (no LO) of increased transmission is seen at a probe photon energy of 1.66 eV (top of Fig. 13). The spectral hole is slightly red-shifted with respect to the excitation spectrum (dashed) and an induced absorption appears around a probing energy of 1.68 eV. This phenomenon is explained as follows. In our experiments, the transmission changes due to the photoexcited carrier distributions are caused by three different mechanisms: (i) The Pauli principle for fermions leads to a decreased absorption of interband transitions that is proportional to the sum of the occupation numbers of holes and electrons blocking the initial and final states. The spectral shape of this Pauli-blocking signal corresponds directly to the combined electron and hole distributions. (ii) The Coulomb attraction between electron and hole leads to excitonic effects influencing the optical properties even in the band-to-band continuum. The resulting Coulomb enhancement factor of the absorption is sensitive to the occupation of charge carriers. The nonlinear optical effects are such that a transmission change is found that is approximately proportional to the first derivative of a carrier distribution with respect to energy (El Sayed and Stanton, 1997). As a result, a quasimonoenergetic distribution of electrons leads to an increased absorption at photon energies corresponding to transitions into states with electron energies above the occupied states and to a bleaching at lower photon energies. (iii) A third contribution arises from the bandgap renormalization due to the occupation of carriers: As the fundamental energy gap decreases, higher joint densities of states shift to transitions at lower photon energies. It has been shown that this effect leads to a small and featureless negative transmission change that is negligible in the energy region probed in the experiments (Fürst et al., 1997a). As a consequence of the combined optical nonlinearities (i) and (ii), the peaks in the transient transmission spectra are red-shifted by approximately 10 meV with respect to the position expected in a situation without Coulomb interaction (pure Pauli blocking, (i)) and an induced absorption appears above the pump photon energy (Fürst et al., 1997a).

It has been explained in Subsection 2 of Section II that during the first part of an ultrafast light pulse, a carrier distribution with an energetic width larger than the pump spectrum is excited. In contrast, we observe an initial bleaching peak that is energetically narrow even at early delay times around $t_D = 0$ fs (Fig. 13). This phenomenon was demonstrated to be a result of the coherent coupling between pump and probe pulses arising when both fields are present in the sample at the same time (Schmenkel et al., 1998). These effects are relevant especially at negative delay times and may be neglected in a qualitative interpretation of the data for $t_D \geq 40$ fs.

We now return to the discussion of the electron–phonon relaxation. At time delays of 40 and 80 fs, a shoulder belonging to the electrons that have emitted one LO phonon arises in the energy range from 1.59 to 1.64 eV. This feature is energetically much broader than the initial bleaching peak. We must emphasize that the energy dispersion of the LO phonons is negligible for the region in the center of the Brillouin zone relevant in our experiment. As a consequence, in the semiclassical picture with its strict energy relaxation one would expect the first phonon replica to exhibit the same energetic width as the initial photoexcited distribution at any time. In contrast, our finding of a broadened shoulder instead of a well-defined peak clearly demonstrates that energy does not have to be conserved in the scattering events on such an early time scale. It must be stressed that in the Boltzmann limit a minimum between the initial distribution and the phonon satellite should exist already at early delay times. In the experiment this minimum builds up only beyond $t_D = 100$ fs at an energy of 1.63 eV. Interestingly, the start of this process coincides roughly with the end of the first LO phonon cycle 115 fs after the maximum of the pump pulse. Looking at the data, one can see the narrowing of the first phonon replica (indicated by the thin converging lines to guide the eye): After 200 fs the -1LO peak at 1.62 eV exhibits a width equal to the excitation maximum. The driving force for this nonclassical behavior of the distribution function is quantum interference. In the early regime, i.e., for delay times less than the phonon cycle, time–energy uncertainty allows scattering into a wide interval of final energies for the electrons. As time proceeds, the wave functions start interfering destructively at energy positions that correspond to scattering events without energy conservation. Constructive interference builds up where the detuning from the original distribution corresponds to exactly one LO phonon energy. As a result, the energetic width of the phonon replica decreases with time approaching the limit given by the original excitation. One might say that the system remembers its history. In theoretical models, exactly these interference phenomena are described by the so-called memory kernels.

Coming back to Fig. 13, we realize that there is still no well-defined separation between the first and the second phonon satellite at a delay time of 200 fs. At a probing energy of 1.59 eV, a minimum between these two maxima cuts in after approximately 300 fs (vertical arrows). This delay corresponds to the formation time of the first replica extended by an additional LO oscillation period. For $t_D = 400$ fs also the second satellite (-2LO) has narrowed. Obviously, *the total time elapsed since the original excitation is not the true criterion for memory effects to be observed.* In a cascading process, such as the subsequent emission of phonons, *the quantum phenomena repeat themselves.* As a result, such features are important on longer time scales than expected from the uncertainty given by the energies exchanged. This result is crucial since it demonstrates a fundamental point regarding the interpretation of the phenomena we find: If one would argue with a simplified picture taking into account only the relationship for the energy uncertainty connected to a certain time delay after excitation, one would expect quantum kinetic effects to be unimportant after a time comparable to the phonon oscillation cycle has elapsed. The fact that these features can continue as long as there is still fast dynamics in the distribution functions can be explained only with the more elaborate model of quantum coherence being a decisive factor during the entire relaxation of the electrons toward the band edge.

Interestingly, the initial bleaching peak remains the most prominent maximum in the spectra even for delay times as late as 0.5 ps (hh), that is, when most electrons should have emitted an LO phonon. At these time delays, the peak originates from the heavy holes that have been excited with an excess energy of 15 meV, making the emission of an LO phonon of an energy of 36 meV impossible. As mentioned earlier, the relaxation of the heavy holes is slow since carrier–carrier scattering is inefficient and the coupling to acoustic phonons via the deformation potential is weak. Nevertheless, a slight broadening of the hh peak within the first 500 fs of delay time is attributed to a combined action of these two processes, even at the low temperatures and densities in our experiment. Obviously, the same broadening does not occur to the features in the electron distribution. At $t_D = 500$ fs, the minimum between the first and the second phonon replica of the electron distribution is very well resolved at a probing energy of 1.59 eV (strong arrow). In contrast, the minimum between the first replica and the initial peak is smoothened out by the spectral broadening of the bleaching by the heavyhole distribution. This difference is explained by the special structure of the heavy-hole band: It is strongly anisotropic with effective masses differing up to a factor of two in different directions of the reciprocal space. As a consequence, even pure momentum scattering of a heavy hole, such as by a long-range Coulomb collision, will change the transition energy that is blocked leading to a broadening of the peak without significant energy relaxation being necessary. In contrast, the conduction band is spherical to a very good approximation. That is why in the case of the electrons, only real energy relaxation is probed in our experiment.

b. *Theoretical Simulations: An Overview*

The first theoretical simulations studying non-Markovian effects in electron–phonon dynamics were published in the late 1980s (Brunetti *et al.*, 1989). In subsequent years several groups contributed new results based on models becoming more realistic and/or including higher orders of electron–phonon correlations (Kuznetsov, 1991; Bányai *et al.*, 1992; Zimmermann, 1992; Hartmann and Schäfer, 1992; Schilp *et al.*, 1994; Heiner, 1996; Kenrow, 1997; Schönhammer and Wöhler, 1997). In response to the experiments presented here, two groups have performed theoretical calculations of the time- and energy-resolved transmission changes based on quantum kinetic models for the electron–phonon interaction (Schmenkel *et al.*, 1998; Zimmermann *et al.*, 1998). In both cases, the transmitted probe field is calculated within the framework of the semiconductor Bloch equations that include the influences of the Coulomb interaction in the nonlinear optical response.

The results obtained by Schmenkel, Bányai, and Haug (1998) are presented in Chapter 5 of this volume. In this work, the coherent dynamics originating from the coherent pump and probe pulses has been simulated in a realistic model: The interband polarization induced during excitation and probing and the coherent interaction of both fields via the semiconductor are accounted for in a three-dimensional system with two parabolic bands. As a result, the calculations fit the experimental data even in the very early time regime ($t_D < 40$ fs). In this domain, the coherent interband polarization induced by the light pulses can interfere resulting in significant deviations of the pump-probe spectra from the situation expected in an incoherent model for the transmission changes. The interaction of the electrons with LO phonons is treated in a quantum kinetic way. The loss of memory by the system is described by an exponential decay superimposed on the interference terms in the memory kernels. A very interesting point is the comparison of the quantum kinetic results with a standard Boltzmann calculation treating, however, the light–semiconductor interaction on the same sophisticated level (Schmenkel et al., 1998): It turns out that the semiclassical simulation already yields well-resolved and sharp phonon replicas at early delay times. This situation is clearly not observed in the experiment. In contrast, the quantum kinetics reproduces the time evolution of the maxima and minima in the measured transmission spectra amazingly well.

It can be shown that the damped memory approach used by Schmenkel et al. (1998) does not lead to a strict conservation of the system energy in the long time limit. This point can be overcome by calculating higher-order correlations between the involved quasiparticles instead of assuming a purely exponential damping of the memory kernels. In one approach, the electron–phonon interaction has been implemented in a fourth-order Born approximation (Zimmermann et al., 1998). The high level of complexity involved in these calculations can be visualized by mentioning that a system of differential equations has to be solved including the following types of independent variables: the semiclassical electron and phonon distributions f_k and n_q, depending both on the particle wave vector as a single argument. In addition, the phonon-assisted density matrices $T_{k,q}$ are calculated, including the lowest level of electron–phonon correlations. As the demanding part, even the three argument correlation functions S^{ep}, S^{ed}, and S^{ee} are taken into account. These equations are exact up to fourth order in the electron–phonon matrix element and lead to a physically meaningful conservation of the total energy. The model for the semiconductor system and its interaction with the laser fields had to be simplified as compared to the work by Schmenkel et al. (1998) to obtain a numerically tractable problem. Nevertheless, the fourth-order Born approximation should give a more reliable description of nature, especially in the long time limit and if stronger electron–phonon coupling than in GaAs is taken into account. However, these expectations have not yet been verified experimentally, since in the present data residual contributions due to carrier–carrier interaction, acoustic phonon scattering, and transitions involving the light-hole band render a comparison to theory difficult on longer time scales.

2. INTERMEDIATE COUPLING REGIME: CdTe

To investigate a material with stronger electron–phonon coupling than GaAs, we have studied the II–VI semiconductor CdTe. This compound exhibits a polaron coupling constant of $\alpha = 0.33 \approx 1$, placing it in the regime of intermediate interaction strength. Several technologically relevant materials, the most prominent example being GaN, exhibit similar conditions. However, the low-temperature bandgap energy of $E_{gap} = 1.6$ eV makes CdTe an ideal candidate for measurements based on the two-color Ti:Sapphire laser system allowing the time resolution and sensitivity needed. In contrast to GaAs, the phonon emission time of approximately 70 fs in the conduction band of CdTe is substantially shorter than the phonon oscillation cycle of 200 fs (see Subsection 1 of Section II). Therefore, the relaxation of the energy distribution of nonequilibrium electrons is expected to be strongly influenced by quantum kinetic effects. A situation that cannot be described by semiclassical Boltzmann kinetics even in a first approximation should be encountered.

In the experiments, free carriers are excited with a pump pulse of a central photon energy of 1.73 eV. In the transition from the heavy-hole band to the conduction band, electrons with an average excess energy of 112 meV are generated while the heavy-hole distribution is centered around 13 meV. The light-hole transition yields electrons of an energy of 70 meV and light holes at 55 meV. Influences of the light-hole band on the transient absorption spectra have been minimized by working with cocircularly polarized pump and probe pulses, as explained in Subsection 3a of Section III. The temporal shape of the excitation pulse fits well to a Gaussian and a pulse duration of 85 fs is deduced from autocorrelation measurements. The time resolution provided by these pulses is sufficient for the LO phonon emission time of 70 fs in CdTe. As a result of the short pulse duration, the spectral bandwidth of the excitation pulses is 19 meV. This value is slightly smaller than the LO phonon energy in CdTe of 21 meV. We have also performed the measurements with pump pulses of a smaller bandwidth of 16 meV and correspondingly longer pulse duration of 107 fs. It turns out that the spectrally more selective excitation yields results comparable to the case with better time resolution presented here.

A low concentration of 3×10^{15} electron–hole pairs is generated per cubic centimeter to keep carrier–carrier scattering inefficient (see Subsection 3b of Section III). The carrier dynamics is probed by a time-delayed test pulse with duration 20 fs and bandwidth 100 meV. The probe spectrum is centered at a photon energy of 1.68 meV and dispersed after the sample with the monochromator resolution set to 6 meV.

The resulting time and energy resolved transmission changes are depicted versus probe photon energy in Fig. 14. At early delay times of $t_D = -40$ fs and $t_D = 0$ fs, a negative transmission change is found close to the center of the pump spectrum (dashed line) and positive bleaching is seen red-shifted by approximately 20 meV. These features are a consequence of the stronger Coulomb effects in CdTe (exciton binding energy: 12 meV) as compared to GaAs (exciton binding energy: 4 meV). At the end of the pump

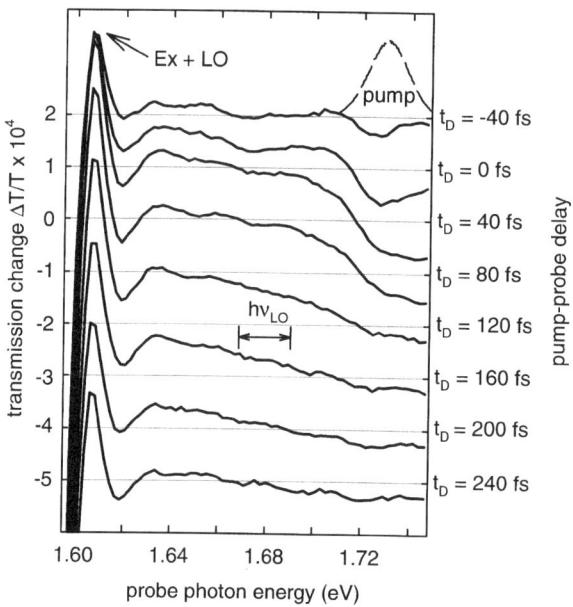

FIG. 14. Spectrally resolved transmission changes $\Delta T/T$ in CdTe measured for different delay times t_D after photogeneration of 3×10^{15} electron–hole pairs per cm³. The lattice temperature is 15 K. The spectrum of the pump pulse is shown as a dashed line. As a result of scattering events without energy conservation, the signals from the relaxing electron distribution show no structures connected to the LO phonon energy. This finding is typical for electron–phonon dynamics in the regime with intermediate coupling ($\alpha = 0.33$ in CdTe). The bleaching peak attributed to an exciton–phonon resonance in the continuum is marked with an arrow (Ex + LO).

pulse ($t_D = 40$ fs), a steplike feature is observed around a probing energy of 1.71 eV, indicating that the photogenerated electron distribution is clearly nonthermal. However, the bleaching feature connected to the electrons is already extended toward the band edge region because of the ultrafast emission of LO phonons. Similar to the results of the simplified model calculation presented in Fig. 3, the electron distribution relaxes to lower energies as an energetically broad wave packet, and no features corresponding to the LO phonon energy in CdTe (indicated by a horizontal bar in Fig. 14) are discernible. This finding may be explained as follows: The LO phonon emission rate in CdTe is so fast that the quantum kinetic memory effects, leading to sharpened phonon replicas in the case of GaAs, become operative only on a time scale where most electrons have already emitted another phonon. Under these circumstances, the concept of energy-conserving scattering events breaks down completely and the relaxation is strongly non-Markovian. No phonon replica are found in the spectra even at delay times of 240 fs, that is, on a time scale where a simple estimate of the time–energy uncertainty would certainly allow for spectral features as broad as 20 meV to become visible. The fact that such structures are still absent has to be seen in analogy to the repeating of the energy broadening in every emission step found in GaAs: As long as there is fast dynamics going on, the quantum kinetics dominates every scattering process and the time delay to the excitation process is not relevant. Instead, the physically important quantities are the phonon emission times and oscillation cycles.

It is clear that a semiclassical simulation of these results would not be able to reproduce the experimental findings, because it leads to phonon satellites in the distribution that are as narrow as the spectrum of the pump pulses convoluted with the energetic width of the excited transitions. Concepts to treat the electron–phonon dynamics on a quantum kinetic level even in the intermediate coupling regime are currently being developed (Zimmermann *et al.*, 1998; Bányai *et al.*, 1998a; Castella and Zimmermann, 1999). However, no simulation of the experimental data presented here is available at the time of this writing, since the full problem turns out to be numerically very challenging.

For probe photon energies below 1.63 eV, the measured transmission changes are no longer closely related to the shape of the photoexcited electron distribution in CdTe. A relatively sharp minimum appears at 1.62 eV already at early delay times, together with a pronounced maximum at 1.61 eV. It turns out that the energy position of this structure is constant and the peak (Ex + LO) is situated exactly one LO phonon energy above the 1s exciton line in our sample. The entire feature resembles a Fano resonance. In CdTe, this exciton–phonon resonance has already been observed in the linear absorption spectrum (Dillinger *et al.*, 1968). In our nonlinear absorption experiment, it appears analogous to the strong bleaching observed at the fundamental exciton resonance due to many-body interactions (see Haug and Koch, 1993). The phenomenon masks the signals due to the carrier occupation close to the band edge in CdTe. The positive transmission change after carrier injection at the Ex + LO position is found to be only one order of magnitude smaller than the signal at the 1s exciton. A much weaker feature with similar shape is also encountered in GaAs in nonlinear transmission spectra (Leitenstorfer *et al.*, 1996) but could not be retrieved in linear spectroscopy (Trallero-Giner *et al.*, 1996). These differences connected to exciton–phonon coupling are another manifestation of the increasing polar–optical coupling strength when going from GaAs to CdTe.

V. Conclusion

In this chapter, femtosecond measurements investigating quantum kinetic aspects of ultrafast electron–phonon collisions in semiconductors have been reviewed. The central findings are summarized in the following.

In interactions between quasiparticles with high scattering rates, the semiclassical assumption of initial and final states with identical energies is no longer a good concept. Instead, local violation of energy conservation is allowed on femtosecond time scales because of the uncertainty principle. As a first example, we have studied the polar–optical LO phonon emission of electrons in GaAs. In this system, the LO phonon emission time of 250 fs is in the same order of magnitude but still longer than the LO phonon oscillation cycle of 115 fs. As a consequence, significant dynamics is happening in a time regime where Fermi's Golden Rule does not apply and scattering events without energy conservation are observed for the first time. However, the electron–phonon coupling is weak enough for these transitions to remain virtual: As time proceeds, destructive quantum interference builds up at energy positions off resonance. In the long time limit, the energetic width of the scattered electron distribution narrows and approaches the shape expected in semiclassical dynamics. This result represents a direct demonstration of a non-Markovian memory effect.

In order to investigate increased electron–phonon coupling, CdTe has been studied where the phonon emission time of approximately 70 fs is even shorter than one phonon oscillation period of 200 fs. It turns out that under these conditions, quantum kinetic features in the scattering events completely dominate the dynamics: The electrons do emit another LO phonon before memory effects governed by quantum interference are able to reestablish the semiclassical limit characterized by local energy conservation. As a consequence, the LO phonon energy does not show up at all in the measured spectra, and the energy-resolved electron distributions remain featureless at any time during relaxation.

An important observation common to both examples is that in a cascading process such as the subsequent emission of LO phonons by the electrons, the quantum phenomena repeat themselves. The finding implies that the time interval elapsed since the preparation of a nonequilibrium state is not relevant for non-Markovian effects to be important. We conclude that the true criterion for quantum kinetic conditions is a high *interaction rate*, which has to be *comparable to or even larger than the frequency of the energy quanta exchanged* in the transitions. As long as this inequality is fulfilled, quantum phenomena and wave mechanics play a dominant role in the dynamics of any system. The incoherent picture of semiclassical particles becomes adequate only when the temporal change in the distribution slows down, for example when a quasi-stationary state is approached.

To experimentally work out these physical relationships as clearly as possible, we have chosen the model system of highly energetic electrons in direct-gap semiconductors interacting with LO phonons via the Fröhlich mechanism. At the end, we want to point out that the quantum kinetic conditions just characterized apply to a large number of ultrafast processes found in nature. Non-Markovian features are expected to play an important role during the relaxation of many kinds of electronic excitations in solids, molecules, molecular liquids, and biological systems.

ACKNOWLEDGMENTS

Our co-workers M. Betz, C. Fürst, and G. Göger have invested a lot of enthusiasm and many ideas that made the investigations presented in this chapter possible. We thank them for all their efforts. The high-quality transmission samples have been grown and processed by G. Böhm, G. Ebert, N. Möllenhoff, E. Sckopke, and G. Tränkle from the Walter-Schottky-Institute (TU München) and by C. R. Becker and K. Ortner (Universität Würzburg). Many interesting discussions with the theorists L. Bányai, H. Haug, L. V. Keldysh, T. Kuhn, F. Rossi, and R. Zimmermann are gratefully acknowledged. We also thank W. Kaiser for many helpful contributions and for his continuing interest in our work. Our studies have been supported financially by the Deutsche Forschungsgemeinschaft within the Schwerpunktprogramm "Quantenkohärenz in Halbleitern."

REFERENCES

Bányai, L., Tran Thoai, D. B., Remling, C., and Haug, H. (1992). "Interband quantum kinetics with LO-phonon scattering in a laser-pulse excited semiconductor," *phys. stat. sol.* (b) **173**, 139 and 149.
Bányai, L., Tran Thoai, D. B., Reitsamer, E., Haug, H., Steinbach, D., Wehner, M. U., Wegener, M., Marschner, T., and Stolz, W. (1995). "Exciton–LO-phonon quantum kinetics: Evidence of memory effects in bulk GaAs," *Phys. Rev. Lett.* **75**, 2188.
Bányai, L., Haug, H., and Gartner, P. (1998a). "Self-consistent PRA retarded polaron Green function for quantum kinetics," *Eur. Phys. J. B* **1**, 209.
Bányai, L., Vu, Q. T., Mieck, B., and Haug, H. (1998b). "Ultrafast quantum kinetics of time-dependent RPA-screened Coulomb scattering," *Phys. Rev. Lett.* **81**, 882.
Bartels, G., Stahl, A., Axt, V. M., Haase, B., Neukirch, U., and Gutowski, J. (1998). "Identification of higher-order electronic coherences in semiconductors by their signature in four-wave-mixing signals," *Phys. Rev. Lett.* **81**, 5880.
Betz, M., Göger, G., Leitenstorfer, A., Ortner, K., Becker, C. R., Böhm, G., and Laubereau, A. (1999). "Ultrafast electron–phonon scattering in semiconductors studied via non-degenerate four-wave-mixing," *Phys. Rev. B* **60**, 11265.
Brudevoll, T., Fjeldly, T. A., Baek, J., and Shur, M. S. (1990). "Scattering rates for holes near the valence band edge in semiconductors," *J. Appl. Phys.* **67**, 7373.
Brunetti, R., Jacoboni, C., and Rossi, F. (1989). "Quantum theory of transient transport in semiconductors: A Monte Carlo approach," *Phys. Rev. B* **39**, 10781.
Camescasse, F. X., Alexandrou, A., Hulin, D., Bányai, L., Tran Thoai, D. B., and Haug, H. (1996). "Ultrafast electron redistribution through Coulomb scattering in undoped GaAs: Experiment and theory," *Phys. Rev. Lett.* **77**, 5429.
Castella, H., and Zimmermann, R. (1999). "Coherent control of a two-level system coupled to phonons," *Phys. Rev. B* **59**, R7801.

Dillinger, J., Konák, C., Prosser, V., Sak, J., and Zvára, M. (1968). "Phonon-assisted exciton transitions in $A^{II}B^{VI}$ semiconductors," *phys. stat. sol.* **29**, 707.
Elsaesser, T., Shah, J., Rota, L., and Lugli, P. (1991). "Initial thermalization of photoexcited carriers in GaAs studied by femtosecond luminescence spectroscopy," *Phys. Rev. Lett.* **66**, 1757.
El Sayed, K., and Stanton, C. J. (1997). "Line-shape analysis of differential transmission spectra in the coherent regime," *Phys. Rev. B* **55**, 9671.
Foing, J.-P., Hulin, D., Joffre, M., Jackson, M. K., Oudar, J.-L., Tanguy, C., and Combescot, M. (1992). "Absorption edge singularities in highly excited semiconductors," *Phys. Rev. Lett.* **68**, 110.
Fürst, C., Leitenstorfer, A., and Laubereau, A. (1996). "Mechanism for self-synchronization of femtosecond pulses in a two-color Ti:sapphire laser," *IEEE J. Sel. Top. Quantum Electron.* **2**, 473.
Fürst, C., Leitenstorfer, A., Laubereau, A., and Zimmermann, R. (1997a). "Quantum kinetic electron–phonon interaction in GaAs: Energy nonconserving scattering events and memory effects," *Phys. Rev. Lett.* **78**, 3733.
Fürst, C., Leitenstorfer, A., Nutsch, A., Tränkle, G., Zrenner, A. (1997b). "Ultrafast Rabi oscillations of free carrier transitions in InP," *phys. stat. sol.* (b) **204**, 20.
Hartmann, M., and Schäfer, W. (1992). "Real time approach to relaxation and dephasing processes in semiconductors," *phys. stat. sol.* (b) **173**, 165.
Haug, H., and Jauho, A.-P. (1996). *Quantum Kinetics in Transport and Optics of Semiconductors.* Springer, Berlin.
Haug, H., and Koch, S. W. (1993). *Quantum Theory of the Optical and Electronic Properties of Semiconductors.* World Scientific, Singapore.
Heberle, A. P., Rühle, W. W., and Ploog, K. (1994). "Quantum beats of electron Larmor precession in GaAs quantum wells," *Phys. Rev. Lett.* **72**, 3887.
Heiner, E. (1996). "Linear response for systems far off equilibrium," *Physica A* **223**, 391.
Kenrow, J. A. (1997). "Quantum kinetic study of the electron–LO-phonon interaction in a semiconductor," *Phys. Rev. B* **55**, 7809.
Kner, P., Schäfer, W., Lövenich, R., and Chemla, D. S. (1998). "Coherence of four-particle correlations in semiconductors," *Phys. Rev. Lett.* **81**, 5386.
Kuznetsov, A. V. (1991). "Interaction of ultrashort light pulses with semiconductors: Effective Bloch equations with relaxation and memory effects," *Phys. Rev. B* **44**, 8721; "Coherent and non-Markovian effects in ultrafast relaxation of photoexcited hot carriers: A model study," *Phys. Rev. B* **44**, 13381.
Langot, P., Tommasi, R., and Vallée, F. (1996). "Nonequilibrium hole relaxation in an intrinsic semiconductor," *Phys. Rev. B* **54**, 1775.
Leitenstorfer, A., Lohner, A., Elsaesser, T., Haas, S., Rossi, F., Kuhn, T., Klein, W., Böhm, G., Tränkle, G., and Weimann, G. (1994). "Coherent optical generation of nonequilibrium electrons studied via band-to-acceptor luminescence in GaAs," *Phys. Rev. Lett.* **73**, 1687; *Phys. Rev. B* **53**, 9876 (1996).
Leitenstorfer, A., Fürst, C., and Laubereau, A. (1995). "Widely tunable two-color mode-locked Ti:sapphire laser with pulse jitter of less than 2 fs," *Opt. Lett.* **20**, 916.
Leitenstorfer, A., Fürst, C., Laubereau, A., Kaiser, W., Tränkle, G., and Weimann, G. (1996). "Femtosecond carrier dynamics in GaAs far from equilibrium," *Phys. Rev. Lett.* **76**, 1545.
Lipavsky, P., Spicka, V., and Velicky, B. (1986). "Generalized Kadanoff–Baym ansatz for deriving quantum transport equations," *Phys. Rev. B* **34**, 6933.
Madelung, O., ed. (1996). *Semiconductors: Basic Data*, 2nd ed. Springer Verlag, Berlin.
Meden, V., Wöhler, C., Fricke, J., and Schönhammer, K. (1995). "Hot electron relaxation: An exactly soluble model and improved quantum kinetic equations," *Phys. Rev. B* **52**, 5624; (1996). "Hot-electron relaxation in one-dimensional models: Exact polaron dynamics versus relaxation in the presence of a Fermi sea," *Z. Phys. B* **99**, 357.

Meier, F. and Zakharchenya, B. P., eds. (1984). *Optical Orientation*. North-Holland, Amsterdam.

Ridley, B. K. (1993). *Quantum Processes in Semiconductors*, 3rd ed. Clarendon Press, Oxford.

Sala, K. L., Kenney-Wallace, G. A., and Hall, G. E. (1980). "CW autocorrelation measurements of picosecond laser pulses, *IEEE J. Quantum Electron.* **QE-16**, 990.

Schilp, J., Kuhn, T., and Mahler, G. (1994). "Electron–phonon quantum kinetics in pulse-excited semiconductors: Memory and renormalization effects," *Phys. Rev. B* **50**, 5435; (1995). "Quantum kinetics of the coupled carrier–phonon system in photoexcited semiconductors," *phys. stat. sol. (b)* **188**, 417.

Schmenkel, A., Bányai, L., and Haug, H. (1998). "Quantum kinetics for differential transmission spectra in GaAs with LO-phonon interaction," *J. Lumin.* **76&77**, 134.

Scholz, R. (1994). "Hole–phonon scattering rates in gallium arsenide," *J. Appl. Phys.* **77**, 3219.

Schönhammer, K., and Wöhler, C. (1997). "Hot electron relaxation: Exact solution for a many-electron model," *Phys. Rev. B* **55**, 13564.

Shah, J. (1999). *Ultrafast Spectroscopy of Semiconductors and Semiconductor Nanostructures*, 2nd ed. Springer, Berlin.

Steinbach, D., Wehner, M. U., Wegener, M., Bányai, L., Reitsamer, E., and Haug, H. (1996). "Exciton–LO-phonon quantum kinetics: three-beam four-wave mixing experiments and theory of bulk GaAs," *Chem. Phys.* **210**, 49.

Trallero-Giner, C., Zimmermann, R., Trinn, M., and Ulbrich, R. (1996). "Exciton–phonon resonance in the continuum absorption of GaAs," in *The Physics of Semiconductors* (M. Scheffler and R. Zimmermann, eds.). World Scientific, Singapore, p. 345.

Vu, Q. T., Bányai, L., Haug, H., Camescasse, F. X., Likforman, J.-P., and Alexandrou, A. (1999). "Screened Coulomb quantum kinetics for resonant femtosecond spectroscopy in semiconductors," *Phys. Rev. B* **59**, 2760.

Wehner, M. U., Ulm, M. H., Chemla, D. S., and Wegener, M. (1998). "Coherent control of electron–LO-phonon scattering in bulk GaAs," *Phys. Rev. Lett.* **80**, 1992.

Young, J. F., Gong, T., Fauchet, P. M., and Kelly, P. J. (1994). "Carrier–carrier scattering rates within nonequilibrium optically injected semiconductor plasmas," *Phys. Rev. B* **50**, 2208.

Zimmermann, R. (1990). "Transverse relaxation and polarization specifics in the dynamical Stark effect," *phys. stat. sol. (b)* **159**, 317.

Zimmermann, R. (1992) "Carrier kinetics for ultrafast optical pulses," *J. Lumin.* **53**, 187; Zimmermann, R., and Wauer, J. (1994). "Non-Markovian relaxation in semiconductors: An exactly soluble model," *J. Lumin.* **58**, 271.

Zimmermann, R., Wauer, J., Leitenstorfer, A., and Fürst, C. (1998). "Observation of memory effects in electron–phonon quantum kinetics," *J. Lumin.* **76&77**, 34.

CHAPTER 2

Spatially and Temporally Resolved Near-Field Scanning Optical Microscopy Studies of Semiconductor Quantum Wires

Christoph Lienau and Thomas Elsaesser

MAX-BORN-INSTITUT FÜR NICHTLINEARE OPTIK UND KURZZEITSPEKTROSKOPIE
BERLIN, GERMANY

I. INTRODUCTION	39
II. SEMICONDUCTOR NANOSTRUCTURES	41
1. General Aspects	41
2. Semiconductor Quantum Wires	43
III. NEAR-FIELD SCANNING OPTICAL MICROSCOPY	46
1. Introduction	46
2. Near-Field Optics	49
3. Theoretical Description of Near-Field Optics of Semiconductor Nanostructures	53
4. Near-Field Probes	56
5. Probe-to-Sample Distance Control	58
6. Low-Temperature Near-Field Microscopy	59
7. Temporally and Spatially Resolved Near-Field Spectroscopy	61
IV. STATIONARY NEAR-FIELD SPECTROSCOPY OF SEMICONDUCTOR NANOSTRUCTURES	65
1. Low-Temperature Near-Field Spectroscopy	65
2. Near-Field Spectroscopy of Quantum Wires on Patterned (311)A GaAs Surfaces	66
V. TIME-RESOLVED NEAR-FIELD SPECTROSCOPY	79
1. Lateral Carrier Transport Studied by Picosecond Near-Field Luminescence Spectroscopy	79
2. Femtosecond Near-Field Pump-and-Probe Spectroscopy	89
VI. OUTLOOK AND CONCLUSIONS	98
REFERENCES	98

I. Introduction

Man-made artificial semiconductor nanostructures allow a spatial confinement of electrons and holes on a length scale comparable to their de Broglie wavelengths, resulting in a quantization of the electronic band structure. Parameters such as dimensionality and size of the structures, material composition, and/or strain of the crystal lattice determine the type of quantization and can be varied by controlling epitaxial growth processes on an atomic scale. In this way, the electronic and optical properties of nanostructures can be tailored for studying the fundamental physics of elementary excitations, including their dynamics, and/or for applying nanostructures in devices such as transistors, semiconductor lasers, and/or optical switches.

Optical studies have provided a wealth of knowledge on the fundamental physical properties of semiconductor nanostructures. The steady-state optical spectra give direct information on the quantized nature of electronic states and on the strength of dipole transitions between those states, as well as on correlation and many-body effects in the carrier system. Time-resolved experiments in the femtosecond domain allow observation of the dynamics of optical excitations in real time and insight into the fundamental nonequilibrium dynamics of optical polarizations and carriers. Detailed theoretical work both on steady-state and transient properties of nanostructures has led to a quantitative theoretical description of optical properties identifying the relevant microscopic interaction processes.

Most optical experiments have been performed under so-called far-field conditions, in which diffraction determines the minimum diameter of the optical beam at the nanostructure. The spatial resolution of such optical studies is limited to about the wavelength of light, as predicted by classical theories of diffraction. This results in a spatial resolution on the order of 1 μm, much larger than the dimensions of nanostructures along their confinement directions, which are on the order of 1 to 100 nm.

As a consequence, optical far-field imaging of individual nanostructures is generally not possible. Consequently, most experiments have addressed the behavior of an ensemble of nanostructures within the optical beam. Size and/or shape fluctuations of the nanostructures in such an ensemble directly affect the quantization of electronic levels and, thus, the optical transition energies, leading to an inhomogeneous broadening of the overall spectra. In addition, the limited spatial resolution makes it difficult to optically resolve the transport of excitations and/or carriers, often occurring on mesoscopic length scales in the submicron range.

The use of experimental techniques with a spatial resolution in the nanometer range is thus crucial for optical studies of single nanostructures. Until now, mainly methods based on the injection of electrons, such as cathodoluminescence spectroscopy, have been used for monitoring the local properties of nanostructures on submicron length scales. However, the initial energy distribution of carriers, its relaxation toward quasiequilibrium, and the transport of carriers in the nanostructure are difficult to control in such measurements. In contrast, all-optical spectroscopy allows for resonant excitation, and thus for generation of carriers under well-defined conditions. To make use of these advantages, one has to overcome the diffraction limit discussed earlier and to push the resolution of optical microscopy toward the 10–100 nm regime. Here, near-field scanning optical microscopy (NSOM) has been introduced as a novel all-optical technique providing a spatial resolution significantly better than the wavelength of the light.

In this paper, the potential of near-field microscopy for the spectroscopy of low-dimensional semiconductor nanostructures is outlined and progress in this area of research is reviewed. The paper is organized as follows. Section II gives a brief introduction to the physics of semiconductor nanostructures with the main emphasis on quasi-one-dimensional systems, so-called quantum wires. In Section III, the principles of near-field optical microscopy as well as theoretical and experimental aspects are discussed with particular emphasis on their relevance for semiconductor spectroscopy. In Section IV, applications of near-field microscopy to the steady-state spectroscopy of quantum wells, quantum wires, and quantum dots are reviewed. Advances of time-resolved near-field spectroscopy providing insight into the dynamics of transport and relaxation phenomena in nanostructures are presented in Section V. Conclusions are given in Section VI.

II. Semiconductor Nanostructures

1. GENERAL ASPECTS

Modern techniques of epitaxial crystal growth such as molecular beam epitaxy (MBE) (Cho and Arthur, 1975; Ploog, 1980) or metal–organic vapor phase epitaxy (MOVPE) (Stringfellow, 1989) allow the controlled growth of semiconductor nanostructures on an atomic scale. Quasi-two-dimensional structures in which carrier motion is restricted to a semiconductor layer have reached a very high degree of perfection using material systems such as GaAs/AlGaAs, GaInAs/InP, GaInAs/AlInAs, Ge/Si, and a number of II–VI and IV–VI materials. The elementary quasi-two-dimensional structures consist of a sequence of nanometer-thick layers of semiconductors with different bandgaps grown on a substrate and forming a sequence of potential wells. The depth of the potential wells is determined by the bandgap discontinuity between the well and the barrier material. The thickness of a potential well is on the order of the de Broglie wavelength of the carriers, $\lambda = h/\sqrt{2m^*E}$ (m^*, E: effective mass and energy of the particles), and thus quantum confinement occurs.

For a simple qualitative description of electronic structure, it is helpful to consider the case of a particle in an n-dimensional potential box with infinitely high potential barriers. In a three-dimensional semiconductor, carriers move freely in all three directions and, within the effective mass approximation, their kinetic energy is

$$E_c = \frac{\hbar^2}{2m^*}(k_x^2 + k_y^2 + k_z^2),$$

where $\hbar$ is Planck's constant, m^* the effective mass, and k the wave vector. In a two-dimensional structure, the carrier motion is confined in one direction leading to a restriction of one component of the wave vector, k_z, to quantized values $k_z = (n\pi/L_z)$, $n = 1, 2, 3\ldots$. This means that motion along the z direction is quantized corresponding to discrete values of carrier energy, whereas carriers can move freely in the plane of the potential well corresponding to a continuum of carrier energies. Simultaneously, the square-root energy dependence of the carrier density of states (DOS) is replaced by a steplike function. This so-called subband structure, which depends directly on the thickness L_z of the quantum well, underlies the strong modification of optical and transport properties compared to bulk semiconductors. Individual potential wells with negligible coupling of the wave functions, that is, thick barriers, are called quantum wells (QWs), whereas coupled potential wells form a superlattice with electronic states delocalized along the quantization direction.

The optical spectra are determined by dipole-allowed transitions between those quantized subbands, both from the valence to the conduction subbands and between consecutive valence or conduction subbands. The shape of those spectra is determined by the joint density of states of the optically coupled subbands, the structural disorder in the potential wells resulting in an additional broadening, many-body effects due to the long-range Coulomb interaction among the carriers, and—in the case of excitonic excitations—by the Coulomb correlation between electrons and holes. The linear and nonlinear optical properties of quasi-two-dimensional semiconductors have been studied thoroughly, both from the theoretical and experimental point of view. There is a number of excellent textbooks and reviews on optical properties in the literature (Chemla et al., 1989; Haug and Koch, 1993; Weisbuch and Vinter, 1991), including aspects of ultrafast dynamics of elementary excitations (Shah, 1999).

Upon further reduction of the dimensionality, the carrier motion also becomes restricted along the y direction for the case of a quantum wire (QWR), and along x and y directions for a quasi-zero-dimensional quantum dot (QD). In a QWR, the particle in a box model predicts a divergence of the DOS at the minimum of the quantized subbands, indicating a $1/E$ (E: energy) dependence of the DOS for each subband in QWRs. For QDs, the states are quantized in all three directions and the DOS is a series of discrete sharp spikes, resembling the case of an atom. So far, growth procedures for QWRs and QDs have not reached the very high degree of structural quality that is available for QWs and superlattices. In particular, fluctuations of size and shape are present in most QWR and QD ensembles, resulting in a substantial inhomogeneous broadening of their optical spectra.

2. SEMICONDUCTOR QUANTUM WIRES

In the following, we concentrate on quasi-one-dimensional QWR structures, as a major part of the near-field studies presented in Sections IV and V are devoted to QWRs. Early work on the fabrication of QWRs was based on a lateral structuring of quasi-two-dimensional nanostructures, both heterojunctions and QWs. GaAs/AlGaAs QWs were structured by chemical etching (Cibert et al., 1986; Kash et al., 1986; Petroff et al., 1982) or by implanting stripes of Ga^+ ions (Scherer et al., 1987). Wirelike geometries of a lateral width between several tens and several hundreds of nanometers were etched directly into single or multiple GaAs QW layers (Fig. 1a). The

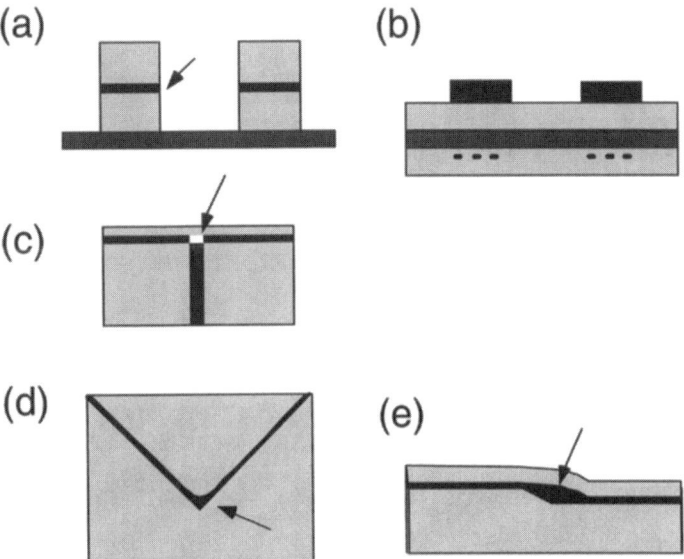

FIG. 1. Schematic of different quantum wire fabrication techniques. (a) Lateral structuring of two-dimensional quantum wells by chemical etching. (b) Lateral structuring of modulation-doped two-dimensional electron gases by low energy ion bombardment. (c) T-shaped quantum wires formed by the cleaved edge overgrowth technique. (d) V-groove quantum wires grown by chemical vapor deposition on prestructured substrates. (e) Sidewall quantum wires grown by molecular beam epitaxy on patterned (311)A GaAs substrates.

main drawback of this technique consists in the introduction of a high density of impurity and trap levels at the etched surfaces, resulting in efficient carrier trapping into those states. This strongly affects both the optical and the transport properties. In a related approach, QWRs were realized in n-type modulation doped QWs by generating a stripelike distortion in the AlGaAs barrier layer that contains the donor atoms (Fig. 1b) (Scherer et al., 1987). The electrostatic interaction between ionized donor atoms and free electrons in the GaAs layer confines the carriers to a quasi-one-dimensional conduction channel of 75-nm width without introducing additional impurities in the GaAs layer. In field effect devices, a metalized contact layer with a grating geometry, that is, an array of parallel lines, has been used to realize quasi-one-dimensional electron channels in an inversion layer beneath the contact (Hansen et al., 1987). In such structures, the lateral confinement potential, which has a period on the order of 200 nm, can be varied by changing the bias voltage on the top contact. For a review of optical and transport properties of those structures, the reader is referred to Hansen (1988).

Epitaxial cleaved edge overgrowth (Pfeiffer et al., 1990) and growth on prestructured substrates (Kapon et al., 1987) represent important techniques for realizing QWRs of high structural quality. In cleaved edge overgrowth, which is based on molecular beam epitaxy, a QW sample is cleaved along a plane perpendicular to the QW plane and a second multilayer system is grown on the cleaved surface. In this T-shaped geometry (Fig. 1c; Chang et al., 1985), the two growth planes define a quasi-one-dimensional intersection along which the confinement potential is modified compared the confinement potential in the QWs. This results in a quasi-one-dimensional carrier confinement along this direction with a relatively small confinement energy of up to 20 meV (Goni et al., 1992; Wegscheider et al., 1993). The optical properties of T-shaped QWRs have been studied both by far-field and near-field techniques (Binnig et al., 1986; Grober et al., 1994; Harris et al., 1996; Wegscheider et al., 1993), and stimulated emission on excitonic transitions has been demonstrated (Wegscheider et al., 1993).

V-groove QWRs represent an important example for quasi-one-dimensional structures grown by chemical vapor deposition on prestructured substrates (Kapon et al., 1987, 1989a, 1989b). V-shaped grooves of a depth of several microns are etched into a (100) GaAs substrates along the [01$\bar{1}$] direction. The deposition of AlGaAs in such a channel results in the formation of a V-shaped groove with nanometer-size facets. At the bottom of the V-groove, a (100) facet is formed between two (311)A facets, whereas (111) surfaces exist on the sidewalls. When GaAs is deposited on such a V-groove, the GaAs growth rate on the (100) and (311)A facets is higher than on the sidewalls, resulting in the formation of a crescent-shaped QWR at the bottom of the V-groove. The growth processes have been characterized in detail (Biasol et al., 1997), and both GaAs and InGaAs (Martinet et

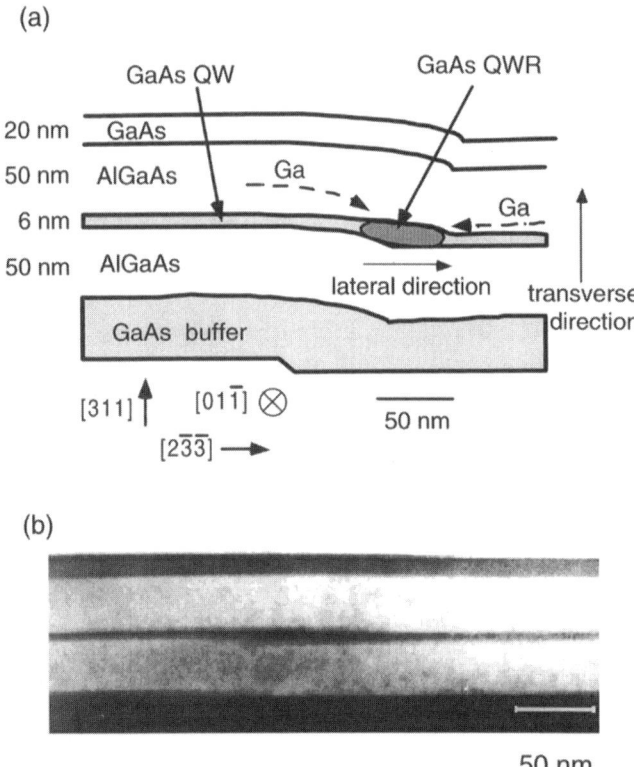

FIG. 2. (a) Schematic of the sidewall quantum wire structure. The sample is grown on a patterned GaAs (311)A substrate with 15-nm-high mesa stripes oriented along the lateral [01$\bar{1}$] direction. Formation of a GaAs quantum wire with a lateral dimension of 50 nm and a vertical dimension of 12 nm is due to the preferential migration of Ga atoms, as indicated by the dashed arrows, during the growth of the 6-nm GaAs quantum well. The GaAs quantum well is sandwiched between two 50-nm $Al_{0.5}Ga_{0.5}As$ layers. The structure is capped with a 20-nm GaAs layer. (b) Cross-sectional TEM image taken along the [01$\bar{1}$] direction.

al., 1998) QWRs have been realized with high structural quality. The lateral width of those QWRs is on the order of 10 to 20 nm with a high confinement energy of about 100 meV. This growth technique has also been used to make QWR stacks.

Epitaxial growth on patterned (311)A GaAs surfaces has been demonstrated to provide high-quality QWRs (Fig. 1e) (Nötzel *et al.*, 1996a, 1998). For QWR formation, (311A) GaAs substrates were patterned by chemical etching and mesa stripes of a height of 15–20 nm were formed along the [00$\bar{1}$] direction (Fig. 2a) (Nötzel *et al.*, 1996b). AlGaAs and GaAs layers grown by molecular beam epitaxy on such patterned substrates develop a smooth convex surface profile near one sidewall of the patterned mesa in the sector toward the next (100) plane. The formation of wirelike regions arises from the preferential migration of Ga atoms within the GaAs layer from the mesa bottom and top toward the sidewall during growth. At the sidewall, the thickness of the GaAs layer is somewhat larger than in the flat part of the structure, resulting in a quasi-one-dimensional confinement of carriers along the sidewall.

In Section IV, we discuss near-field experiments on such a QWR sample that consists of a nominally 6-nm-thick GaAs QW layer embedded between 50-nm $Al_{0.5}Ga_{0.5}As$ barriers. A 50-nm GaAs buffer is grown between the patterned substrate and the lower AlGaAs barrier, and the upper barrier is separated from the surface by a 20-nm GaAs cap layer. Cross-sectional TEM images (Nötzel et al., 1996b) (Fig. 2b) indicate that—as a consequence of the migration process—the thickness of the GaAs QW near the sidewall increases from 6 to 13 nm. This change in QW thickness results in a quasi-one-dimensional confinement over a lateral width of 50 nm and a confinement energy on the order of several tens of milli–electron volts.

Until now, optical experiments have mainly concentrated on ensembles of a large number of QWRs. Evidence for a quasi-one-dimensional subband structure has been found in photoluminescence and photoluminescence excitation studies on T-shaped (Harris et al., 1996) and V-groove QWRs (Vouilloz et al., 1997), as well as in studies of far-infrared intersubband absorption in etched QWRs (Hansen et al., 1987). Optical studies of sidewall QWRs are presented in Section IV. Stimulated optical emission on interband transitions has been demonstrated both with V-groove (Kapon et al., 1989a) and T-shaped QWRs (Wegscheider et al., 1993). Nonlinear optical properties and carrier dynamics in quasi-one-dimensional structures have been studied by Oestreich et al. (1993) and Ryan et al. (1996). The quasi-one-dimensional band structure and the optical properties of QWRs have been studied in a number of theoretical papers (see, e.g., Saar et al., 1996; Rossi and Molinari, 1996; and references therein), partly including an analysis of experimental data.

III. Near-Field Scanning Optical Microscopy

1. INTRODUCTION

The technique of near-field scanning optical microscopy (NSOM) is a member of the family of scanning probe microscopies (Bonnell, 1993; Wiesendanger and Güntherodt, 1994–1996). As for the other members of this family, such as scanning tunneling microscopy (Binnig et al., 1992), atomic force microscopy (Binnig et al., 1986), or ballistic electron emission microscopy (Bell and Kaiser, 1988), a nanoscopic probe is raster-scanned

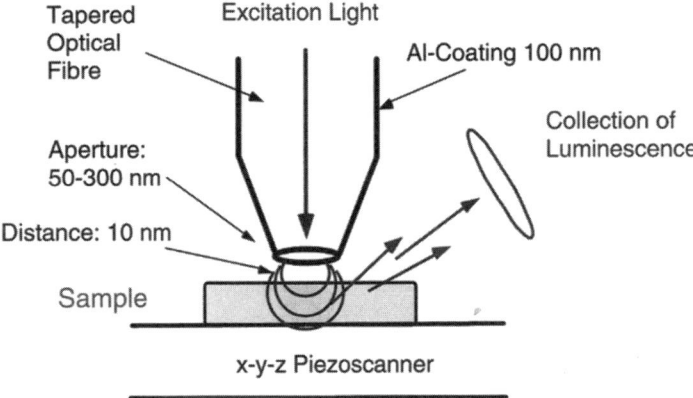

FIG. 3. Schematic of a near-field scanning optical microscope in an illumination-mode geometry. The sample is locally excited by transmitting light through a nanometer-sized aperture at the end of a metal-coated tapered single-mode optical fiber. Piezoelectric actuators position the aperture within a few nanometers of the sample surface and raster-scan the aperture relative to the sample. A distance regulation setup maintains the aperture-to-sample distance at a few nanometers while scanning. Near-field images are generated by recording the photodetector signal, for example, the sample luminescence as a function of near-field probe position.

across the surface of the sample, selectively testing specific local sample properties with nanometer resolution.

In near-field microscopy, the probe is a nanoscopic light emitter or collector, such as an illuminated aperture with diameter of a fraction of the wavelength of the incident light. In the direct vicinity of the aperture, that is, in its *near field*, the spatial resolution is defined by the dimension of the aperture rather than by diffraction and can thus be reduced by sufficiently decreasing the size of this aperture. This idea was originally suggested as early as 1928 by Synge, rediscovered by O'Keefe in 1956, and rediscovered again and demonstrated at microwave frequencies by Ash and Nicholls in 1972. In 1984, Pohl *et al.* and Lewis and co-workers demonstrated, in independent work, the first near-field scanning optical microscopes for visible light. Since then, the technique has found widespread applications in numerous different research areas such as single-molecule spectroscopy, high-density data storage, biology, high-resolution lithography, and semiconductor spectroscopy. For reviews, the reader is referred to Courjon and Bainier (1994) and Paesler and Moyer (1996).

A schematic of a typical NSOM setup is shown in Fig. 3. The main parts are (i) a near-field probe, such as a nanometer-sized aperture located at the end of a metal-coated tapered single-mode optical fiber, (ii) piezoelectric actuators that position the aperture within a few nanometers of the sample

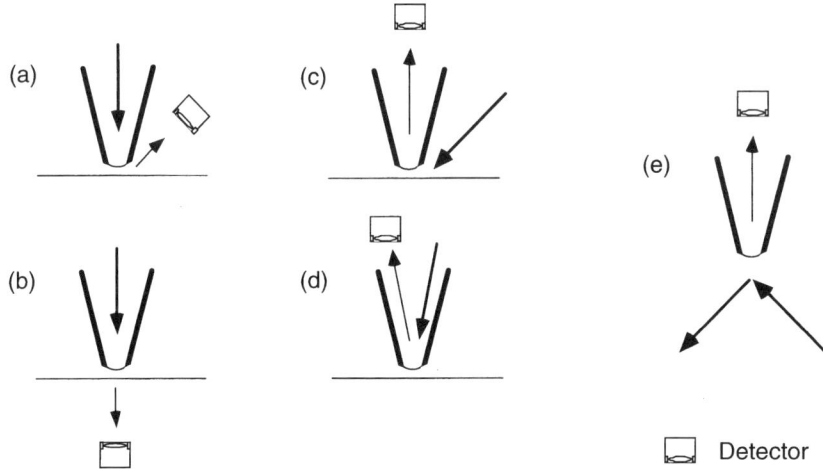

FIG. 4. Contrast mechanisms in near-field microscopy. (a) Illumination mode; detection in a reflection geometry. (b) Illumination mode; detection in a transmission geometry. (c) Collection mode. (d) Illumination/collection mode. (e) Dark-field excitation.

surface and raster-scan the aperture relative to the sample along the surface, (iii) a distance regulation setup, to keep the aperture-to-sample distance constant while scanning, (iv) an excitation light source, and (v) suitable photodetectors. Near-field images are generated by recording the photodector signal as a function of near-field probe position as the probe is scanned across the sample.

As with conventional far-field light microscopy, near-field microscopy offers a variety of different contrast mechanisms (Fig. 4). In the so-called *illumination*-mode geometry, light is transmitted through the aperture, locally exciting the object under study. The light that is reflected off (a) or transmitted through (b) the sample is then collected with a conventional far-field microscope either in a transmission or reflection geometry. Important spectroscopic information can be obtained by collecting the luminescence emission that is locally excited. In an alternative, *collection-mode* geometry (c), the sample is conventionally excited in the far field and scattered light or luminescence is locally detected by a near-field probe. In the combined *illumination/collection* mode (d), the excitation light is transmitted through the near-field aperture and the light from the sample is collected through the same aperture. In a *dark-field* geometry, the incident light is made to totally internally reflect from the substrate surface. The local evanescent field in the near field of the sample is then disturbed by a near-field probe, such as an uncoated fiber tip, and the scattered light is collected through the probe.

Each of these geometries has its particular range of applications. In the illumination mode, it is important to note that the collected light is not necessarily generated at the point of excitation. The geometry is particularly sensitive to carrier transport processes within the sample that may pose an upper limit on the spatial resolution that is obtained. In contrast, in the collection mode the position of light generation is probed with a resolution that is ideally defined by the aperture diameter.

In addition to these different geometrical configurations, a number of further contrast mechanisms have been realized in near-field microscopy. Among the most important is the selection of the excitation wavelength and/or detection wavelength, as well as polarization and temporal contrast. In combination with time-resolved excitation and detection schemes, near-field microscopy has the potential to give direct access to processes on ultrashort time and length scales, as discussed in Section V.

2. NEAR-FIELD OPTICS

In conventional far-field optical microscopy a fundamental limit is imposed on the spatial resolution by diffraction. Two pointlike objects can only be resolved if their distance is larger than $0.61\lambda/\mathrm{NA} = 0.61\lambda/(n\sin\delta)$, with NA being the numerical aperture of the microscope objective, δ the half-angle of the maximum cone of light collected by this lens, and n the refractive index of the immersing medium between object and lens. Commercial lenses have maximum numerical apertures of about 0.9 for air and 1.4 for oil immersion objectives. Thus, the resolution is typically limited to about one-half of the wavelength of the light. Further improvement may be obtained by solid immersion lenses (Terris *et al.*, 1994) with refractive indices of up to 3.2 (Vollmer *et al.*, 1999).

In near-field microscopy, the diffraction limit is overcome by working with *evanescent* electromagnetic waves $E_0 \exp(-(\omega t - \vec{k}\vec{r}))$ with

$$k_z^2 = k^2 - k_x^2 - k_y^2 < 0 \quad (k = |\vec{k}| = \sqrt{k_x^2 + k_y^2 + k_z^2} = 2\pi n/\lambda = \omega/c).$$

We take the object plane as (x, y) and z as the optical axis. The electric field of these evanescent waves decays exponentially with distance as $\exp(-|k_z|z)$. Such nonradiative evanescent waves are always generated if light is diffracted by finite-size objects, as can be seen for the case of diffraction of a plane electromagnetic wave at a one-dimensional slit of width d (Vigoreux *et al.*, 1992).

In the experiments that are reported in this article, aperture near-field probes are mainly used for generating the evanescent waves. For the interpretation of such experiments, it is often important to have a detailed understanding of the transmission properties of such nanoapertures. An exact calculation of the electromagnetic field is generally quite involved because of the complex shape of the experimentally used near-field probes and the vector nature of the electromagnetic field. It is thus instructive to consider the diffraction of light through subwavelength-sized apertures with a simplified geometry. Here, an important model system is the case of a subwavelength-size circular aperture of radius a in a perfectly conducting plane screen illuminated from one side by a plane electromagnetic wave $E_0 \hat{x} \exp(-i(\omega t - kz))$. This circular case gives an indication of several important features common to small apertures of various shapes, in particular the polarization dependence of the electromagnetic field, and its dependence on the distance from the aperture. A first solution to this problem was given by Bethe (1944); later, an improved analytical expression for the field in the aperture plane was given by Bouwkamp (1950a, 1950b).

Those results show that the power density of the electric field is strongly confined to the size of the aperture and that its spatial variation exhibits a strong direction dependence due to the different boundary conditions of the tangential and normal component of the electric field vector at the rim. While the electric field amplitude $|\vec{E}|$ along the x direction, parallel to the incident field vector, shows a square-root divergence at the rim of the aperture, the maximum along the y direction is at the center of the aperture and the amplitude decreases monotonically with distance from the center. Strong support for the very pronounced polarization dependence that is predicted by the solution of Bouwkamp (1950a, 1950b) has been given in near-field images of individual dye molecules (Betzig and Chichester, 1993; Veerman et al., 1998) where the dipole of randomly oriented molecules was used to map the polarization dependence of the electric field distribution in a near-field aperture.

For semiconductor studies it is of particular importance to understand how the electric field distribution varies with increasing distance from the aperture, as most of today's nanostructures are buried below the sample surface, since they are capped with a thin protection layer. Grober et al. (1996) emphasized that the solution given by Bouwkamp (1950a, 1950b) for the electric field behind the aperture holds only in the immediate neighborhood of the hole, and a modified expression for the z component of the electric field in the aperture plane was proposed. This solution preserves the vector nature of the electromagnetic field and recovers both the behavior of the near field in the limit of $z \to 0$ and the nature of the far field that is that of a magnetic dipole $m = -(4a^3/3\pi)$ oriented along the y direction. Figures 5a and 5b show the variation of the x component of the electric field vector along the x and the y direction, respectively, in planes of constant values of z for an aperture radius a of 30 nm and a wavelength of 800 nm. Figure 6 compares the electric field intensities on the z axis for different refractive indices of $n = 1$ (open circles) and $n = 3.5$ (closed circles) of the dielectric

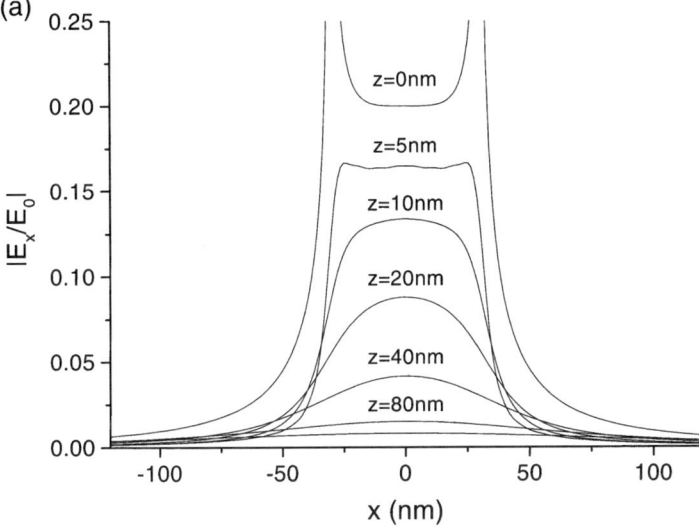

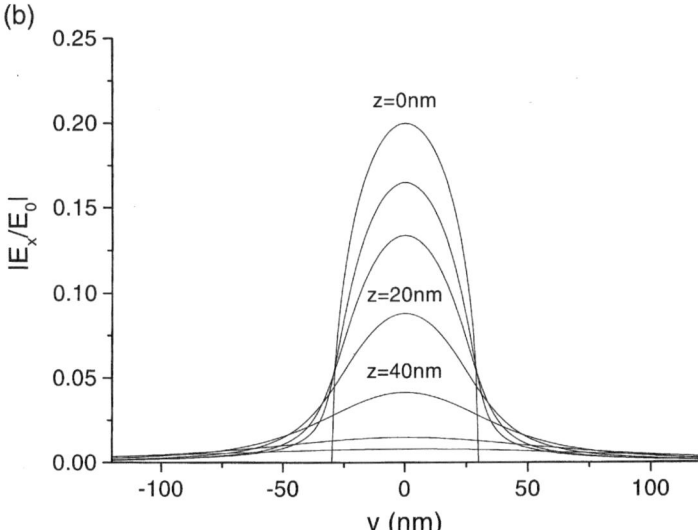

FIG. 5. Amplitude of the x-component of the electric field vector $|E_x(x, y, z)|$ behind an aperture of diameter 60 nm and a wavelength of 800 nm in planes of constant distance $z = 0$, 5, 10, 20, 40, 80, and 120 nm. (a) Variation of $|E_x(x, 0, z)|$ along the x direction. (b) Variation of $|E_x(0, y, z)|$ along the y direction.

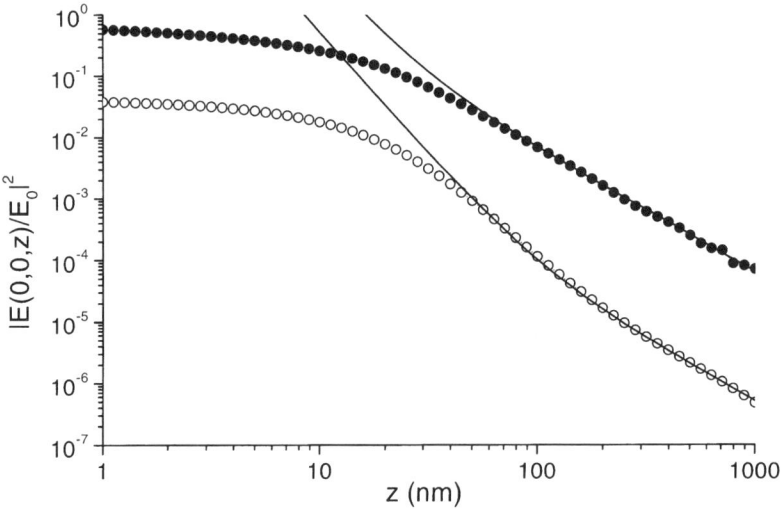

FIG. 6. Electric field intensity $|E^2|$ along the z axis for an aperture diameter of 60 nm and a wavelength of 800 nm. The open circles are calculated for a refractive index of 1, and the closed circles for $n = 3.5$ of the dielectric behind the aperture. The solid lines are the corresponding field intensities of a magnetic dipole oriented along the y axis.

behind the aperture. The electric field remains confined to a dimension of the aperture diameter over a distance of about one aperture radius. For larger distances, the field diverges similar to that of a magnetic dipole, for which the full-width-at-half-maximum of the power density profile, given by

$$|\vec{S}| = \frac{c}{4\pi} |\vec{E} \times \vec{H}|,$$

at a fixed value of z, corresponds to about twice the value of z, independent of the refractive index of the material. This shows that a high, aperture-size-limited resolution can only be achieved if the sample is placed in close proximity to the aperture. The pronounced difference in the electric field profiles along the x and y directions, $|E_x(x, 0, z)|$ and $|E_x(0, y, z)|$, is only maintained for distances z of about one aperture diameter, whereas the profiles become very similar for larger distances z. Very close to the aperture, $z < \lambda/100$, the double logarithmic plot of $|E(0, 0, z)|^2$ vs z (Fig. 6) shows a weak field fall-off with distance that evolves into that of a magnetic dipole at larger distances. Here, the on-axis electric field intensity $|E(0, 0, z)|^2$ decreases approximately exponentially with increasing distance, with a decay length of about $a/2$. In general, the results indicate that an exact calculation of the vectorial near-field distribution is important if the object is placed less than one aperture diameter away from the tip, whereas outside this region the near field of a magnetic dipole seems to represent a fairly good representation of the field distribution and its distance dependence.

The model system of a circular aperture in a perfectly conducting screen is certainly very instructive for analyzing the generation and the spatial dependence of the near-field behind a sub-wavelength aperture. However, it cannot fully reproduce the transmission properties of experimentally used near-field probes that are generally of quite complex shape. A number of authors thus attempted to model the transmission through near-field probes of a more realistic geometry (Christensen, 1995; Roberts, 1987, 1989, 1991). Detailed calculations for both two- and three-dimensional configurations have been presented by Novotny *et al.* (1994, 1995a, 1995b, 1995c, 1997), Pohl and Novotny (1994), and Pohl *et al.* (1996) using the so-called multiple-multipole-method (MMP) (Hafner, 1990; Hafner and Bomholt, 1993). The transmission through a variety of different near-field probes, such as uncoated dielectric probes, metal-coated aperture probes, entirely metal-coated fiber probes, and plasmon probes, has been investigated within this formalism. For a metal-coated dielectric waveguide, both the polarization and spatial dependence of the transmitted field are similar to what is found in the model calculations for a circular aperture in a perfectly conductive screen. In the plane of polarization, the fields are strongly enhanced at the edges of the aperture, and the near field consists of two peaks with a distance equal to the aperture diameter, as in the Bouwkamp solution. In addition, the calculations highlight the rapid decay of the field intensity inside the fiber in the region where the core diameter becomes smaller than $\lambda/2$. This decay in the evanescent region of the waveguide leads to a strong decrease of the transmitted power for small taper angle. A large portion of the incident power is reflected back from the aluminum coating, leading to a pronounced standing-wave pattern inside the taper. Because of the finite conductivity of the aluminum coating, the field also penetrates into the cladding, leading to considerable power dissipation and consequently heating of the metal coating. The skin depth of 6.5 nm of aluminum is an upper limit for the field confinement and gives the theoretical limit for the resolution that can be achieved with an aperture probe. Thus, if the cladding is chosen too thin, light is leaking out from the sides of the taper, resulting in a strong decrease of the attainable spatial resolution.

3. Theoretical Description of Near-Field Optics of Semiconductor Nanostructures

A theoretical description of NSOM experiments requires not only the knowledge of the electromagnetic near-field distribution in the vicinity of the fiber probe, but in particular a correct description of the interaction between this field distribution and the object under investigation. In general, such a description requires a full, self-consistent solution of Maxwell's equations for the coupled probe–object system (*global* theories). Often, the description is simplified by sequential calculation of the electromagnetic field distribution of the near-field probe and the interaction of this field distribution with the object (*nonglobal* theories) (Girard and Dereux, 1996).

An example of an application of the nonglobal approach to the field of semiconductor spectroscopy is the recent work on the near-field excitation of semiconductor QDs (Bryant, 1998). In this paper, the near-field incident on the sample is described by the Bethe–Bouwkamp model (Bouwkamp, 1950a, 1950b). Optical transition rates for dots that are excited with such a near-field distribution are determined within the Fermi Golden Rule approximation. The transition matrix elements between confined electron and hole states $\Psi^{(e,h)}$ are calculated in the electric dipole approximation. These rates depend sensitively on the overlap of the dot state envelope wave functions with the electric near field and on the atomic transition matrix element between the electron and hole states. In the limit that the QD dimensions are smaller than the tip aperture, the transition rates are a very sensitive probe of the vectorial character of the near field. In particular, the ratio of the transition rates from valence states with predominantly heavy-hole character relative to those from light-hole valence states is varying strongly when the tip is scanned across the dot. In this model this effect is a consequence of the strong polarization dependence of the Bouwkamp solution. It is shown that light-hole transitions can be selectively enhanced in the vicinity of the probe aperture. On the contrary, if the tip aperture is smaller than the dot size, it is predicted that the near-field technique allows one to directly map the spatial variation of the electron–hole pair wave functions. This prediction is supported by a theoretical study of the local absorption properties of a coupled QWR structure (Mauritz *et al.*, 1999).

A first theoretical study of the near-field optics of semiconductor nanostructures that goes beyond the perturbative approach just described has been presented by Hanewinkel *et al.* (1997). The authors describe the near-field response of optically excited semiconductor QDs in a fully self-consistent manner by solving the vector wave equation for the electric field,

$$\vec{\nabla}^2 \vec{E} - \vec{\nabla}\vec{\nabla}\vec{E} - \frac{1}{c^2}\frac{\partial^2}{\partial t^2} \vec{E} = \frac{4\pi}{c_0^2}\frac{\partial^2}{\partial t^2} \vec{P}, \qquad (1)$$

applying an efficient Green function method (Girard and Dereux, 1994; Girard *et al.*, 1994, 1995, 1997; Martin *et al.*, 1995). Assuming a local response of the medium and no anisotropy, the frequency-dependent polarization is $\vec{P}(\vec{r}, \omega) = \chi(\vec{r}, \omega)\vec{E}(r, \omega)$. The QDs are treated as pointlike particles, localized in space. A Lorentzian line shape with a polarization decay rate Γ_{ij} and a center frequency ω_{ij} is assumed for the susceptibility of the optical transitions of the individual QDs:

$$\chi(\vec{r}, \omega) = \frac{i}{\Omega_0} \sum_{i,j,k} \frac{|d_{ij}|^2 f_{ij}}{\Gamma_{ij} + i(\omega_{ij} - \omega)} \delta(\vec{r} - \vec{r}_k).$$

Here, i, j denote the different quantum states in the QD; k, placed at $\vec{r}_k \cdot d_{ij}$, are the matrix elememts of the electric dipole operator; and Ω_0 is the volume of the elementary cell. f_{ij} denotes a static band filling factor that equals unity for weak excitation. Comparing the fully self-consistent solution with a situation where the interaction of the dot and the tip material is neglected, it can be seen that—including the near-field interaction—the signal from the QD that is observed in an illumination-mode experiment is strongly enhanced by about one order of magnitude. This highlights the important role of the tip–sample interaction in near-field experiments. The tip–sample interaction gradually becomes weaker as the distance between tip and sample increases, and the length scale for this decrease is given by the decay length of the evanescent field of about $\lambda/2\pi$. Calculations done for a set of different susceptibilities $\chi_0 = 0.1, 1, 10$ predict that the spatial shape of the illumination mode signal that is obtained when the tip is scanned across the dot depends significantly on the value of the susceptibility. For small susceptibilities, the main signal contribution is peaked at the QD position where the induced dipole field is strongest. For large susceptibilities, an interference pattern is formed by the near field of the tip and the induced dipole of the dot, resulting in a double peak structure with maxima at the tip walls. Although the signal strength and the spatial shape in illumination mode experiments are affected by the tip–sample interaction, there is only a minor effect on the line shape of the susceptibility. The results also show that magnetic dipole interactions, which are generally negligible in far-field experiments, are enhanced in the near field. Yet their contribution is still small for the typical size of the apertures that are used experimentally. An extension of this work to the near-field response of QW and bulk excitons was presented by the same group (Knorr et al., 1998) Concerning the transition rates of heavy- and light-hole excitons in a GaAs QW, the authors reach conclusions similar to those drawn by Bryant (1998). The paper extends the theoretical description of semiconductor near-field optics to spatial propagation effects (Knorr et al., 1998) showing that ballistic transport of exciton wave packets over a length of several micrometers should occur on a picosecond time scale and should be observable experimentally under conditions where the relevant scattering times in the system are long. Recent calculations show the pronounced influence of disorder on the propagation of the excitonic wave packets (Hanewinkel et al., 1999). In addition, the authors give a first discussion of the temporal dynamics of the far-field emission from a locally excited excitonic wave packet. In the presence of disorder, strong interference effects between the emissions from excitons in different localization sites are predicted. Presently, a fully self-consistent theoretical discussion of semiconductor near-field optics has only been given for a few model systems. It is expected that such theoretical descriptions will greatly enhance the insight into near-field spectroscopy and also into the spatiotemporal dynamics of excitons and free carriers in semiconductor nanostructures.

4. NEAR-FIELD PROBES

The central elements in every near-field experiment are the near-field probes. A method for producing such a probe is based on metal-coated sharply tapered single-mode fiber and was introduced by Betzig et al. (1991). The fiber is heated with a CO_2 laser and drawn to a sharp taper in a micropipette puller. The resulting taper has an approximately conical shape with opening angle of typically 10° and a flat end face. The diameter of the flat end face (<30 to 300 nm) and the exact geometry of the taper can be varied by careful adjustment of the pulling parameters (Madsen et al., 1998). After angled evaporation of the sides of the taper with ~ 100 nm of aluminum, apertures ranging from <20 to >500 nm in diameter are reproducibly formed (Fig. 7). With such probes, a resolution of 12 nm ($\lambda/43$)

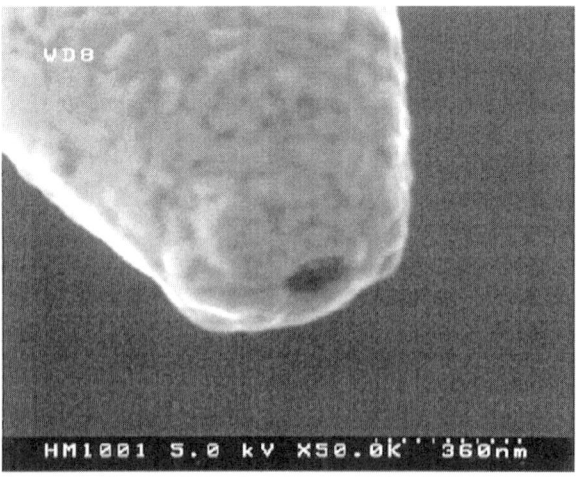

FIG. 7. Scanning electron microscope image of a metal-coated laser pulled near-field aperture probe with a tip diameter of 160 nm.

was deduced from scans of single aluminum particles on a glass substrate (Betzig et al., 1991). The transmission efficiencies of pulled fiber probes are typically on the order of 10^{-5} for an aperture diameter of 100 nm and decrease to less than 10^{-6} for diameters of less than 50 nm (Saiki et al., 1996), limited by the small taper angle of about $10°$ and the correspondingly long evanescent field region inside the taper, where the fiber diameter decreases below the wavelength of the transmitted light. The maximum power that can be launched into these probes is mainly limited by light absorption and heating of the metal coating, resulting in tip destruction for incident powers of more than a few milliwatts (Kavaldjiev et al., 1995; Larosa et al., 1995; Lienau et al., 1996; Stahelin et al., 1996). The polarization of the incoming light is mainly preserved within these fibers (Betzig et al., 1992; Jalocha et al., 1995; Lacoste et al., 1998). The properties, as well as the manufacturing and coating of the pulled fiber tips, have been carefully investigated (Hollars and Dunn, 1998; Madsen et al., 1998; Valaskovic et al., 1995). Recent attempts to form the apertures in pulled fiber probes in a more controlled way rely on focused ion beam milling or drilling (Lacoste et al., 1998; Muranishi et al., 1997; Veerman et al., 1998). With this method, aperture diameters as small as 20 nm are formed and an excellent polarization behavior is obtained.

Strategies for increasing the probe transmissivity rely on increasing the taper angle in order to decrease the evanescent field region within the taper. Fiber probes with cone angles around $30°$ were fabricated by wet chemical etching of the fiber (Bukofsky and Grober, 1997; Hoffman et al., 1995; Islam et al., 1997; Zeisel et al., 1996). Transmission efficiencies of about 10^{-3} were reported for fibers giving a spatial resolution of ~ 100 nm, an improvement by two orders of magnitude over pulled fiber probes. It was found that the surface roughness of the etched probes and thus the reproducibility of the manufacturing process can be improved by performing the etching through the plastic jacket of the fiber (Fig. 8) (Lambelet et al., 1998). Etched fiber probes that are particularly designed for illumination/collection mode studies of semiconductor nanostructures have been reported (Mononobe et al., 1998; Muranishi et al., 1997).

It is believed that the ultimate resolution that can be reached with these aperture probes is limited by the finite skin depth and grain size of the metal coating to about 30 nm. This resolution limit may be overcome by apertureless near-field techniques (Zenhausern et al., 1994, 1995), which use a sharp dielectric or metallic tip instead of the aperture probe. Here, one focuses a light beam in the gap between sample surface and tip apex and uses the strongly enhanced electric field at the tip apex as the excitation source (Aigony et al., 1999; Bachelot et al., 1995; Grésillon et al., 1999). In particular, near-field microscopy based on two-photon excitation of sharp metal tips has been proven both theoretically (Novotny et al., 1997) and experimentally (Sanchez et al., 1999) to present a promising novel approach for high-resolution luminescence imaging.

(a)

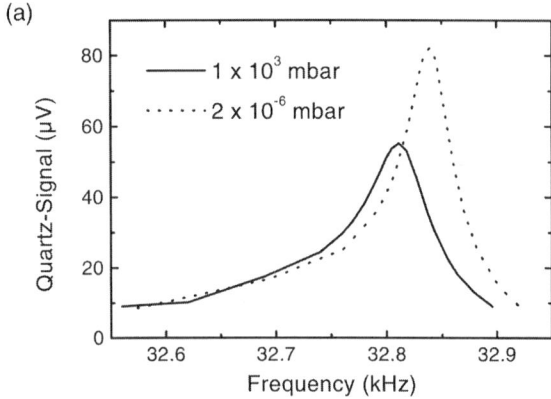

(b)

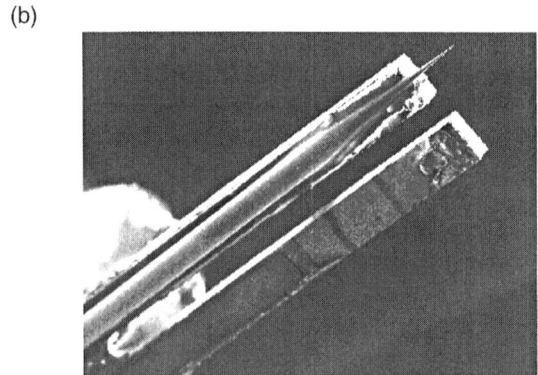

FIG. 8. (a) Resonance of a pulled near-field fiber probe mounted on a quartz tuning fork at ambient pressure and in vacuum at a pressure of 2×10^{-6} mbar. The excitation amplitude of the dither piezo is 10 mV. The quality factor of the tuning fork/tip oscillator is about 400 at ambient and increases to 530 in vacuum. (b) Scanning electron microscope image of a pulled near-field fiber probe mounted on one arm of a quartz tuning fork. The dimensions of one arm of the tuning fork are 3.98 mm × 0.625 mm × 0.125 mm.

5. Probe-to-Sample Distance Control

As pointed out in Subsection 2, the evanescent field in the vicinity of a near-field probe decays on a length scale of $\lambda/2\pi$, that is, typically 30 to 100 nm. To record a high-resolution near-field image, the tip-to-sample distance should be less than this value. Keeping the electric field within the sample constant poses an even stricter requirement on the probe-to-sample distance and suggests that it should not vary by more than a few nanometers. Consequently, the probe-to-sample distance is usually sensed and regulated during scanning to permit the probe to track the topography of the sample and maintain a constant gap width between probe tip and sample. In most near-field microscopes, this control and regulation is currently achieved with the noncontact shear-force technique (Betzig et al., 1992; Toledo-Crow et al., 1992). In this technique, a small lateral tip vibration, parallel to the sample surface, is excited at or near the mechanical resonance frequency of the near-field probe. The vibration amplitude is much less than the expected optical resolution, typically a few angstroms to nanometers; and most near-field probes have a fairly sharp mechanical resonance with a frequency in the 10- to 100-kHz range and a quality factor of about 100 to 1000 (Fig. 8a). As the probe approaches the sample surface, the resonant vibration is damped by shear forces between tip and sample. The shear-force damping sets in at a tip–sample distance of about 20 nm and leads to a monotonic decrease of the vibration amplitude with decreasing distance. Therefore, sensing the vibration amplitude and/or phase enables one to regulate the z piezo voltage and thus the probe-to-sample separation via a feedback loop and to scan the sample or the tip in a constant shear-force mode. Different techniques have been studied for sensing the dither amplitude, such as optical (Betzig et al., 1992; Froehlich and Milster, 1994; Wei et al., 1995), capacitance (Leong and Williams, 1995), impedance (Hsu et al., 1995), piezoelectric (Atia and Davis, 1997; Barenz et al., 1996; Brunner et al., 1997; Debarre et al., 1997; Decca et al., 1997; Drabenstedt et al., 1996; Karrai and Grober, 1995; Salvi et al., 1998), or interferometric (Pfeffer et al., 1997; Toledo-Crow et al., 1992) detection. A nonoptical distance control technique that is now in common use in a number of near-field microscopes is the tuning fork setup (Karrai and Grober, 1995). Here, the fiber probe is attached to one arm of a crystal quartz tuning fork and the tip vibration is sensed through the oscillating piezoelectric potential at the electrical contacts of the tuning fork (Fig. 8b).

The physical origin of the "shear-force" interaction has been a controversial subject in the literature. Several possible mechanisms have been proposed, such as viscous damping, van der Waals forces, capillary forces, and direct contact between tip and sample (knocking) (Froehlich and Milster, 1994, 1997; Gregor et al., 1996; Hollricher et al. 1998; Toledo-Crow et al., 1992; Wei and Fann, 1998).

6. LOW-TEMPERATURE NEAR-FIELD MICROSCOPY

Optical spectroscopy on semiconductors, in particular luminescence spectroscopy, is mostly performed at cryogenic temperatures. Particular reasons for going to low temperatures are (i) the increase in the luminescence quantum yield as nonradiative recombination processes are suppressed and (ii) the strong reduction of thermally induced line broadening processes.

This makes it desirable to perform near-field spectroscopy at temperatures close to or even below that of liquid helium.

Two different classes of instruments have been developed. In Grober et al. (1994), the microscope head is designed as an insert for a liquid helium bath cryostat, and both sample and scan head are cooled by direct contact with either gaseous or liquid helium at operating temperatures between 1.5 and 300 K. Modified designs following the same concept have been reported by several research groups (Ghaemi et al., 1995; Gohde et al., 1997; Levy et al., 1996).

Such a design has several drawbacks that complicate the operation of the instrument: (i) The scan range of the x–y–z piezos is strongly temperature dependent and decreases at 4 K to about 20% of its room temperature value. This limits the attainable maximum scan range to typically $5 \times 5 \times 1\,\mu$m and requires careful scanner calibration for every operation temperature. (ii) Hardware linearization of the piezo nonlinearity is generally not available. (iii) Cool-down of the entire microscope setup is necessary and thermal equilibration occurs on a time scale of several hours. (iv) Changes in operation temperature change the resonance frequency of the shear-force setup.

Some of these problems may be circumvented by placing the microscope head and the sample in a high vacuum chamber (Behme et al., 1997). This allows cooling of the sample by connecting it to the cold finger of a helium flow cryostat while leaving the temperature of the NSOM scan head uncontrolled and close to room temperature. A schematic diagram of such an instrument is shown in Fig. 9. The entire setup is mounted inside a vacuum chamber that is evacuated to a pressure of 2×10^{-7} mbar with a magnetic-bearing turbo pump and the sample temperature is actively controlled in the range between 8 and 330 K with a stability of better than 0.1 K. A tuning fork setup (Fig. 8b) is used for distance regulation. For fine positioning of the tip perpendicular to the sample, a piezoelectric translator with a vertical scan range of 10 μm is used, whereas in the x–y plane parallel to the sample surface fine positioning is achieved with a hardware-linearized, vacuum-compatible x–y monolithic scan stage. The maximum scan area of the stage is 100 μm $\times$ 100 μm. Coarse positioning is accomplished by three linear translation stages that permit motion in the x, y, and z directions over a range of up to 16 mm. The translation stages are driven by motorized actuators with a minimum step size of 0.5 μm. A long-working-distance microscope objective with a numerical aperture of 0.4 permits light collection, far-field excitation, and visual inspection of the tip–sample region. Near-field photoluminescence spectra of semiconductor nanostructures taken with this instrument give no indication of a significant increase of the local sample temperature directly below the tip relative to that of the cold finger of the cryostat, despite the presence of the near-field tip, which is at room temperature. Local sample temperatures down to 10 K are reached.

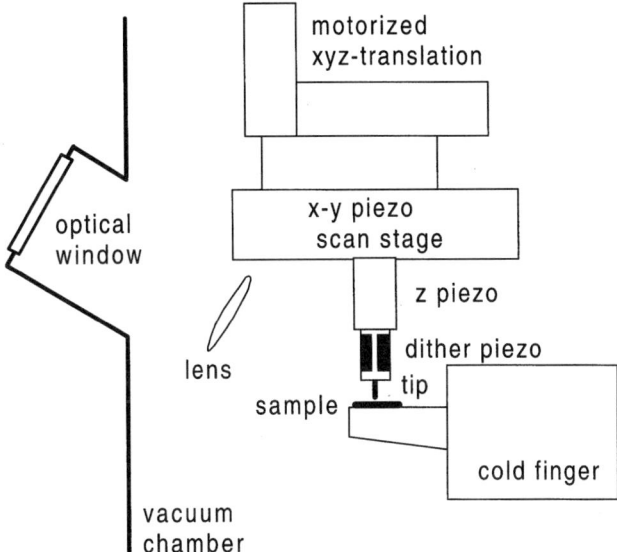

FIG. 9. Schematic diagram of the NSOM scan head according to the design of Behme et al. The sample is mounted on a sample holder connected to the cold finger of a continuous flow helium cryostat. The temperature of this cold finger is actively controlled in the range between 8 and 330 K. The NSOM scan head consists of a near-field fiber tip, a tuning fork shear-force tip-to-sample distance regulator, a 10-μm piezoelectric actuator for z motion, a 100 μm × 100 μm linearized x-y piezo scan stage, motorized x-y-z translation stages for coarse positioning, and a 0.4 NA microscope objective for far-field illumination or luminescence collection. The entire setup is mounted inside a vacuum chamber at a base pressure of 3×10^{-7} mbar.

Experimental results obtained with this instrument are discussed in the following sections.

7. TEMPORALLY AND SPATIALLY RESOLVED NEAR-FIELD SPECTROSCOPY

Time-resolved optical spectroscopy, in particular with femtosecond resolution, has provided a wealth of information on dynamic processes such as the dephasing, scattering, and energy relaxation of free carriers and excitons in semiconductor nanostructures and has led to a detailed understanding of these processes (Shah, 1999). In general, however, such studies give only indirect information on the ballistic and diffusive real-space transfer of photogenerated carriers or their trapping into low-dimensional nanostructures, as these processes involve the real-space motion of carriers on typical length scales of 10–1000 nm, that is, beyond the spatial resolution of far-field spectroscopy. Also, far-field resolution is often not sufficient to isolate the ultrafast response of a single nanostructure, resulting in investigations of inhomogeneously broadened ensembles of nanostructures. This triggered considerable experimental efforts toward combining the techniques of time-resolved spectroscopy and near-field microscopy. The most straightforward approach is to time resolve the laser-induced luminescence using electronic techniques such as time-correlated single-photon counting or synchroscan streak camera detection. In particular, the first method is characterized by a extremely high detection sensitivity. The time resolution of this technique is limited by the available single-photon counting detectors to about 20 ps for microchannel-plate photomultipliers and about 100 ps for avalanche photodiodes. To further improve the temporal resolution into the femtosecond domain, nonlinear pump-and-probe techniques have been combined with NSOM. In a typical femtosecond pump-and-probe experiment, the sample is excited by a first pump pulse. The time-dependent response of the sample to this excitation pulse is monitored via the pump-induced changes (e.g., in sample transmission or reflectivity) with a second time-delayed probe pulse. Ideally, the temporal resolution in such an experiment is only limited by the duration of laser pulses and may be as short as a few femtoseconds. Subwavelength spatial resolution in such an experiment is achieved by transmitting pump and/or probe laser pulses through a near-field aperture probe.

A number of ultrafast near-field pump-and-probe setups have been described in the literature. The pump-induced transmission changes in GaAs microdisks have been investigated in a far-field pump/near-field probe configuration (Stark *et al.*, 1996). Smith *et al.*, (1998) chose a degenerate equal-pulse correlation technique sending both pump and probe beams through a fiber probe to study transmission changes. Strain-induced reflectivity changes in a patterned gold nanostructure were measured using far-field pump and probe pulses and collecting the reflected probe light with the near-field probe (Vertikov *et al.*, 1996). Levy *et al.* (1996a, 1996b) analyzed the spin-dependent diffusion of excitons at low temperatures in patterned magnetic semiconductors by a luminescence intensity autocorrelation technique. Studies of carrier diffusion in ion-implanted patterned GaAs QWs by measuring local differential transmission changes induced by a far-field pump pulse have been reported (Achermann *et al.*, 1999; Nechay *et al.*, 1999a, 1999b).

When combining femtosecond techniques with near-field spectroscopy, a number of experimental issues have to be addressed.

(i) Femtosecond pulses propagating through the fiber probe experience a substantial broadening in time because of the group velocity dispersion in the fiber. For a bandwidth-limited Gaussian input pulse of duration τ_{in}, the output pulse length τ_{out} after propagation through a dispersive material of length L is given by $\tau_{out} = \tau_{in}\sqrt{1+4|k_d|^2 L^2/\tau_{in}^4}$ (Diels and Rudolph, 1996), where the group velocity dispersion (GVD) parameter $k_d = \partial^2 k/\partial \omega^2$, with k

being the wave vector and ω the angular frequency. Higher order dispersion is neglected in this formula. For quartz single-mode fibers at laser wavelength around 800 nm, $k_d \sim 100$ ps^2/km. In 50 cm of quartz single-mode fiber, a 50-fs input pulse is thus stretched to a chirped pulse about 2 ps long. Thus, a GVD precompensation setup using a prism and/or grating sequence must be introduced before the fiber in order to add a negative GVD that compensates the positive GVD of the fiber. For a 50-cm-long fiber, pulse widths of about 100 fs are routinely obtained with standard prism compressors, whereas third-order dispersion should also be compensated to reach shorter pulse widths. In addition to GVD, self-phase modulation of ultrashort pulses due to the third-order nonlinearity of the fiber material may result in a substantial distortion of the pulse spectrum. With unamplified laser pulses taken from mode-locked oscillators, these effects are generally small as long as the average power that is launched into the fiber is kept below 10 mW (pulse energies of about 0.1 nJ). As the optical powers that can be launched into metal-coated fiber probes are limited to some few milliwatts because of tip heating, self-phase modulation is generally of little concern.

(ii) Typical transmission efficiencies of metal-coated aperture probes are on the order of 10^{-5}–10^{-4}, corresponding to transmitted, far-field detected power levels of about 10–100 nW. Typical differential transmission or reflection changes in single semiconductor nanostructures are on the order of 10^{-4} or even less. This means that in order to detect such a small differential transmission change within an integration time of about 1 s, the noise level on the photodetector for the transmitted probe beam should typically be less than 1 pW. This can be achieved with pin or avalanche photodiodes in combination with low-noise preamplifiers. Apart from detector shot noise, a major noise source is background scattered light of the pump laser. In particular in far-field pump/near-field probe geometries, the pump power that is reflected off the fiber tip is typically several orders of magnitude higher than that of the probe, requiring efficient pump suppression. In quasi-two-color configurations, spectral filtering is effective and is generally combined with other methods such as polarization or spatial filtering. Also, dual-frequency modulation of pump and probe beams and signal detection at either the difference or sum frequencies is an efficient means of background suppression. Yet, care has to be taken to avoid a mixing of pump and probe beams due to the photodetector nonlinearity, which may give rise to a time-delay-dependent background signal.

A schematic of a setup for a quasi-two-color near-field pump-and-probe experiment is shown in Fig. 10. Both pump and probe pulses are derived from a mode-locked Ti:sapphire oscillator providing 20- to 40-fs pulses that are tunable in the wavelength range from 810 to 870 nm. The laser works at a repetition rate of 80 MHz and gives an average power of up to several

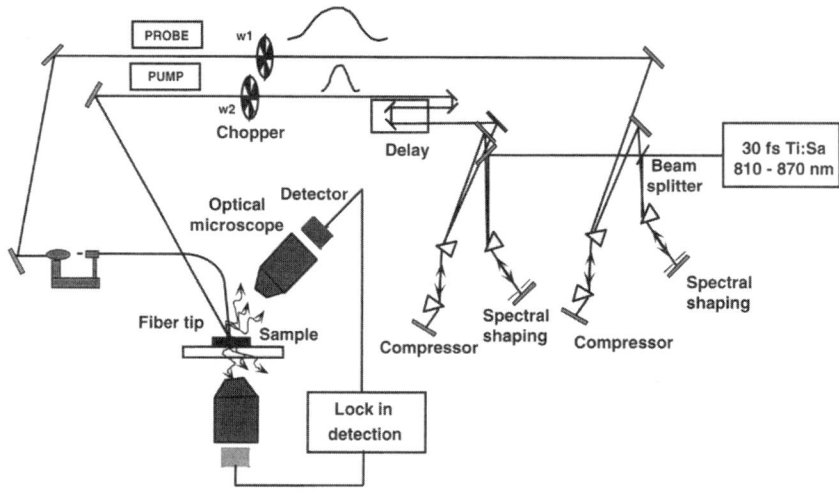

FIG. 10. Schematic of the experimental setup for femtosecond near-field spectroscopy. Pump and probe pulses are derived from a mode-locked Ti:sapphire laser using two independent arrangements for spectral shaping and compression. The pump excites a large area of the sample, whereas the probe is fed into a near-field probe, allowing for a spatially resolved detection of ultrafast transmission or reflection changes. The nonlinear changes of transmission or reflection are measured with the help of a lock-in technique.

hundreds of milliwatts. The laser output is split into a pump and a probe beam, each of which is traveling through a separate prism setup for spectral selection and precompensation of group velocity dispersion. The pump beam is chopped at a frequency f_1. The pump can either be focused onto the sample by a far-field $f = 8$ cm graded index lens, resulting in a spot size on the sample of about 30 μm, or be coupled into the fiber probe, resulting in a subwavelength excitation spot. The probe beam is chopped at a frequency f_2, coupled into the fiber, and transmitted onto the sample through the subwavelength aperture at the end of the 50-cm-long fiber probe. The time delay between pump and probe is adjusted with a delay stage in the pump arm. After interaction with the sample, the transmitted or reflected probe light is collected through standard long-working-distance microscope objectives and detected by a photodiode in conjunction with a lock-in amplifier. For suppression of background signals, the lock-in amplifier detects the amplified photodiode signal at the at the sum frequency $f_1 + f_2$ or difference frequency $f_1 - f_2$. Either acoustooptical modulators with modulation frequencies in the megahertz range or standard mechanical choppers with kilohertz chopping frequencies may be used. High-frequency modulation of pump and probe beams in the megahertz range significantly reduces thermal effects due to the periodic heating and cooling of the tip that occurs on longer millisecond time scales (Stahelin et al., 1996). In addition, the difference frequency can be chosen in the 50–100 kHz range to reduce the effects of laser noise.

IV. Stationary Near-Field Spectroscopy of Semiconductor Nanostructures

1. Low-Temperature Near-Field Spectroscopy

The potential of near-field scanning optical microscopy for the spectroscopy of semiconductor nanostructures was first demonstrated by Hess et al. (1994). Combining low-temperature near-field microscopy and luminescence spectroscopy, the emission from individual luminescence sites in single GaAs QWs was spatially and spectrally resolved. Monolayer fluctuations of the QW thickness result in local variations of the QW confinement potential. Photogenerated electron–hole pairs are trapped in local minima of the confinement potential. The spatially resolved luminescence spectra consist of a pattern of sharp lines originating from the different sites. In a similar experiment, the effects of electron localization in a two-dimensional electron gas produced by modulation doping of a QW structure were clarified (Eytan et al., 1998). Here, the presence of ionized donor atoms in the remote dopant layers gives rise to random potential fluctuations in the plane of the two-dimensional electron gas. These fluctuations could be mapped by recording the spatial fluctuations of the luminescence intensity of negatively charged excitons. Potential fluctuations also arise from disorder in compound semiconductors. This was investigated in a near-field photoluminescence (PL) study of partially ordered (GaIn)P where a spatial anticorrelation of distinct photoluminescence bands was found (Gregor et al., 1995).

In the first near-field spectrosocpic study of quasi-one-dimensional QWR structures, the spatial origin of different emission peaks was analyzed in an array of GaAs/AlGaAs QWRs made by cleaved edge overgrowth (Grober et al., 1994b). In a later study of strained layer QWRs (Harris et al., 1996); single QWRs were resolved and their photoluminescence excitation (PLE) spectrum was measured. A factor of three enhancement in the absorption strength of the QWR with respect to that of a comparable QW was estimated.

Considerable effort has been devoted to the near-field spectrosopic investigation of single QDs. First experiments focused on spatially imaging single QDs (Saiki et al., 1998; Toda et al., 1996, 1998a, 1998b). In a near-field PL study of single self-assembled InAs/GaAs QDs under high magnetic fields, the Zeeman splitting of individual emission lines of single dots was observed and spin-flip transitions were studied using magneto-PLE spectroscopy (Toda et al., 1998b). Most recently, a near-field study of the PLE spectrum of a single InGaAs/GaAs self-assembled QD was presented (Toda et al., 1999). The PLE spectrum shows, in addition to a number of sharp lines, signatures of two-dimensional continuum states. The presence of these states gives rise to an efficient intradot relaxation that is suggested to proceed via the relaxation of carriers within the continuum states and subsequent transitions to the excitonic ground state by resonant emission of localized phonons. It should be noted that single QDs have also been studied by other techniques, for example, by using metallic masks with subwavelength pinholes on top of the samples (Bonadeo et al., 1998a, 1998b; Gammon et al., 1996a, 1996b).

2. NEAR-FIELD SPECTROSCOPY OF QUANTUM WIRES ON PATTERNED (311)A GaAs SURFACES

In this section, we review a detailed near-field study of single GaAs QWRs grown on (311)A GaAs substrates (Lienau et al., 1998a, 1998b; Richter et al., 1997a, 1997b, 1997c, 1998a). This QWR geometry was explained in Subsection 2 of Section II, and the experiments discussed here were performed with the sample schematically shown in Fig. 2. Prior to performing near-field experiments on this sample, the sample was characterized by microphotoluminescence (μ-PL) spectroscopy with a spatial resolution of about 1.5 μm. For excitation on the flat part of the mesa structure, a μ-PL spectrum recorded at 8 K (dashed line) shows a single peak at 1.602 eV originating from the 6-nm-thick QW (Fig. 11, upper panel). At room temperature (solid line), the reduced bandgap results in a red shift of the emission spectrum. On the sidewall, however, the low-temperature μ-PL spectrum shows two well separated peaks at 1.540 and 1.605 eV (Fig. 11, lower panel). The peak at 1.540 eV is assigned to QWR emission, and the one at 1.605 eV to emission from the adjacent QW. The large energy difference between the QW and QWR emission lines of about 60 meV indicates a pronounced confinement of quasi-one-dimensional states. This is confirmed by the observation of a strong QWR emission peak, spectrally well separated from the QW emission, over the whole temperature range from 8 K up to 300 K. In the sidewall spectrum, taken at room temperature, the QWR emission at 1.46 eV is still well separated from the heavy-hole and light-hole QW peaks at 1.526 and 1.555 eV, respectively.

Detailed information on the local energetics of the QWR nanostructure is obtained by combining PLE spectroscopy and near-field microscopy for mapping the local luminescence properties of the QWR structure in real space with subwavelength resolution (Lienau, 1999; Lienau et al., 1998b; Richter et al., 1997c). In such an experiment, the sample is locally excited through the NSOM probe with a tunable narrowband Ti:sapphire laser and

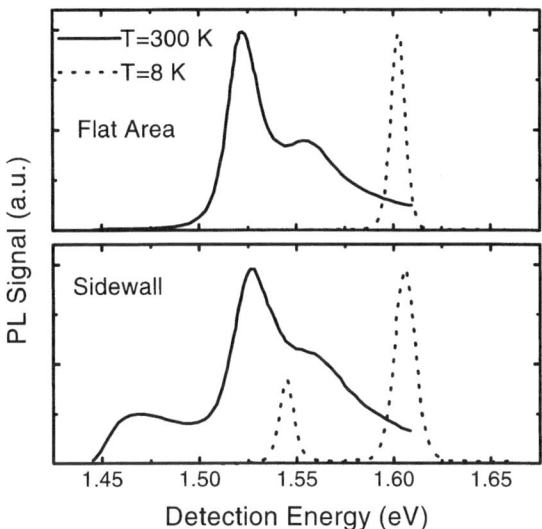

FIG. 11. Micro-PL spectra taken at room temperature (solid line) and 8 K (dotted line) from the flat area of the mesa (upper part) and at the sidewall (lower part). The peak in the lower spectra at 1.47 eV (300 K) and 1.545 eV (8 K) is due to the sidewall quantum wire structure.

the QWR luminescence is collected in the far field. The luminescence is spectrally dispersed in a 0.22-m double monochromator and detected with a single photon counting silicon avalanche photodiode. The experiments were carried out in a home-built low-temperature near-field microscope as described in Subsection 6 of Section III (Behme et al., 1997). In these measurements, the metal-coated tapered fiber tip is scanned along the lateral y direction perpendicular to the QWR located at $y = 0$. In Fig. 12a, the intensity of the QWR luminescence at a detection energy of 1.544 eV (spectral resolution 2 meV), that is, slightly below the maximum of the QWR emission, is plotted as a function of excitation energy (abscissa) and of lateral distance of the exciting fiber tip from the QWR position at $y = 0$ (ordinate). The PLE spectrum is spatially confined to the QWR region. For excitation energies below 1.61 eV, the width of this region has a value of about 250 nm, as is evident from a spatial cross section along the y direction (Fig. 13). This represents the spatial resolution of the experiment, which is limited by the finite distance of 70 nm between the buried QWR and sample surface (cf. Fig. 2) and the finite aperture size of about 200 nm used in this measurement. In Fig. 12b, a cross section through the image of Fig. 12a along the energy (x) axis for $y = 0$ is shown. This PLE spectrum displays several maxima below the onset of QW absorption at 1.63 eV. Such peaks are due to the quasi-one-dimensional subband structure of the QWR and reflect the QWR absorption spectrum (Rossi and Molinari, 1996). Calcula-

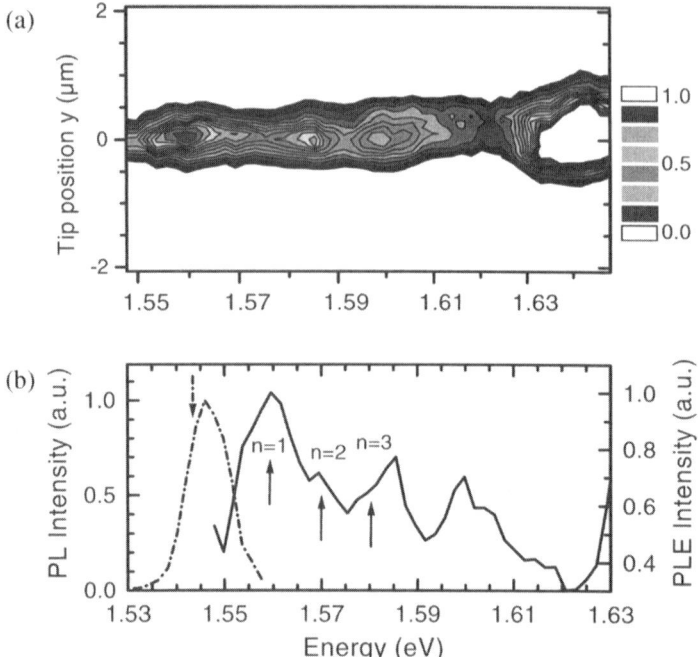

FIG. 12. (a) Near-field PLE spectrum recorded for spatially resolved excitation of the QWR sample through the NSOM fiber probe at a temperature of 10 K. The intensity of the QWR luminescence at 1.544 eV is plotted as a function of excitation energy (abscissa) and of lateral distance between the QWR located at $y = 0$ and the fiber tip (ordinate). Only carriers excited at the QWR contribute to the signal. (b) Cross-section through (a) at $y = 0$ (solid line) and QWR near-field luminescence spectrum (dash–dotted line).

tions of the 1D subband structure within the adiabatic approximation (Kapon et al., 1989a) indicate that the first peaks are due to the $n = 1$ (1.558 eV), $n = 2$ (1.570 eV), and $n = 3$ (1.583 eV) excitonic heavy-hole to conduction band transitions (arrows in Fig. 12b). The splitting between the $n = 1$ and $n = 2$ transition is on the order of 12 meV. This is in good agreement with the width of this QWR structure of about 50 nm, which was estimated from TEM images and observed diamagnetic shifts (Nötzel et al., 1996c). The strong intensities of the third peak (at 1.583 eV) and the fourth peak (at about 1.60 eV) suggest a significant contribution from excitonic transitions between light-hole valence and conduction band states. The role of such transitions has been addressed both experimentally (Vouilloz et al., 1997) and theoretically (Goldoni et al, 1997) in investigations of valence band mixing effects on the linear optical properties of QWRs. We also note that the shape of the spectrum, in particular the strong decrease of the PLE

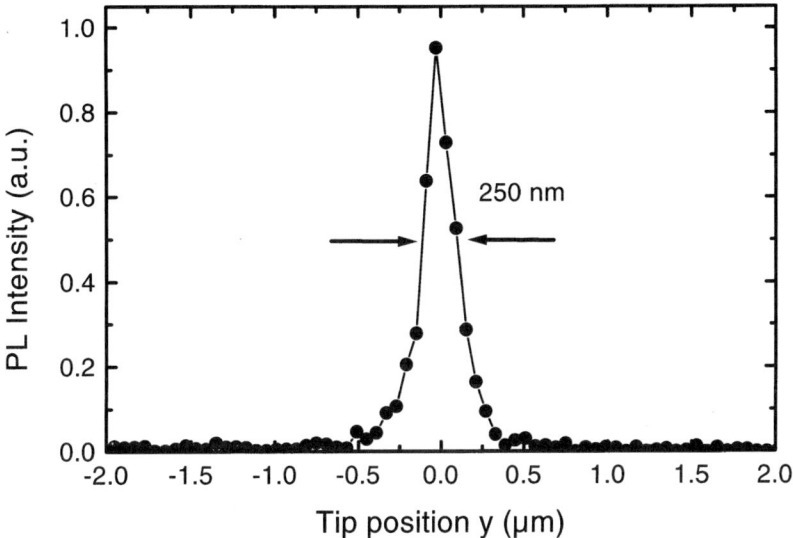

FIG. 13. QWR luminescence intensity as a function of excitation tip position. Spatially resolved excitation of the QWR sample at 1.56 eV is performed by transmitting the excitation laser through the NSOM fiber probe. The tip is scanned along the lateral y direction, perpendicular to the QWR axis. The sample temperature is 10 K and the detection energy 1.54 eV. The spatial resolution of about 250 nm is mainly limited by the finite distance of 70 nm between QWR and sample surface.

intensity around 1.62 eV, indicates that the observed peaks are mainly excitonic. Support for the pronounced modification of the absorption spectra by Coulomb correlation effects has been given in a theoretical analysis (Rossi and Molinari, 1996).

The small inhomogeneous broadening of the spectra of about 10 meV reflects the structural properties of the corrugated (311)A surface. The absorption line comprises contributions from different localized excitons, as observed for QWs on low-index GaAs surfaces (Gammon et al., 1996a, 1996b; Hess et al., 1994). On (311)A surfaces the typical lateral size of monolayer-high islands is much smaller than on (100) surfaces (Gammon et al., 1996a), where these islands can extend to diameters of more than 100 nm. (311)A surfaces have been shown to spontaneously form a corrugation with a lateral periodicity of 3.2 nm and a step height of two monolayers (0.34 nm). On a larger, 20–40 nm scale, STM images of these surfaces show mesoscopic scale roughness of few monolayers (Wassermeier et al., 1995). Because of this short corrugation length, components from individual localized excitons are not resolved even on a 100-nm length scale, and the NSOM PLE spectrum shows significant inhomogeneous broadening.

Similar excitation spectra of the QWR luminescence are observed for detection on the high-energy side of the QWR spectrum at 1.550 eV (Lienau et al., 1998a) and at different spatial positions along the wire axis. This points to a predominantly inhomogeneously broadened QWR PL spectrum. The rise at energies above 1.63 eV is due to the onset of QW absorption. From its separation to the first QWR peak, one derives a confinement energy of 80 meV. Unlike for QWR absorption, the spatial dependence of the QW absorption peak is not limited by the spatial resolution of the experiment. Its spatial intensity variation is reasonably described by a Gaussian profile with a width of about 800 nm. This width is mainly given by the lateral confinement potential in the vicinity of the QWR wire, as discussed later.

The spatially resolved PLE spectrum changes drastically as the sample temperature is increased to 77 K. In Fig. 14, the luminescence intensity detected near the maximum of the QWR emission at 1.533 eV is plotted as a function of excitation energy and position as the tip is scanned perpendicular to the QWR. Again, the excitation light is transmitted through the near-field probe and the luminescence from the sample is collected in the far

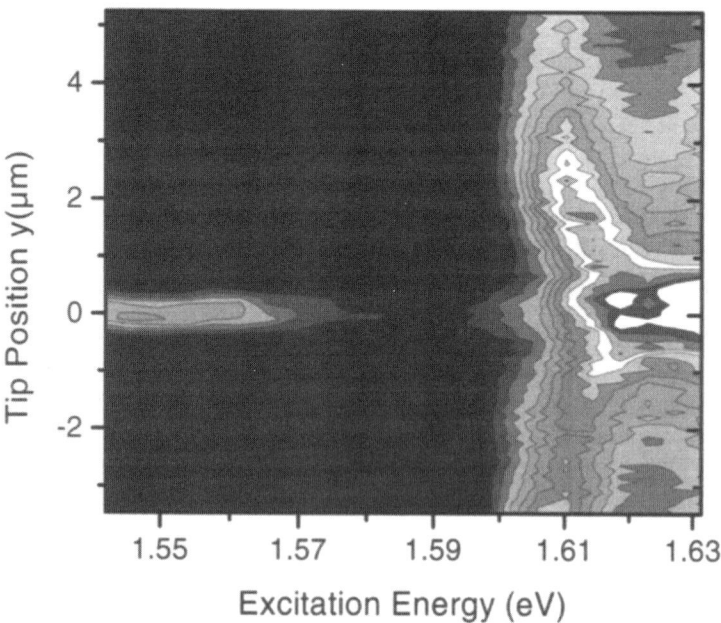

FIG. 14. Near-field PLE spectrum at a sample temperature of 77 K and a detection energy of 1.533 eV, corresponding to the maximum of the QWR emission spectrum. Notice the pronounced spatial blue shift of the excitonic QW absorption peak as the excitation tip approaches the QWR at $y = 0$.

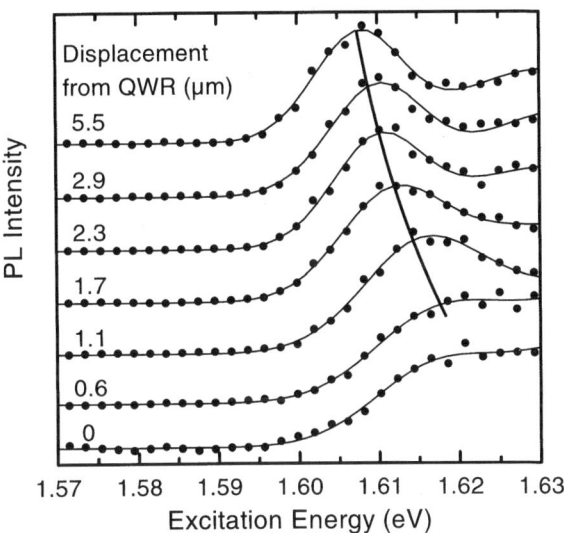

FIG. 15. Cross sections through Fig. 14. Near-field mesa top QW PLE spectra at fixed excitation position relative to the QWR location at $y = 0$ at a sample temperature of 77 K and a detection energy of 1.533 eV. Experimental data points are shown as solid circles; the solid lines represent fits of the data to the Elliot equation.

field. The QWR PLE spectrum appears again as a spatially narrow peak around $y = 0$, slightly red-shifted with respect to the 10 K spectrum. This red shift reflects the decrease of the GaAs bandgap with lattice temperature. The subband-related modulation of the spectrum is washed out because of increased thermal broadening. More important and in contrast to the low-temperature spectrum, strong QWR luminescence is now detected for QW excitation at energies higher than 1.6 eV and at distances y of several micrometers away from the QWR location. The spectral shape of the PLE spectrum for $y \neq 0$ changes strongly with excitation position. Figure 15 shows cross sections through the image of Fig. 14 for different separations from the QWR on the mesa top ($y > 0$) (symbols). The peak in these spectra, which is due to the excitonic enhancement of QW absorption, shifts by about 14 meV to higher energies as one moves from a distance of $y = 5.5\,\mu m$ to the location of the QWR. On the mesa bottom ($y < 0$), this shift is even more pronounced with PLE maxima at 1.610 eV for $y = -3.5\,\mu m$ and at 1.628 eV for $y = -0.4\,\mu m$. The data demonstrate that carriers that are locally generated within the QW undergo real-space transfer to the location of the QWR, where they are trapped into quasi-one-dimensional QWR states and contribute to the QWR emission. The time scale on which real-space transfer occurs is given by the recombination lifetime of carriers within the QW. Time-resolved measurements under the same excitation conditions give a lifetime of QW PL of 1.2 ns at 77 K. This time scale is much longer than typical thermalization and cooling times of carriers, which occur on the femto- and picosecond time scales (Shah, 1999). As a result, the carriers undergoing real-space transfer form a quasiequilibrium distribution with a temperature close to lattice temperature. This is discussed in detail in Section V.

The PLE spectra recorded on the QW are closely related to the local QW absorption spectrum at the specific excitation position. The overall intensity at a given position is proportional to the fraction of photoexcited carriers trapped into the QWR. The results in Fig. 15 reveal (i) a pronounced excitonlike feature in the absorption of the QW and (ii) a blue shift of this maximum with decreasing separation from the QWR. To analyze this behavior in a more quantitative way, we used Elliott's formula (Haug and Koch, 1993) for the absorption spectrum of a quasi-two-dimensional semiconductor. The dominant features are the lowest heavy-hole exciton absorption peak and the continuum absorption at higher energies. At each excitation position, the QW absorption spectrum in Fig. 15 was fitted to Elliot's formula using a Gaussian profile to account for the inhomogeneous broadening of the exciton peak and the continuum absorption that is probably due to interface roughness of the QW. The solid lines in Fig. 15 represent the calculated spectra and are in good agreement with the experimental results. The calculation gives the local bandgap energy $E_g(y)$ as a function of excitation position y. A plot of the lateral variation of the QW bandgap vs y is shown in Fig. 16. We note that the change in QW thickness near the QWR is correlated with an increase in inhomogeneous broadening. This is directly observed as an increase in the spectral width of the exciton absorption peak.

On each side of the QWR, the potential shows two pronounced maxima, separated by approximately 1 μm. On the mesa bottom, $y < 0$, the bandgap increases by as much as 18 meV over a length scale of about 2 μm as approaching the location of the QWR, while on the mesa top, $y > 0$, the bandgap shift is slightly less pronounced and amounts to about 14 meV. The change in bandgap energy originates mainly from a change in the average local thickness of the QW, which decreases from about 5.6 nm on the flat-area mesa top and bottom to 4.8 nm ($y < 0$), respectively 5.0 nm ($y > 0$), in the vicinity of the QWR. This thinning is a consequence of the Ga atom migration toward the sidewall during the growth process and is determined by the specific MBE growth parameters, in particular the substrate temperature. The asymmetry in the bandgap profile on mesa top ($y > 0$) and bottom ($y < 0$) is most likely related to the wet chemical etching procedure that is used to pattern the mesa structure, and the resulting changes of Ga migration.

Based on these results, the complete confinement potential of the quantum-well embedded QWR is derived. Two different spatial regions are

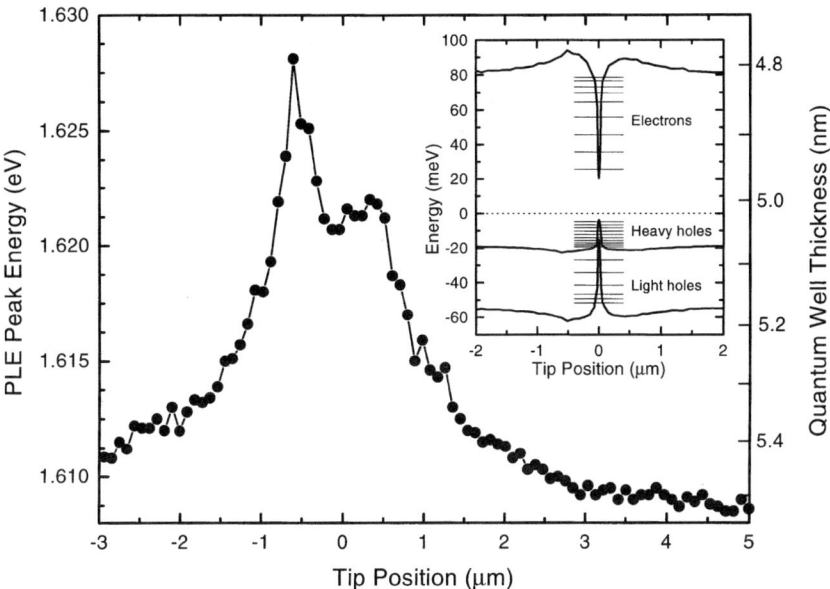

FIG. 16. Lateral bandgap profile as obtained from an analysis of the QW PLE spectrum shown in Fig. 14. The lateral shift of the excitonic QW absorption peak corresponds to a decrease of the average QW thickness in the region close to the QWR. Inset: QWR confinement potential as derived from the experimental data and calculated energies of the 1D subbands (lines).

distinguished. In the region $|y| > 100$ nm, the potential can be taken directly from the results of Fig. 16, whereas in the central 100 nm around $y = 0$ the potential is derived from the cross-sectional TEM images (Fig. 2), supported by the PLE spectrum at 10 K. In the inset of Fig. 16, the potential is plotted as a function of y for the whole interval between -2 and $+2$ μm. Using this information and a ratio of 2:1 for the conduction to valence band offset energy in the 2D GaAs/Al$_{0.5}$Ga$_{0.5}$As structure, the quasi-one-dimensional subband structure and the interband transition energies of the QWR can now be calculated within the adiabatic approximation. We find a splitting of 10 meV between the two lowest conduction band subbands, in agreement with the low-temperature PLE spectrum. We note that Coulomb correlation effects that affect both the position and shape of the peaks in the PLE spectrum are neglected in these calculations. Since the height of the local barriers of about 15 meV is only a small fraction of the QWR confinement energy and since the barriers are separated by about 1 μm, the influence of the barriers on the energetic position of the lowest QWR subbands is rather small. Nonetheless, the presence of the barriers strongly influences the carrier transport within the sample, as is evidenced from the temporally and spatially resolved PL measurements discussed later.

In addition to the effect of the local variation of the QW thickness on the lateral confinement potential, the height and shape of the energetic barriers separating QWR and QW may be influenced by the local strain distribution, which is induced by growing the QWR on a prestructured mesalike substrate. We have recorded a series of low-temperature (10 K) excitation spectra of the QW luminescence to clarify this issue. In Fig. 17, spatially resolved PLE spectra are presented. QW luminescence was detected in the far field at 1.598 eV, and the photon energy of the near-field excitation was tuned between 1.605 and 1.675 eV. The relatively strong intensity fluctuations in those spectra are due to the localization of QW excitons at low temperature, which corresponds to local variations of the luminescence transition energies. For excitation energies between 1.605 and 1.63 eV, one finds a continuous blue shift of the onset of QW absorption with decreasing separation from the QWR, in agreement with the data of Fig. 14. The central dark area around $y = 0$ corresponds to the region between the two

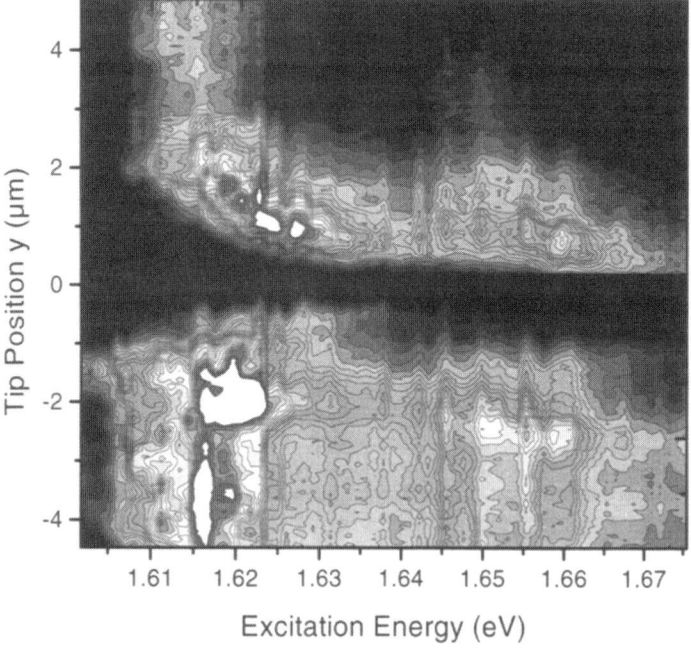

FIG. 17. Near-field PLE spectrum of the quantum well for a lattice temperature of 10 K (detection energy 1.598 eV). The data taken for photon energies between 1.605 and 1.63 eV reveal the continuous blue shift of the onset of quantum well absorption with decreasing separation from the quantum wire, in agreement with the data of Fig. 14.

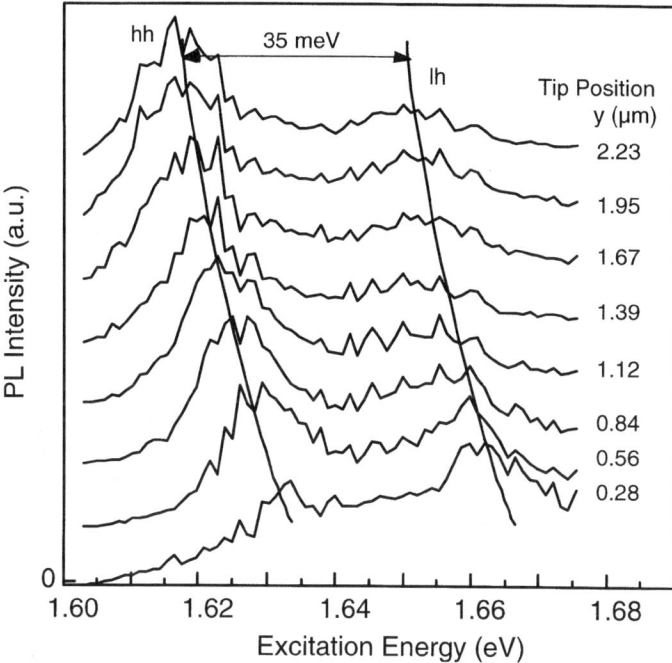

FIG. 18. Cross sections through Fig. 17 for different separations y between excitation tip and quantum wire. Both the heavy- and the light-hole excitonic peaks of the quantum well PLE spectrum show a blue shift when approaching the quantum wire. In contrast, the energy separation of the heavy- and light-hole peaks is nearly the same for all y values.

shallow barriers where carriers are trapped very efficiently into the QWR and consequently do not contribute to the QW luminescence. Figure 18 gives cross sections through Fig. 17 for different separations from the QWR position. Both the heavy- and the light-hole peaks in the QW excitation spectra are resolved. The two peaks show a continuous blue shift with decreasing distance from the QWR, whereas the energy separation of the two peaks, that is, the heavy-hole light-hole splitting, remains essentially unchanged. The absolute value of this splitting of about 35 meV is in good agreement with theoretical estimates for an unstrained 6-nm-wide GaAs QW. We conclude from this agreement and from the constant heavy–light hole splitting that strain plays a minor role for the observed blue shift of the excitonic peaks. Instead, the shift is related to the local thinning of the QW in the range of the shallow barriers.

In addition to providing detailed information on the local energetics of nanostructures, near-field spectroscopy gives insight into the spatial homogeneity of such structures and in particular on carrier transport occurring

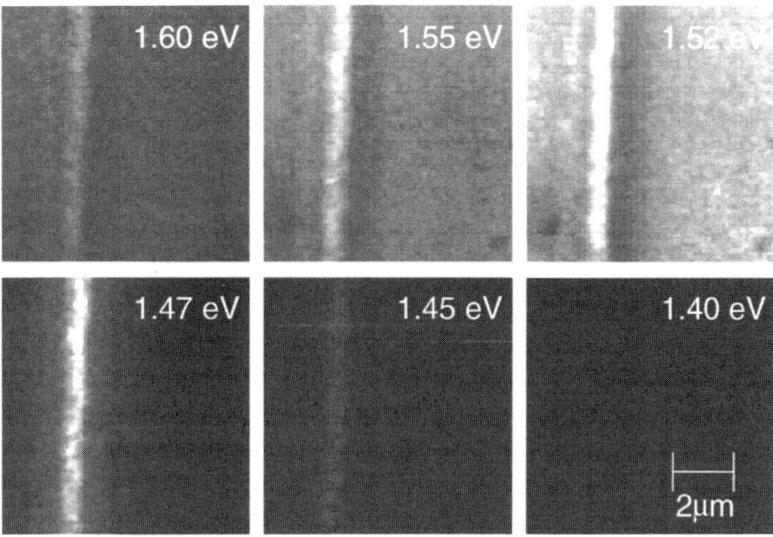

FIG. 19. Two-dimensional near-field photoluminescence images taken at room temperature for detection energies between 1.60 and 1.40 eV. The excitation energy was chosen as 1.959 eV. The scan range in these images is 8 μm × 8 μm. Note the strong quantum wire emission at 1.47 eV.

on subwavelength length scales. To investigate the sample homogeneity along the wire axis, we recorded two-dimensional near-field photoluminescence images at room temperature (Fig. 19). Here, the sample was locally excited with a HeNe laser at 1.959 eV that was transmitted through an aluminum-coated fiber probe. This excitation generates carriers in highlying continuum states in both the QW and QWR, but not in the AlGaAs barriers. The photoluminescence of the sample was detected for various detection energies between 1.40 and 1.60 eV. In Fig. 19, the detected PL intensity is plotted as a function of the tip position. The scan range in these images is 8 μm × 8 μm with a pixel size of 0.1 μm. All images have been recorded under identical excitation conditions. In these images, the black and white areas correspond to zero and maximum PL intensity, respectively. For detection energies between 1.45 and 1.60 eV, the images clearly show the correlation of the QWR emission with the location of the QWR. The emission appears homogeneous over the entire length of the QWR, confirming the high structural quality of the sample. A detailed analysis shows slight nonuniformities in the emission pattern that are related to nonuniformities of the etching process by which the (311)A substrate was prepared. At energies between 1.5 and 1.6 eV, spatially homogeneous emission from the embedding QW is observed in addition to QWR luminescence.

Steady-state near-field spectra that are recorded in an illumination configuration provide important information on carrier transport phenomena in semiconductor nanostructures. In this experimental configuration, only the excitation position is fixed, while real-space transfer of carriers may occur during the lifetime of the optically excited carriers. Since the luminescence is collected in the far field from a sample area of about 10 μm diameter, the experiment is sensitive to carrier transport on a length of about 100 nm to 10 μm. In spatially inhomogenous samples, the position where the emission occurs may be fixed by a spectrally selecting a specific emission wavelength. Thus, in near-field PL spectra, both the carrier generation and the recombination point may be fixed by combining a technique with inherently high spatial resolution with an optical marker concept. This potential of this concept is demonstrated in near-field photoluminescence spectra of the QWR sample that are recorded over a wide range of temperatures between 10 and 300 K (Fig. 20). At 300 K, QWR emission appears as a spectrally and spatially broad peak around 1.47 eV. The cross section through the QWR PL spectrum shows (i) a narrow peak at $y = 0$ arising from direct QWR excitation and (ii) broad tails that decay exponentially over a length of several micrometers in the QW area. The occurrence of QWR luminescence after localized QW excitation at distances of up to several micrometers to the QWR involves carrier drift diffusion within the QW and subsequent trapping into the QWR. At room temperature, the thermal energy is larger than the local energy barriers, with a height of about 15 meV, that are found to separate the QWR from the flat area QW (Fig. 16). Their effect is thus small and the transport will be mainly diffusive. Correspondingly, the QWR PL signal $I_{QWR}(y)$ is expected to decrease exponentially with increasing distance from the QWR, that is, $I_{QWR}(y) \propto \exp(-|y|/L_d)$. For a tip separation $|y| > 1$ μm, the data in Fig. 20 show this behavior. A quantitative analysis gives a diffusion length of $L_d = 1.6$ μm. Using the relations $D = L_d/t_{rec}$ and $\mu_{eff} = De/(kT)$ for the diffusion coefficient D and the effective mobility μ_{eff} (t_{rec}, 2.1 ns QW recombination time; e, electron charge; kT, thermal energy), one derives values of $D = 12.2$ cm^2/s and $\mu_{eff} = 490$ cm^2/Vs for $T = 300$ K. Those numbers are characteristic for an ambipolar diffusion process of electron–hole pairs in GaAs QWs in which the overall mobility is determined by hole diffusion.

At lower temperatures, the lateral transport is strongly influenced by the barriers in the vicinity of the QWR (Fig. 20). At 10 K, carrier transfer from the QW into the QWR is fully suppressed, as the thermal carrier energy is much less than the barrier height. This is also evident from the spatially confined shape of the 10 K PLE spectrum in Fig. 12. A more detailed discussion of lateral transport is given in Subsection 1 of Section V.

In summary, these experiments show that the high spatial resolution of the NSOM technique allows separation of the optical spectra of QWR and QW, provides evidence for the low-dimensional carrier confinement, and

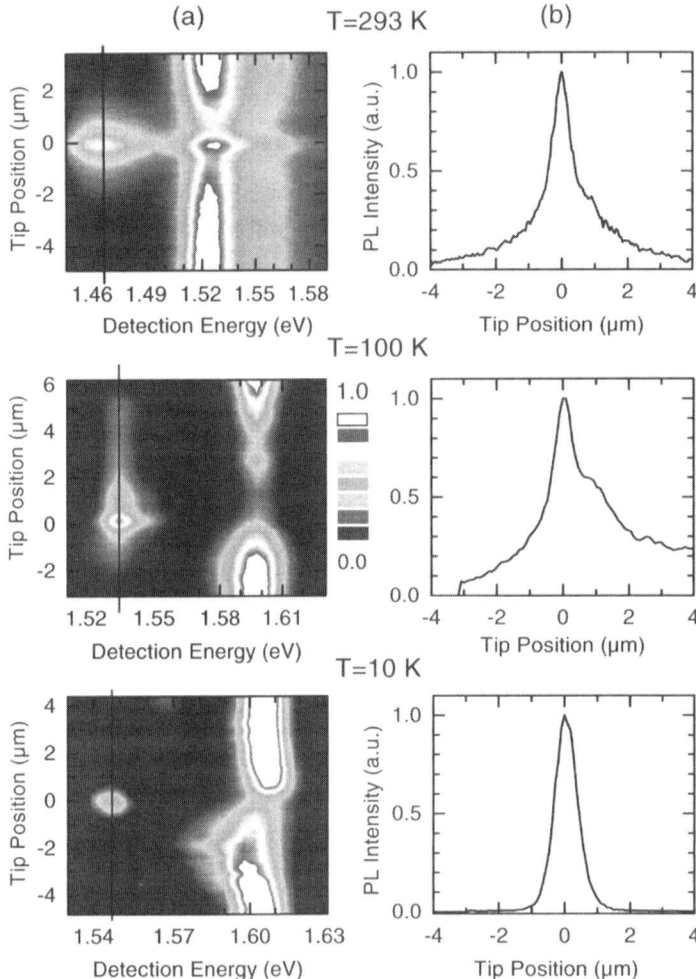

FIG. 20. (a) Near-field photoluminescence spectra of the QWR sample for localized excitation at 1.595 eV at three different sample temperatures of 300, 100, and 10 K. The luminescence intensity is plotted as a function of detection energy (abscissa) and of lateral distance between the QWR located at $y = 0$ and the fiber tip (ordinate). (b) Spatial cross sections through the near-field PL spectra at the energy of the QWR emission maximum.

permits mapping of the local confinement potential in real space. This gives direct access to growth-related properties of nanostructures, as demonstrated for the QWR investigated here. Furthermore, the technique provides a particularly sensitive tool for imaging carrier transport phenomena in real space.

V. Time-Resolved Near-Field Spectroscopy

1. LATERAL CARRIER TRANSPORT STUDIED BY PICOSECOND NEAR-FIELD LUMINESCENCE SPECTROSCOPY

In recent experiments on QWRs on patterned (311)A GaAs surfaces, low-temperature near-field microscopy and time-resolved luminescence spectroscopy were combined to study the influence of the local band structure, as described in the previous paragraph, on the microscopic carrier transport dynamics within the nanostructure (Richter *et al.*, 1998b, 1999).

Femtosecond excitation pulses from a mode-locked Ti:sapphire laser were transmitted through a metal-coated NSOM fiber probe with an aperture of 200 nm to generate a transient electron–hole pair distribution within a subwavelength spot on the GaAs QW. The dimensions of this spot are mainly given by the aperture diameter. The dynamics of this carrier distribution, in particular their real-space transfer toward the QWR and trapping into 1D QWR states, can be monitored by time-resolving the QWR luminescence using time-correlated single-photon counting detection with a temporal resolution of 260 ps.

First, experiments at a sample temperature of 100 K are discussed. The PLE spectra discussed earlier (Fig. 14) and stationary PL experiments (Fig. 20) show that in this temperature range carriers generated in the embedding QW up to distances of several micrometers from the QWR undergo real-space transfer to the QWR and are effectively trapped into the QWR. Moreover, the energy of the potential barriers separating QWR and QW is similar to the thermal energy at this temperature so that we expect a considerable influence of the barrier potential on the carrier dynamics. In Fig. 21, the intensity of the time-resolved QWR emission at a sample temperature of 100 K is shown as a function of the tip position along the lateral y axis (abscissa) and of the delay time. In this experiment, the 16-meV broad spectrum of the excitation laser is centered at 1.614 eV and overlaps the excitonic absorption band of the QW. QWR PL at 1.54 eV, the maximum of the QWR PL spectrum, is collected in the far field, dispersed in a 0.22-m monochromator with a spectral resolution of 1.2 nm, and time resolved. Cross sections through Fig. 21, that is, the time evolution of the luminescence for fixed excitation positions $|y| = 0$ (QWR position), 0.4, 1, and 2 μm are shown in Figs. 22a and 22b for excitation on mesa top and mesa bottom, respectively. In the insets, these data are plotted on a logarithmic intensity scale. For excitation at $y = 0$, we find a rise of the QWR luminescence that is limited by the temporal resolution of the experiment of 260 ps and a decay of the QWR luminescence that is single exponential over two orders of magnitude with a decay time constant of 1.5 ns. This temporal dependence of the QWR luminescence remains

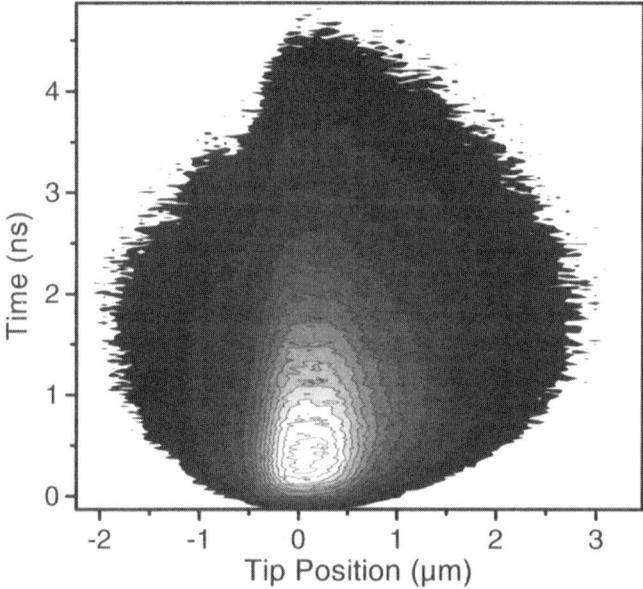

FIG. 21. Spatially and temporally resolved photoluminescence of the QWR at a temperature of 100 K after localized excitation in the embedding quantum well ($E_{ex} = 1.614$ eV, detection energy 1.54 eV). The luminescence intensity is plotted as a function of the lateral distance y between excitation and QWR position and as a function of time.

unchanged for $|y| < 320$ nm, that is, for excitation between the barriers separating QWR and embedding QW. For a distance $|y| > 320$ nm, the onset of the QWR luminescence is delayed with respect to that at $y = 0$. The rise of the QWR luminescence becomes monotonically slower with increasing $|y|$, as is highlighted in Fig. 23, where normalized QWR luminescence profiles for fixed excitation positions of $|y| = 0$, 0.9, and 1.8 µm are compared.

This delayed rise manifests itself in a shift of the maximum of the PL intensity in time with increasing y. In Fig. 24, the temporal position of the maximum PL intensity, $t_{max} - t_0$, is plotted as a function of the excitation position $y(t_0 \approx 0.4$ ns for $|y| < 0.32$ µm). For distances $|y| > 0.32$ µm, the value of $t_{max} - t_0$ increases almost linearly with increasing distance y, with a slope of 0.73 ns/µm for $y < -0.32$ µm and 0.62 ns/µm for $y > 0.32$ µm. The maximum intensity of QWR luminescence after excitation on the mesa bottom ($y < 0$) is significantly smaller than for mesa top ($y > 0$) excitation, that is, fewer carriers are trapped into the QWR. This is evident from the amplitude values in Figs. 22a and 22b and from the shape of the plot in Fig. 21 at positive and negative y.

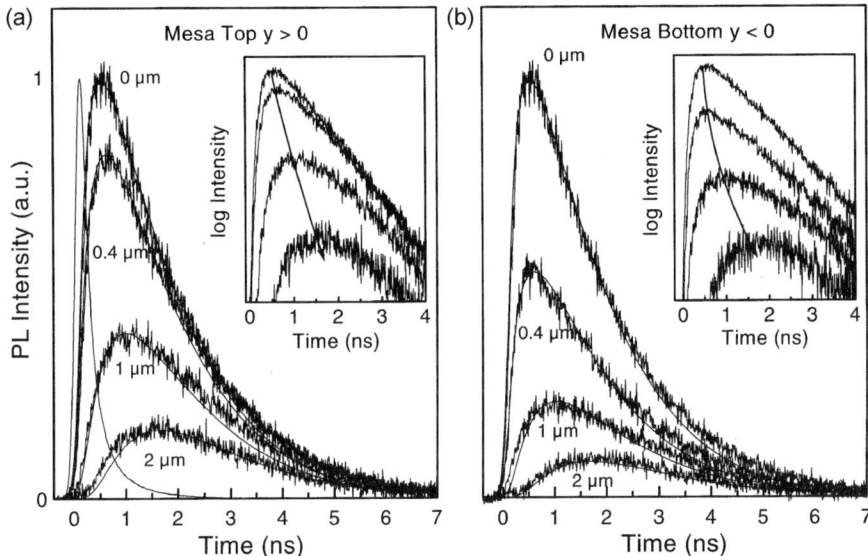

FIG. 22. Time evolution of the QWR luminescence after excitation on (a) the mesa top and (b) the mesa bottom of the QWR structure. The luminescence intensity at 1.54 eV is plotted vs time for different excitation positions y and shows an increasingly delayed rise with increasing y that is due to the transport of excitons from the excitation spot to the QWR. Insets: Data plotted on a logarithmic intensity scale. Solid lines: Results of a numerical calculation based on a drift-diffusion model for exciton transport.

In the following, the luminescence behavior found for different excitation positions y is discussed. For $|y| < 0.32$ μm, the single exponential decay of the QWR PL gives a population lifetime of the QWR (Akiyama et al., 1994; Gershoni et al., 1994) of 1.5 ns, slightly longer than the PL decay time of the embedding QW of 1.35 ns. The slow single exponential decay of the QWR PL is indicative of the high structural quality of the QWR sample. The fast rise of the QWR PL after excitation at $|y| < 0.32$ μm demonstrates that carrier relaxation from QW continuum states into localized QWR states occurs within the time resolution of the experiment of 260 ps. This relaxation involves both trapping into and energy relaxation within the QWR. For our very low excitation density of $\leqslant 10^5$ cm^{-1}, carrier–carrier scattering is negligible and emission of longitudinal optical phonons represents the dominant process populating low-lying QWR states (Ryan et al., 1996).

The onset of the QWR luminescence for excitation at $|y| > 0.32$ μm, that is, for photogeneration of carriers in the QW, occurs with a delay of several hundred picoseconds. This delay increases with y (Fig. 23), reflecting the

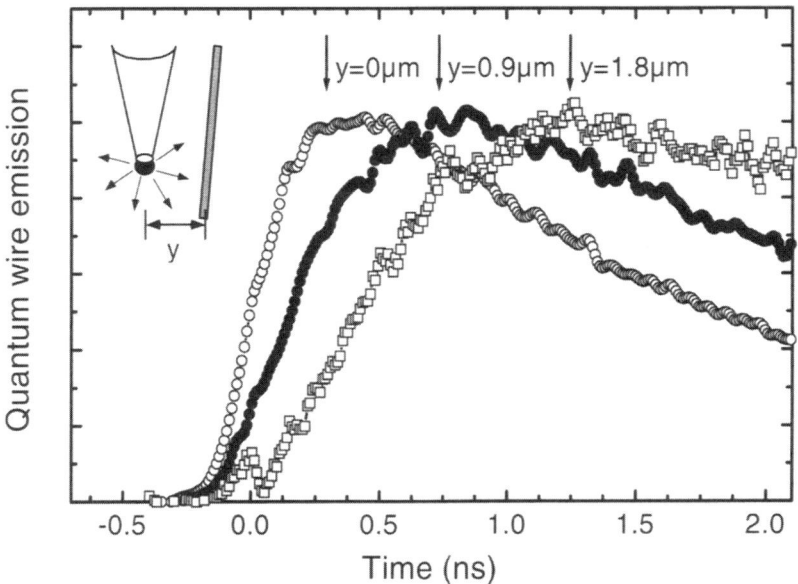

FIG. 23. Temporal profiles of the QWR luminescence for fixed excitation position y for mesa top QW excitation as in 21. Exciton transport within the GaAs QW from the generation point toward the QWR location causes a delayed rise of the QWR luminescence. The maximum of the QWR emission for each excitation position has been normalized to unity. Shown as a solid line is the response function of the single photon counting detection.

traveling time of carriers from the excited QW area to the QWR location. In our experiment, the optical excitation is resonant to the heavy-hole exciton transition of the QW, which is about 10 meV below the onset of the band-to-band continuum.

For a lattice temperature of 100 K and weak excitation, the real-space transfer in the QW is dominated by excitonic transport (Hillmer *et al.*, 1992). The time scale of transport is several hundred picoseconds, much longer than the formation and energy relaxation times of excitons, that is, excitons undergoing real-space transfer form a quasiequilibrium energy distribution close to the lattice temperature. In such a case, an isothermal drift-diffusion model is appropriate to describe the spatially resolved exciton dynamics. A purely diffusive transport model is clearly inappropriate as it does not account for (i) the constant decay curves for $|y| < 0.32\,\mu m$ and (ii) the strong difference in QWR PL intensities for mesa top and bottom excitation.

Exciton transport is described by the two-dimensional particle current density $\vec{j}(\vec{r}, t) = \vec{j}_{\text{diff}}(\vec{r}, t) + \vec{j}_{\text{drift}}(\vec{r}, t)$, being the sum of a diffusion term $\vec{j}_{\text{diff}}(\vec{r}, t) = -D_{\text{ex}} \vec{\nabla} n(\vec{r}, t)$ induced by the gradient of the exciton concentra-

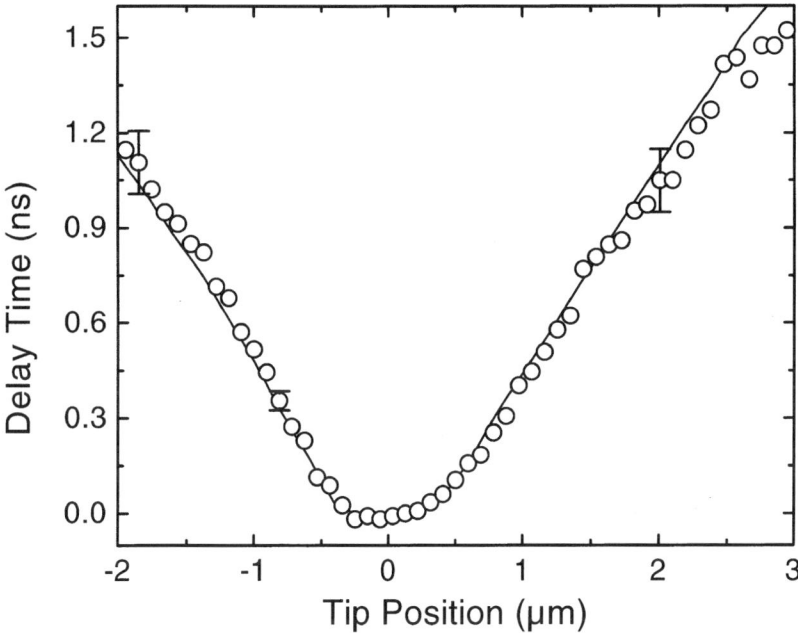

FIG. 24. Temporal position of the luminescence maxima of Fig. 23 as a function of the lateral separation between excitation spot and QWR (symbols). Solid line: Temporal positions calculated from the drift-diffusion model for exciton transport.

tion $n(\vec{r}, t)$ ($r = x, y$); D_{ex} is the exciton diffusion coefficient) and a drift term $\vec{j}_{drift}(\vec{r}, t) = -\mu_{ex} n(\vec{r}, t) \vec{\nabla} U(\vec{r})$, induced by the action of the local bandgap gradient $\vec{\nabla} U(\vec{r})$ on the exciton center of mass motion (μ_{ex} is the equivalent charged particle mobility of the exciton) (Tamor and Wolfe, 1980). Under the conditions of the experiment, μ_{ex} is linked to the diffusion coefficient D_{ex} by the Einstein relation $\mu_{ex} = eD_{ex}/kT$ (e, electron charge; k, Boltzmann constant). Taking the spatial variation of the band gap $U(\vec{r})$ along the lateral y direction directly from the confinement potential in Fig. 16, the two-dimensional evolution of the exciton concentration $n(\vec{r}, t)$ was calculated from the drift-diffusion equation. The spatiotemporal evolution of the exciton concentration $n(\vec{r}, t)$ is described by the two-dimensional continuity equation, including generation, diffusion, drift, and recombination terms:

$$\frac{\partial n(\vec{r}, t)}{\partial t} = g(\vec{r}, \vec{r}_0, t) + D_{ex}\Delta n(\vec{r}, t) + \mu_{ex} n(\vec{r}, t)\Delta U(\vec{r}) + \mu_{ex}\vec{\nabla} n(\vec{r}, t)\vec{\nabla} U(\vec{r}) - n(\vec{r}, t)/\tau(\vec{r}). \qquad (2)$$

For the generation term $g(\vec{r}, t)$, we use a Gaussian shape in time (FWHM of 1 ps) and space (FWHM of 300 nm) centered at $\vec{r} = (0, y)$. The intensity of the time-resolved QWR luminescence is then proportional to

$$I(t) = \int dx \int dy \, \partial n(\vec{r}, t)/\partial t,$$

where the integration has to be performed over the length of the QWR along the x axis and over the width of the QWR region along the y axis. For comparison with experiment, $I(t)$ is convoluted with the temporal response of the photodetector.

In Fig. 22, the results of such simulations for different tip positions y are compared to the experimental results. Good agreement of the simulation with the experiment for excitation on both the mesa top and the mesa bottom is obtained with an exciton diffusion coefficient $D_{ex} = 13 \text{ cm}^2/\text{s}$. This diffusion coefficient corresponds to an excitonic mobility $\mu_{ex} = 1500 \text{ cm}^2/\text{Vs}$, given mainly by the mobility of quasi-two-dimensional holes and limited in this temperature range by LO phonon scattering. In particular, the model calculation correctly describes the influence of the lateral bandgap variation $U(\vec{r})$ on the exciton transport. In the region outside the barriers, $U(\vec{r})$ exerts a force on the excitonic center of mass motion that opposes the diffusive real-space transfer toward the QWR. This is manifested in the experimental data as a significantly weaker QWR luminescence for mesa bottom ($y < -0.4 \,\mu\text{m}$) than mesa top ($y > 0.4 \,\mu\text{m}$) excitation due to the higher barrier on the mesa bottom. On both mesa top and bottom, the calculated t_{max} are in good agreement with the measured values (see Fig. 22), indicating that the exciton diffusion constants are similar on both sides of the wire. In the region inside the barriers, the bandgap variation accelerates the exciton transport toward the QWR. This directly explains the fast rise of the PL decay curve within the time resolution of the present experiment.

The quantitative agreement between model simulations and experiment indicates that the classical excitonic drift-diffusion model provides a good description of the transport processes that are monitored in this experiment. To illustrate these dynamics more clearly, simulations of the full spatio-temporal evolution of the two-dimensional exciton density $n(\vec{r}, t)$ are shown in Figs. 25 and 26. The simulations are obtained by direct numerical integration of the drift-diffusion equation with $\mu_{ex} = 1500 \text{ cm}^2/\text{Vs}$ using a generation term $g(\vec{r}, t)$ as described earlier. For excitation at the QWR location, $y = 0 \,\mu\text{m}$ (Fig. 26), the presence of the potential barriers leads to a rapid trapping of practically all generated excitons into the QWR. This results in a spatial narrowing of the exciton distribution along the y axis, perpendicular to the QWR axis. Diffusion of trapped QWR excitons results in a spreading of the QWR exciton distribution along the wire axis x. In the simulations, the exciton mobility along the wire axis was taken to be the

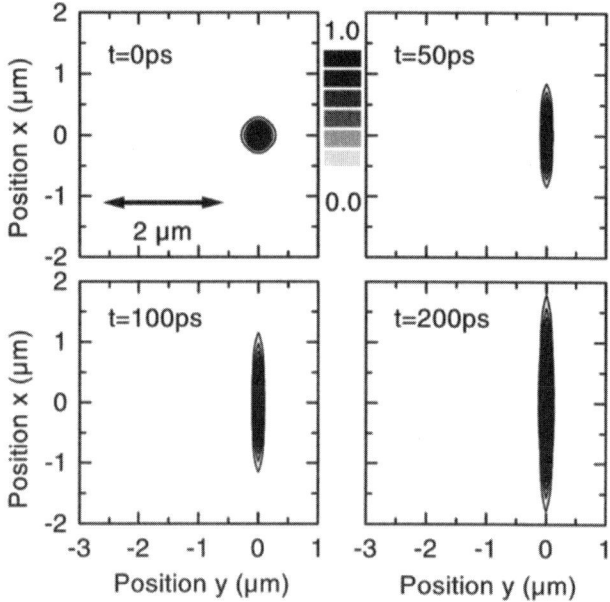

FIG. 25. Spatiotemporal evolution of the exciton density as calculated within the drift-diffusion model for localized excitation at the QWR position, $y = 0$, and at a sample temperature of 100 K. For each delay time, the maximum of the exciton density is normalized to 1.

same as the QW mobility even though theoretical models predict a higher 1D mobility due to a change in scattering rates (Sakaki, 1980). For excitation outside the barriers, for example, at $y = -1\,\mu$m (Fig. 26), the initially narrow exciton distribution spreads out within the QW plane. As observed experimentally, the buildup of the QWR exciton population is delayed by roughly 100 ps and the QWR exciton population continues to increase within almost 1 ns, leading to a concomitant depletion of the QW exciton population.

The pronounced influence of the drift current relative to the diffusion current becomes even more important at lower temperatures, where the ratio between barrier height and thermal energy increases. This is demonstrated in a similar time-resolved experiment at a sample temperature of 10 K (Fig. 27). Again the sample is locally excited by transmitting femtosecond pulses that are resonant to the excitonic QW absorption band, and time-resolved QW luminescence is detected at the center of the QW emission band.

At this temperature QWR luminescence is only observed for excitation in a narrow region around $y = 0$. The intensity of the time-integrated QWR

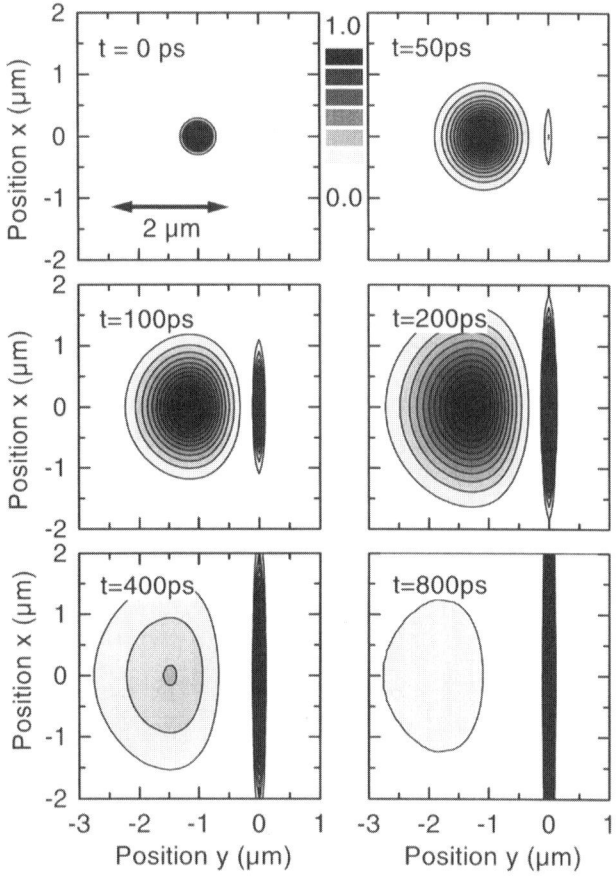

FIG. 26. Spatiotemporal evolution of the exciton density as calculated within the drift-diffusion model for localized excitation at $y = -1\,\mu$m with respect to the QWR at $y = 0$. For each delay time, the maximum of the exciton density is normalized to 1.

luminescence is found to decrease monotonously with increasing separation of excitation tip and QWR. The spatial y-dependence of the time-integrated luminescence can be described by a Gaussian profile with a full width at half maximum of 570 nm. This width is more than a factor of 2 larger than the spatial resolution in the experiment, which was independently determined by mapping the QWR luminescence for near-resonant QWR excitation. At each excitation position, the temporal dependence of the QWR luminescence can be described by a single-exponential decay with a decay constant of 0.9 ns, the population lifetime of the QWR. No shift of the temporal position of the maximum PL intensity is observed within the accuracy of the experiment (± 50 ps). This indicates that at low temperatures ($T = 10$ K)

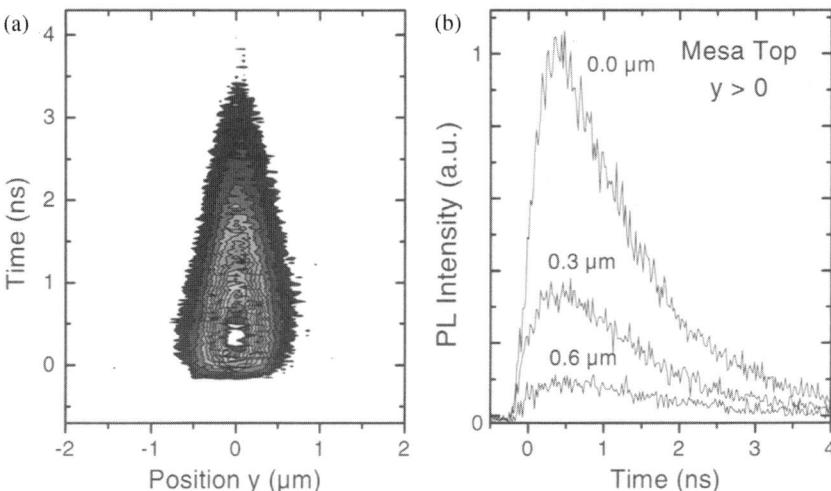

FIG. 27. (a) Temporally resolved QWR photoluminescence at a temperature of 10 K after localized excitation in the QW. Excitation conditions as in Fig. 21. The luminescence intensity is plotted as a function of the lateral distance y between excitation and QWR position and as a function of time. (b) Time evolution of the QWR luminescence after excitation on the mesa top. The luminescence intensity at 1.545 eV is plotted vs time for different excitation positions.

exciton transfer from outside the barriers into the QWR is almost fully suppressed, in agreement with the steady-state PLE spectrum in Fig. 12. QWR luminescence is mainly observed for excitation between the maxima of the energetic barriers separating QWR and QW. The fact that no temporal shift of the PL maximum is observed suggests that at this temperature diffusive exciton transport is of minor importance for real-space transfer and that the trapping of excitons generated in the region between the barriers is greatly accelerated because of the bandgap-induced drift motion. Experiments at slightly higher temperatures, $T = 30$ K (Fig. 28a and $T = 60$ K (Fig. 28b) highlight the influence of the asymmetry of the lateral bandgap variation on the exciton transport into the wire. At both temperatures strong QWR luminescence is found for excitation around the QWR, $|y| < 0.4\,\mu\text{m}$, showing a resolution-limited rise time. QWR luminescence is practically fully suppressed for mesa bottom excitation ($y < -0.4\,\mu\text{m}$, high barrier), whereas the QWR luminescence intensity that is observed for excitation on the mesa top increases strongly with increasing temperature.

As the confinement potential is experimentally known, the remaining free parameter in the drift-diffusion equation is the two-dimensional exciton diffusion coefficient D_{ex}. A simulation of the time-resolved near-field luminescence profiles within the drift-diffusion model thus allows one to

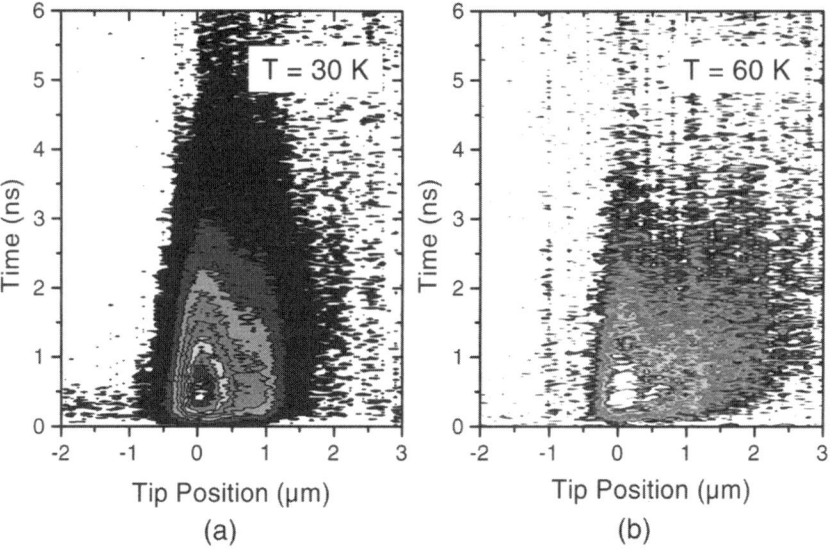

FIG. 28. As in 27, for sample temperatures of (a) 30 K and (b) 60 K.

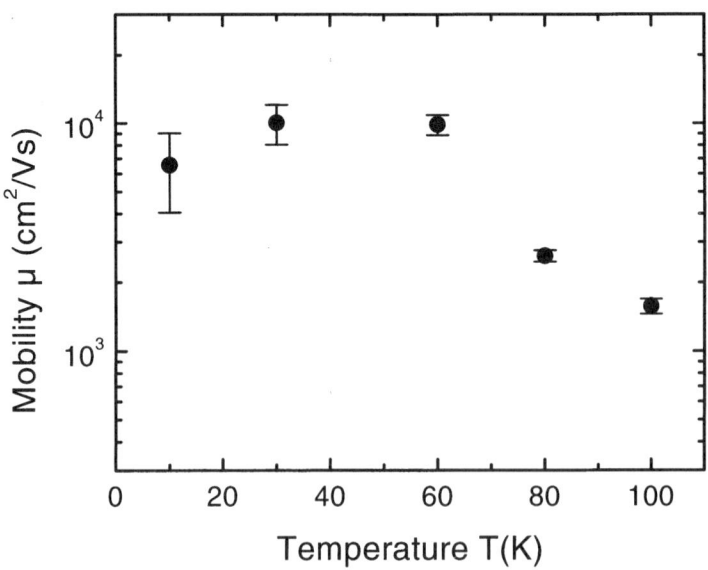

FIG. 29. QW exciton mobilities as a function of sample temperature as extracted from time-resolved near-field PL experiments.

extract the D_{ex} and, with the help of the Einstein equation, the QW exciton mobility μ_{ex} (Fig. 29). At the lowest temperature of this study of 10 K, we find a QW mobility of 6000 cm^2/Vs, in good agreement with results of previous experiments on (100) GaAs QWs (Hillmer et al., 1989). At low temperatures, the exciton mobility is mainly limited by interface-roughness scattering. Since interface-roughness scattering rates are decreasing with increasing temperatures, this leads to an increase of the exciton mobility at temperatures of up to 50 K. At higher temperatures, acoustic-deformation potential scattering and polar-optical scattering (above 100 K) become dominant and result in the pronounced decrease of the exciton mobility with increasing temperature for $T > 50$ K.

The experiments demonstrate that through combination of near-field spectroscopy with time-resolved detection and excitation schemes, direct access to the spatiotemporal carrier dynamics in single nanostructures becomes possible. In particular, the extension of the temporal resolution toward the femtosecond regime offers new perspectives for gaining direct insight into transport and relaxation processes in low-dimensional semiconductors.

2. Femtosecond Near-Field Pump-and-Probe Spectroscopy

The experiments discussed in the previous section provide detailed information about the quasi-two-dimensional exciton transport in our samples. Yet, their temporal resolution in the 100-ps range is clearly not sufficient for resolving the carrier relaxation in an individual nanostructure. One possible approach for increasing the temporal resolution by three orders of magnitude while maintaining subwavelength spatial resolution is to combine near-field microscopy with nonlinear femtosecond pump-and-probe techniques. In such experiments, carrier dynamics in a single QWR is monitored by probing the temporal variation of carrier-induced changes in the transmission and/or reflection of the nanostructure.

The first experiments along this direction are described in this section (Guenther et al., 1999). In the following, we first analyze the different components contributing to pump-probe signals in such an experiment. We then demonstrate that specific information on local carrier relaxation phenomena on a 100-fs time scale and on carrier transport along the QWR on a picosecond time scale can be extracted from those measurements.

Spatially resolved femtosecond studies were performed with a quasi-two-color near-field pump and probe setup as described in Section III. The pump and probe pulses were derived from a mode-locked 80-MHz Ti:sapphire oscillator tunable between 810 and 870 nm. Two different optical configurations were used. In a near-field pump/near-field probe geometry, both pump and probe pulses were transmitted through the same near-field fiber probe. In a far-field pump/near-field probe geometry, only the probe pulse was sent through the fiber while the pump was focused onto the sample by a far-field graded index lens, resulting in a spot size on the sample of about 30 μm. As the substrate of the QWR structure (Fig. 2) is not transparent in the wavelength range of interest, all experiments were performed in a reflection geometry. The probe light reflected from the sample was collected in the far field through a microscope objective, spatially and spectrally filtered, and detected with a photodiode. For suppression of background signals, pump and probe beams were mechanically chopped at frequencies $f_1 = 1.2$ kHz and $f_2 = 2.1$ kHz. All data were taken at a sample temperature of 300 K.

First, far-field pump/near-field probe spectra were recorded on the mesa top of the sample, that is, in the QW region. The pump pulses, with a bandwidth of 40 meV, are centered around the QW absorption resonance at 1.52 eV. The excitation density was about 3×10^{11} cm^{-2}. The pump-induced change in reflectivity was measured with tunable probe pulses of 11-meV bandwidth. For photon energies E_{pr} of the probe between 1.51 and 1.55 eV, the creation of electron–hole pairs results in an increase in reflectivity, whereas a decrease of reflectivity is found for $E_{pr} < 1.51$ eV. In Fig. 30a, the reflectivity change $\Delta R/R_0(t_d) = (R(t_d) - R_0)/R_0$ for a time delay t_d of 10 ps between pump and probe is plotted as a function of E_{pr} (R, R_0: sample reflectivity with and without pump). This spectrum shows a pronounced maximum at $E_{pr} = 1.515$ eV and follows — for $E_{pr} \geqslant 1.515$ eV — the shape of the QW photoluminescence spectrum (inset of Fig. 30a). For excitation densities between 10^{10} and 3×10^{11} cm^{-2}, the magnitude of $\Delta R/R_0$ is proportional to the pump intensity. The reflectivity changes rise within the time resolution of the experiment of 200 fs. For $E_{pr} \geqslant 1.515$ eV, the signal decays by carrier recombination on a nanosecond time scale. For $E_{pr} < 1.50$ eV, a fast partial decay is found within the first 3 ps, followed by the long-lived component.

Next, we present data taken in the range of the QWR resonance around 1.46 eV (cf. inset of Fig. 30a). In this measurement, both pump and probe were transmitted through the fiber probe. Excitation at 1.48 eV generates carriers exclusively in the QWR that give rise to a local change of reflectivity. This is evident from the image in Fig. 30b, where the change of reflectivity at $E_{pr} = 1.45$ eV and a delay time of 10 ps is plotted as a function of two spatial coordinates in the QW plane. The pronounced local change of reflectivity occurs along a line that coincides with the QWR position at the sidewall of the [01-1] mesa stripes. This position is independently identified from simultaneously recorded shear-force topography images. The local reflectivity change is superimposed on a spatially slowly varying background signal that is due to the GaAs cap and substrate layers. In Fig. 30c (solid circles), the quantity $\Delta R(y)/R_0$ is plotted along the y direction, that is, perpendicular to the QWR axis ($y = 0$: QWR position). The local

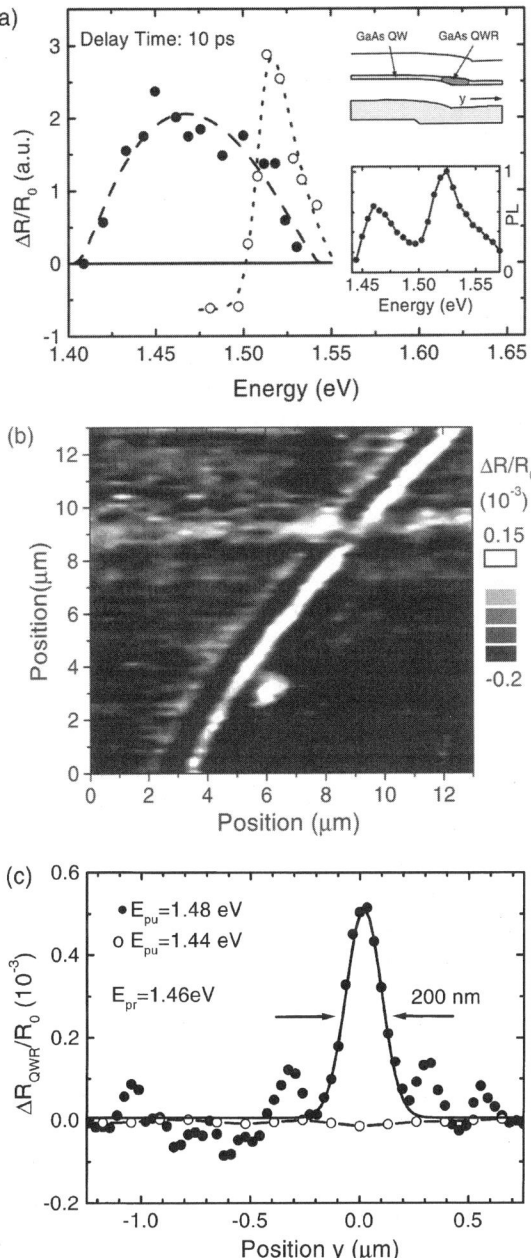

FIG. 30. (a) Probe wavelength dependence of the pump-induced quantum wire reflectivity change $\Delta R_{QWR}(t_d = 10\,ps)/R_0$ (closed circles) and of the reflectivity change in the mesa top region of the sample (open circles). $E_{pu} = 1.52$ eV. Insets: Schematic of the QWR sample and room temperature near-field photoluminescence spectrum. (b) Spatial map of the pump-induced reflectivity change $\Delta R/R_0$ at a delay time t_d of 10 ps. The pump and probe lasers are set to 1.51 and 1.45 eV, respectively. (c) Spatial variation of $\Delta R/R_0$ along a line perpendicular to the wire axis at a probe wavelength of $E_{pr} = 1.46$ eV. Closed circles, $E_{pu} = 1.48$ eV; open circles, $E_{pu} = 1.44$ eV.

reflectivity peak is measured with a spatial resolution of 200 nm, which is limited by the finite distance between the buried QWR and the sample surface of about 70 nm. The variation of the reflectivity change $\Delta R_{QWR}/R_0 = (\Delta R(y = 0) - \Delta R(|y| > 1\ \mu m))/R_0$ with probe energy E_{pr} is presented in Fig. 30a (solid circles) for a delay of 10 ps ($\Delta R(|y| > 1\ \mu m)$: background signal). The line shape resembles the shape of the room-temperature PLE spectrum. For excitation at a lower photon energy of 1.44 eV where the QWR absorption is negligible, the reflectivity change $\Delta R_{QWR}/R_0$ vanishes completely (Fig. 30c, open circles). This result demonstrates that the local reflectivity change ΔR_{QWR} is due to a change in the QWR susceptibility.

To gain insight into the carrier dynamics of the QWR, we follow the temporal variation of $\Delta R_{QWR}(t_d)$ for different excitation conditions. First, carriers are generated with a 50-fs far-field pulse with a spot size of 30 μm centered at 1.52 eV (Fig. 31c, inset). At this energy, a spatially homogeneous carrier distribution is generated in the GaAs cap layer, the QW and high-lying states of the QWR, and the GaAs substrate. The density generated by the pump pulse was 5×10^{11} cm^{-2}, corresponding to nondegenerate excitation conditions ($T = 300$ K). Probe pulses at 1.475 eV, near the QWR resonance, were transmitted through the fiber probe. On both sides of the QWR (Fig. 31a), a spatially homogeneous transient reflectivity decrease is observed that decays on a time scale of several picoseconds (Fig. 31b, solid circles). This signal is dominated by the transient carrier-induced reflectivity change of the GaAs cap layer, and the picosecond decay of the change of reflectivity reflects the trapping of carriers into surface states (Baumberg et al., 1997). In contrast, a decrease in reflectivity with a smaller amplitude occurs at the QWR position $y = 0$ (Fig. 31b, open circles). The QWR contribution, ΔR_{QWR}, is extracted by taking the difference of the two transients in Fig. 31b. The time evolution of ΔR_{QWR} (Fig. 31c) shows a steplike behavior with an ultrafast rise within 200 fs, the time resolution of the experiment, and a constant value up to delay times of 50 ps. Similar transients are measured with pump pulses attenuated by a factor of 3. A similar temporal evolution of ΔR_{QWR} is observed for resonant far-field excitation of the QWR ($E_{pu} = 1.47$ eV) and spatially resolved probing at energies of high-lying QWR states around 1.51 eV.

Different dynamics are observed with both pump and probe pulses transmitted through the fiber. For excitation at 1.52 eV and probing at 1.46 eV, a cap-layer-induced decrease in reflectivity occurs on both sides of the QWR. This signal decays with a time constant of 2.5 ps (Fig. 32a, open circles). In contrast, at the QWR position, there is an increase in reflectivity for all positive delay times. The difference between both signals, ΔR_{QWR}, decays by more than 35% on a time scale of 20 ps (Fig. 32b). Similar dynamics are observed for several different probe energies between 1.48 and 1.43 eV (Fig. 33). The excitation density in these experiments was 5×10^{11} cm^{-2}, that is, close to the one chosen for far-field excitation.

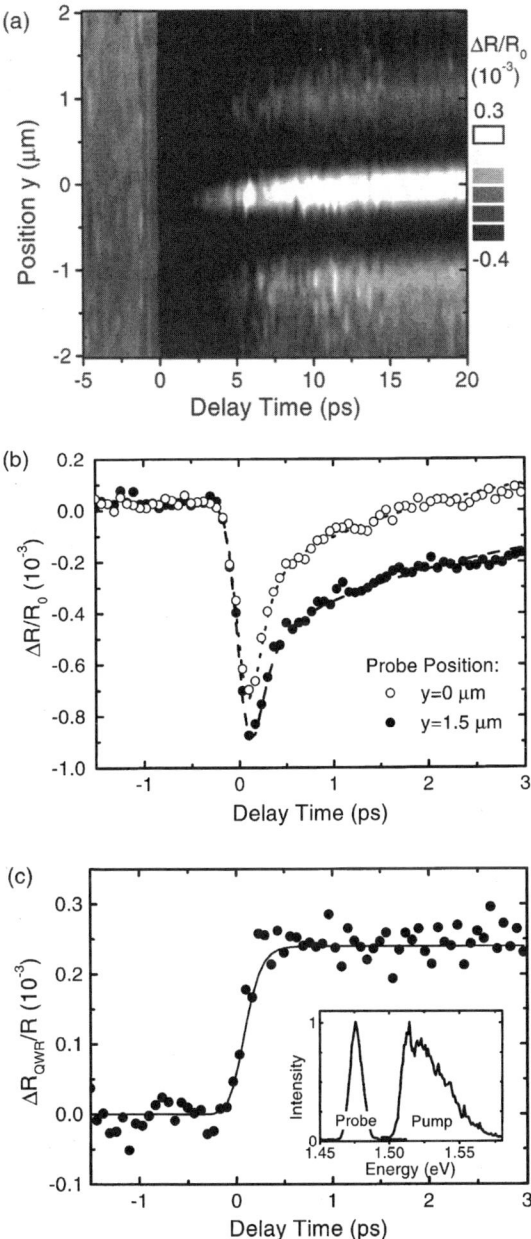

FIG. 31. (a) Pump-induced reflectivity change $\Delta R/R_0$ as a function of tip position along the y axis, perpendicular to the quantum wire axis, and as a function of delay time t_d. $E_{pu} = 1.51$ eV, $E_{pr} = 1.475$ eV. The excitation is performed with a far-field spot having a diameter of 30 μm. (b) Temporal variation of $\Delta R/R_0$ at fixed spatial probe positions of $y = 0$ (QWR position, open circles) and on the mesa top part at $y = 1.5$ μm. (c) $\Delta R_{QWR}(t_d)/R_0 = (\Delta R(y=0, t_d) - \Delta R(y=1.5 \,\mu m, t_d))/R_0$. Inset: Spectra of pump and probe laser.

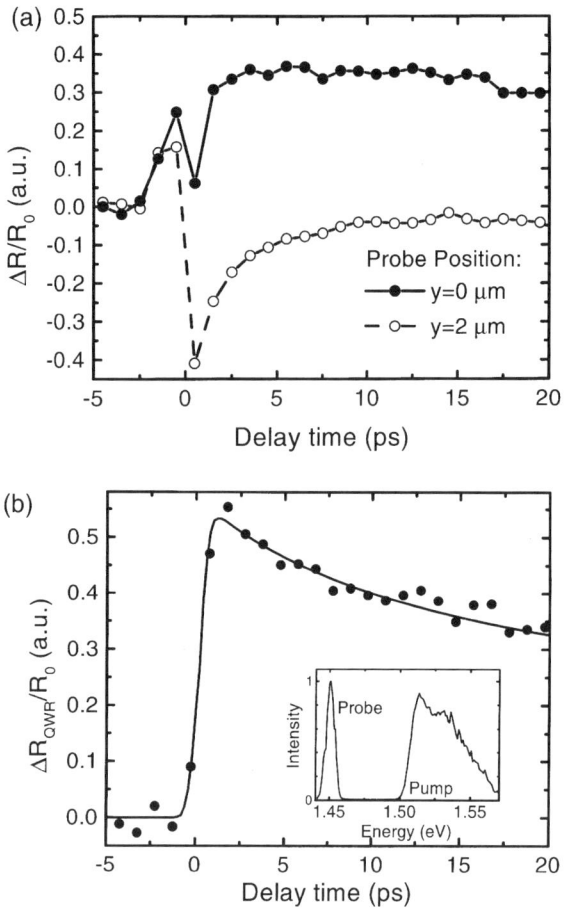

FIG. 32. (a) Temporal variation of $\Delta R/R_0$ at fixed spatial probe positions of $y = 0$ (QWR position, closed circles) and on the mesa top part at $y = 2\,\mu m$. Localized excitation and probing is achieved by sending both pump and probe laser through the near-field probe. (b) Temporal variation of the pump-induced change of the QWR reflectivity $\Delta R_{QWR}(t_d)/R_0 = (\Delta R(y = 0, t_d) - \Delta R(y = 2\,\mu m, t_d))/R_0$. Solid line: Simulation assuming one-dimensional diffusion along the wire axis with a diffusion coefficient $D_{QWR} = 70\,cm^2/s$. Inset: Spectra of pump and probe lasers.

In general, reflectivity spectra of multilayer semiconductor nanostructures depend sensitively on the details of the layer structure (Gurioli et al., 1999). To interpret our experiments in terms of carrier dynamics, we performed an analysis of the pulse propagation through our multilayer nanostructure. The structure (Fig. 2) consists of the GaAs cap layer, the GaAs QW in which the QWR is embedded, the GaAs substrate, and—in between—the AlGaAs barriers. For photon energies $E_{pump} < 1.44$ eV, only the cap layer and the

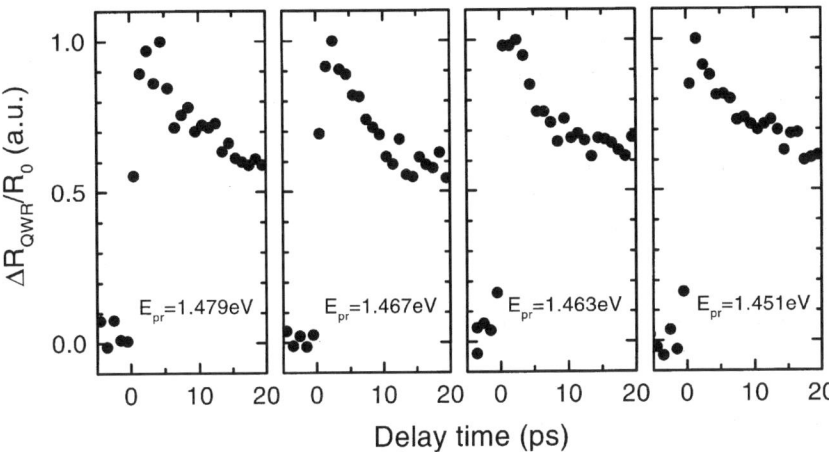

FIG. 33. Probe energy dependence of the temporal variation of the pump-induced change of the QWR reflectivity $\Delta R_{QWR}(t_d)/R_0 = (\Delta R(y = 0, t_d) - \Delta R(y = 2\,\mu m, t_d))/R_0$. Localized excitation and probing is achieved by sending both pump and probe laser through the near-field probe. Excitation conditions as in 32.

substrate are excited; for 1.44 eV < E_{pump} < 1.50 eV, those layers and the QWR. For higher photon energies, carriers are also created in the QW. Carrier generation leads to a modification of the optical susceptibility $\chi(\omega, n_{ex})$ of each layer with carrier density n_{ex}. This results in a change of reflectivity measured at the different probe wavelengths. A numerical simulation of these changes assuming thermalized carrier distributions for both electrons and holes was performed by using the numerical maxtrix inversion method (Schmitt-Rink et al., 1986). The assumption of thermalized distributions is reasonably justified, as our time resolution is less than the sub–100 fs dephasing and thermalization times of free carriers in quasi-two-dimensional QWs at room temperature (Bigot et al., 1991; Kim et al., 1992; Knox et al., 1988). The model accounts quantitatively for the influence of phase-space filling, screened Coulomb interaction (here treated in a static approximation), and bandgap renormalization on $\chi(\omega, n_{ex})$. It gives absolute values for the density dependent absorption coefficient $\alpha(\omega, n_{ex})$ and refractive index $n_r(\omega, n_{ex})$ of each layer. A transfer matrix formalism is then used to derive the reflectivity $R(\omega, n_{ex})$ and the carrier-induced change in reflectivity $\Delta R(\omega, n_{ex}) = R(\omega, n_{ex}) - R(\omega, 0)$ for the multilayer structure.

This analysis suggests the following interpretation: For E_{pump} > 1.5 eV, excitation of the QW leads to a decrease of interband absorption around the QW absorption resonance at E_{pr} = 1.52 eV, that is, of the imaginary part of the susceptibility. This results in an increase in reflectivity (open circles in Fig. 30a) with a spectral dependence close to that of the QW PLE spectrum. At probe energies $E_{pr} < 1.5$ eV, the reflectivity signal of opposite sign is mainly due to the cap layer. The change in sign occurs because the cap layer signal is most sensitive to the change in refractive index, that is, the real part of the susceptibility. This effect is due to the large jump in refractive index at the sample surface. The observed decay of the pump-induced reflectivity change on a time scale of a few picoseconds — which is observed in all experiments for probe positions outside the wire region — reflects the trapping of carriers in the cap layer into surface states (Baumberg et al., 1997).

Now we consider the signals that are measured with the fiber probe at the QWR. Here, the additional contribution due to carriers in the QWR results in a smaller amplitude of the change of reflectivity (open circles in Fig. 31b). This contribution is described by the quantity ΔR_{QWR}. The spectrum of ΔR_{QWR} in Fig. 30a resembles the PLE spectrum of the QWR (inset). If the photon energy of the pump pulses is below the onset of QWR absorption, this signal disappears.

The QWR signal ΔR_{QWR} is sensitive to the concentration and energy distribution of carriers in the QWR. For $E_{pump} = 1.52$ eV (Fig. 31), carriers are excited both to high-lying QWR states and into the surrounding QW. Probing at $E_{pr} = 1.46$ eV, that is, at the bottom of the QWR, one observes a rise of ΔR_{QWR} within the time resolution of 200 fs and a constant amplitude up to the maximum delay of 50 ps. There are two mechanisms that contribute to this behavior: (i) screening of the QWR absorption resonance by the optically injected carriers and (ii) bandfilling due to a carrier population in the low-lying states of the QWR. In QWs both effects are of similar magnitude (Stark et al., 1992). Because of the short-range nature of the exciton screening (Schmitt-Rink et al., 1985), the strength of both effects are a sensitive probe of the *local* carrier concentration within the QWR region. The first mechanism rises with the optically generated carrier concentration, that is, it follows instantaneously the time integral of the pump pulse. In contrast, the second mechanism requires carrier redistribution from optically excited high-lying states to the bottom of the QWR. The absence of any slower dynamics on the transient in Fig. 31c suggests that this redistribution process occurs within the first 200 fs. This relaxation involves a complex scattering scenario between high-lying QWR states and states at the bottom of the QWR, involving both carrier–LO phonon and carrier–carrier scattering. This conclusion is strongly supported by the ultrafast rise in reflectivity that is observed when the pump laser is tuned in resonance with the QWR absorption and the probe wavelength is set to probe high-lying QWR states.

The experiments discussed so far were performed with far-field excitation (30 μm spot size), and — within the excitation spot — a homogeneous carrier distribution is generated along the QWR. Thus, a diffusive motion of carriers along the QWR is not visible for far-field excitation and the reflectivity signal decays with the electron–hole recombination time of 1.9 ns, identical to the luminescence decay time.

A different behavior occurs for near-field excitation through the fiber probe. Here, the excitation and probe pulse are spatially overlapping and localized to a small spot of about 400 nm dimension. The diffusion of carriers out of the excitation volume results in a decrease of the local carrier concentration within the probe spot, and thus to a reduction of the pump-induced reflectivity change. This carrier diffusion is contributing to the decay of ΔR_{QWR} on a 10-ps time scale plotted in Figs. 32 and 33. In general, the transport of carriers out of the excitation volume can occur both along the QWR and in the lateral direction, that is, from the QWR into the QW. In these experiments, performed at a pump photon energy of 1.52 eV, carriers are generated in a 400-nm spot in high-lying QWR and in QW states. The data in Figs. 31b and 31c suggest that a quasiequilibrium population of both QWR and QW states is formed within about 100 fs after excitation. In such a distribution, consisting of a total of 5×10^{11} electron–hole pairs per square centimeter, the population of high-lying states from which carriers can undergo diffusive motion in the lateral direction, that is, into the unexcited embedding QW, is very small, even at 300 K. Taking this small fraction of carriers and taking into account the ambipolar diffusion coefficient of 12 cm^2/s measured in our sample (cf. Fig. 20), one finds a negligible contribution of lateral transport to the transients in Figs. 32 and 33 for the first 20 ps. This estimate is confirmed by a numerical simulation of the ambipolar drift diffusion of carriers within the experimentally obtained confinement potential, as outlined in Subsection 1 of Section V. We conclude that the experiment is mainly sensitive to the carrier transport along the QWR. Assuming a drift-diffusion model for transport along the QWR, the observed decay of the pump-induced change in QWR reflectivity is well described by taking an ambipolar diffusion coefficient $D_{QWR} = 70 \pm 30$ cm^2/s. This value is substantially higher than the diffusion coefficient $D_{QW} = 12$ cm^2/s found for transport within the QW. This value corresponds to a mobility $\mu = 3000$ cm^2/Vs, which is close to the room temperature mobility of electrons in GaAs. This suggests that, in our experiments, the carrier transport along the QWR axis is strongly different from a conventional ambipolar diffusive transport regime, with a spatially and temporally correlated motion of electrons and holes and a mobility that is hole limited. Instead, the data suggest that we are observing a rapid motion of electrons out of the excitation volume, along the QWR axis, with a mobility that is basically unaffected by the hole distribution within the QWR (Emiliani et al., 2000). Experiments at cryogenic temperature and as a function of excitation energy will provide additional information.

In conclusion, we have demonstrated how femtosecond near-field spectroscopy can be applied to directly image the carrier dynamics in a single semiconductor nanostructure. By combining ultrahigh spatial and temporal resolution, we have resolved ultrafast relaxation of carriers into a single QWR on a 100-fs time scale at room temperature. First evidence for the diffusive transport of photogenerated carriers along the wire axis on a picosecond time and 100-nm length scale was given.

VI. Outlook and Conclusions

During the past years, near-field scanning optical microscopy has emerged as a novel tool for the spectroscopy of semiconductor nanostructures. It offers many of the different contrast mechanisms of conventional far-field optical microscopy and combines them with unprecedented spatial resolution in the 100-nm, sometimes even sub-100-nm, range. This makes the technique an ideal choice for selectively studying the optical and electronic properties of *single* semiconductor nanostructures, in particular by combining near-field microscopy at low temperatures and photoluminescence or photoluminescence excitation spectroscopy.

In addition, the combination of near-field microscopy with time-resolved optical spectroscopy in the pico- and femtosecond range gives direct access to the spatiotemporal carrier dynamics in nanostructures on ultrashort time and length scales. With continuing technological progress in this newly emerging area of research, this combination promises to give important new insight into the transport and relaxation dynamics of photoexcited carriers in real space and time.

ACKNOWLEDGMENTS

The near-field work in our laboratory would have been impossible without the commitment and dedication of Gerd Behme, Valentina Emiliani, Tobias Günther, Francesca Intonti, Alexander Richter, Marko Süptitz, and Monika Tischer. It is a pleasure to acknowledge Richard Nötzel, Manfred Ramsteiner, and Klaus Ploog from the Paul-Drude-Institut in Berlin. Financial support by the Deutsche Forschungsgemeinschaft (SFB 296) and the European Union (EFRE program) is gratefully acknowledged.

References

Achermann, M., Nechay, B. A., Morier-Genoud, F., Schertel, A., Siegner, U., and Keller, U. (1999). Direct experimental observation of different diffusive transport regimes in semiconductor nanostructures. *Physical Review B — Condensed Matter* 60, 2101–2105.

Aigony, L., Lahrech, A., Gresillon, S., Cory, H., Boccara, A. C., and Rivoal, J. C. (1999). Polarization effects in apertureless scanning near-field optical microscopy: an experimental study. *Optics Letters* 24, 187–189.

Akiyama, H., Koshiba, S., Someya, T., Wada, K., Noge, H., Nakamura, Y., Inoshita, T., Shimizu, A., and Sakaki, H. (1994). Thermalization effect on radiative decay of excitons in quantum wires. *Physical Review Letters* 72, 924–927.

Ash, E. A., and Nicholls, G. (1972). Super-resolution aperture scanning microscope. *Nature* 237, 510–513.

Atia, W. A., and Davis, C. C. (1997). A phase-locked shear-force microscope for distance regulation in near-field optical microscopy. *Applied Physics Letters* 70, 405–407.

Bachelot, R., Gleyzes, P., and Boccara, A. C. (1995). Near-field optical microscope based on local perturbation of a diffraction spot. *Optics Letters* 20, 1924–1926.

Barenz, J., Hollricher, O., and Marti, O. (1996). An easy-to-use non-optical shear-force distance control for near-field optical microscopes. *Review of Scientific Instruments* 67, 1912–1916.

Baumberg, J. J., Williams, D. A., and Köhler, K. (1997). Ultrafast acoustic phonon ballistics in semiconductor heterostructures. *Physical Review Letters* 78, 3358–3361.

Behme, G., Richter, A., Suptitz, M., and Lienau, C. (1997). Vacuum near-field scanning optical microscope for variable cryogenic temperatures. *Review of Scientific Instruments* 68, 3458–3463.

Bell, L. D., and Kaiser, W. J. (1988). Observation of interface band structure by ballistic-electron-emission microscopy. *Physical Review Letters* 61, 2368–2371.

Bethe, H. A. (1944). Theory of diffraction by small holes. *Physical Review* 66, 163–182.

Betzig, E., and Chichester, R. (1993). Single molecules observed by near-field scanning optical microscopy. *Science* 262, 1422–1462.

Betzig, E., Trautman, J. K., Harris, T. D., Weiner, J. S., and Kostelak, R. L. (1991). Breaking the diffraction barrier: optical microscopy on a nanometric scale. *Science* 251, 1468–1470.

Betzig, E., Finn P. L., and Weiner, J. S. (1992a). Combined shear force and near-field scanning optical microscopy. *Applied Physics Letters* 60, 2484–2487.

Betzig, E., Trautman, J. K., Weiner, J. S., Harris, T. D., and Wolfe, R. (1992b). Polarization contrast in near-field scanning optical microscopy *Applied Optics* 31, 4563–4568.

Biasol, G., Reinhardt, F., Gustafsson, A., and Kapon, E. (1997). Self-limiting growth of GaAs surfaces on nonplanar substrates. *Applied Physics Letters* 71, 1831–1833.

Bigot, J.-Y., Portella, M. T., Schoenlein, R. W., Cunningham, J. E., Shank, C. V. (1991). Two-dimensional carrier-carrier screening in a quantum well. *Physical Review Letters* 67, 636–639.

Binnig, G., Rohrer, H., Gerber, Ch., and Weibel, E. (1982). Tunneling through a controllable vacuum gap. *Physical Review Letters* 49, 57–60.

Binnig, G., Quate, C. F., and Gerber, Ch. (1986). Atomic force microscope. *Physical Review Letters.* 56, 930–933.

Bonadeo, N. H., Chen, G., Gammon, D., Katzer, D. S., Park, D., Steel, D. G. (1998a). Nonlinear nano-optics: Probing one exciton at a time. *Physical Review Letters* 81, 2759–2761 (1998).

Bonadeo, N. H., Erland, J., Gammon, D., Park, D., Katzer, D. S., Steel, D. G. (1998b). Coherent optical control of a single quantum dot. *Science* 282, 1473–1476.

Bonnell, D. A., ed. (1993). *Scanning Tunneling Microscopy and Spectroscopy: Theory, Techniques and Applications.* VCH, New York.

Bouwkamp, C. J. (1950a). On Bethe's theory of diffraction by small holes. *Philips Research Reports* 5, 321–332.

Bouwkamp, C. J. (1950b). On the diffraction of electromagnetic waves by small circular disks and holes. *Philips Research Reports* 5, 401–422.

Brunner, R., Bietsch, A., Hollricher, O., and Marti, O. (1997). Distance control in near-field optical microscopy with piezoelectrical shear-force detection suitable for imaging in liquids. *Review of Scientific Instruments* 68, 1769–1772.

Bryant, G. W. (1998). Probing quantum nanostructures with near-field optical microscopy and vice versa. *Applied Physics Letters* 72, 768–770.
Bukofsky, S. J., and Grober, R. D. (1997). Video rate near-field scanning optical microscopy. *Applied Physics Letters* 71, 2749–2751.
Chang, Y. C., Chang, L. L., and Esaki, L. (1985). A new one–dimensional quantum well structure. *Applied Physics Letters* 47, 1324–1326.
Chemla, D. S., Schmitt-Rink, S., and Miller, D. A. B. (1989). Linear and nonlinear optical properties of semiconductor quantum wells. *Advances in Physics* 38, 89–188.
Cho, A. Y., and Arthur, J. R. (1975). Molecular beam epitaxy. *Progress in Solid State Chemistry* 10, 157.
Christensen, D. A. (1995). Analysis of near field tip patterns including object interaction using finite-difference time-domain calculations. *Ultramicroscopy* 57, 189–195.
Cibert, J., Petroff, P. M., Dolan, G. J., Pearton, S. J., Gossard, A. C., and English, J. H. (1986). Optically detected carrier confinement to one and zero dimension in GaAs quantum well wires and boxes. *Applied Physics Letters* 49, 1275–1277.
Courjon, D., and Bainier, C. (1994). Near field microscopy and near field optics. *Reports on Progress in Physics* 57, 989–1028.
Debarre, A., Richard, A., and Tchenio, P. (1997). High-contrast piezoelectric fiber resonance detection for near-field optical microscopy. *Review of Scientific Instruments* 68, 4120–4123.
Decca, R. S., Drew, H. D., and Empson, K. L. (1997). Mechanical oscillator tip-to-sample separation control for near-field optical microscopy. *Review of Scientific Instruments* 68, 1291–1295.
Diels, J.-C., and Rudolph, W. (1996). *Ultrashort Laser Pulse Phenomena*. Academic Press, San Diego.
Drabenstedt, A., Wrachtrup, J., and Vonborczyskowski, C. (1996). A distance regulation scheme for scanning near-field optical microscopy. *Applied Physics Letters* 68, 3497–3499.
Emiliani, V., Guenther, T., Lienau, C., Nötzel, R., and Ploog, K. H. (2000). Ultrafast near-field spectroscopy of quasi-one-dimensional transport in a single quantum wire. *Physical Review B—Condensed Matter* 61, R10583–R10586.
Eytan, G., Yayon, Y., Rappaport, M., Shtrikman, H., and Barjoseph, I. (1998). Near-field spectroscopy of a gated electron gas: A direct evidence for electron localization. *Physical Review Letters* 81, 1666–1669.
Froehlich, F. F., and Milster, T. D. (1994). Minimum detectable displacement in near-field scanning optical microscopy. *Applied Physics Letters* 65, 2254–2256.
Froehlich, F. F., and Milster, T. D. (1997). Mechanical resonance behavior of near-field optical microscope probes. *Applied Physics Letters* 70, 1500–1502.
Gammon, D., Snow, E. S., Shanabrook, B. V., Katzer, D. S., and Park, D. (1996a). Fine structure splitting in the optical spectra of single GaAs quantum dots. *Physical Review Letters* 76, 3005–3008.
Gammon, D., Snow, E. S., Shanabrook, B. V., Katzer, D. S., and Park, D. (1996b). Homogeneous linewidths in the optical spectrum of a single gallium arsenide quantum dot. *Science* 273, 87–90.
Gershoni, D., Katz, M., Wegscheider, W., Pfeiffer, L. N., Logan, R. A., and West, K. (1994). Radiative lifetimes of excitons in quantum wires. *Physical Review B—Condensed Matter* 50, 8930–8933.
Ghaemi, H., Cates, C., and Goldberg, B. B. (1995). Low temperature near field spectroscopy and microscopy. *Ultramicroscopy* 57, 165–168.
Girard, C., and Dereux, A. (1994). Optical spectroscopy of a surface at the nanometer scale: a theoretical study in real space. *Physical Review B—Condensed Matter* 49, 11344–11351.
Girard, C., and Dereux, A. (1996). Near-field optics theories. *Reports on Progress in Physics* 59, 657–699.
Girard, C., Dereux, A., Martin, O. J. F., and Devel, M. (1994). Importance of confined fields in

near-field optical imaging of subwavelength objects. *Physical Review B — Condensed Matter* 50, 14467–14473.

Girard, C., Martin, O. J. F., and Dereux, A. (1995). Molecular lifetime changes induced by nanometer scale optical fields. *Physical Review Letters* 75, 3098–3101.

Girard, C., Weeber, J. C., Dereux, A., Martin, O. J. F., and Goudonnet, J. P. (1997). Optical magnetic near-field intensities around nanometer-scale surface structures. *Physical Review B — Condensed Matter* 55, 16487–16497.

Gohde, W., Tittel, J., Basche, T., Brauchle, C., Fischer, U. C., and Fuchs, H. (1997). A low-temperature scanning confocal and near-field optical microscope. *Review of Scientific Instruments* 68, 2466–2474.

Goldoni, G., Rossi, F., Molinari, E., and Fasolino, A. (1997). Band structure and optical anisotropy in V-shaped and T-shaped semiconductor quantum wires. *Physical Review B — Condensed Matter* 55, 7110–7123.

Goni, A. R., Pfeiffer, L. N., West, K. W., Pinczuk, A., Baranger, H. U., and Stormer, H. L. (1992). Observation of quantum wire formation at intersecting quantum wells. *Applied Physics Letters* 61, 1956–1958.

Gregor, M. J., Blome, P. G., Ulbrich, R. G., Grossmann, P., Grosse, S., Feldmann, J., Stolz, W., Gobel, E. O., Arent, D. J., Bode, M., Bertness, K. A., and Olson, J. M. (1995). Near-field optical characterization of the photoluminescence from partially ordered (GaIn)P. *Applied Physics Letters* 67, 3572–3574.

Gregor, M. J., Blome, P. G., Schofer, J., and Ulbrich, R. G. (1996). Probe surface interaction in near-field optical microscopy: The nonlinear bending force mechanism. *Applied Physics Letters* 68, 307–309.

Grésillon, S., Aigouy, L., Boccara, A. C., Rivoal, J. C., Quelin, X., Desmarest, C., Gadenne, P., Shubin, V. A., Sarychev, A. K., and Shalaev, V. M. (1999). Experimental observation of localized optical excitations in random metal-dielectric films, *Physical Review Letters* 82, 4520–4523.

Grober, R. D., Harris, T. D., Trautman, J. K., and Betzig, E. (1994a). Design and implementation of a low temperature near-field scanning optical microscope. *Review of Scientific Instruments* 65, 626–631.

Grober, R. D., Harris, T. D., Trautman, J. K., Betzig, E., Wegscheider, W., Pfeiffer, L., and West, K. (1994b). Optical spectroscopy of a GaAs/AlGaAs quantum wire structure using near-field scanning optical microscopy. *Applied Physics Letters* 64, 1421–1423.

Grober, R. D., Rutherford, T., and Harris, T. D. (1996). Modal approximation for the electromagnetic field of a near-field optical probe. *Applied Optics* 35, 3488–3495.

Guenther, T., Emiliani, V., Intonti, F., Lienau, C., Elsaesser, T., Nötzel, R., and Ploog, K. H. (1999). Femtosecond near-field spectroscopy of a single GaAs quantum wire. *Applied Physics Letters* 75, 3500–3502.

Gurioli, M., Piantelli, S., Colocci, M., and Franchi, S. (1999). Disorder characterization by means of radiatively coupled quantum wells. *Applied Physics Letters* 74, 3365–3367.

Hafner, Ch. (1990). *The Generalized Multiple Multipole Technique for Computational Electrodynamics*. Artech, Boston MA.

Hafner, Ch., and Bomholt, L. H. (1993). *The 3D Electrodynamic Wave Simulator*. John Wiley & Sons, Chichester.

Hanewinkel, B., Knorr, A., Thomas, P., and Koch, S. W. (1997). Optical near-field response of semiconductor quantum dots. *Physical Review B — Condensed Matter* 55, 13715–13725.

Hanewinkel, B., Knorr, A., Thomas, P., and Koch, S. W. (1999). Near-field dynamics of excitonic wave packets in semiconductor quantum wells. *Physical Review B — Condensed Matter* 60, 8975–8983.

Hansen, W. (1988). Quasi-one-dimensional electron systems on GaAs/AlGaAs heterojunctions. *Advances in Solid State Physics* 28, 121–140.

Hansen, W., Horst, M., Kotthaus, J. P., Merkt, U., Sikorski, Ch., and Ploog, K. (1987).

Intersubband resonance in quasi one-dimensional inversion channels. *Physical Review Letters* 58, 2586–2589.

Harris, T. D., Gershoni, D., Grober, R. D., Pfeiffer, L., West, K., and Chand, N. (1996). Near-field optical spectroscopy of single quantum wires. *Applied Physics Letters* 68, 988–990.

Haug, H., and Koch, S. W. (1993). *Quantum Theory of the Optical and Electronic Properties of Semiconductors.* World Scientific, Singapore.

Hess, H. F., Betzig, E., Harris, T. D., Pfeiffer, L. N., and West, K. W. (1994). Near-field spectroscopy of the quantum constituents of a luminescent system. *Science* 264, 1740–1745.

Hillmer, H., Forchel, A., Hansmann, S., Morohashi, M., Lopez, E., Meier, H. P., and Ploog, K. (1989). Optical investigations on the mobility of two-dimensional excitons in GaAs/ Ga_{1-x}/Al_x/As quantum wells. *Physical Review B — Condensed Matter* 39, 10901–10912.

Hillmer, H., Forchel, A., and Tu, C. W. (1992). Enhancement of electron-hole pair mobilities in thin GaAs/Al_x/Ga_{1-x}/As quantum wells. *Physical Review B — Condensed Matter* 45, 1240–1245.

Hoffmann, P. Dutoit, B., and Salath, R.-P. (1995). Comparison of mechanically drawn and protection layer chemically etched optical fiber tips. *Ultramicroscopy* 61, 165.

Hollars, C. W., and Dunn, R. C. (1998). Evaluation of thermal evaporation conditions used in coating aluminum on near-field fiber-optic probes. *Review of Scientific Instruments* 69, 1747–1752.

Hollricher, O., Brunner, R., and Marti, O. (1998). Piezoelectrical shear-force distance control in near-field optical microscopy for biological applications. *Ultramicroscopy* 71, 143–147.

Hsu, J. W. P., Lee, M., and Deaver, B. S. (1995). A nonoptical tip–sample distance control method for near-field scanning optical microscopy using impedance changes in an electromechanical system. *Review of Scientific Instruments* 66, 3177–3181.

Islam, M. N., Zhao, X. K., Said, A. A., Mickel, S. S., and Vail, C. F. (1997). High-efficiency and high-resolution fiber-optic probes for near field imaging and spectroscopy. *Applied Physics Letters* 71, 2886–2888.

Jalocha, A., Moers, M. H. P., Ruiter, A. G. T., and Vanhulst, N. F. (1995). Multi-detection and polarisation contrast in scanning near-field optical microscopy in reflection. *Ultramicroscopy* 61, 221–226.

Kapon, E., Tamargo, M. C., and Hwang, D. M. (1987). Molecular beam epitaxy of GaAs/AlGaAs superlattice heterostructures on nonplanar surfaces. *Applied Physics Letters* 50, 347–349.

Kapon, E., Hwang, D. M., and Bhat R. (1989a). Stimulated emission in semiconductor quantum wire heterostructures. *Physical Review Letters* 63, 430.

Kapon, E., Simhony, S., Bhat, R., and Hwang, D. M. (1989b). Single quantum wire semiconductor lasers, *Applied Physics Letters* 55, 2715–2717.

Karrai, K., and Grober, R. D. (1995). Piezoelectric tip–sample distance control for near field optical microscopes. *Applied Physics Letters* 66, 1842–1844.

Kash, K., Scherer, A., Worlock, J. M., Craighead, H. G., and Tamargo, M. C. (1986). Optical spectroscopy of ultrasmall structures etched from quantum wells. *Applied Physics Letters* 49, 1043–1045.

Kavaldjiev, D. I., Toledocrow, R., and Vaeziravani, M. (1995). On the heating of the fiber tip in a near-field scanning optical microscope. *Applied Physics Letters* 67, 2771–2773.

Kim, D. S., Shah, J., Cunningham, J. E., Damen, T. C., Schäfer, W., Hartmann, M., Schmitt-Rink, S. (1992). Giant excitonic resonance in time-resolved four-wave mixing in quantum wells. *Physical Review Letters* 68, 1006–1009.

Knorr, A., Steininger, F., Hanewinkel, B., Kuckenburg, S., Thomas, P., and Koch, S. W. (1998). Theory of ultrafast spatio-temporal dynamics in semiconductor quantum wells: Electronic wavepackets and near-field optics. *Physica Status Solidi B — Basic Research* 206, 139–151.

Knox, W. H., Chemla, D. S., Livescu, G., Cunningham, J. E., and Henry, J. E. (1988).

Femtosecond carrier thermalization in dense Fermi seas. *Physical Review Letters* 61, 1290–1293.

Lacoste, T., Huser, T., Prioli, R., and Heinzelmann, H. (1998). Contrast enhancement using polarization-modulation scanning near-field optical microscopy (PM-SNOM). *Ultramicroscopy* 71, 333–340.

Lambelet, P., Sayah, A., Pfeffer, M., Philipona, C., and Marquis-Weible, F. (1998). Chemically etched fiber tips for near-field optical microscopy: a process for smoother tips. *Applied Optics* 37, 7289–7292.

Larosa, A. H., Yakobson, B. I., and Hallen, H. D. (1995). Origins and effects of thermal processes on near-field optical probes. *Applied Physics Letters* 67, 2597–2599.

Leong, J. K., and Williams, C. C. (1995). Shear force microscopy with capacitance detection for near-field scanning optical microscopy. *Applied Physics Letters* 66, 1432–1434.

Levy, J., Nikitin, V., Kikkawa, J. M., Awschalom, D. D., and Samarth, N. (1996a). Femtosecond near-field spin microscopy in digital magnetic heterostructures. *Journal of Applied Physics* 79, 6095–6100.

Levy, J., Nikitin, V., Kikkawa, J. M., Cohen, A., Samarth, N., Garcia, R., and Awschalom, D. D. (1996b). Spatiotemporal near-field spin microscopy in patterned magnetic heterostructures. *Physical Review Letters* 76, 1948–1951.

Lewis, A., Isaacson, M., Harootunian, A., and Murray, A. (1984). Development of a 500Å resolution light microscope. *Ultramicroscopy* 13, 227–231.

Lienau, C. (1999). Near-field scanning optical spectroscopy of semiconductor nanostructures. *Advances in Solid State Physics* 38, 325–339.

Lienau, C., Richter, A., and Elsaesser, T. (1996). Light-induced expansion of fiber tips in near-field scanning optical microscopy. *Applied Physics Letters* 69, 325–327.

Lienau, C., Richter, A., Behme, G., Suptitz, M., Elsaesser, T., Ramsteiner, M., Notzel, R., and Ploog, K. H. (1998a). Mapping of the local confinement potential in semiconductor nanostructures by near-field optical spectroscopy. *Physica Status Solidi B — Basic Research* 206, 153–166.

Lienau, C., Richter, A., Behme, G., Suptitz, M., Heinrich, D., Elsaesser, T., Ramsteiner, M., Notzel, R., and Ploog, K. H. (1998b). Nanoscale mapping of confinement potentials in single semiconductor quantum wires by near-field optical spectroscopy. *Physical Review B — Condensed Matter* 58, 2045–2049.

Madsen, S., Holme, N. C. R., Ramanujam, P. S., Hvilsted, S., Hvam, J. M., and Smith, S. J. (1998). Optimizing the fabrication of aluminum-coated fiber probes and their application to optical near-field lithography. *Ultramicroscopy* 71, 65–71.

Martin, O. J. F., Girard, C., and Dereux, A. (1995). Generalized field propagator for electromagnetic scattering and light confinement. *Physical Review Letters* 74, 526–529.

Martinet, E., Reinhardt, F., Gustafsson, A., Biasiol, G., and Kapon, E. (1998). Self-ordering and confinement in strained InGaAs/AlGaAs V-groove quantum wires grown by low-pressure organometallic chemical vapor deposition. *Applied Physics Letters* 72, 701–703.

Mauritz, O., Goldoni, G., Rossi, F., and Molinari, E. (1999). Local optical spectroscopy in quantum confined systems: a theoretical description. *Physical Review Letters* 82, 847–850.

Mononobe, S., Saiki, T., Suzuki, T., Koshihara, S., and Ohtsu, M. (1998). Fabrication of a triple tapered probe for near-field optical spectroscopy in UV region based on selective etching of a multistep index fiber. *Optics Communications* 146, 45–48.

Muranishi, M., Sato, K., Hosaka, S., Kikukawa, A., Shintani, T., and Ito, K. (1997). Control of aperture size of optical probes for scanning near-field optical microscopy using focused ion beam technology. *Japanese Journal of Applied Physics Part 2 — Letters* 36, L942–L944.

Nechay, B. A., Siegner, U., Achermann, M., Bielefeldt, H., and Keller, U. (1999a). Femtosecond pump-probe near-field optical microscopy, *Review of Scientific Instruments* 70, 2758–2764 (1999).

Nechay, B. A., Siegner, U., Morier-Genoud, F., Schertel, A., and Keller, U. (1999b). Femtosecond near-field optical spectroscopy of implantation patterned semiconductors. *Applied Physics Letters* 74, 61–63.

Nötzel, R., Menniger, J., Ramsteiner, M., Ruiz, A., Schönherr, H.-P., and Ploog, K. H. (1996a). Selectivity of growth on patterned GaAs (311)A substrates. *Applied Physics Letters* 68, 1132–1134.

Nötzel, R., Ramsteiner, M., Menniger, J., Trampert, A., Schonherr, H. P., Däweritz, L., and Ploog, K. H. (1996b). Micro-photoluminescence study at room temperature of sidewall quantum wires formed on patterned GaAs (311)A substrates by molecular beam epitaxy. *Japanese Journal of Applied Physics Part 2—Letters* 35, L297–L300.

Nötzel, R., Ramsteiner, M., Menniger, J., Trampert, A., Schönherr, H.-P., Däweritz L., and Ploog, K. H. (1996c). Patterned growth on high-index GaAs (n11) substrates: Application to sidewall quantum wires. *Journal of Applied Physics* 80, 4108–4111.

Nötzel, R., Niu, Z. C., Ramsteiner, M., Schönherr, H.-P., Trampert, A., Däweritz, L., and Ploog, K. H. (1998). Uniform quantum-dot arrays formed by natural self-faceting on patterned substrates. *Nature* 392, 56–58.

Novotny, L., Pohl, D. W., and Regli, P. (1994). Light propagation through nanometer-sized structures—The 2-dimensional-aperture scanning near-field optical microscope. *Journal of the Optical Society of America A—Optics Image Science and Vision* 11, 1768–1779.

Novotny, L., Pohl, D. W., and Hecht, B. (1995a). Light confinement in scanning near-field optical microscopy. *Ultramicroscopy* 61, 1–9.

Novotny, L., Pohl, D. W., and Hecht, B. (1995b). Scanning near-field optical probe with ultrasmall spot size. *Optics Letters* 20, 970–972.

Novotny, L., Pohl, D. W., and Regli, P. (1995c). Near-field, far-field and imaging properties of the 2D aperture SNOM. *Ultramicroscopy* 57, 180–188.

Novotny, L., Bian, R. X., and Xie, X. S. (1997). Theory of nanometric optical tweezers. *Physical Review Letters* 79, 645–648.

Oestreich, M., Ruhle, W. W., Lage, H., Heitmann, D., and Ploog, K. (1993). Reduced exciton–exciton scattering in quantum wires. *Physical Review Letters* 70, 1682–1684.

O'Keefe, J. A. (1956). Resolving power of visible light. *Journal of the Optical Society of America A* 46, 359–366.

Paesler, M. A., and Moyer, P. J. (1996). *Near-field Optics, Theory, Instrumentation and Applications*. Wiley, New York.

Petroff, P. M., Gossard, A. C., Logan, R. A., and Wiegmann, W. (1982). Toward quantum well wires: Fabrication and optical properties. *Applied Physics Letters* 41, 635–637.

Pfeffer, M., Lambelet, P., and Marquiswelbe, F. (1997). Shear-force detection based on an external cavity laser interferometer for a compact scanning near field optical microscope. *Review of Scientific Instruments* 68, 4478–4482.

Pfeiffer, L., West, K. W., Stoermer, H. L., Eisenstein, J. P., Baldwin, K. W., Gershoni, D., and Spector, J. (1990). Formation of a high-quality two-dimensional electron gas on cleaved GaAs. *Applied Physics Letters* 58, 1697–1699.

Ploog, K. (1980). In *Crystals: Growth, Properties and Applications* (H. C. Freyhardt, ed.), p. 75. Springer, Berlin.

Pohl, D. W., and Novotny, L. (1994). Near-field optics—Light for the world of NANO. *Journal of Vacuum Science & Technology B* 12, 1441–1446.

Pohl, D. W., Denk, W., and Lanz, M. (1984). Optical stethoscopy: image recording with resolution $\lambda/20$. *Applied Physics Letters* 44, 651–653.

Pohl, D. W., Novotny, L., Hecht, B., and Heinzelmann, H. (1996). Radiation coupling and image formation in scanning near-field optical microscopy. *Thin Solid Films* 273, 161–167.

Richter, A., Suptitz, M., Lienau, C., Elsaesser, T., Ramsteiner, M., Notzel, R., and Ploog, K. H. (1997a). near-field optical spectroscopy of single GaAs quantum wires. *Surface and Interface Analysis* 25, 583–592.

Richter, A., Behme, G., Suptitz, M., Lienau, C., Elsaesser, T., Ramsteiner, M., Notzel, R., and

Ploog, K. H. (1997b). near-field optical spectroscopy of carrier exchange between quantum wells and single GaAs quantum wires. *Physica Status Solidi B — Basic Research* 204, 247–250.

Richter, A., Behme, G., Suptitz, M., Lienau, C., Elsaesser, T., Ramsteiner, M., Notzel, R., and Ploog, K. H. (1997c). Real-space transfer and trapping of carriers into single GaAs quantum wires studied by near-field optical spectroscopy. *Physical Review Letters* 79, 2145–2148.

Richter, A., Süptitz, M., Lienau, C., Elsaesser, T., Ramsteiner, M., Notzel, R., and Ploog, K. H. (1998a). Carrier trapping into single GaAs quantum wires studied by variable temperature near-field spectroscopy. *Ultramicroscopy* 71, 205–212.

Richter, A., Suptitz, M., Heinrich, D., Lienau, C., Elsaesser, T., Ramsteiner, M., Notzel, R., and Ploog, K. H. (1998b). Exciton transport into a single GaAs quantum wire studied by picosecond near-field optical spectroscopy. *Applied Physics Letters* 73, 2176–2178.

Richter, A., Süptitz, M., Lienau, C., Elsaesser, T., Ramsteiner, M., Nötzel, R., and Ploog, K. H. (1999). Time-resolved near-field optics: Exciton transport in semiconductor nanostructures. *Journal of Microscopy* 194, 393–400.

Roberts, A. (1987). Electromagnetic theory of diffraction by a circular aperture in a thick, perfectly conducting screen. *Journal of the Optical Society of America A — Optics Image Science and Vision* 4, 1970–1983.

Roberts, A. (1989). Near-zone fields behind circular apertures in thick, perfectly conducting screens. *Journal of Applied Physics* 65, 2896–2899.

Roberts, A. (1991). Small-hole coupling of radiation into a near-field probe. *Journal of Applied Physics* 70, 4045–4049.

Rossi, F., and Molinari, E. (1996). Coulomb-induced suppression of band-edge singularities in the optical spectra of realistic quantum-wire structures. *Physical Review Letters* 76, 3642–3645.

Ryan, J. F., Maciel, A. C., Kiener, C., Rota, L., Turner, K., Freyland, J. M., Marti, U., Martin, D., Moriergemoud, F., and Reinhart, F. K. (1996). Dynamics of electron capture into quantum wires. *Physical Review B — Condensed Matter* 53, R4225–R4228.

Saar, A., Calderon, S., Givant, A., Benshalom, O., Kapon, E., and Caneau, C. (1996). Energy subbands, envelope states, and intersubband optical transitions in one-dimensional quantum wires: The local-envelope-states approach. *Physical Review B — Condensed Matter* 54, 2675–2684.

Saiki, T., Mononobe, S., Ohtsu, M., Saito, N., and Kusamo, J. (1996). Tailoring a high-transmission fiber probe for photon scanning tunneling microscope. *Applied Physics Letters* 68, 2612–2614.

Saiki, T., Nishi, K., and Ohtsu, M. (1998). Low temperature near-field photoluminescence spectroscopy of InGaAs single quantum dots. *Japanese Journal of Applied Physics Part 1 — Regular Papers Short Notes & Review Papers* 37, 1638–1642.

Sakaki, H. (1980). Scattering suppression and high-mobility effect of size-quantized electrons in ultrafine semiconductor wire structures. *Japanese Journal of Applied Physics Part 2 — Letters* 19, L735–738. (1980).

Salvi, J., Chevassus, P., Mouflard, A., Davy, S., Spajer, M., Courjon, D., Hjort, K., and Rosengren, L. (1998). *Review of Scientific Instruments* 69, 1744.

Sanchez, E. J., Novotny, L., and Xie, X. S. (1999). Near-field fluorescence microscopy based on two-photon excitation with metal tips. *Physical Review Letters* 82, 4014–4017.

Scherer, A., Roukes, M. L., Craighead, H. G., Ruthen, R. M., Beebe, E. D., and Harbison, J. P. (1987). Ultranarrow conducting channels defined in GaAs–AlGaAs by low-energy ion damage. *Applied Physics Letters* 51, 2133–2135.

Schmitt-Rink, S., Chemla, D. S., and Miller, D. A. B. (1985). Theory of transient excitonic optical nonlinearities in semiconductor quantum-well structures. *Physical Review B — Condensed Matter* 32 6601–6609.

Schmitt-Rink, S., Ell, C., and Haug, H. (1986). Many-body effects in the absorption, gain, and luminescence spectra of semiconductor quantum-well structures. *Physical Review B — Condensed Matter* 33, 1183–1189.

Shah, J. (1999). *Ultrafast Spectroscopy of Semiconductors and Semiconductor Nanostructures*, 2nd ed., Springer Verlag, Berlin.

Smith, S., Holme, N. C. R., Orr, B., Kopelman, R., and Norris, T. (1998). Ultrafast measurement in GaAs thin films using NSOM. *Ultramicroscopy* 71, 213–223.

Stahelin, M., Bopp, M. A., Tarrach, G., Meixner, A. J., and Zschokkegranacher, I. (1996). Temperature profile of fiber tips used in scanning near-field optical microscopy. *Applied Physics Letters* 68, 2603–2605.

Stark, J. B., Knox, W. H., and Chemla, D. S. (1992). Femtosecond circular dichroism study of nonthermal carrier distributions in two- and zero-dimensional semiconductors. *Physical Review Letters* 68, 3080–3083.

Stark, J. B., Mohideen, U., Betzig, E., and Slusher, R. E. (1996). Time-resolved nonlinear near-field optical microscopy of semiconductor microdisks. In *Ultrafast Phenomena IX* (P. F. Barbara, W. H. Knox, G. A. Mourou, and A. H. Zewail, eds.). Springer Series in Chemical Physics, Berlin, pp. 349–350.

Stringfellow, G. B. (1989). *Organometallic Vapor Phase Epitaxy: Theory and Practice*. Academic Press, New York.

Synge, E. A. (1928). A suggested method for extending microscopic resolution into the ultra-microscopic region. *Philosophical Magazine* 6, 356–358.

Tamor, M. A., and Wolfe, J. P. (1980). Drift and diffusion of free excitons in Si. *Physical Review Letters* 44, 1703–1706.

Terris, B. D., Mamin, H. J., Rugar, D., Studenmund, W. R., and Kino, G. S. (1994). Near-field optical data storage using a solid immersion lens. *Applied Physics Letters* 65, 388–390.

Toda, Y., Kourogi, M., Ohtsu, M., Nagamune, Y., and Arakawa, Y. (1996). Spatially and spectrally resolved imaging of GaAs quantum-dot structures using near-field optical technique. *Applied Physics Letters* 69, 827–829.

Toda, Y., Shinomori, S., Suzuki, K., and Arakawa, Y. (1998a). near-field magneto-optical spectroscopy of single self-assembled InAs quantum dots. *Applied Physics Letters* 73, 517–519.

Toda, Y., Shinomori, S., Suzuki, K., and Arakawa, Y. (1998b). Polarized photoluminescence spectroscopy of single self-assembled InAs quantum dots. *Physical Review B — Condensed Matter* 58, R10147–10150.

Toda, Y., Moriwaki, O., Nishioka, M., and Arakawa, Y. (1999). Efficient carrier relaxation mechanism in InGaAs/GaAs self-assembled quantum dots based on the existence of continuum states. *Physical Review Letters* 82, 4114–4117.

Toledo-Crow, R., Yang, P. C., Chen, Y., and Vaez-Irvani, M. (1992). Near-field differential scanning optical microscope with atomic force regulation. *Applied Physics Letters* 60, 2957–2959.

Valaskovic, G. A., Holton, M., and Morrison, G. H. (1995). Parameter control, characterization, and optimization in the fabrication of optical fiber near field probes. *Applied Optics* 34, 1215–1228.

Veerman, J. A., Otter, A. M., Kuipers, L., and Vanhulst, N. F. (1998). High definition aperture probes for near-field optical microscopy fabricated by focused ion beam milling. *Applied Physics Letters* 72, 3115–3117.

Vertikov, A., Kuball, M., Nurmikko, A. V., and Maris, H. J. (1996). Time-resolved pump-probe experiments with subwavelength lateral resolution. *Applied Physics Letters* 60, 2456–2458.

Vigoureux, J. M., Depasse, F., and Girard, C. (1992). Superresolution of near-field optical microscopy defined from properties of confined electromagnetic waves. *Applied Optics* 31, 3036–3045.

Vollmer, M., Giessen, H., Stolz, W., Ghislain, L., and Elings, V. (1999). Ultrafast nonlinear subwavelength solid immersion spectroscopy at T = 8 K. *Applied Physics Letters* 74, 1791–1793.

Vouilloz, F., Oberli, D. Y., Dupertuis, M. A., Gustafsson, A., Reinhardt, F., and Kapon, E. (1997). Polarization anisotropy and valence band mixing in semiconductor quantum wires. *Physical Review Letters* 78, 1580–1583.

Wassermeier, M., Sudijono, J., Johnson, M. D., Leung, K. T., Orr, B. G., Daweritz, L., and Ploog, K. (1995). Reconstruction of the GaAs(311) A surface. *Physical Review B — Condensed Matter* 51, 14721–14724.

Wegscheider, W., Pfeiffer, L. N., Dignam, M. M., Pinczuk, A., West, K. W., Mccall, S. L., and Hull, R. (1993). Lasing from excitons in quantum wires. *Physical Review Letters* 71, 4071–4074.

Wei, P. K., and Fann, W. S. (1998). The probe dynamics under shear force in near-field scanning optical microscopy. *Journal of Applied Physics* 83, 3461–3468.

Wei, C.–C., Wei, P.-K., and Fann, W. (1995). Direct measurements of the true vibrational amplitudes in shear force microscopy. *Applied Physics Letters* 67, 3835–3537.

Weisbuch, C., and Vinter, B. (1991). *Quantum Semiconductor Structures: Fundamentals and Applications*. Academic Press, Boston.

Wiesendanger, R., and Güntherodt, H.-J., eds. (1994–1996). *Scanning Tunneling Microscopy I–III*. Springer, Berlin (1994–1996).

Zeisel, D., Nettesheim, S., Dutoit, B., and Zenobi, R. (1996). Pulsed laser-induced desorption and optical imaging on a nanometer scale with scanning near-field microscopy using chemically etched fiber tips. *Applied Physics Letters* 68, 2491–2492.

Zenhausern, F., Oboyle, M. P., and Wickramasinghe, H. K. (1994). Apertureless near-field optical microscope. *Applied Physics Letters* 65, 1623–1625.

Zenhausern, F., Martin, Y., and Wickramasinghe, H. K. (1995). Scanning interferometric apertureless microscopy: optical imaging at 10 angstrom resolution. *Science* 269, 1083.

CHAPTER 3

Ultrafast Dynamics in Wide Bandgap Wurtzite GaN

K. T. Tsen

DEPARTMENT OF PHYSICS AND ASTRONOMY
ARIZONA STATE UNIVERSITY
TEMPE, ARIZONA

I. INTRODUCTION 109
II. RAMAN SPECTROSCOPY IN SEMICONDUCTORS 110
  1. Theory of Raman Scattering from Carriers in Semiconductors 110
  2. Theory of Raman Scattering by Lattice Vibrations in Semiconductors 122
III. SAMPLES, EXPERIMENTAL SETUP, AND APPROACH 125
IV. PHONON MODES IN THE WURTZITE STRUCTURE GaN 128
V. EXPERIMENTAL RESULTS 128
  1. Studies of Electron–Phonon Interactions in Wurtzite GaN by Nonlinear
     Laser Excitation Processes 128
  2. Nonequilibrium Electron Distributions and Electron–Longitudinal Optical
     Phonon Scattering Rates in Wurtzite GaN Studied by an Ultrashort,
     Ultraviolet Laser 132
  3. Anharmonic Decay of the Longitudinal Optical Phonons in Wurtzite GaN
     Studied by Subpicosecond Time-Resolved Raman Spectroscopy 139
VI. CONCLUSIONS AND FUTURE EXPERIMENTS 147
   REFERENCES 148

I. Introduction

The recent surge of activity in wide bandgap semiconductors (Pankove and Moustakas, 1997) has arisen from the need for electronic devices capable of operation at high power levels, high temperatures and caustic environments, and separately, from a need for optical materials, especially emitters, that are active in the blue and ultraviolet wavelengths. Electronics based upon the existing semiconductor device technologies of Si and GaAs cannot tolerate greatly elevated temperatures or chemically hostile environments. The wide bandgap semiconductors with their excellent thermal conductivities, large breakdown fields, and resistance to chemical corrosion will be the materials of choice for these applications. Among the wide bandgap semiconductors, the III–V nitrides have long been viewed as a very promising semiconductor system for device applications in the blue and ultraviolet wavelengths. The wurtzite polytypes of GaN, AlN, and InN form a continuous alloy system whose direct bandgaps range from 1.9 eV for InN, to 3.4 eV for GaN, to 6.2 eV for AlN. Thus, III–V nitrides could potentially be fabricated into optical devices that are active at wavelengths ranging from the red all the way into the ultraviolet. Because of lower ohmic contact resistances and larger predicted electron saturation velocity (Ferry, 1975), research on GaN is of particular interest.

Although much progress has been made in device-oriented applications with wide bandgap semiconductors, very little information concerning their dynamical properties has yet been obtained. Knowledge of both carrier and phonon dynamical properties is indispensable for device engineers to design better and faster devices. For example, carrier energy loss rate is primarily determined by electron–phonon scattering rates (Shah, 1996); electron relaxation may be greatly influenced by hot phonon effects (Hamaguchi and Inoue, 1992), which in turn are governed by the population relaxation time of optical phonons. In this chapter, electron–phonon and phonon–phonon interactions and nonequilibrium electron distributions in wurtzite GaN are studied by using subpicosecond time-resolved Raman spectroscopy.

II. Raman Spectroscopy in Semiconductors

We first present a comprehensive theory of Raman scattering from carriers in semiconductors, that will be particularly useful for situations where electron distributions are non-equilibrium; and then we give thorough discussions on the theory of Raman scattering by lattice vibrations in semiconductors.

1. THEORY OF RAMAN SCATTERING FROM CARRIERS IN SEMICONDUCTORS

a. A Simple Model

In order to understand how Raman spectroscopy can be used to probe electron distribution function in semiconductors, we start with the simplest physical concept — Compton scattering. As shown in Fig. 1, let's consider that an incident photon with wave vector $\vec{k}_i$ and angular frequency ω_i is interacting with an electron of mass m_e^* traveling at a velocity $\vec{V}$. After the scattering event, the scattered photon is characterized with wave vector $\vec{k}_s$ and angular frequency ω_s. The scattered electron is then moving at a velocity $\vec{V}'$. From the conservation of energy and momentum, we can write the following equations:

$$\hbar\omega_i + \tfrac{1}{2}m_e^*\vec{V}^2 = \hbar\omega_s + \tfrac{1}{2}m_e^*\vec{V}'^2 \qquad (1)$$

$$\hbar\vec{k}_i + m_e^*\vec{V} = \hbar\vec{k}_s + m_e^*\vec{V}'. \qquad (2)$$

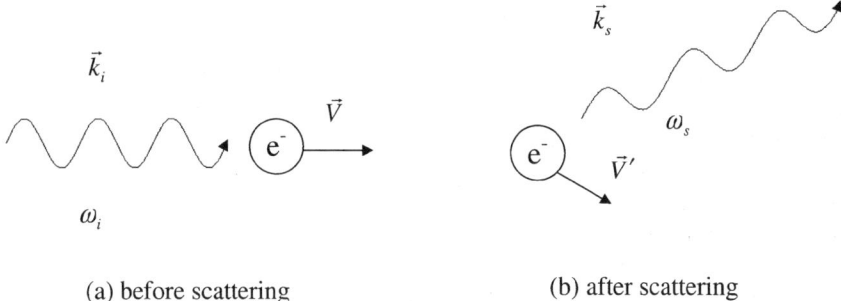

FIG. 1. A simple model—Compton scattering, demonstrating how the electron distribution function in semiconductors can be probed by Raman spectroscopy.

If we define the energy transfer and the wave vector transfer of the photon to be $\omega \equiv \omega_i - \omega_s$ and $\vec{q} \equiv \vec{k}_i - \vec{k}_s$, respectively, then from Eqs. (1) and (2) we have

$$\omega = \vec{V} \cdot \vec{q} + \frac{\hbar \vec{q}^2}{2m_e^*}. \tag{3}$$

This important equation states that the energy transfer of incident photon is (apart from a constant term $\hbar \vec{q}^2 / 2m_e^*$) directly proportional to the electron velocity along the direction of wave-vector transfer. In other words, it implies that Raman scattering intensity, measured at an angular frequency ω, is proportional to the number of electrons that have a velocity component along the direction of wave-vector transfer given by Eq. (3), irrespective of their velocity components perpendicular to $\vec{q}$.

Therefore, if electron distribution function is Maxwell–Boltzmann-like, then the lineshape of Raman scattering spectrum will be Gaussian-like centered around $\omega \cong 0$, whereas a drifted Maxwell–Boltzmann distribution with an electron drift velocity $\vec{V}_d$ will result in a Raman scattering spectrum that is a shifted Gaussian centered around $\omega \cong \vec{q} \cdot \vec{V}_d$.

However, we note that, strictly speaking, this simple picture is only correct for a system of noninteracting electron gas in vacuum. For an electron gas in a semiconductor such as GaAs or GaN, many-body effects and the effects of band structure have to be considered. The former are usually taken into account by the random-phase approximation (RPA) (Platzman and Wolff, 1973) and the latter by sophisticated band structure calculations such as $\vec{k} \cdot \vec{p}$ approximation (Ferry, 1991; Ridley, 1993; Kane, 1957).

b. A Full Quantum-Mechanical Approach

We now use a quantum-mechanical method to calculate the Raman scattering cross section or single-particle scattering (SPS) spectrum for a single-component plasma in a direct bandgap semiconductor such as GaAs or GaN, probed by an ultrafast laser having pulse width t_p. For simplicity, we assume that the probe pulse is a square pulse from $-t_p/2$ to $+t_p/2$ and the electron elastic scattering is the dominant scattering process in the solid-state system. For the more general case in that inelastic scattering plays a role in the electron scattering processes, see Chia et al. (1993).

We start with a typical electron–photon interaction Hamiltonian, which has been shown in the equilibrium case to be (Jha, 1969)

$$H = H_0 + \sum_i \left[\frac{-e}{2m_e^* c} [\vec{p}_i \cdot \vec{A}(\vec{r}_i) + \vec{A}(\vec{r}_i) \cdot \vec{p}_i] + \frac{e^2}{2m_e^* c} \vec{A}^2(\vec{r}_i) \right]$$
$$\equiv H_0 + H_1 + H_2, \quad (4)$$

where H_0 is the total Hamiltonian of the system in the absence of radiation field; e is the charge of an electron; c is the speed of light; $\vec{p}$ is the electron momentum; Σ_i refers to summation over electrons; the second and third terms describe the interactions of electrons with the radiation field; and $\vec{A}$ is the vector potential of the radiation field,

$$\vec{A}(\vec{r}_i) \equiv \frac{1}{\sqrt{V}} \sum_j \left(\frac{2\pi\hbar c^2}{\omega_j} \right)^{1/2} (e^{i\vec{k}_j \cdot \vec{r}_i} b_{\vec{k}_j} + e^{-i\vec{k}_j \cdot \vec{r}_i} b_{\vec{k}_j}^+) \hat{e}_j; \quad (5)$$

where V is the volume of the semiconductor; $b_{\vec{k}_j}$, $b_{\vec{k}_j}^+$ are photon annihilation and creation operators, respectively. Since the vector potential $\vec{A}$ is a linear combination of photon creation and annihilation operators, and the Raman scattering process involves the annihilation of an incident photon and the creation of a scattered photon, $\vec{p} \cdot \vec{A}$ and $\vec{A} \cdot \vec{p}$ terms in Eq. (4) will contribute to the scattering matrix element in the second order and $\vec{A}^2$ terms contribute in the first order in the perturbation-theory calculations of the Raman scattering cross section.

The transition amplitude from an electron–photon state $|I\rangle$ to a state $|F\rangle$ is given by

$$c_{IF} = \frac{1}{i\hbar} \int_{-t_p/2}^{t_p/2} e^{i\omega_{FI} t'} dt' \langle F| \left[H_2 - \frac{i}{\hbar} \int_0^{t+t_p/2} H_1 e^{-\frac{i}{\hbar} H_0 t''} H_1 e^{\frac{i}{\hbar} H_0 t''} dt'' \right] |I\rangle, \quad (6)$$

where $\omega_{FI} \equiv (\varepsilon_F - \varepsilon_I)/\hbar$. For convenience, we have chosen the zero of time so that the probe pulse runs from $-t_p/2$ to $+t_p/2$.

We let the initial state $|I\rangle$ be a product of an electron/phonon state $|i\rangle$ and the phonon state $|n_i, 0_s\rangle$ consisting of n_i incident photons and zero scattered photons. The final state $|F\rangle$ is the product of an electron/phonon state $|f\rangle$ and the photon state $|n_i - 1, 1_s\rangle$. Thus, we can write the transition amplitude in a more convenient form,

$$c_{IF} = \frac{1}{i\hbar} \int_{-t_p/2}^{t_p/2} e^{i\omega_{FI}t'} C(\omega_i, \omega_s) M_{fi}(t') dt' \tag{7}$$

with

$$C(\omega_i, \omega_s) \equiv \frac{-e^2}{m_e^* c^2} \sqrt{n_i} \left[\frac{2\pi \hbar c^2}{(\omega_i \omega_s)^{1/2}} \right] \tag{8}$$

and

$$M_{fi}(t') \equiv (\hat{e}_i \cdot \hat{e}_s)\langle f|n_{\vec{q}}^c|i\rangle + \frac{1}{i\hbar m_e^*}$$
$$\times \Bigg[\int_{-t'-t_p/2}^{0} dt'' e^{i\omega_s t''} \langle f|\hat{e}_i \cdot \vec{j}(\vec{k}_i)\vec{j}(-\vec{k}_s, t'') \cdot \hat{e}_s|i\rangle$$
$$+ \int_{0}^{t'+t_p/2} dt'' e^{i\omega_s t''} \langle f|\hat{e}_s \cdot \vec{j}(-\vec{k}_s, t'')\vec{j}(\vec{k}_i) \cdot \hat{e}_i|i\rangle \Bigg]. \tag{9}$$

The matrix element $M_{fi}(t')$ contains in order three terms represented by three Feynman diagrams; $\vec{k}_i, \vec{k}_s$ are the wave vectors of incident and scattered photons, $\hat{e}_i, \hat{e}_s$ are the polarization vectors of incident and scattered photons, and ω_i, ω_s are the angular frequencies of incident and scattered photons, respectively. $\vec{q} \equiv \vec{k}_i - \vec{k}_s$ is the momentum transfer of the photon to the electronic system. The operator $n_{\vec{q}}^c \equiv \Sigma_{j=1}^{N} e^{i\vec{q}\cdot\vec{r}_j}$ is the Fourier transform of the density operator of electron in the conduction band. In the second quantized notation, the current operator $\vec{j}(\vec{k})$ is (Mooradian and McWhorter, 1969; Hamilton and McWhorter, 1969)

$$\vec{j}(\vec{k}) = \sum_{n,n',\vec{k}',\alpha,\alpha'} \langle \vec{k}+\vec{k}', n, \alpha|e^{i\vec{k}\cdot\vec{r}}\vec{p}|\vec{k}', n', \alpha'\rangle a^+_{\vec{k}+\vec{k}',n,\alpha} a_{\vec{k}',n',\alpha'} \tag{10}$$

where $a_{\vec{p},n,\sigma}, a^+_{\vec{p},n,\sigma}$ are the annihilation and creation operators for a Bloch state $|\vec{p}, n, \sigma\rangle$, respectively.

We assume that energy is nearly conserved so that $\omega_i \cong \omega_s$. We define

$$\vec{j}(\vec{k},t) \equiv e^{\frac{i}{\hbar}H_e^0 t} \vec{j}(\vec{k}) e^{-\frac{i}{\hbar}H_e^0 t} \tag{11}$$

where H_e^0 is the Hamiltonian for the unperturbed electronic system.

We are specifically interested in the case where we scatter off the electrons in the conduction band and hole contributions are neglected. The neglect of the hole contributions is a good approximation, because the coefficient $C(\omega_i, \omega_s)$ has a factor $e^2/m_e^*c^2$, which means that the larger the effective mass, the smaller the Raman scattering cross section. Since the effective mass of electrons in the conduction band of III–V semiconductors is typically much smaller than that of holes in the valence band, it is justified to neglect the contributions of the holes.

We use Kane's two-band model (Kane, 1957) with a full set of valence bands and also use the free electron approximation between different bands, that is, for $n \neq n'$,

$$a^+_{\vec{p},n,\sigma}(t) a_{\vec{p}',n',\sigma'}(t) = e^{\frac{i}{\hbar}(\varepsilon_{p,n,\sigma} - \varepsilon_{p',n',\sigma'})t} a^+_{\vec{p},n,\sigma} a_{\vec{p}',n',\sigma'}, \qquad (12)$$

where $\varepsilon_{p,n,\omega}$ is the energy of the Bloch state $|\vec{p},n,\sigma\rangle$. With this approximation and letting $\vec{k} \to 0$, and $\omega_i \cong \omega_s$, we have

$$M_{fi}(t) = \langle f| \sum_{\vec{p},\alpha,\beta} \gamma_{\alpha\beta}(\vec{p},t,t_p,\omega_i) a^+_{\vec{p}+\vec{q},\alpha} a_{\vec{p},\beta} |i\rangle, \qquad (13)$$

where

$$\gamma_{\alpha\beta}(\vec{p},t,t_p,w_i) \equiv (\hat{e}_i \cdot \hat{e}_s)\delta_{\alpha\beta} + \frac{1}{\hbar m_e^*}$$

$$\times \sum_{n,\sigma} [\langle c,\alpha|\vec{p}\cdot\hat{e}_i|n,\sigma\rangle\langle n,\sigma|\vec{p}\cdot\hat{e}_s|c,\beta\rangle\Theta(t) + \langle c,\alpha|\vec{p}\cdot\hat{e}_s|n,\sigma\rangle\langle n,\sigma|\vec{p}\cdot\hat{e}_i|c,\beta\rangle\phi(t)]. \qquad (14)$$

Equation (14) is the generalization of the time-independent $\gamma_{\alpha\beta}$ derived by Hamilton and McWhorter (1969) for the equilibrium case. The summation is taken over the three valence bands (heavy hole, light hole, and split-off hole bands), and c represents the conduction band. The spin indices are α, β, and σ. The steplike functions $\Theta(t)$ and $\phi(t)$ appear in the expression because of the finite pulse width of the probe pulse and are given by

$$\Theta(t) = \frac{1}{i}\int_{-t-t_p/2}^{0} dt' e^{i\omega_i t'} e^{\frac{i}{\hbar}(\varepsilon_{p,n} - \varepsilon_{p,c})t'}; \quad \phi(t) = \frac{1}{i}\int_{0}^{t+t_p/2} dt' e^{i\omega_i t'} e^{\frac{i}{\hbar}(\varepsilon_{p,c} - \varepsilon_{p,n})t'}. \qquad (15)$$

For simplicity, we evaluate $\gamma_{\alpha\beta}(\vec{p},t,t_p,\omega_i)$ near the center of the Brillouin zone with $\vec{k}\cdot\vec{p}$ wave functions provided by Kane, that yields

$$\gamma_{\alpha\beta}(\vec{p},t,t_p,\omega_i) = \hat{e}_i \cdot \vec{\vec{C}}_p(t,\omega_i) \cdot \hat{e}_s \delta_{\alpha\beta} + i(\hat{e}_i \times \hat{e}_s) \cdot \vec{\vec{S}}_p(t,\omega_i) \cdot \vec{\sigma}_{\alpha\beta}, \qquad (16)$$

where $\vec{\sigma}_{\alpha\beta}$ are Pauli-matrices, and $\vec{C}_p(t, \omega_i)$ and $\vec{S}_p(t, \omega_i)$ are time-dependent dyadic tensors, involving charge density, energy-density fluctuations, and spin-density fluctuations, respectively. They can be written as

$$\vec{C}_p(t, \omega_i) \equiv \vec{I} \left\{ 1 + \frac{2P^2}{3m_e^*} \right.$$

$$\times \sum_{n=1}^{3} A_c(n) \cdot \frac{E_{g_n} - e^{\frac{i}{\hbar}E_{g_n}(t+t_p/2)}\{E_{g_n}\cos[\omega_i(t+t_p/2)] - i\hbar\omega_i \sin[\omega_i(t+t_p/2)]\}}{(E_{g_n} - i\Gamma_n)^2 - (\hbar\omega_i)^2} \right\}$$

$$- \left(\hat{p}\hat{p} - \frac{\vec{I}}{3} \right) \frac{P^2}{m_e^*}$$

$$\times \sum_{n=1}^{2} B_c(n) \cdot \frac{E_{g_n} - e^{\frac{i}{\hbar}E_{g_n}(t+t_p/2)}\{E_{g_n}\cos[\omega_i(t+t_p/2)] - i\hbar\omega_i \sin[\omega_i(t+t_p/2)]\}}{(E_{g_n} - i\Gamma_n)^2 - (\hbar\omega_i)^2}$$

(17)

$$\vec{S}_p(t, \omega_i) \equiv -\vec{I} \left\{ \frac{P^2}{3m_e^*} \right.$$

$$\times \sum_{n=1}^{3} A_s(n) \cdot \frac{\hbar\omega_i - e^{\frac{i}{\hbar}E_{g_n}(t+t_p/2)}\{\hbar\omega_i \cos[\omega_i(t+t_p/2)] - iE_{g_n}\sin[\omega_i(t+t_p/2)]\}}{(E_{g_n} - i\Gamma_n)^2 - (\hbar\omega_i)^2} \right\}$$

$$- \left(\hat{p}\hat{p} - \frac{\vec{I}}{3} \right) \frac{P^2}{m_e^*}$$

$$\times \sum_{n=1}^{2} B_s(n) \cdot \frac{\hbar\omega_i - e^{\frac{i}{\hbar}E_{g_n}(t+t_p/2)}\{\hbar\omega_i \cos[\omega_i(t+t_p/2)] - iE_{g_n}\sin[\omega_i(t+t_p/2)]\}}{(E_{g_n} - i\Gamma_n)^2 - (\hbar\omega_i)^2},$$

(18)

where

$$A_c(1) = A_c(2) = A_c(3) = B_c(1) = -B_c(2) = 1,$$
$$A_s(1) = A_s(2) = B_s(1) = -B_s(2) = 1, \text{ and } A_s(3) = -2.$$

$\Gamma_1, \Gamma_2, \Gamma_3$ are the damping constants involved in the Raman scattering processes. The $E_{g_1}, E_{g_2}, E_{g_3}$ are the energy difference between the conduction band and the heavy-hole, light-hole, and split-off-hole bands evaluated at wave vector $\vec{k}$, respectively. $P \equiv -i\langle S|p_z|Z\rangle$ is the momentum matrix element between the conduction and valence bands at the Γ-point in Kane's notations.

The Raman scattering cross section is proportional to the transition probability $|c_{IF}|^2$ averaged over all the possible initial states (that might be far from equilibrium). Since the coefficient $C(\omega_i, \omega_s)$ is a constant if $\omega_i \cong \omega_s$, we finally obtain

$$\frac{d^2\sigma}{d\omega d\Omega} \propto \left\langle \sum_f \left| \int_{-t_p/2}^{t_p/2} dt e^{i\omega_{FI}t} \langle f | \sum_{\vec{p},\alpha,\beta} \gamma_{\alpha\beta}(\vec{p},t,t_p,\omega_i) a^+_{\vec{p}+\vec{q},\alpha} a_{\vec{p},\beta} | i \rangle \right|^2 \right\rangle$$

$$= \int_{-t_p/2}^{t_p/2} dt \int_{-t_p/2-t}^{t_p/2-t} dt' e^{i\omega t'} \sum_{\vec{p},\alpha,\beta} \gamma_{\alpha\beta}(\vec{p},t,t_p,\omega_i) \sum_{\vec{p}',\alpha',\beta'} \gamma^+_{\alpha'\beta'}(\vec{p}',t'+t,t_p,\omega_i)$$

$$\times \langle a^+_{\vec{p}',\beta'}(t') a_{\vec{p}'+\vec{q},\alpha'}(t') a^+_{\vec{p}+\vec{q},\alpha} a_{\vec{p},\beta} \rangle. \tag{19}$$

Our result for Raman scattering cross section shows that the cross section is a double time integral of a product of two $\gamma_{\alpha\beta}(\vec{p},t,t_p,\omega_i)$ terms (which contain information about the polarization of incident and scattered photons, the probe pulse width, and the characteristics of the band structure near the Γ-point and a dynamical electron correlation function. We note that $\gamma_{\alpha\beta}(\vec{p},t,t_p,\omega_i)$ is completely known from Eqs. (17) and (18), once the polarizations of incident and scattered photons, the probe pulse width, and the near-bandgap energies are given. The only unknown quantity in Eq. (19) is the time-dependent electron correlation function

$$\langle a^+_{\vec{p}',\beta'}(t') a_{\vec{p}'+\vec{q},\alpha'}(t') a^+_{\vec{p}+\vec{q},\alpha} a_{\vec{p},\beta} \rangle,$$

which we now determine for a nonequilibrium electron distribution.

To evaluate the electron correlation function

$$\langle a^+_{\vec{p}',\beta'}(t') a_{\vec{p}'+\vec{q},\alpha'}(t') a^+_{\vec{p}+\vec{q},\alpha} a_{\vec{p},\beta} \rangle,$$

we need to use the equation of motion (Pines, 1962) and the RPA. The electrons of interest are in the conduction band, and the time-dependence of the operators is for times during the probe pulse of duration t_p. It is convenient to find the correlation function by defining its frequency Fourier transform function, that is,

$$\int_{-\infty}^{+\infty} dt' e^{i\omega' t'} \langle a^+_{\vec{p}',\beta'}(t') a_{\vec{p}'+\vec{q},\alpha'}(t') a^+_{\vec{p}+\vec{q},\alpha} a_{\vec{p},\beta} \rangle$$

$$\equiv -i\hbar [g^+_{\alpha',\beta',\alpha,\beta}(\vec{p}',\vec{p},\omega',\vec{q}) + g^-_{\alpha',\beta',\alpha,\beta}(\vec{p}',\vec{p},\omega',\vec{q})]. \tag{20}$$

In Eq. (20), we have defined the retarded and advanced correlation function,

$$g^+_{\alpha',\beta',\alpha,\beta}(\vec{p}',\vec{p},t',\vec{q}) \equiv \frac{i}{\hbar}\varphi(t')\langle a^+_{\vec{p}',\beta'}(t')a_{\vec{p}'+\vec{q},\alpha'}(t')a^+_{\vec{p}+\vec{q},\alpha}a_{\vec{p},\beta}\rangle \quad (21)$$

and

$$g^-_{\alpha',\beta',\alpha,\beta}(\vec{p}',\vec{p},t',\vec{q}) \equiv \frac{i}{\hbar}\varphi(-t')\langle a^+_{\vec{p}',\beta'}(t')a_{\vec{p}'+\vec{q},\alpha'}(t')a^+_{\vec{p}+\vec{q},\alpha}a_{\vec{p},\beta}\rangle, \quad (22)$$

where $\varphi(t)$ is a steplike function with $\varphi(0) = \frac{1}{2}$. The relationship between the Fourier transform of the retarded and advanced functions can be found to be

$$g^-_{\alpha',\beta',\alpha,\beta}(\vec{p}',\vec{p},\omega',\vec{q}) = -[g^+_{\alpha',\beta',\alpha,\beta}(\vec{p}',\vec{p},\omega',\vec{q})]^*. \quad (23)$$

We use the equation of motion technique within the RPA to derive the time dependence of $g^+_{\alpha',\beta',\alpha,\beta}(\vec{p}',\vec{p},\omega',\vec{q})$. From the equation of motion in Heisenberg representation, we have

$$(\hbar\omega' + \varepsilon_{\vec{p}'} - \varepsilon_{\vec{p}'+\vec{q}} + i\hbar\delta)g^+_{\alpha',\beta',\alpha,\beta}(\vec{p}',\vec{p},\omega',\vec{q})$$
$$= -\langle a^+_{\vec{p}',\beta'}a_{\vec{p}'+\vec{q},\alpha'}a^+_{\vec{p}+\vec{q},\alpha}a_{\vec{p},\beta}\rangle$$
$$+ \delta_{\alpha'\beta'}\left[\frac{4\pi e^2}{q^2\varepsilon(\omega)}\right][n(\vec{p}') - n(\vec{p}'+\vec{q})]\sum_{\vec{k},\sigma}g^+_{\sigma,\sigma,\alpha,\beta}(\vec{k},\vec{p},\omega',\vec{q})$$
$$- \frac{i\hbar}{\tau}\left[g^+_{\alpha',\beta',\alpha,\beta}(\vec{p}',\vec{p},\omega',\vec{q}) - \oint\frac{d\Omega_{\vec{p}'}}{4\pi}g^+_{\alpha',\beta',\alpha,\beta}(\vec{p}',\vec{p},\omega',\vec{q})\right], \quad (24)$$

where δ is an infinitesimal positive quantity, and

$$\varepsilon(\omega) = \varepsilon_\infty + \frac{(\varepsilon_0 - \varepsilon_\infty)\omega_T^2}{\omega_T^2 - \omega^2}$$

is the frequency-dependent dielectric function. ε_0, ε_∞ are the static and high-frequency dielectric constants, respectively; ω_T is the angular frequency of transverse optical phonon. Here, we have phenomenologically added the last term of Eq. (24), which conserves the particle number, spin, and energy. This way of introducing a phenomenological collision time is exactly the same as the work of Hamilton and McWhorter (1969). It essentially introduces an elastic collision time τ in the electronic system, because the correlation function relaxes to its average value at a constant energy $\varepsilon_{\vec{p}}$.

Specifically, τ includes electron–electron, electron–impurity, and electron–acoustic phonon scattering processes.

Equation (24) allows us to find the closed form of the retarded electron correlation function in frequency domain. The Raman scattering cross section is proportional to the imaginary part of $g^+_{\alpha',\beta',\alpha,\beta}(\vec{p}',\vec{p},\omega',\vec{q})$ through Eqs. (19), (20), and (21), so the cross section becomes

$$\frac{d^2\sigma}{d\omega d\Omega} \propto \int_{-\infty}^{\infty} d\omega' \int_{-t_p/2}^{t_p/2} dt \int_{-t_p/2-t}^{t_p/2-t} dt' e^{i(\omega-\omega')t'} \sum_{\vec{p},\alpha,\beta} \gamma_{\alpha\beta}(\vec{p},t,t_p,\omega_i)$$

$$\times \sum_{\vec{p}',\alpha',\beta'} \gamma^+_{\alpha',\beta'}(\vec{p}',t'+t,t_p,\omega_i) \cdot \text{Im}\left\{\left[-\langle a^+_{\vec{p}',\beta'} a_{\vec{p}'+\vec{q},\alpha'} a^+_{\vec{p}+\vec{q},\alpha} a_{\vec{p},\beta}\rangle \right.\right.$$

$$\left. + \delta_{\alpha'\beta'}\left[\frac{4\pi e^2}{q^2\varepsilon(\omega)}\right][n(\vec{p}') - n(\vec{p}'+\vec{q})]\sum_{\vec{k},\sigma} g^+_{\sigma,\sigma,\alpha,\beta}(\vec{k},\vec{p},\omega',\vec{q})\right]$$

$$\left.\times \frac{1}{\hbar\omega'+\varepsilon_{\vec{p}'}-\varepsilon_{\vec{p}'+\vec{q}}+i\hbar/\tau'}\cdot\left[1-\frac{i\hbar}{\tau}\left\langle\frac{1}{\hbar\omega'+\varepsilon_{\vec{p}'}-\varepsilon_{\vec{p}'+\vec{q}}+i\hbar/\tau'}\right\rangle_{\Omega_{\vec{p}'}}\right]^{-1}\right\},$$

(25)

where

$$\left\langle\frac{1}{\hbar\omega'+\varepsilon_{\vec{p}'}-\varepsilon_{\vec{p}'+\vec{q}}+i\hbar/\tau'}\right\rangle_{\Omega_{\vec{p}'}}$$

represents the average over the solid angle $\Omega_{\vec{p}'}$, and

$$\frac{1}{\tau'} = \frac{1}{\tau} + \delta.$$

$n(\vec{p})$ is the electron distribution function. "Im" means taking the imaginary part of.

We would like to stress that our derivation has not used the fluctuation-dissipation theorem, so Eq. (25) is valid irrespective of the nature of electron distribution, that is, equilibrium or nonequilibrium.

The Raman scattering cross section given by Eq. (25) has three parts, namely, the charge-density fluctuations (CDF), energy-density fluctuations (EDF), and spin-density fluctuations (SDF). The spin of the electron is coupled with the incident light through the second-order $\vec{p}_i \cdot \vec{A}$ perturbation terms; SDF is the dominant contribution for higher electron concentrations. In addition, EDF contribution dominates when CDF contribution is screened at high electron densities. These three contributions are related to different polarization selection rules as pointed out later.

Applying the RPA to the zero-time electron correlation function

$$\langle a^+_{\vec{p}',\beta'} a_{\vec{p}'+\vec{q},\alpha'} a^+_{\vec{p}+\vec{q},\alpha} a_{\vec{p},\beta} \rangle,$$

we obtain a simple and useful relation:

$$\langle a^+_{\vec{p}',\beta'} a_{\vec{p}'+\vec{q},\alpha'} a^+_{\vec{p}+\vec{q},\alpha} a_{\vec{p},\beta} \rangle = n(\vec{p})[1 - n(\vec{p}+\vec{q})] \delta_{\vec{p}\vec{p}'} \delta_{\alpha\alpha'} \delta_{\beta\beta'}. \tag{26}$$

From Eqs. (24) and (26), we have

$$\sum_{\vec{k},\sigma} g^+_{\sigma,\sigma,\alpha,\beta}(\vec{k},\vec{p},\omega',\vec{q}) = \dfrac{-\dfrac{n(\vec{p})[1 - n(\vec{p}+\vec{q})]}{\hbar\omega' + \varepsilon_{\vec{p}} - \varepsilon_{\vec{p}+\vec{q}} + i\hbar\delta} \cdot \delta_{\alpha\beta}}{1 - 2\sum_{\vec{k}} \dfrac{\left[\dfrac{4\pi e^2}{q^2 \varepsilon(\omega)}\right][n(\vec{k}) - n(\vec{k}+\vec{q})]}{\hbar\omega' + \varepsilon_{\vec{k}} - \varepsilon_{\vec{k}+\vec{q}} + i\hbar\delta}}. \tag{27}$$

The Kronecker deltas in Eqs. (26) and (27) are very important in separating the contributions of SDF from those of CDF and EDF.

The multiplication of $\gamma_{\alpha\beta}(\vec{p}, t, t_p, \omega_i)$ and $\gamma^+_{\alpha'\beta'}(\vec{p}', t'+t, t_p, \omega_i)$ that appears in Eq. (25) is in general very complicated. For simplicity, we drop the anisotropic part, that is, we ignore the part containing $(\hat{p}\hat{p} - \bar{I}/3)$ in Eqs. (17) and (18). Then

$$\gamma_{\alpha\beta}(\vec{p}, t, t_p, \omega_i)\gamma^+_{\alpha'\beta'}(\vec{p}', t'+t, t_p, \omega_i)$$
$$\cong [(\hat{e}_i \cdot \hat{e}_s)C_p(t, \omega_i)\delta_{\alpha\beta} + i(\hat{e}_i \times \hat{e}_s)\cdot \vec{\sigma}_{\alpha\beta} S_p(t, \omega_i)]$$
$$\times [(\hat{e}_i \cdot \hat{e}_s)C^*_{p'}(t'+t, \omega_i)\delta_{\alpha\beta} + i(\hat{e}_i \times \hat{e}_s)\cdot \vec{\sigma}^+_{\alpha\beta} S^*_{p'}(t'+t, \omega_i)], \tag{28a}$$

where $C_p(t, \omega_i)$ and $S_p(t, \omega_i)$ are the remaining scalars occurring from the isotropic $(\bar{I})$ part of Eqs. (17) and (18), respectively. We notice that Eq. (28a) gives rise not only to terms containing $(\hat{e}_i \cdot \hat{e}_s)$ and $(\hat{e}_i \times \hat{e}_s)$, but also to cross terms involving $(\hat{e}_i \cdot \hat{e}_s)(\hat{e}_i \times \hat{e}_s)$. Fortunately, the cross terms exactly vanish when the Kronecker deltas and a property of Pauli matrices $(\Sigma_\alpha \bar{\sigma}_{\alpha,\alpha} = 0)$ are used. Equation (28a) becomes

$$\gamma_{\alpha\beta}(\vec{p}, t, t_p, \omega_i)\gamma^+_{\alpha'\beta'}(\vec{p}', t'+t, t_p, \omega_i)$$
$$\cong (\hat{e}_i \cdot \hat{e}_s)^2 \delta_{\alpha\beta}\delta_{\alpha'\beta'} C_p(t, \omega_i) C^*_{p'}(t'+t, \omega_i)$$
$$+ (\hat{e}_i \times \hat{e}_s)\cdot \vec{\sigma}_{\alpha\beta}(\hat{e}_i \times \hat{e}_s)\cdot \vec{\sigma}^+_{\alpha'\beta'} S_p(t, \omega_i) S^*_{p'}(t'+t, \omega_i). \tag{28b}$$

We now combine the results of Eqs. (26), (27), and (28) to obtain the desired form of Raman scattering cross section. The contribution proportional to $(\hat{e}_i \cdot \hat{e}_s)^2$ corresponds to that of CDF and EDF and is given by

$$\left(\frac{d^2\sigma}{d\omega d\Omega}\right)_{CDF,EDF} \propto \sum_{\vec{p}} -n(\vec{p})[1-n(\vec{p}+\vec{q})](\hat{e}_i \cdot \hat{e}_s)^2$$

$$\times \int_{-\infty}^{\infty} d\omega' \int_{-t_p/2}^{t_p/2} dt \int_{-t_p/2-t}^{t_p/2-t} dt' e^{i(\omega-\omega')t'} \sum_{\vec{p}} [C_p(t,\omega_i) C_{p'}^*(t'+t,\omega_i) \delta_{\vec{p}\vec{p}'}$$

$$+ 2C_p(t,\omega_i) C_{p'}^*(t'+t,\omega_i) \cdot \frac{\left[\frac{4\pi e^2}{q^2 \varepsilon(\omega)}\right]\left[\frac{n(\vec{p}')-n(\vec{p}'+\vec{q})}{\hbar\omega'+\varepsilon_{\vec{p}'}-\varepsilon_{\vec{p}'+\vec{q}}}\right]}{1 - 2\sum_{\vec{k}} \frac{\left[\frac{4\pi e^2}{q^2 \varepsilon(\omega)}\right][n(\vec{k}')-n(\vec{k}'+\vec{q})]}{\hbar\omega'+\varepsilon_{\vec{k}'}-\varepsilon_{\vec{k}'+\vec{q}}}}$$

$$\times \mathrm{Im}\left\{\frac{1}{\hbar\omega'\varepsilon_{\vec{p}'}-\varepsilon_{\vec{p}'+\vec{q}}+i\hbar/\tau} \cdot \left[1 - \frac{i\hbar}{\tau}\left\langle\frac{1}{\hbar\omega'+\varepsilon_{\vec{p}'}-\varepsilon_{\vec{p}'+\vec{q}}+i\hbar/\tau}\right\rangle_{\Omega_{\vec{p}'}}\right]^{-1}\right\}.$$

(29)

The CDF and EDF contributions are very complicated. They contain both the screening effects of Coulomb interaction and the electron–phonon interaction through the dielectric constant. The double time integrals in Eq. (29) show the effects of probing with ultrashort laser pulse-broadening of the Raman spectra; the imaginary part, which contains the collision time, gives rise to the narrowing of Raman spectra.

The other contributions to the Raman scattering cross section is from the SDF. The SDF contributions are proportional to $(\hat{e}_i \times \hat{e}_s)^2$ and are independent of both the Coulomb interaction and electron–phonon interactions. It is therefore much simpler than either CDF or EDF and is most conveniently used to probe electron distribution functions in semiconductors. The expression for SDF is given by

$$\left(\frac{d^2\sigma}{d\omega d\Omega}\right)_{SDF} \propto \sum_{\vec{p}} -n(\vec{p})[1-n(\vec{p}+\vec{q})](\hat{e}_i \cdot \hat{e}_s)^2$$

$$\times \int_{-\infty}^{\infty} d\omega' \int_{-t_p/2}^{t_p/2} dt \int_{-t_p/2-t}^{t_p/2-t} dt' e^{i(\omega-\omega')t'} S_p(t,\omega_i) S_p^*(t'+t,\omega_i)$$

$$\times \mathrm{Im}\left\{\frac{1}{\hbar\omega'\varepsilon_{\vec{p}}-\varepsilon_{\vec{p}+\vec{q}}+i\hbar/\tau} \cdot \left[1 - \frac{i\hbar}{\tau}\left\langle\frac{1}{\hbar\omega'+\varepsilon_{\vec{p}}-\varepsilon_{\vec{p}+\vec{q}}+i\hbar/\tau}\right\rangle_{\Omega_{\vec{p}}}\right]^{-1}\right\}.$$

(30)

Equations (29) and (30) represent results for the CDF, EDF, and SDF contributions to the Raman scattering cross section, respectively. These equations are useful for a direct bandgap semiconductor for which elastic scattering processes such as electron–electron, electron–hole, electron–impurity, and electron–acoustic phonon scattering are dominant. These results are derived from a full time-dependent quantum mechanical treatment for a square probing laser pulse of duration t_p and frequency ω_i. In general, the effects of probing with short laser pulses are to broaden the SPS spectrum, and the effects of electron scattering processes are to narrow the SPS spectrum. In particular, the fluctuation-dissipation theorem has not been used in our derivation of Eqs. (29) and (30) so that the results are applicable to both equilibrium and nonequilibrium systems. A phenomenological electron scattering time τ has been introduced in a similar way to that used in the equilibrium case. We note that in the limit of very long probe pulse ($t_p \to \infty$) and equilibrium electron distributions, our results can be shown to reduce to expressions previously given for the Raman scattering cross section in the equilibrium case (Hamilton and McWhorter, 1969).

It is very instructive to note that if we assume that the pulse width of the probe pulse is sufficiently wide, collision effects are negligible, the electron distribution function is nondegenerate, and the term involving matrix elements S_p does not depend upon the electron momentum, then Eq. (30) can be shown to become

$$\left(\frac{d^2\sigma}{d\omega d\Omega}\right)_{SDF} \propto \int d^3p \cdot n(\vec{p}) \cdot \delta\left[\omega - \vec{V}\cdot\vec{q} - \frac{\hbar q^2}{2m_e^*}\right]. \qquad (31)$$

Here, the δ function ensures that both the energy and momentum are conserved.

We note that Eq. (31) has come to the same conclusion as that we have obtained previously by simply considering the conservation of energy and momentum in the Compton scattering process. It shows that the measured Raman scattering cross section at a given solid angle $d\Omega$ (which determines $\vec{q}$) provides *direct* information about the electron distribution function in the direction of wave-vector transfer $\vec{q}$.

An intriguing point for probing carrier distributions with Raman spectroscopy is that since Raman scattering cross section is inversely proportional to the effective mass of the carrier, it preferentially probes electron transport even if holes are simultaneously present. This unique feature makes the interpretation of electron distribution in Raman scattering experiments much simpler than those of other techniques.

2. Theory of Raman Scattering by Lattice Vibrations in Semiconductors

Consider an incident laser beam of angular frequency ω_i that is scattered by a semiconductor and the scattered radiation is analyzed spectroscopically, as shown in Fig. 2. In general, the scattered radiation consists of a laser beam of angular frequency ω_i accompanied by weaker lines of angular frequencies $\omega_i \pm \omega$. The line at an angular frequency $\omega_i - \omega$ is called a Stokes line, whereas that at an angular frequency $\omega_i + \omega$ is usually referred to as an anti-Stokes line. The important aspect is that the angular frequency shifts ω are independent of ω_i. In this way, this phenomenon differs from that of luminescence, in that it is the angular frequency of the emitted light that is independent of ω_i. The effect just described is called the Raman effect. It was predicted by Smekal (1923) and is implicit in the radiation theory of Kramers and Heisenberg (1925). It was discovered experimentally by Raman (1928) and by Landsberg and Mandel'shtam (1928) in 1928. It can be understood as an inelastic scattering of light in which an internal form of motion of the scattering system is either excited or absorbed during the process.

a. A Simple Classical Theory

Let us imagine that we have a crystalline lattice having an internal mode of vibration characterized by a normal coordinate

$$Q = Q_0 \cos \omega t. \tag{32}$$

The electronic polarizability α is generally a function of Q and, since in general, $\omega \ll \omega_i$, at each instant we can regard Q as fixed compared with the

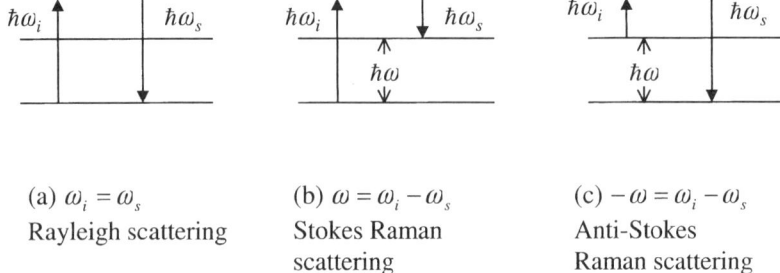

(a) $\omega_i = \omega_s$
Rayleigh scattering

(b) $\omega = \omega_i - \omega_s$
Stokes Raman scattering

(c) $-\omega = \omega_i - \omega_s$
Anti-Stokes Raman scattering

FIG. 2. A diagram showing (a) Rayleigh scattering process; (b) Stokes Raman scattering process; and (c) anti-Stokes Raman scattering process.

variation of the external field $\vec{E}$, that is, at angular frequency ω_i the induced dipole moment $\vec{P}$ is

$$\vec{P} = \alpha\vec{E} = \alpha(Q)\vec{E}. \tag{33}$$

Let $\alpha_0 = \alpha(0)$ be the polarizability in the absence of any excitation. We can write

$$\alpha(Q) = \alpha_0 + \left(\frac{\partial\alpha}{\partial Q}\right)_0 Q + \frac{1}{2}\left(\frac{\partial^2\alpha}{\partial Q^2}\right)_0 Q^2 + \cdots$$

$$= \alpha_0 + \alpha_1 Q + \frac{1}{2}\alpha_2 Q^2 + \cdots, \tag{34}$$

where

$$\left[\frac{\partial\alpha}{\partial Q}\right]_0 \equiv \alpha_1; \quad \left[\frac{\partial^2\alpha}{\partial Q^2}\right]_0 \equiv \alpha_2$$

and the derivative is to be evaluated at zero excitation field.

If we assume that $\vec{E} = \vec{E}_0 \cos\omega_i t$, we find that

$$\vec{P}(t) = \left(\alpha_0\vec{E}_0 + \frac{1}{4}\alpha_2 Q_0^2 \vec{E}_0\right)\cos\omega_i t + \frac{\vec{E}_0}{2}\alpha_1 Q_0[\cos(\omega_i + \omega)t + \cos(\omega_i - \omega)t]$$

$$+ \frac{1}{8}\alpha_2 Q_0^2 \vec{E}_0[\cos(\omega_i + 2\omega) + \cos(\omega_i - 2\omega)] + \cdots. \tag{35}$$

For an oscillating dipole moment, the magnetic and electric fields of emitted electromagnetic wave are given by (Jackson, 1975)

$$\vec{B} = \frac{1}{c^2 r}\left[\frac{\partial^2 \vec{P}(t - r/c)}{\partial^2 t}\right] \times \hat{n} \tag{36a}$$

and

$$\vec{E} = \vec{B} \times \hat{n}, \tag{36b}$$

where $\vec{r}$ is the position vector connecting the center of dipole moment to the point of observation, and $\hat{n} = \vec{r}/|\vec{r}|$.

Therefore, the first term in Eq. (35) gives rise to Rayleigh scattering; the second term gives the anti-Stokes and Stokes first-order Raman lines, respectively; the third term takes into account the anti-Stokes and Stokes second-order Raman lines, and so on. We notice that in Eq. (35), the intensities of the Stokes and anti-Stokes lines are equal; this is because all classical theories neglect the possibility of spontaneous emission.

b. A Quantum-Mechanical Theory

In the quantum-mechanical treatment of scattering of light by lattice vibrations, we consider the total Hamiltonian of the system, including the radiation field;

$$H = H'_0 + H_{el-ph} + H', \tag{37}$$

where H'_0 includes contributions from the electronic system, lattice vibrations (or phonons) and radiation field; $H_{el-ph} = -e\varphi(\vec{r}_i)$ describes the interaction of electrons with phonons; $\varphi(\vec{r}_i)$ is the potential due to, say, deformation potential and/or Fröhlich interactions; and

$$\begin{aligned}H' &= \sum_i \frac{-e}{2m_e c}[\vec{A}(\vec{r}_i) \cdot \vec{p}_i + \vec{p}_i \cdot \vec{A}(\vec{r}_i)] + \sum_i \frac{e^2}{2m_e c^2} \vec{A}^2(\vec{r}_i) \\ &= \sum_i \frac{-e}{m_e c}[\vec{p}_i \cdot \vec{A}(\vec{r}_i)] + \sum_i \frac{e^2}{2m_e c^2} \vec{A}^2(\vec{r}_i) \equiv H'_1 + H'_2 \end{aligned} \tag{38}$$

takes into account the electron–photon interactions, where $\vec{A}(\vec{r}_i)$ is the vector potential of radiation field given by Eq. (5).

We notice that for a typical Raman scattering process in which $\omega_i \gg \omega$, photons do not interact directly with phonons, but through electron–phonon interactions, that is, the H_{el-ph} term in the total Hamiltonian. Since the Raman scattering process involves the annihilation of an incident photon and the creation of a scattered photon, $\vec{p} \cdot \vec{A}$ and $\vec{A} \cdot \vec{p}$ terms in Eq. (38) will contribute to the scattering matrix element in the third order and $\vec{A}^2$ terms contribute in the second order in the perturbation-theory calculations of Raman scattering cross section. If we neglect nonlinear processes, then only $\vec{p} \cdot \vec{A}$ and $\vec{A} \cdot \vec{p}$ terms in Eq. (38) are important and need to be considered.

From the time-dependent perturbation theory and Fermi Golden Rule, we obtain for the scattering probability (that is proportional to Raman scattering cross section) for a one-phonon Stokes Raman process (Yu and Cardona, 1996):

$$\begin{aligned}P(\omega_s) = \frac{2\pi}{\hbar} \bigg| &\sum_{n,n'} \frac{\langle i|H'_1|n\rangle\langle n|H_{el-ph}|n'\rangle\langle n'|H'_1|i\rangle}{[\hbar\omega_i - (E_n - E_i)][\hbar\omega_i - \hbar\omega - (E_{n'} - E_i)]} \\ + &\sum_{n,n'} \frac{\langle i|H'_1|n\rangle\langle n|H'_1|n'\rangle\langle n'|H_{el-ph}|i\rangle}{[\hbar\omega_i - (E_n - E_i)][\hbar\omega_i - \hbar\omega_s - (E_{n'} - E_i)]}\end{aligned}$$

$$+ \sum_{n,n'} \frac{\langle i|H'_1|n\rangle\langle n|H_{el-ph}|n'\rangle\langle n'|H'_1|i\rangle}{[-\hbar\omega_s - (E_n - E_i)][-\hbar\omega_s - \hbar\omega - (E_{n'} - E_i)]}$$

$$+ \sum_{n,n'} \frac{\langle i|H'_1|n\rangle\langle n|H'_1|n'\rangle\langle n'|H_{el-ph}|i\rangle}{[-\hbar\omega_s - (E_n - E_i)][-\hbar\omega_i - \hbar\omega - (E_{n'} - E_i)]}$$

$$+ \sum_{n,n'} \frac{\langle i|H_{el-ph}|n\rangle\langle n|H'_1|n'\rangle\langle n'|H'_1|i\rangle}{[-\hbar\omega - (E_n - E_i)][-\hbar\omega + \hbar\omega_i - (E_{n'} - E_i)]}$$

$$+ \left. \sum_{n,n'} \frac{\langle i|H_{el-ph}|n\rangle\langle n|H'_1|n'\rangle\langle n'|H'_1|i\rangle}{[-\hbar\omega - (E_n - E_i)][-\hbar\omega - \hbar\omega_s - (E_{n'} - E_i)]} \right|^2$$

$$\times \delta(\hbar\omega_i - \hbar\omega_s - \hbar\omega) \tag{39}$$

where $|i\rangle$ is the initial state of the system and E_i is its energy; $|n\rangle$, $|n'\rangle$ are intermediate states with energies E_n, $E_{n'}$, respectively.

We note that there are three processes involved in one-phonon Raman scattering: The incident photon is annihilated; the scattered photon is emitted; and a phonon is annihilated (or created). Since these three processes can occur in any time order in the time-dependent perturbation-theory calculations of scattering probability, we expect that there will be six terms or contributions to $P(\omega_s)$, that is consistent with Eq. (39). The δ function here ensures that energy is conserved in the Raman scattering process.

III. Samples, Experimental Setup, and Approach

The wurtzite structure GaN sample used in our Raman experiments was a 2-μm-thick undoped GaN (with a residual electron density of $n \cong 10^{16}$ cm^{-3}) grown by molecular beam epitaxy on a (0001)-oriented sapphire substrate. The z axis of this wurtzite structure GaN is perpendicular to the sapphire substrate plane.

The experimental setup for transient/time-resolved picosecond/subpicosecond Raman spectroscopy is shown in Figs. 3 and 4, respectively (Tsen et al., 1989, 1991, 1996a, 1996b, 1998a; Ruf et al., 1993; Tsen, 1993; Grann et al., 1994, 1995a, 1995b, 1996, 1997). The output of the higher harmonic of a cw mode-locked Ti-sapphire laser is used to either carry out transient Raman measurements or perform pump/probe experiments. In the case of transient experiments, the output from the higher harmonics is directed into the sample, whereas for time-resolved Raman experiments, the output is split into two equally intense, but perpendicularly polarized beams. One of them is used to excite the electron–hole pair density in the sample and the other, after being suitably delayed, is used to probe the evolution of

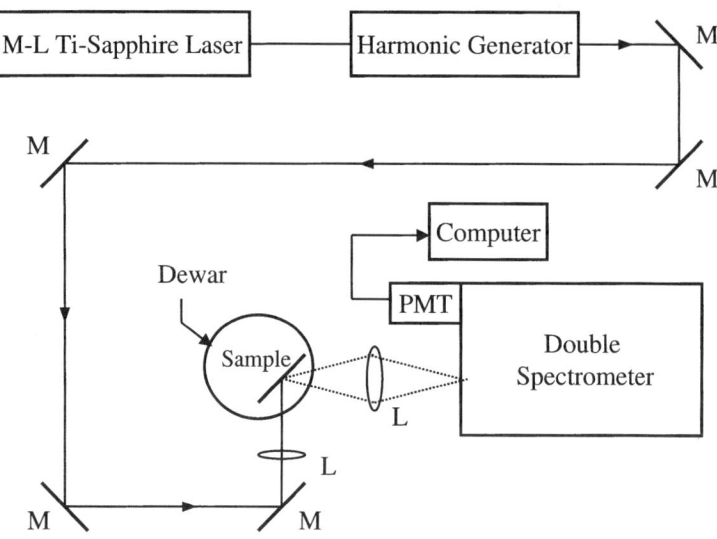

Fig. 3. Experimental setup for transient Raman scattering experiments where a single laser pulse is used to both excite and probe nonequilibrium excitations. M, mirror; L, lens; PMT, photomultiplier tube.

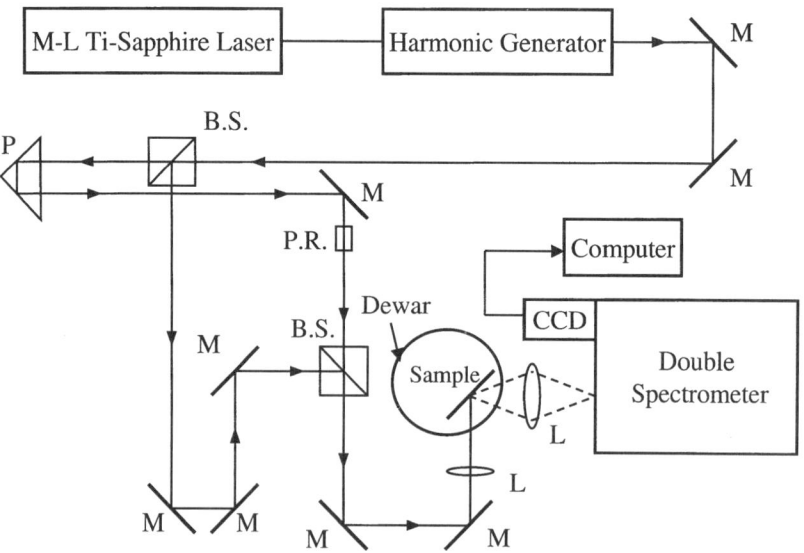

Fig. 4. Experimental setup for time-resolved Raman scattering experiments where a laser pulse is used to excite nonequilibrium excitations, whereas the other time-delayed pulse is employed to probe the time evolution of nonequilibrium excitations. M, mirror; L, lens; B.S., beam splitter; P.R., polarization rotator; P, variable delay line; CCD, charge coupled device.

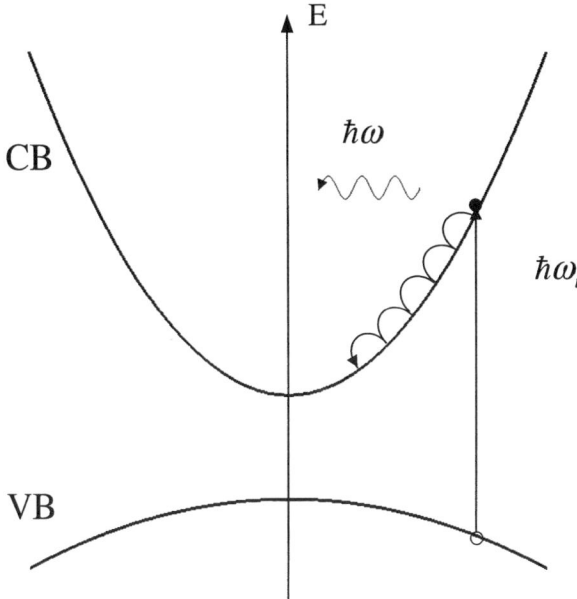

FIG. 5. A two-band diagram showing how nonequilibrium excitations such as phonons can be created and studied by transient/time-resolved Raman spectroscopy. CB, conduction band; VB, valence band; $\hbar\omega$, the energy of generated nonequilibrium excitations, in this case LO phonons; $\hbar\omega_i$, energy of the excitation photon.

nonequilibrium electron/phonon distributions. The 90° and/or backward-scattered Raman signal is collected and analyzed by a standard Raman system that consists of a double spectrometer, a photomultiplier tube, and/or a CCD detector.

One important advantage of probing nonequilibrium excitations with Raman spectroscopy in semiconductors is that since it detects Raman signal only when excitation photons are present, its time resolution is essentially limited by the pulse width of the excitation laser and not by the response of the detection system. This explains why our detection system has a time resolution on the order of a nanosecond, whereas the time resolution in our Raman experiments is typically on the subpicosecond scale.

We use picosecond/subpicosecond time-resolved Raman spectroscopy for studying nonequilibrium electron distributions and electron–phonon and phonon–phonon interactions in wurtzite GaN. As shown in Fig. 5, electron–hole pairs are photoexcited by the excitation photons across the bandgap of GaN either directly (in the case of above-bandgap excitation) or through nonlinear processes (in the case of below-bandgap excitations). These energetic electron–hole pairs will relax to the bottom of the conduction band (for electrons) and to the top of the valence band (for holes) by emitting nonequilibrium phonons through electron–phonon interactions. By monitoring the occupation number of these emitted nonequilibrium phonons, information such as the strength of electron–phonon interactions and phonon–phonon interactions can be readily obtained.

IV. Phonon Modes in the Wurtzite Structure GaN

The phonon modes in the wurtzite structure GaN are complicated because the structure contains a large number of atoms in the unit cell. We here limit our discussions to the phonon modes at the zone center that are directly accessible by Raman spectroscopy. Furthermore, only the optical phonon modes are illustrated. The acoustic modes are, since the wave vector is zero, just translations. The space group at the Γ point is represented by C_{6v}. This full, reducible representation can be decomposed into irreducible representations according to (Hayes and Loudon, 1978)

$$\Gamma_{opt} = A_1(Z) + 2B_1 + E_1(X, Y) + 2E_2, \quad (40)$$

where the X, Y, and Z in parentheses represent the polarization directions: $X = (100)$, $Y = (010)$, $Z = (001)$. The atomic vibrations of various modes are illustrated in Fig. 6. The E_2 modes are Raman active; A_1, E_1 modes are both Raman and infrared active; and B_1 modes are silent. It is worthwhile pointing out that there is a close relation between the cubic and hexagonal wurtzite structure — the difference in the neighboring atoms begins only in the third shell. However, the unit cell in the wurtzite structure contains twice as many atoms as that in the zincblende structure. This results in a folding of the Brillouin zone of the zincblende structure in the $\Gamma \to L$ direction, and then in a doubling of the number of modes. Therefore, instead of the TO mode in the zincblende structure, we have in the wurtzite structure two modes: E_2^1 and TO modes. At the Γ point of the Brillouin zone, the latter is split into A_1 and E_1 modes.

V. Experimental Results

1. STUDIES OF ELECTRON–PHONON INTERACTIONS IN WURTZITE GaN BY NONLINEAR LASER EXCITATION PROCESSES

In this section (Shieh *et al.*, 1995), a below-bandgap picosecond laser system was used to excite electron–hole pairs through two-photon absorption process, the highly energetic electrons and holes then relaxed by emitting

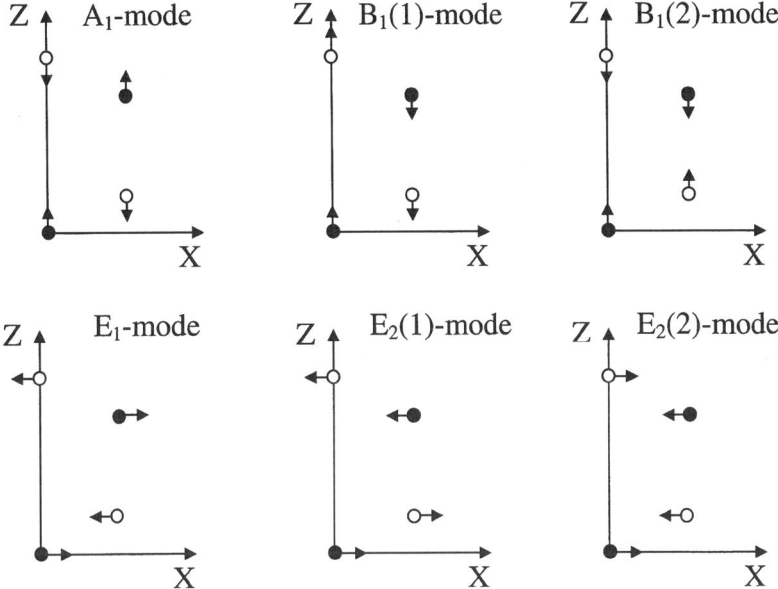

FIG. 6. The atomic vibrations of various phonon modes in wurtzite GaN. Solid circles represent Ga atoms and open ones correspond to N atoms.

various phonon modes. The same laser was employed to detect the emitted nonequilibrium phonon populations by Raman spectroscopy. From the analysis of experimental results, we conclude that high-energy electrons relax primarily through the emission of longitudinal optical phonons and that Fröhlich interaction is much stronger than deformation potential interaction in wurtzite structure GaN.

a. Samples and Experimental Technique

The sample used in this work is described in detail in Section III. Very intense ultrashort laser pulses were generated by the second harmonic of a cw mode-locked Ti-sapphire laser operating at a repetition rate of about 80 MHz. The photon energy was about 2.8 eV. The average output power of the laser was about 80 mW. The laser has a pulse width of about 1 ps. A laser power density as high as 1 GW/cm^2 can be achieved by focusing the laser beam with a microscope objective. We note that because of the wide bandgap of GaN, the photons in our laser beam do not have sufficient energy to directly excite electron-hole pairs across the bandgap. A two-photon absorption process was used to create electron-hole pairs with large excess energies ($\geqslant 2$ eV). The same high-power laser beam was also used to study the strength of electron–phonon interaction by monitoring the emitted nonequilibrium phonon populations with Raman scattering, as depicted in Fig. 3. Polarized Raman scattering experiments were carried out in a variety of scattering geometries as specified later. The sample was kept in contact with the cold finger tip of a closed-cycle refrigerator. The temperature of the sample was estimated to be about $T \cong 25$ K. The scattered signal was collected and analyzed with a standard Raman setup equipped with a CCD detection system.

b. *Experimental Results and Analysis*

Figures 7a and 8a show two polarized Stokes Raman spectra of a GaN sample taken at $T \cong 25$ K and at a laser power density of 1 GW/cm², and in $Z(X, X)\bar{Z}$, $Y(X, X)\bar{Y}$ scattering configurations, respectively, where $X = (100)$, $Y = (010)$, $Z = (001)$. Similar cw Raman spectra had been reported by Azuhata *et al.* (1995) at $T = 300$ K. In Fig. 7a, the sharp peak around 757 cm⁻¹ comes from scattering of light by the E_g phonon mode of sapphire; the shoulder close to 741 cm⁻¹ belongs to the A_1(LO) phonon mode in GaN; on the other hand, in Fig. 8a, the sharp peak around 574 cm⁻¹ corresponds to the E_2 phonon mode in GaN; the peak centered about 538 cm⁻¹ represents the A_1(TO) phonon mode in GaN; the small structure around 422 cm⁻¹ is from the A_{1g} phonon mode of sapphire.

The anti-Stokes Raman spectra corresponding to Figs. 7a and 8a are shown in Figs. 7b and 8b, respectively. Since at a sample temperature as low as $T \cong 25$ K, thermal occupations of phonons are varnishingly small, any Raman signal observed in the anti-Stokes Raman spectra must arise from nonequilibrium phonon modes. With the help of the Stokes Raman spectra in Figs. 7a and 8a, we can make the following identifications: The broad structure centered around -741 cm⁻¹ comes from Raman scattering from the nonequilibrium A_1(LO) phonon mode. Strikingly, we notice that within the experimental accuracy there are no detectable nonequilibrium phonon populations for either A_1(TO) or E_2 phonon modes of GaN and for E_g as well as A_{1g} phonon modes of sapphire.

We have also carried out similar Raman experiments in scattering geometries of $A(Z, X)Y$ and $Y(Z, X)\bar{Y}$, where E_1(LO) and E_1(TO) phonon modes could be observed; here, $A = (\sin \theta, -\cos \theta, 0)$ with $\theta \cong 108°$. These experimental results indicated that a substantial occupation of nonequilibrium phonons was observed for the E_1(LO) mode but not for the E_1(TO).

Therefore, our experimental results demonstrate directly that hot electrons thermalize primarily through the emission of polar longitudinal optical phonons in wurtzite structure GaN. Electrons interact with LO phonons through Fröhlich interaction as well as deformation potential

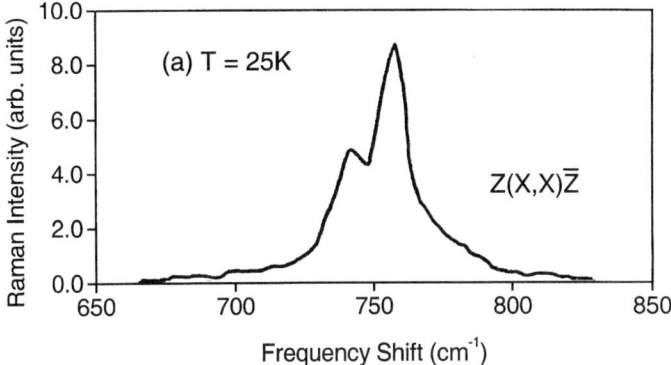

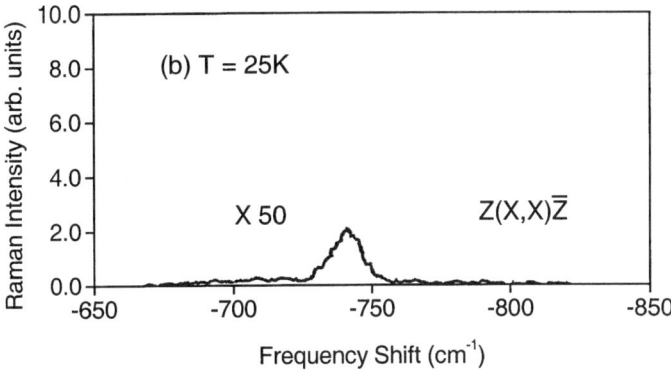

FIG. 7. Transient (a) Stokes and (b) anti-Stokes Raman spectra for a wurtzite GaN sample, taken by an ultrashort pulse laser having a pulse width of 1 ps and photon energy of 2.8 eV. The scattering configuration is $Z(X, X)\bar{Z}$. The anti-Stokes signal has been magnified by a factor of 50.

interaction, and they interact with TO phonons via deformation potential interaction only. Since only the LO phonon modes are driven out of equilibrium, our experimental results show that Fröhlich interaction is much stronger than deformation potential interaction in this wide bandgap semiconductor.

From the measured Raman spectra we can estimate that the phonon occupation number of the observed nonequilibrium $A_1(LO)$ and $E_1(LO)$ phonon modes at the laser power density of $1\,GW/cm^{-1}$ is about $\Delta n \cong 0.16 \pm 0.01$, 0.14 ± 0.01, respectively. This suggests that electrons interact almost as strongly with $A_1(LO)$ as with $E_1(LO)$ phonon modes in wurtzite structure GaN.

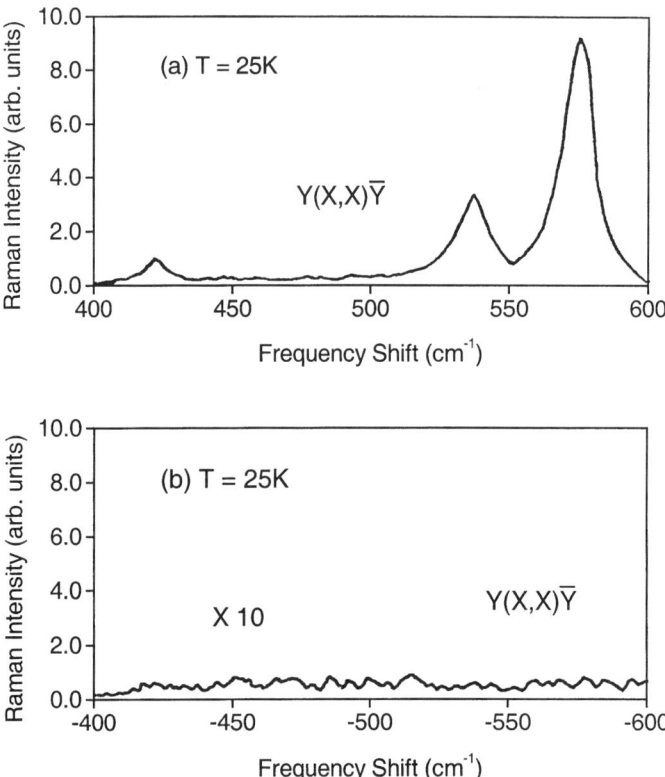

FIG. 8. Transient (a) Stokes and (b) anti-Stokes Raman spectra for a wurtzite GaN sample, taken by an ultrashort pulse laser having a pulse width of 1 ps and photon energy of 2.8 eV. The scattering configuration is $Y(X, X)\bar{Y}$. The anti-Stokes signal has been magnified by a factor of 10.

Figure 9 shows the measured nonequilibrium population of the $A_1(\text{LO})$ phonon mode as a function of the square of excitation laser power density. The experimental data can be very well fit by a straight line. The fact that the observed nonequilibrium phonon population increases linearly with the square of laser power density confirms that the hot electron–hole pairs are photoexcited by the two-photon absorption process.

2. NONEQUILIBRIUM ELECTRON DISTRIBUTIONS AND
ELECTRON–LONGITUDINAL OPTICAL PHONON SCATTERING RATES IN
WURTZITE GaN STUDIED BY AN ULTRASHORT, ULTRAVIOLET LASER

In this section (Tsen et al., 1996c, 1997), an ultrashort, ultraviolet laser system is employed to investigate the nature of nonequilibrium electron

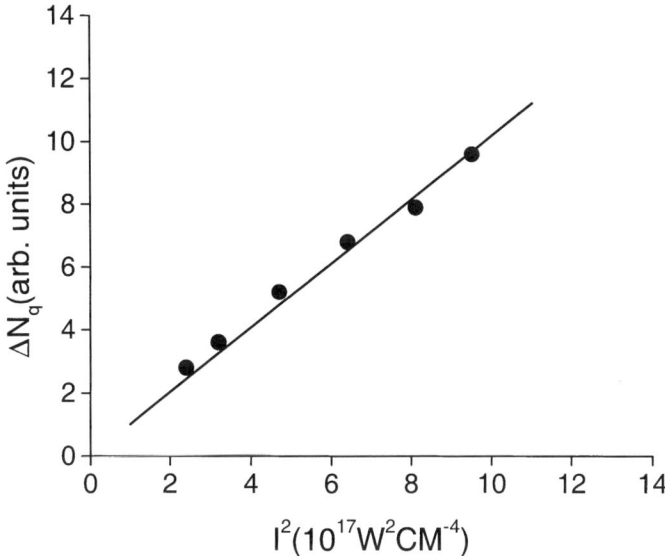

FIG. 9. The nonequilibrium A_1(LO) phonon population as a function of the square of excitation laser intensity. The fact that data are fit very well by a straight line suggests that the nonequilibrium phonons are generated by a nonlinear process, e.g., two-photon absorption process.

distributions as well as to directly measure electron–longitudinal phonon scattering rates in wurtzite GaN. The total electron–LO phonon scattering rate is determined to be $(4 \pm 0.8) \times 10^{13}$ sec^{-1}, which is about an order of magnitude larger than that found in GaAs. The very larger electron–LO phonon scattering rate is mostly due to the large ionicity in wurtzite GaN. In addition, we show that for electron densities $n \geqslant 5 \times 10^{17}$ cm^{-1}, as a result of efficient electron–electron scattering, nonequilibrium electron distributions photoexcited in wurtzite GaN can be very well described by Fermi–Dirac distribution functions with effective electron temperatures much higher than the lattice temperature.

a. Samples and Experimental Technique

The wurtzite structure GaN used in this study has been described in detail in Section III. The third harmonic of a cw mode-locked Ti-sapphire laser was used to photoexcite electron–hole pairs across the bandgap of GaN. For the study of nonequilibrium electron distributions, a photon energy of 4.5 eV was chosen so that electron–hole pairs were photoexcited with an excess energy of $\cong 1$ eV. On the other hand, in the investigation of electron–

LO phonon interactions, a photon energy of 5 eV was selected so that electron–hole pairs possess an excess energy of $\cong 1.5$ eV. The pulse width of the laser could be varied from about 50 fs to 2 ps. For interrogating nonequilibrium electron distributions, the same laser pulse was used to both excite and probe nonequilibrium electron distributions, as shown in Fig. 3; therefore, the experimental results presented here represent an average over the laser pulse duration. In the pump/probe configuration, which was used to study electron–phonon interactions, the laser beam was split into two equal but perpendicular polarized beams, as depicted in Fig. 4. One was used to excite electron–hole pairs and the other to probe nonequilibrium LO phonon populations through Raman scattering. The density of photo-excited electron–hole pairs was determined by fitting the time-integrated luminescence spectrum and was about $n \cong 10^{16}$ cm^{-3} for the experiments involving the measurements of electron–LO phonon scattering rates. The sample was kept in contact with the cold finger tip of a closed-cycle refrigerator. The temperature of the sample was estimated to be $T \cong 25$ K. The backward scattered Raman signal was collected and analyzed with a standard Raman setup equipped with a photomultiplier tube (for the detection of electron distributions) and a charge-coupled device (for the detection of phonons) detection systems.

b. *Experimental Results and Discussions*

Figures 10a and 10b show two transient single-particle scattering (SPS) spectra of a wurtzite structure GaN sample taken at $T \cong 25$ K and by an ultrafast laser system with a pulse width of about 600 fs and for electron densities $n \cong 3 \times 10^{18}$ cm^{-3}, 5×10^{17} cm^{-3}, respectively. Here the scattering geometry of $Z(X, Y)\bar{Z}$ was used, which ensures that the SPS signal comes from scattering of light associated with spin-density fluctuations. One intriguing aspect of using the Raman scattering technique to probe carrier distribution functions needs to be emphasized here: Raman scattering cross section is inversely proportional to the square of the effective mass; since the effective mass of electrons is usually much smaller than that of holes in semiconductors, Raman spectroscopy effectively probes electron distributions even if holes are simultaneously present.

In order to obtain better insight, we have used Eq. (30) that we developed in Subsection 1 of Section II to fit the line shapes of these SPS spectra. We note that in fitting the experimental results, three parameters were used: Γ, τ, and T_{eff} (the effective electron temperature). In addition, because the photon energy used in the experiments is sufficiently far from any relevant energy gaps in GaN, the fitting processes are not sensitive to the detailed band structure of GaN. The solid curves in Figs. 10a and 10b are theoretical calculations based on Eq. (30) with Fermi–Dirac distribution functions and

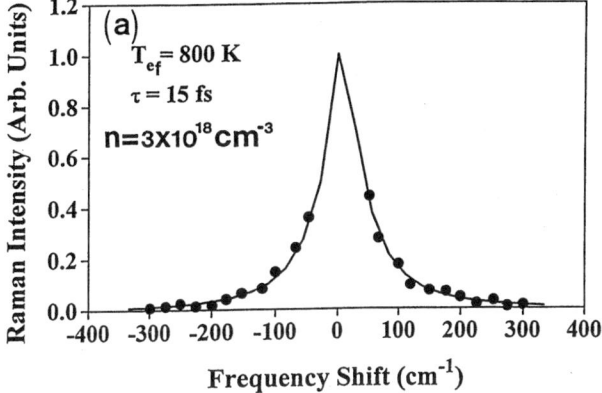

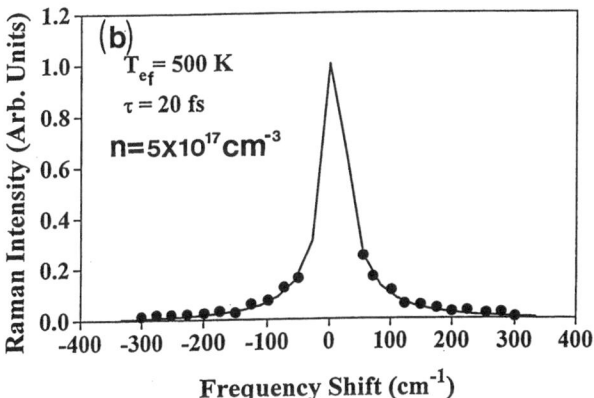

FIG. 10. Transient single-particle scattering spectra for a wurtzite GaN sample, taken by an ultrafast laser having a pulse width of 600 fs and for electron–hole pair densities of (a) $n \cong 3 \times 10^{18}$ cm^{-3}; (b) $n \cong 5 \times 10^{17}$ cm^{-3}, respectively. The solid circles are data. The curves are theoretical fit based on Eq. (30).

parameter sets that best fit the data: $T_{\mathrm{eff}} = 800$ K, $\tau = 15$ fs, $\Gamma = 20$ meV for $n \cong 3 \times 10^{18}$ cm^{-3}; and $T = 500$ K, $\tau = 20$ fs, $\Gamma = 20$ meV for $n \cong 5 \times 10^{17}$ cm^{-3}. The effective temperatures of electrons have been found to be much higher than that of the lattice, as expected. In addition, as the electron density increases, the effective electron temperature increases and the electron collision time becomes shorter. We also notice that the value of damping constant $\Gamma(=20 \text{ meV})$ involved in the Raman scattering process is very close to that ($\cong 13$ meV) obtained from an analysis of resonance Raman profile under the equilibrium conditions (Pinczuk et al., 1979). From the quality of the fit, we conclude that for electron densities

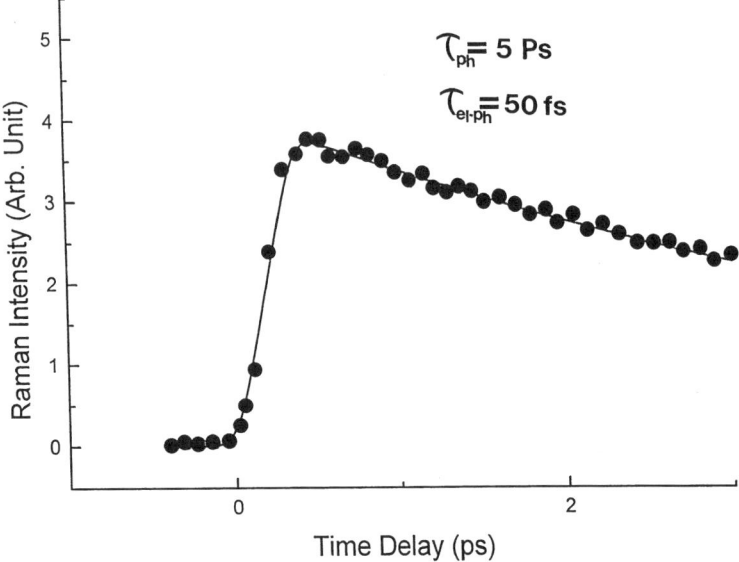

FIG. 11. Nonequilibrium populations of A_1(LO) phonon mode as a function of time delay. The solid circles are data. The curve is based on a simple-cascade model with electron–phonon scattering rate as the only adjustable parameter.

$n \geqslant 5 \times 10^{17}$ cm^{-3}, as a result of efficient momentum randomization, electron distribution functions in wurtzite GaN can be very well described by Fermi–Dirac distributions with the effective temperature of electrons much higher than that of the lattice.

Figure 11 shows time-resolved nonequilibrium populations of A_1(LO) phonon mode in wurtzite GaN taken at $T \cong 25$ K and with an ultrafast laser having a pulse width of 100 fs. The rise and the decrease of the signal reflect the generation and decay, respectively, of nonequilibrium A_1(LO) phonon mode.

We have used an electron cascade model (Tsen and Morkoc, 1988) to fit the experimental data in Fig. 11. The use of an electron cascade model under our experimental conditions is appropriate because (1) the electron–hole pair density excited is low ($n \cong 10^{16}$ cm^{-3}), and (2) the electrons do not have sufficient excess energy to scatter to other satellite valleys in GaN (Bloom et al., 1974).

In this model, the nonequilibrium LO phonon occupation number $n_{ph}(t)$ is given by

$$\frac{dn_{ph}(t)}{dt} = G(t) + \frac{n_{ph}(t)}{\tau_{ph}}, \qquad (41)$$

where $G(t)$ is the LO phonon generation rate by the excitation pulse laser; τ_{ph} is the LO phonon lifetime.

For our current experimental conditions, the LO phonon generation rate is given by

$$G(t) = \frac{f(t)}{\tau_{el-ph}} \qquad (42)$$

where $f(t) = 1$ for $0 \leqslant t \leqslant m\tau_{el-ph}$; and $f(t) = 0$ for $t \leqslant 0$ or $t \geqslant m\tau_{el-ph}$; τ_{el-ph} is the electron–LO phonon scattering time; m is an integer determined by the photon energy, LO phonon energy, bandgap, and band structure of wurtzite GaN.

For wurtzite GaN (Strite and Morkoc, 1992; Pearton et al., 1999), if we take $m_e^* = 0.2m_e$; $m_h^* = 0.8m_e$; $\hbar\omega_i = 5$ eV; $E_g = 3.49$ eV; the index of refraction $= 2.8$; $\hbar\omega_{LO} = 0.09263$ eV (corresponding to A_1(LO) phonon mode energy), then the phonon wave vector probed in our backscattered Raman experiments is $q = 2nk_i = 1.42 \times 10^6$ cm^{-1} and electron excess energy is given by

$$\Delta E_e = (\hbar\omega_i - E_g)\left(\frac{m_h^*}{m_e^* + m_h^*}\right) = 1.21 \text{ eV}.$$

This means that the energetic electrons are capable of emitting 13 LO phonons during their thermalization to the bottom of the conduction band. However, because of the conservation of both energy and momentum for the electron–LO phonon interaction process, there exists a range of LO phonon wave vectors that electrons can emit. As depicted in Fig. 12, for an electron with wave vector $\vec{k}_e$ and excess energy ΔE_e, the minimum and maximum LO phonon wave vectors it can interact with are given by

$$k_{min} = \frac{\sqrt{2m_e^*}}{\hbar}(\sqrt{\Delta E_e} - \sqrt{\Delta E_e - \hbar\omega_{LO}}) \qquad (43)$$

and

$$k_{max} = \frac{\sqrt{2m_e^*}}{\hbar}(\sqrt{\Delta E_e} + \sqrt{\Delta E_e - \hbar\omega_{LO}}). \qquad (44)$$

Because of the nature of the energy–wavevector relationship of the electron, the lower the electron's energy, the larger the k_{min} and the smaller the k_{max}. Therefore, at some electron energy, the k_{min} of the LO phonon will be larger than the wave vector probed by our Raman scattering experiments $q = 1.42 \times 10^6$ cm^{-1}. When that happens during the relaxation process, the

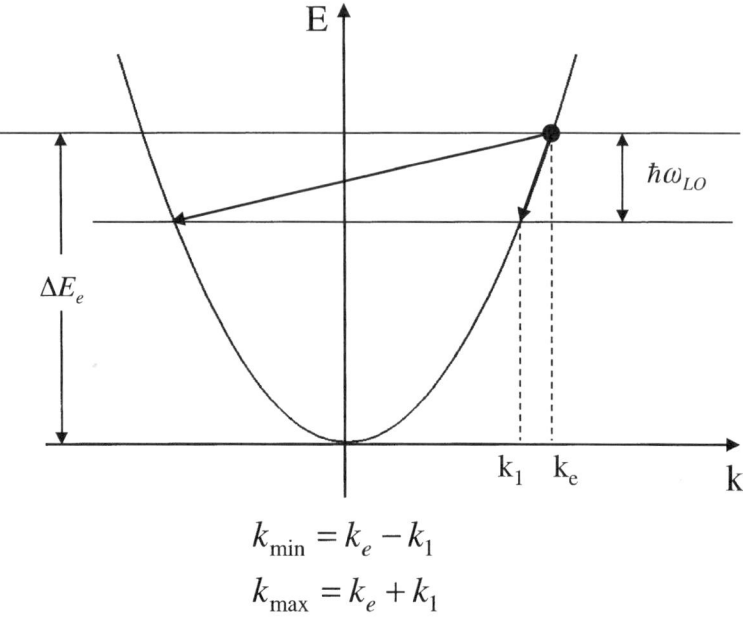

FIG. 12. A diagram demonstrating that because of the conservation of both energy and momentum, the minimum and maximum phonon wave vectors that an electron can interact with during its thermalization toward the bottom of the conduction band are given by Eqs. (43) and (44), respectively.

energetic electrons can no longer emit LO phonons with wave vector detectable in our Raman scattering experiments. By taking this into consideration, we have found that although in principle the energetic electrons are capable of emitting 13 LO phonons during their thermalization to the bottom of the conduction band, only 7 of them can be detected in our Raman experiments because of the conservation of both energy and momentum; in other words, $m = 7$ under our current experimental conditions.

Therefore, there are two adjustable fitting parameters in this electron cascade model — phonon lifetime (τ_{ph}) and electron–phonon scattering time (τ_{el-ph}). Since from the decaying part of the Raman signal, the LO phonon lifetime τ_{ph} can be independently measured, the electron–LO phonon scattering time τ_{el-ph} is used as the only adjustable parameter in the fitting process.

We have found that $\tau_{el-ph} = 50 \pm 10$ fs, which gives rise to an electron–LO phonon scattering rate of $\Gamma_{el-ph} = 1/\tau_{el-ph} = (2.2 \pm 0.4) \times 10^{13}$ sec^{-1} and provides the best fit to the experimental data in Fig. 11.

Similar experiments for the E_1(LO) phonon mode were also carried out. Our experimental results in this case show that the electron–LO phonon scattering rate for E_1(LO) phonon mode is given by $\tau_{\text{el-ph}} = (2.4 \pm 0.4) \times 10^{13}$ sec^{-1}, which is very close to the value found for A_1(LO) phonon mode.

Consequently, the total electron–LO phonon scattering rate is given by $\Gamma_{\text{total}} = (4.6 \pm 0.8) \times 10^{13}$ sec^{-1}. Since the electron–LO phonon scattering rate in GaAs is about 5×10^{12} sec^{-1}, the observed total electron–LO phonon scattering rate in wurtzite GaN is almost one order of magnitude larger than that in GaAs. The much larger LO–TO phonon energy splitting in wurtzite GaN provides a clue to this mystery. We attribute this enormous increase of electron–LO phonon scattering rate in GaN to its much larger ionicity. In general, the strength of electron–LO phonon coupling is set by the lattice-dipole interactions, that is expressed by (Conwell, 1967)

$$\frac{1}{\gamma} = \omega_{\text{LO}}^2 \left[\frac{1}{\varepsilon_\infty} - \frac{1}{\varepsilon_0} \right] = \left[\frac{\omega_{\text{LO}}^2 - \omega_{\text{TO}}^2}{\varepsilon_\infty} \right]. \tag{45}$$

This splitting is directly proportional to the lattice polarization and is a measure of the effective charge. In GaAs, the LO–TO energy splitting is about 3 meV, whereas it is about 25 meV in wurtzite GaN. However, in the two-mode system of hexagonal GaN, this must be split between the two modes to compute the contribution of each of the dielectric functions. This, together with the much smaller dielectric constant of GaN, leads to an expected increase of a factor of 9 in the electron–LO phonon scattering strength, that is quite close to that observed experimentally. Phillips (1973) estimates that the effective charge of GaN is about twice that of GaAs, while the dielectric constant is almost a half. These two factors together would suggest a factor of 8 increase in the electron–LO phonon scattering strength. This provides a consistent interpretation of the increase of electron–LO phonon scattering rate as due to the ionicity of wurtzite GaN relative to GaAs.

3. Anharmonic Decay of the Longitudinal Optical Phonons in Wurtzite GaN Studied by Subpicosecond Time-Resolved Raman Spectroscopy

In this section (Tsen et al., 1998b), decay of the longitudinal optical phonons in wurtzite GaN has been studied by subpicosecond time-resolved Raman spectroscopy. In contrast to the 2LA decay channel usually believed for LO phonons in other semiconductors (Kash and Tsang, 1991), our experimental results show that, among the various possible decay channels, the LO phonons in wurtzite GaN decay primarily into a large wave vector

TO and a large wave vector LA or TA phonon. These experimental results are consistent with theoretical calculations of the phonon dispersion curves in wurtzite structure GaN.

a. Samples and Experimental Technique

The GaN sample used in this study is described in detail in Section III. In our time-resolved Raman experiments, ultrashort, ultraviolet laser pulses were generated by a frequency-quadrupled cw mode-locked Ti-sapphire laser system, as shown in Fig. 4. The laser pulse width could be varied from 50 fs to about 2 ps. The photon energy was chosen to be about 5 eV so that electron–hole pairs could be excited with an excess energy of more than 1 eV. In the investigation of phonon–phonon interactions in GaN, a pump/probe configuration was employed. The laser beam was split into two equal intensity but perpendicularly polarized beams. One is used to excite electron–hole pairs and the other to probe nonequilibrium LO phonon populations through Raman scattering. The density of the photoexcited electron–hole pairs was determined by fitting the time-integrated luminescence spectrum and was about $n \cong 5 \times 10^{16} \text{ cm}^{-3}$. The sample was kept in contact with the cold finger tip of a closed-cycle refrigerator. The temperature of the laser-irradiated sample was estimated to be $T \cong 25 \text{ K}$. The scattering geometry was taken to be $Z(X, X)\bar{Z}$. The backward scattered Raman signal was collected and analyzed with a standard Raman setup equipped with a CCD detection system.

b. Experimental Results and Discussions

Figures 13a and 13b show a typical set of transient polarized Stokes and anti-Stokes Raman spectra for a GaN sample taken at $T \cong 25 \text{ K}$ with an ultrafast laser having a 2 ps pulse width and a photon energy of 5 eV, in $Z(X, X)\bar{Z}$ scattering geometry. These spectra are very similar to Figs. 7a and 7b, where a photon energy below the bandgap of GaN was used for Raman scattering experiments. We demonstrate here again that nonequilibrium LO phonon mode can be generated with an above-bandgap laser excitation of hot electron–hole pair density. Similar to arguments made in Subsection 1b of Section V, we can conclude that the broad structure around -741 cm^{-1} in the anti-Stokes Raman spectrum of Fig. 13b comes from scattering of light by the nonequilibrium $A_1(\text{LO})$ phonon mode in wurtzite GaN.

In order to obtain better insight into the generation of nonequilibrium optical phonons in semiconductors, we have used a simple-cascade, two-

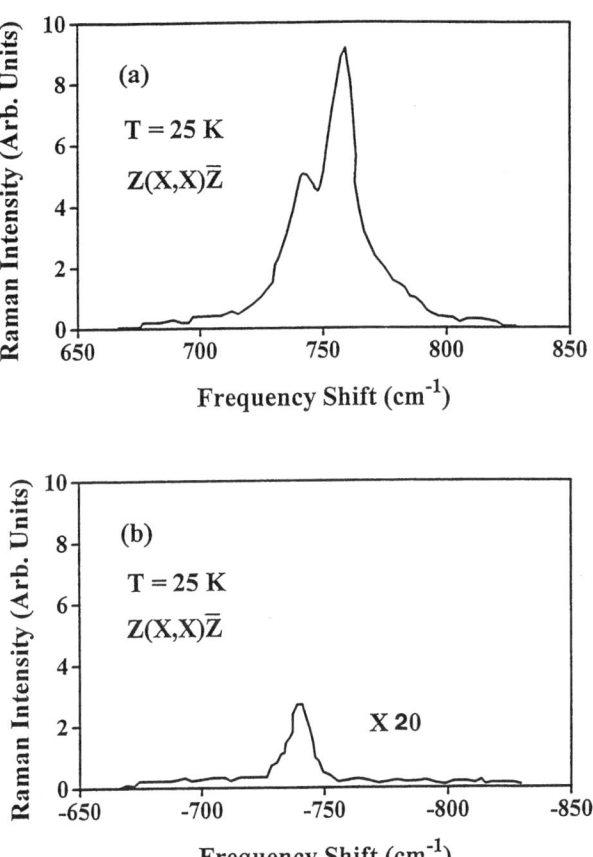

FIG. 13. Transient (a) Stokes and (b) anti-Stokes Raman spectra for a wurtzite GaN sample, taken by an ultrashort pulse laser having a pulse width of 2 ps and photon energy of 5.0 eV. The scattering configuration is $Z(X, X)\bar{Z}$. The anti-Stokes signal has been magnified by a factor of 20.

parabolic-band model (Colins and Yu, 1984) to calculate the nonequilibrium phonon occupation number as a function of phonon wave vector under our experimental conditions. Although both electrons and holes are excited, we assume that the electron mass is much smaller than the hole mass so that most of the excess energy of the photon goes to the electron. For simplicity, we neglect the nonequilibrium phonons emitted by the relaxation of the holes. With these assumptions we have only two populations to consider: $f_{\vec{k}}(t)$, the occupation number of electrons with wave vector $\vec{k}$, and $n_{\vec{q}}(t)$, the occupation number of LO phonons with wave vector $\vec{q}$. The two distributions can be calculated as a function of time by solving two Boltzmann equations:

$$\frac{df_{\vec{k}}}{dt} = G_{\vec{k}} + \left(\frac{\partial f_{\vec{k}}}{\partial t}\right)_{LO} - \frac{f_{\vec{k}}}{\tau_{\vec{k}}} \qquad (46)$$

$$\frac{dn_{\vec{q}}}{dt} = \left(\frac{\partial n_{\vec{q}}}{\partial t}\right)_{e} - \frac{n_{\vec{q}}}{\tau_{LO}}. \qquad (47)$$

In Eq. (46), $G_{\vec{k}}$ is the rate of generation of electrons due to optical excitation; $(\partial f_{\vec{k}}/\partial t)_{LO}$ is the rate of change in electron occupation number due to scattering with LO phonons; and $\tau_{\vec{k}}$ is the electron lifetime due to any other decaying processes, such as radiative recombination. In Eq. (47), $(\partial n_{\vec{q}}/\partial t)_e$ is the rate of change of the LO phonon occupation number due to its interaction with the electron; τ_{LO} is the lifetime of LO phonons due to their decay into lower energy phonons. For simplicity, τ_{LO} is assumed to be independent of $\vec{q}$. The rates $(\partial f_{\vec{k}}/\partial t)_{LO}$ and $(\partial n_{\vec{q}}/\partial t)_e$ can be evaluated from time-dependent perturbation theory as follows.

It has been shown that the electron–LO phonon interaction in polar semiconductors such as GaAs, GaN is dominated by the Fröhlich interaction $H_{\text{ele-ph}} = -e\varphi(\vec{r}_i)$. If we represent the matrix element of $H_{\text{ele-ph}}$ for scattering an electron in state $\vec{k}$ to $\vec{k} + \vec{q}$ (or $\vec{k} - \vec{q}$) with absorption (or emission) of a LO phonon with wave vector $\vec{q}$ as

$$M_{\vec{k},\vec{k} \pm \vec{q}} = |\langle \vec{k}|H_{\text{ele-ph}}|\vec{k} \pm \vec{q}\rangle|^2, \qquad (48)$$

then using Fermi's Golden Rule, we can write $(\partial f_{\vec{k}}/\partial t)_{LO}$ as

$$\begin{aligned}\left(\frac{\partial f_{\vec{k}}}{\partial t}\right)_{LO} = \frac{2\pi}{\hbar} \sum_{\vec{q}} [&-M_{\vec{k},\vec{k}-\vec{q}}(n_{\vec{q}}+1)f_{\vec{k}}(1-f_{\vec{k}-\vec{q}})\delta(E_{\vec{k}}-E_{\vec{k}-\vec{q}}-\hbar\omega_{LO})\\ &- M_{\vec{k},\vec{k}+\vec{q}}n_{\vec{q}}f_{\vec{k}}(1-f_{\vec{k}+\vec{q}})\delta(E_{\vec{k}}-E_{\vec{k}+\vec{q}}+\hbar\omega_{LO})\\ &+ M_{\vec{k}+\vec{q},\vec{k}}(n_{\vec{q}}+1)(1-f_{\vec{k}})f_{\vec{k}+\vec{q}}\delta(E_{\vec{k}+\vec{q}}-E_{\vec{k}}-\hbar\omega_{LO})\\ &+ M_{\vec{k}-\vec{q},\vec{k}}n_{\vec{q}}(1-f_{\vec{k}})f_{\vec{k}+\vec{q}}\delta(E_{\vec{k}-\vec{q}}-E_{\vec{k}}+\hbar\omega_{LO})].\end{aligned} \qquad (49)$$

The first two terms in Eq. (47) arise from the decay of the electron in state $\vec{k}$ with emission or absorption of a LO phonon. The last two terms are due to scattering of electrons in states $\vec{k} \pm \vec{q}$ into the state $\vec{k}$ through emission or absorption of a LO phonon.

Similarly, we find that

$$\begin{aligned}\left(\frac{\partial n_{\vec{q}}}{\partial t}\right)_{e} = \frac{2\pi}{\hbar} \sum_{\vec{k}} [&M_{\vec{k},\vec{k}-\vec{q}}(n_{\vec{q}}+1)f_{\vec{k}}(1-f_{\vec{k}-\vec{q}})\delta(E_{\vec{k}}-E_{\vec{k}-\vec{q}}-\hbar\omega_{LO})\\ &- M_{\vec{k},\vec{k}+\vec{q}}n_{\vec{q}}f_{\vec{k}}(1-f_{\vec{k}+\vec{q}})\delta(E_{\vec{k}}-E_{\vec{k}+\vec{q}}+\hbar\omega_{LO})],\end{aligned} \qquad (50)$$

where

$$E_{\vec{k}} = \hbar^2 k^2/2m_e^*; \quad M_{\vec{k},\vec{k}\pm\vec{q}} = \frac{2\pi\hbar^2 e}{Vm_e^* q^2}\left[\left(\frac{m_e^* e\omega_{LO}}{\hbar}\right)\left(\frac{1}{\varepsilon_\infty} - \frac{1}{\varepsilon_0}\right)\right],$$

with ε_∞, ε_0 being high-frequency and static dielectric constants.

By assuming that the electron distribution is nondegenerate and that

$$\left(\frac{\partial f}{\partial E}\right)\hbar\omega_{LO} \ll f(E)$$

(Levinson and Levinsky, 1975), we have

$$\frac{df(E,t)}{dt} = \frac{e}{(2m_e^* E)^{1/2}}\left(\frac{m_e^* e\omega_{LO}}{\hbar}\right)\left(\frac{1}{\varepsilon_\infty} - \frac{1}{\varepsilon_0}\right)$$

$$\times \left[f(E^+, t)\ln\left[\frac{(E^+)^{1/2} + E^{1/2}}{(E^+)^{1/2} - E^{1/2}}\right] - f(E,t)\ln\left[\frac{E^{1/2} + (E^-)^{1/2}}{E^{1/2} - (E^-)^{1/2}}\right]\right]$$

$$+ G(E,t) - \frac{f(E,t)}{\tau(E)} \tag{51}$$

and

$$\frac{dn_{\vec{q}}}{dt} = -\frac{n_{\vec{q}}}{\tau_{LO}} + \frac{2m_e^* e}{\hbar^3 q^3}\left(\frac{m_e^* e\omega_{LO}}{\hbar}\right)\left(\frac{1}{\varepsilon_\infty} - \frac{1}{\varepsilon_0}\right)\int_{E_{max}}^{\infty} f(E,t)dE \tag{52}$$

where

$$E_{max} = \left(\frac{q'}{2} - \frac{\hbar\omega_{LO}}{2q'}\right)^2 + \hbar\omega_{LO}; \quad E_{min} = \left(\frac{q'}{2} - \frac{\hbar\omega_{LO}}{2q'}\right)^2; \quad q' = \frac{\hbar q}{\sqrt{2m_e^*}};$$

$$E^+ = E + \hbar\omega_{LO}; \quad E^- = E - \hbar\omega_{LO}.$$

Equations (51) and (52) can be solved very easily to obtain both electron distributions and nonequilibrium phonon distributions.

Figure 14 shows the calculated $A_1(LO)$ phonon occupation number as a function of phonon wave vector for a GaN sample, taken with a ultrafast laser having a pulse width of 300 fs and at a time delay of 300 fs after the peak of the excitation laser pulse. The excitation photon energy is 5 eV. The electron–hole pair density is 5×10^{16} cm^{-3}. The dashed vertical line indicated the wave vector probed by Raman spectroscopy. The sharp cutoff in phonon occupation number at around 8×10^5 cm^{-1} is due to the

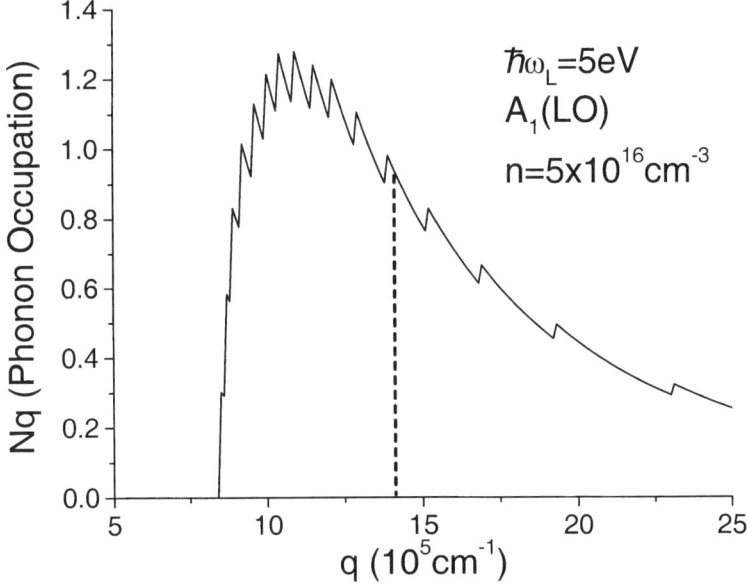

FIG. 14. The calculated A_1(LO) phonon occupation number as a function of phonon wave vector for a GaN sample. The simulated ultrafast laser has a pulse width of 300 fs and the phonon occupation is monitored at a time delay of 300 fs. The excitation photon energy is 5 eV. The photoexcited electron–hole pair has a density of $n \cong 5 \times 10^{16}$ cm^{-3}. The dashed vertical line indicates the value of wave vector probed by our Raman scattering experiments.

conservation of both energy and momentum during the electron–phonon interaction processes.

Figure 15 shows several time-resolved nonequilibrium populations of A_1(LO) phonon mode in wurtzite GaN, taken at the lattice temperature of $T = 10$, 150, and 300 K, respectively, and with an ultrafast laser having a pulse width of 300 fs and a photon energy of 5 eV. From the decaying parts of the Raman signals, the population relaxation time of A_1(LO) phonons as a function of the lattice temperature ranging from 10 to 300 K can be obtained, which is shown in Fig. 16.

Klemens (1966) and Ridley (1996) used perturbation theory to show that the temperature-dependence part of the decay of LO phonon population $n_{ph}(\omega, T)$ in semiconductors can be expressed as

$$\frac{dn_{ph}(\omega, T)}{dt} = -n_{ph}(\omega, T)[1 + n_{ph1}(\omega_1, T) + n_{ph2}(\omega_2, T)]/\tau_0, \quad (53)$$

where τ_0 is the decay time of LO phonon at $T = 0$ K (in this case, τ_0 has

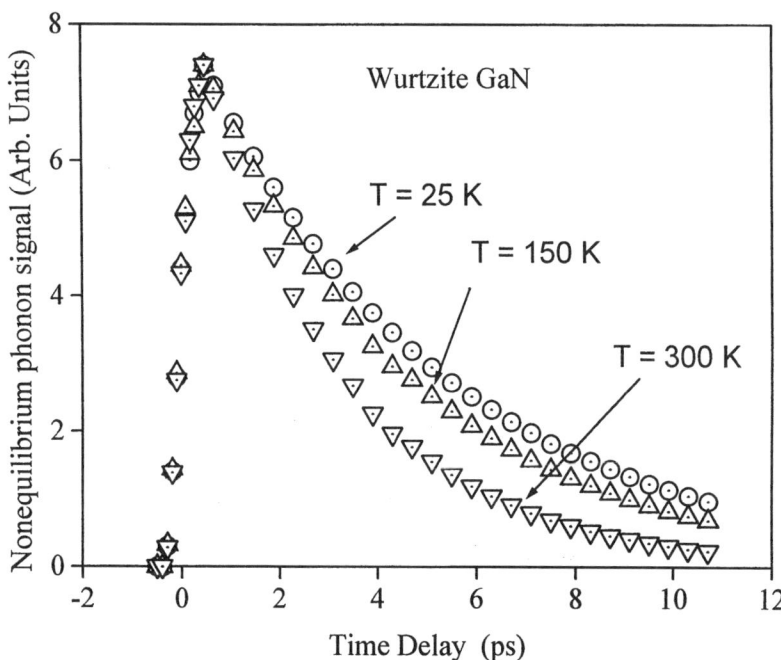

FIG. 15. A typical set of nonequilibrium populations of A_1(LO) phonon mode as a function of time delay, taken with an ultrafast laser having a pulse width of 300 fs and photon energy of 5 eV, at the lattice temperatures of $T \cong 10$, 150, 300 K, respectively.

been measured to be about 5 ps); $\omega = \omega_1 + \omega_2$; and $n_{\text{ph}1}(\omega_1, T) = 1/(e^{\hbar\omega_1/k_B T} - 1)$, $n_{\text{ph}2}(\omega_2, T) = 1/(e^{\hbar\omega_2/k_B T} - 1)$ are the occupation numbers of decayed lower-energy phonons at the lattice temperature T; and k_B is Boltzmann's constant. Equation (53) says that the temperature dependence of LO phonon lifetime is given by

$$\frac{1}{\tau(T)} = \frac{1}{\tau_0}[1 + n_{\text{ph}1}(\omega_1, T) + n_{\text{ph}2}(\omega_2, T)]. \tag{54}$$

We notice that the population relaxation time for LO phonons in GaAs is about 8 ps at $T = 10$ K; on the other hand, that for LO phonons in wurtzite GaN is about 5 ps at $T = 10$ K. This decrease of LO phonon lifetime might be due to either defects and dislocations present in the GaN sample or a larger density-of-state for the decay of LO phonons in GaN. Further investigation is now under way to clarify this point.

There are various possible channels for the decay of zone-center LO phonons in wurtzite GaN. First of all, from our experimental data in

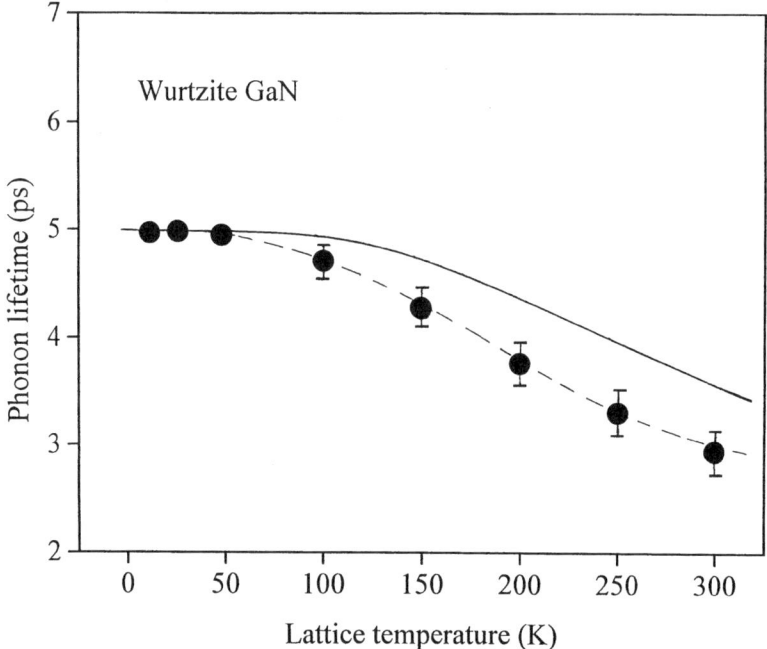

FIG. 16. The temperature-dependence of the population relaxation time of A_1(LO) phonon mode for a GaN sample. The solid circles are experimental data. The solid curve is for the decay channel $\omega_{LO} \rightarrow 2\omega_{LA(TA)}$ with $\hbar\omega_{LA(TA)} = 370\,\text{cm}^{-1}$; the dashed curve is for the decay channel $\omega_{LO} \rightarrow \omega_{TO} + \omega_{LA(TA)}$, as predicted by Eq. (54). The energies of the TO and LA or TA phonons involved in the fitting process are $\omega_{TO} = 540\,\text{cm}^{-1}$, $\omega_{LA(TA)} = 201\,\text{cm}^{-1}$, respectively.

Fig. 16, the decay of zone-center LO phonons into a small wave vector LO and a small wave vector LA or TA phonons does not seem to be likely, because this channel predicts a much larger temperature dependence of the lifetime of LO phonons than the experimental data indicate. The solid curve in Fig. 16 represents the decay channel $\omega_{LO} \rightarrow 2\omega_{LA(TA)}$ with $\hbar\omega_{LA(TA)} = 370\,\text{cm}^{-1}$, which is half of the energy of zone-center LO phonons, as predicted by Eq. (54). We have found that, within our experimental uncertainty, the zone-center LO phonons can not decay into large-wave-vector, equal-energy but opposite-momentum LA or TA phonons, which is usually assumed in the decay of LO phonons in other III–IV compound semiconductors. On the other hand, we have found that the decay channel of zone-center LO phonons in wurtzite GaN into a large-wave-vector TO phonon (assuming that the TO phonon dispersion curve is relatively flat across the Brillouin zone) and a large-wave-vector LA or TA phonon fits our experimental data very well. The dotted curve corresponds to such a decay channel: $\omega_{LO} \rightarrow \omega_{TO} + \omega_{LA(TA)}$, as predicted by

Eq. (54). The energies of the TO and LA or TA phonons involved in the fitting process are $\omega_{TO} = 540\,\text{cm}^{-1}$, $\omega_{LA(TA)} = 201\,\text{cm}^{-1}$, respectively.

Azuhata et al. (1996) have calculated the phonon dispersion curve of wurtzite GaN by using a rigid-ion model. Their calculations indicate that the zone-center LO phonons in wurtzite GaN cannot decay into two large-wave-vector lower-branch LA or TA phonons, because $2\omega_{LA(TA)} < \omega_{LO}$, whereas it is possible for a zone-center LO phonon to decay into a large-wave-vector TO and a large-wave-vector LA or TA phonon. These theoretical predictions agree very well with our experimental analysis.

VI. Conclusions and Future Experiments

We have presented experimental results on dynamical properties of wurtzite structure GaN such as electron–phonon scattering rates, phonon–phonon interaction times, and nonequilibrium electron distributions. A comprehensive theory of Raman scattering in semiconductors is also given, which is particularly useful under the circumstances of nonequilibrium conditions. We demonstrate that the total electron–LO phonon scattering rate in wurtzite GaN is $(4 \pm 0.8) \times 10^{13}\,\text{sec}^{-1}$, which is about an order of magnitude larger than that found in GaAs. The very much larger electron–LO phonon scattering rate is mostly due to the large ionicity in wurtzite GaN. In addition, we show that for electron densities $n \geqslant 5 \times 10^{17}\,\text{cm}^{-3}$, as a result of efficient electron–electron scattering, nonequilibrium electron distributions photoexcited in wurtzite GaN can be very well described by Fermi–Dirac distribution functions with effective electron temperature much higher than the lattice temperature. We have found that the population relaxation time of LO phonons in wurtzite GaN is about 5 ps, a value comparable to that in GaAs. In contrast to the usually believed 2LA decay channel for LO phonons in other semiconductors, our experimental results show that, among the various possible decay channels, the LO phonons in wurtzite GaN decay primarily into a large-wave-vector TO and a large-wave-vector LA or TA phonon. These experimental results are consistent with recent theoretical calculations of the phonon dispersion curves in wurtzite structure GaN.

Although these experimental results have helped establish a solid foundation for our understanding of dynamical properties in nitride-based III–V semiconductors, there are still many important aspect that remain to be explored. One of them is the investigation of transient nonequilibrium electron transport in nitride-based III–V semiconductors. Because of the much larger electron effective mass in nitride-based III–V semiconductors, the drift velocity (and therefore) device operation speed of electronic devices built upon these nitride materials is inherently much smaller than those of other semiconductor devices such as GaAs. This will tremendously limit the application range of these materials. In order to enhance the speed of devices based upon these nitride materials, good use must be made of novel transient transport properties such as electron velocity overshoot and ballistic electron transport that we have learned of in the case of GaAs and InP. This made the study of transient electron transport properties very interesting and important.

ACKNOWLEDGMENTS

I thank D. K. Ferry for his helpful discussions and H. Morkoc for GaN samples. This work was supported by the National Science Foundation under Grant No. DMR-9301100.

REFERENCES

Azuhata, T., Matsunaga, T., Shimada, K., Yoshida, K., Sota, T., Suzuki, K., and Nakamura, S. (1996). *Physica B: Condens. Matter* **219**, 493.
Azuhata, T., Sota, T., Suzuki, K., Nakamura, S. (1995). *J. Phys. Condens. Matter* **7**, L129.
Bloom, S., Harbeke, G., Meier, E., Ortenburger, I. B. (1974). *Phys. Stat. Solidi* **B66**, 161.
Chia, C., Sankey, O. F., Tsen, K. T. (1993). *Mod. Phys. Lett.* **B7**, 331.
Colins, C. L., Yu, P. Y. (1984). *Phys. Rev. B* **30**, 4501.
Conwell, E. M. (1967). *High Field Transport in Semiconductors.* Academic Press, New York.
Ferry. D. K. (1975). *Phys. Rev. B* **12**, 2361.
Ferry, D. K. (1991). *Semiconductors.* MacMillan, New York, Chapter 13.
Grann, E. D., Shieh, S. J., Chia, C., Tsen, K. T., Sankey, O. F., Gunser, S., Ferry, D. K., Maracus, G., Droopad, R., Salvador, A., Botchkarev, A., Morkoc, H. (1994). *Appl. Phys. Lett.* **64**, 1230.
Grann, E. D., Shieh, S. J., Tsen, K. T., Sankey, O. F., Gunser, S. E., Ferry, D. K., Salvador, A., Botchkarev, A., Morkoc, H. (1995a). *Phys. Rev. B* **51**, 1631.
Grann, E. D., Tsen, K. T., Sankey, O. F., Ferry, D. K., Salvador, A., Botchkarev, A., Morkoc, H. (1995b). *Appl. Phys. Lett.* **67**, 1760.
Grann, E. D., Tsen, K. T., Ferry, D. K., Salvador, A., Botchkarev, A., Morkoc, H. (1996). *Phys. Rev. B* **53**, 9838.
Grann, E. D., Tsen, K. T., Ferry, D. K., Salvador, A., A. Botchkarev, A., Morkoc, H. (1997). *Phys. Rev. B* **56**, 9539.
Hamaguchi, C., Inoue, M. (1992). *Proc. of the 7th Int. Conf. on Hot Carriers in Semiconductors.* Adam Hilger, New York.
Hamilton, D. C., McWhorter, A. L. (1969). In *Light Scattering Spectra of Solids*, edited by G. B. Wright. Springer, New York, p. 309.
Hayes, W., Loudon, R. (1978). *Scattering of Light by Crystals.* John Wiley & Sons, New York.
Jackson, J. D. (1975). *Classical Electrodynamics*, 2nd ed. John Wiley & Sons, New York, pp. 391–397.
Jha, S. S. (1969). *Nuovo Cimento* **63B**, 331.

Kane, E. O. (1957). *J. Phys. Chem. Solids* **1**, 249.
Kash, J. A., Tsang, J. C. (1991).In *Light Scattering in Solids VI*, edited by M. Cardona and G. Guntherodt. Springer, New York, p. 423.
Klemens, P. G. (1966). *Phys. Rev.* **148**, 845.
Kramers, H. A., Heisenberg, W. (1925). *Zeit. F. Physik* **31**, 681.
Landsberg, G., Mandel'shtam, L. (1928). *Naturwissensch.* **16**, 57; *ibid*, **16**, 772.
Levinson, Y. B., Levinsky, B. N. (1975). *Solid State Commun.* **16**, 713.
Mooradian, A., McWhorter, A. L. (1969). In *Light Scattering Spectra of Solids,* edited by G. B. Wright. Springer, New York, p. 297.
Pankove, J. I., Moustakas, T. D. (1997). *Gallium Nitride*, Vol. 50 of *Semiconductors and Semimetals*, edited by R. K. Willardson and E. R. Weber. Academic Press, New York.
Pearton, S. J., Zolper, J. C., Shul, R. J., Ren, F. (1999). *Appl. Phys. Rev.* **86**, 1.
Pinczuk, A., Abstreiter, G. A., Trommer, R., Cardona, M. (1979). *Solid State Commun.* **30**, 429.
Phillips, J. C. (1973). *Bonds and Bands in Semiconductors*. Academic Press, New York.
Pines, D. (1962). *The Many-Body Problem*. Benjamin, New York, p. 44.
Platzman, P. M., Wolff, P. A. (1973). *Waves and Interactions in Solid State Plasmas*, Solid State Supplement, Vol. **13**. Academic Press, New York.
Raman, C. V. (1928). *Ind. J. Phys.* **2**, 387.
Ridley, B. K. (1993). *Quantum Processes in Semiconductors*, 3rd ed. Oxford Press, London.
Ridley, B. K. (1996). *J. Phys. Condens. Matter* **8**, L511.
Ruf, T., Wald, K. R., Yu, P. Y., Tsen, K. T, Morkoc, H. K. T., Chan, K. T. (1993). *Superlattices and Microstructures* **13**, 203.
Shah, J. (1996). *Ultrafast Spectroscopy of Semiconductors and Semiconductor Nanostructures*, Vol. 115 in Springer Series in Solid-State Sciences. Springer, New York.
Shieh, S. J., Tsen, K. T., Ferry, D. K., Botchkarev, A., Sverdlov, B., Salvador, A., Morkoc, H. (1995). *Appl. Phys. Lett.* **67**, 1757.
Smekal, A. (1923). *Naturwissensch.* **11**, 873.
Strite, S., Morkoc, H. (1992). *J. Vac. Sci. Technol. B* **10**, 1237.
Tsen, K. T., Morkoc, H. (1988). *Phys. Rev. B* **38**, 5615.
Tsen, K. T., Halama, G., Sankey, O. F., Tsen, S.-C. Y., Morkoc, H. (1989). *Phys. Rev. B* **40**, 8103.
Tsen, K. T., Wald, K. R., Ruf, T., Yu, P. Y., Morkoc, H. (1991). *Phys. Rev. Lett.* **67**, 2557.
Tsen, K. T. (1993). *Int. J. Modern Phys. B* **7**(25), 4165.
Tsen, K. T., Grann, E. D., Guha, S., Menendez, J. (1996a). *Appl. Phys. Lett.* **68**, 1051.
Tsen, K. T., Ferry, D. K., Wang, J. S., Huang, C. H., Lin, H. H. (1996b). *Appl. Phys. Lett.* **69**, 3575.
Tsen, K. T., Joshi, R. P., Ferry, D. K., Botchkarev, A., Sverdlov, B., Salvador, A., Morkoc, H. (1996c). *Appl. Phys. Lett.* **68**, 2990.
Tsen, K. T., Ferry, D. K., Botchkarev, A., Sverdlov, B., Salvador, A., Morkoc, H. (1997). *Appl. Phys. Lett.* **71**, 1852.
Tsen, K. T., Ferry, D. K., Salvador, A., Morkoc, H. (1998a). *Phys. Rev. Lett.* **80**, 4807.
Tsen, K. T., Joshi, R. P., Ferry, D. K., Botchkarev, A., Sverdlov, B., Salvador, A., Morkoc, H. (1998b). *Appl. Phys. Lett.* **72**, 2132.
Yu, P. Y., Cardona, M. (1996). *Fundamentals of Semiconductors*. Springer, New York.

CHAPTER 4

Ultrafast Dynamics and Phase Changes in Highly Excited GaAs

J. Paul Callan, Albert M.-T. Kim, Christopher A. D. Roeser, and Eric Mazur

DEPARTMENT OF PHYSICS AND DIVISION OF ENGINEERING AND APPLIED SCIENCES
GORDON MCKAY LABORATORY
HARVARD UNIVERSITY
CAMBRIDGE, MASSACHUSETTS

I. INTRODUCTION	152
II. MEASURING ULTRAFAST PHENOMENA WITH LIGHT	154
1. *Ultrafast Measurements: The Pump-Probe Technique*	154
2. *Linear Optical Properties and the Dielectric Function*	154
3. *Microscopic Theory of the Dielectric Function*	157
4. *The Dielectric Function of Crystalline Solids: Interband Contributions*	158
5. *The Dielectric Function of Crystalline Solids: Intraband Contributions*	160
6. *Lattice Structure Effects on the Dielectric Function*	162
7. *Electronic Configuration Effects on the Dielectric Function*	163
8. *Why We Measure the Dielectric Function*	165
III. DYNAMICS OF ELECTRONS AND ATOMS FROM FEMTOSECONDS TO MICROSECONDS	166
1. *Mechanisms of Carrier Excitation*	167
2. *Carrier Redistribution, Thermalization, and Cooling*	169
3. *Carrier–Lattice Thermalization*	172
4. *Carrier Recombination*	172
5. *Carrier Diffusion*	174
6. *Structural Effects*	175
7. *Summary*	177
IV. MEASURING THE TIME-RESOLVED DIELECTRIC FUNCTION	179
1. *White-Light Probe Generation*	180
2. *Measuring Transient Reflectivity Spectra*	180
3. *Extracting the Transient Dielectric Function*	181
V. TIME-RESOLVED DIELECTRIC FUNCTION OF HIGHLY EXCITED GaAs	182
1. *Low Fluence Regime ($F < 0.5$ kJ/m^2)*	183
2. *Medium Fluence Regime ($F = 0.5$–0.8 kJ/m^2)*	185
3. *High Fluence Regime ($F > 0.8$ kJ/m^2)*	189
VI. CARRIER AND LATTICE DYNAMICS	191
1. *Carrier Dynamics: Excitation, Scattering, and Relaxation*	192
2. *Structural Dynamics: Lattice Heating, Disordering, and Phase Transitions*	197
VII. CONCLUSIONS	200
REFERENCES	202

I. Introduction

The progress of science in modern times is often driven by improvements in the precision of observations of natural systems or in the scientist's ability to control systems in experiments. Femtosecond lasers, developed about 20 years ago, are a technology that has both increased the precision of observations and allowed us to explore the behavior of materials under novel extreme conditions. These two capabilities have opened up new scientific fields—ultrafast phenomena and laser-induced phase transformations—that span the traditional disciplines of physics, chemistry, materials science, and biology. The first, and most obvious, characteristic of femtosecond lasers is that they produce pulses of light shorter than five femtoseconds (fs) (Sutter *et al.*, 1999). These short pulses can be used to observe phenomena with a time resolution 1,000 to 10,000 times greater than possible in any other experiment. Many important ultrafast phenomena can now be observed directly, for example, energy transfer from electrons to the lattice in solids (Shah, 1996) and chemical reactions in gases (Zewail, 1994).

The second key characteristic of femtosecond pulses is that they can deliver an enormous amount of power with just a modest quantity of total energy, because the energy is packed into such a short time. We use 100-fs pulses with up to 1-mJ of energy, giving a peak power of 10 GW. That is three times the peak power output of the entire Irish national grid, which was 3.3 GW during a particularly cold day in December 1997 (Electricity Supply Board of Ireland, 1999). When such a laser pulse is focused to a small spot, the intensity and electric field produced are truly enormous; the electric field can reach values millions of times greater than the field with which the nucleus of an atom attracts its electrons. It is thus possible to generate a small plasma of ions and free electrons in practically any material, and to do so with very little total pulse energy. Furthermore, the laser-induced plasma is created on a timescale much shorter than the time for electrons to transfer energy to the lattice in a solid. Consequently, the electrons in a solid reach extremely high temperatures while the atomic structure is still cold; this is a unique nonequilibrium excitation of a solid and can cause novel and unusual phase transitions.

In this chapter, we explore some of these remarkable new phase transitions, specifically those occurring in the semiconductor gallium arsenide (GaAs). In inducing these phase transitions, we use the enormous peak power in femtosecond laser pulses to modify the solid. We observe the phase transitions as they occur with femtosecond time resolution. Our specific approach, which is to measure the time-resolved dielectric function spectrum, provides us with the first ever detailed picture of the carrier and lattice dynamics in GaAs at very high excited carrier densities (1–20% of the valence electrons).

We structure the chapter as follows. In Section II, we explain how the optical properties of the material are linked to its structure and electronic configuration. In order to determine ultrafast structural and electronic dynamics — that is, how the material changes — we measure changes in some optical property. The basic technique, called the pump-probe method, involves exciting a material with one pulse, the pump pulse, and measuring an optical property of the material with a second pulse, the probe pulse, as a function of the variable time delay between pump and probe pulses. We measure the dielectric function because it is the fundamental linear optical property of a material and the property that best identifies changes in the phase following excitation.

In Section III, the rich variety of ultrafast processes in semiconductors is laid out. The litany of mechanisms takes one through the first few picoseconds (ps) after excitation: excitation of charge carriers (electrons and holes) by photons from the femtosecond pulse, scattering among carriers, transfer of energy from the carriers to the lattice via carrier–phonon scattering, loss of energy via carrier recombination, and finally mesoscopic and macroscopic changes to the lattice such as ablation, melting, evaporation, and resolidification. Many different pump-probe techniques have been put into action to learn about these processes. Most of these experiments were conducted for low ($<10^{-17}$ cm) and medium ($<10^{-17}$–10^{-20} cm) excited-carrier densities. By contrast, the experiments described here study the dyanmics for excited-carrier densities in the range from 10^{-21}–10^{-22} cm. At such densities, it is not possible to design experiments similar to those used for low-density studies to investigate one single process or another.

Measurement of the time-resolved dielectric function makes possible the task of ascertaining which processes dominate the electronic and structural dynamics of materials at extremely high excited-carrier densities. The experimental methods for directly measuring the time-resolved spectral dielectric function of a material are the subject of Section IV.

In Section V, we detail our measurements of the spectral dielectric function of GaAs following intense femtosecond laser excitation. The data enable us to infer the phase changes caused by laser excitations of different strengths. As we shall see in Section VI, close analysis of the time-resolved dielectric function provides the most comprehensive description to date of electronic and structural dynamics at these high excited-carrier densities. For different excitation strengths, different ultrafast processes dominate the dynamics, and the timescales for these dominant processes can often be calculated from the changes in the dielectric function.

II. Measuring Ultrafast Phenomena with Light

Optical properties tell us a great deal about the state of a material. The air is transparent but coal is not. Transparent solids and liquids, such as glass and water, produce reflections that distinguish them from air. Metals are more shiny than insulators such as wood or plastic. Changes in optical properties call our attention to changes in state or phase transformations in materials. For example, when a metal is heated strongly, it turns red; indeed, in English the term "red hot" denotes this change in state.

What determines the material's optical properties such as reflectivity, transmissivity, or light emission (luminescence)? Furthermore, can one work back from the optical properties to gain an understanding of the material itself? The linear optical properties we observe — reflectivity, transmissivity, etc. — are determined by one fundamental property of the material, called the dielectric function $\varepsilon(\omega)$, and depend on the light frequency ω. The dielectric function can be derived from the atomic lattice structure and electronic configuration of the material, that is, the allowed energy states and the occupation of those states.

1. Ultrafast Measurements: The Pump-Probe Technique

To track ultrafast phase changes in the structure or electronic configuration, we must measure changes in some optical property using the *pump-probe technique*: A *pump pulse* induces changes in the material. After a certain time delay, a weaker, *probe pulse* is incident on the excited area of the sample. A detector or camera measures the amount of the probe that is reflected, transmitted, converted to second-harmonic frequency, or whatever. The optical property of the material is measured for different time delays after the pump pulse, revealing how the material changes in response to the excitation. Figure 1 shows a schematic setup for a pump-probe experiment to measure the transient transmission following excitation.

It remains to select an optical property to track with the probe pulses in our experiments. In order to do so, we must first understand the links between the microscopic state of a material and its optical properties.

2. Linear Optical Properties and the Dielectric Function

Light is an electromagnetic wave, and thus its interaction with a material depends on the response of the medium to electric and magnetic fields. When the optical fields are weak compared to those exerted by atoms on electrons, the response is linear and described by the refractive index

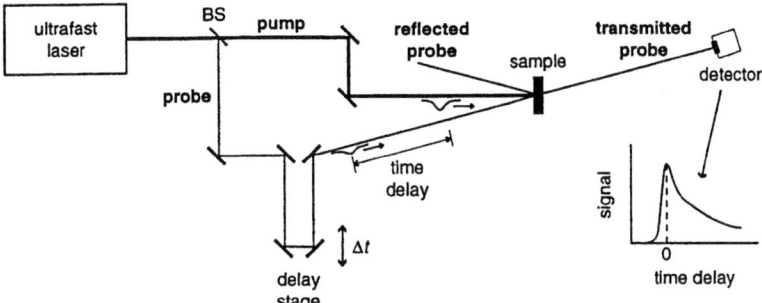

FIG. 1. Schematic representation of a pump-probe experiment. In the configuration illustrated here, transmitted light is detected. However, the detector could measure the reflected probe instead. BS = beamsplitter; Δt = pump-probe time delay. Many optical components are not shown, such as lenses or mirrors to focus pump and probe beams on the sample.

(Jackson, 1975)

$$n(\omega) = \sqrt{\varepsilon(\omega)\mu(\omega)}. \tag{1}$$

The dielectric function $\varepsilon(\omega)$ describes the material's response to applied electric fields, and the magnetic permeability $\mu(\omega)$ encapsulates the linear response to applied magnetic fields. For optical frequencies we can take $\mu = 1$ (Landau et al., 1984), and the optical response of a medium is determined by the light's electric field. In general, $n = \eta + i\kappa$ is a complex quantity, and the imaginary part causes absorption of light during propagation governed by the light's wave equation (Jackson, 1975),

$$\mathbf{E}(x, t) = \mathbf{E}_0\, e^{i(kx - \omega t)} = \mathbf{E}_0\, e^{-\kappa(\omega/c)x}\, e^{i[\eta(\omega/c)x - \omega t]}, \tag{2}$$

because $\omega = (c/n)k$.

Consider now the most common linear optical properties, namely reflection and transmission of light at an interface. These are functions of the geometry (i.e., the angle of incidence θ_i and light polarization relative to the interface) as well as of the materials on either side of the interface. For the case where the polarization is in the plane of incidence (p-polarization), we find that in the notation of Fig. 2a (Jackson, 1975),

$$r_p = \frac{E_r}{E_i} = \frac{n_2^2 \cos \theta_i - n_1\sqrt{n_2^2 - n_1^2 \sin^2 \theta_i}}{n_2^2 \cos \theta_i + n_1\sqrt{n_2^2 - n_1^2 \sin^2 \theta_i}} \tag{3}$$

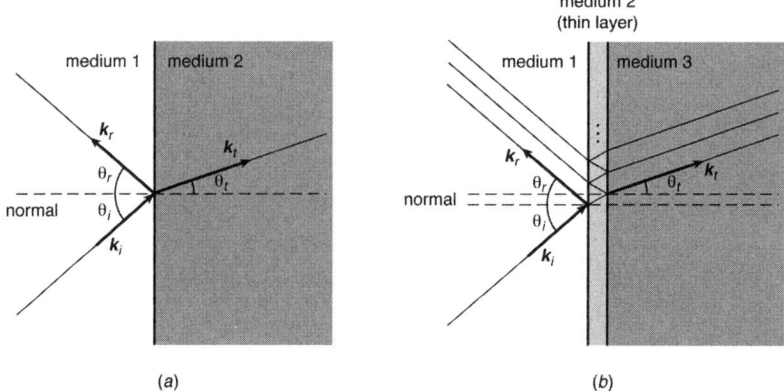

FIG. 2. Incident (labeled with subscripts i), reflected (r), and transmitted (t) beams at (a) a simple interface between two media, (b) an interface with a thin layer.

and

$$t_p = \frac{E_t}{E_i} = \frac{2n_1 n_2 \cos \theta_i}{n_2^2 \cos \theta_i + n_1 \sqrt{n_2^2 - n_1^2 \sin^2 \theta_i}}. \qquad (4)$$

The reflectivity R and transmissivity T are defined as the ratio of reflected and transmitted fluxes, respectively, to the incident flux on the interface. They are given by (Hecht, 1987)

$$R = |r|^2, \qquad T = \frac{n_2 \cos \theta_t}{n_1 \cos \theta_i} |t|^2. \qquad (5)$$

If the sample has multiple layers, the expressions for R and T are more complex. For example, in measuring the reflectivity of many materials in air, one must account for a thin oxide layer, and the multiple reflections that can occur in the layer as shown in Fig. 2b. The total reflectivity for such a three-layer system is (Born and Wolf, 1980)

$$R = |r_{\text{total}}|^2 = \left| \frac{r_{12} + r_{23} e^{i2\beta}}{1 + r_{12} r_{23} e^{i2\beta}} \right|^2, \qquad (6)$$

where r_{12} and r_{23} are Fresnel factors calculated from Eq. (3) or the equivalent formula for light polarized perpendicular to the plane of incidence (s-polarization) for each of the interfaces and

$$\beta = \frac{\omega}{c} d \sqrt{\varepsilon_2 - \varepsilon_1 \sin^2 \theta_i} \qquad (7)$$

is related to the thickness d of the oxide layer.

In all cases — reflection, transmission, or absorption — the optical properties depend on a combination of geometrical factors and the materials. The optical response of a material is described by the refractive index, or equivalently the dielectric function. We now turn to the task of understanding the microscopic origins of the dielectric function, and derive an expression for the dielectric function in terms of the allowed quantum energy states in a material.

3. Microscopic Theory of the Dielectric Function

The response of a material at optical frequencies is determined by the outer, or valence, electrons. We use a semiclassical approach in which we treat the electrons in the material using quantum mechanics and treat the external electric field classically. The perturbing potential due to the light is taken to be $H_1(t) = (-e)\mathbf{x} \cdot \mathbf{E}(t)$ if we neglect the effect of the light's magnetic field. This is called the *electric dipole approximation*. Based on linear response theory, or time-dependent perturbation theory, one can show that the dielectric function of a solid is given in SI units by (Yu and Cardona, 1996)

$$\varepsilon(\omega) = 1 + \frac{e^2}{\varepsilon_0 \hbar} \sum_{v,c} |x_{cv}|^2 \cdot \left(\frac{\mathscr{P}}{\omega - \omega_{cv}} - i\pi\delta(\omega - \omega_{cv}) + \frac{\mathscr{P}}{\omega + \omega_{cv}} - i\pi\delta(\omega + \omega_{cv}) \right) \tag{8}$$

where v and c label occupied and unoccupied states, respectively, $x_{cv} = \langle c|x|v \rangle$ is the matrix element of the position operator between the states c and v, and $\omega_{cv} = \omega_c - \omega_v$. In this equation, $\delta(\omega - \omega_0)$ is the Dirac delta function about ω_0, and $\mathscr{P}$ denotes the principal part of the integral over ω, that is, the integral leaving out an infinitesimally small region about ω_0. Physically, the first two terms in Eq. (8) correspond to absorption, in which photons of energy $\hbar\omega$ are taken from the electromagnetic field of the light in order to excite electrons to higher energy states. The delta function requires that the energy gained by an electron must equal the energy of the absorbed photon and thus incorporates the principle of conservation of energy. The imaginary part of the dielectric function, which determines in large part the absorption by the medium, has resonances whenever the frequency of the light is such that the photon energy matches the separation between electronic energy levels. The second pair of terms in the equation, involving $\omega + \omega_{cv}$, describes stimulated emission (gain). The delta function requires that ω_{cv} be negative for resonance, that is, the presence of the electromagnetic field induces electrons to drop from higher energy states to lower energy states and release their energy as photons. Because electrons typically occupy the lowest energy states, absorption is usually more significant than stimulated emission.

4. THE DIELECTRIC FUNCTION OF CRYSTALLINE SOLIDS: INTERBAND CONTRIBUTIONS

For a crystal, the allowed electronic wave functions are periodic plane waves of the Bloch form (Ashcroft and Mermin, 1976),

$$|\psi\rangle_{b,\mathbf{k}} = u_{b,\mathbf{k}}(\mathbf{x})\, e^{i\mathbf{k}\cdot\mathbf{x}}, \qquad (9)$$

where **k** here is the *crystal momentum*, b is the *band index*, the quantum number that distinguishes different bands, and $u_{b,\mathbf{k}}(\mathbf{x})$ has the same periodicity as the lattice. The allowed energy levels $E_b(\mathbf{k})$ corresponding to these eigenstates of the crystal Hamiltonian can be plotted as a function of crystal momentum, showing the various bands. This is called the *band structure* of the crystal. Figure 3a displays the band structure for the crystalline semiconductor gallium arsenide (c-GaAs) along two directions in **k**-space. The directions are shown in the inset, which displays the first Brillouin zone for a face-centered cubic (fcc) crystal such as c-GaAs. Each band contains two valence electrons (i.e., outer, as opposed to core, electrons) per **k** value, and the number of **k** values equals the number of unit cells (the unit which is repeated in the lattice). Crystalline GaAs has an average of four valence electrons per atom, and two atoms (one gallium and one arsenic) per unit cell. For each of the eight electrons per unit cell, there are two states, one bonding and one antibonding state. Hence, there must be 16 states per **k** value, which requires a total of eight bands. In the ground state of GaAs, the bottom four bands (the valence bands or bonding states) are completely full and the top four bands (the conduction bands or antibonding states) are

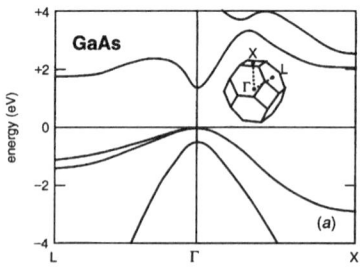
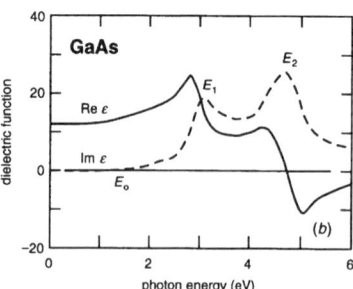

FIG. 3. (a) Band structure (Cohen and Chelikowsky, 1989) and (b) dielectric function (Palik, 1985) of crystalline GaAs.

completely empty. There is a gap between them, and this is what makes GaAs a semiconductor. The top three valence bands and the lowest two conduction bands appear in Fig. 3a.

When we calculate the matrix element x_{cv} in Eq. (8) with the crystalline wave functions, integrating over the exponentials produces the requirement that

$$\mathbf{k}_c - \mathbf{k}_v = 0 \quad \Rightarrow \quad \mathbf{k}_c = \mathbf{k}_v. \tag{10}$$

Thus, the transitions between bands induced by light have to be *vertical* when looking at band structures such as that in Fig. 3a. Under this constraint, the dielectric function reduces to

$$\varepsilon(\omega) = 1 - \frac{e^2}{\varepsilon_0} \sum_{v,c} \int \frac{d^3k}{8\pi^3} |x_{cv}(\mathbf{k})|^2 \left[\frac{\mathscr{P}}{E_{cv}(\mathbf{k}) - \hbar\omega} - i\pi\delta(E_{cv}(\mathbf{k}) - \hbar\omega) \right] \tag{11}$$

if the material is isotropic (Yu and Cardona, 1996). Here, the summation is now over conduction and valence bands labeled by indices c and v; $x_{cv}(\mathbf{k})$ is the position matrix element between states in bands c and v at the same crystal momentum $\mathbf{k}$; and $E_{cv}(\mathbf{k}) = E_c(\mathbf{k}) - E_v(\mathbf{k})$ is the difference in energies between the conduction and valence bands at a given $\mathbf{k}$. Note that this expression only includes the absorption term; there is also a stimulated emission term in which $E_{cv}(\mathbf{k}) - \hbar\omega$ is replaced by $E_{cv}(\mathbf{k}) + \hbar\omega$. Position and momentum matrix elements between eigenstates of the crystal Hamiltonian are related by

$$\langle \psi_1 | x | \psi_2 \rangle = \frac{\langle \psi_1 | p | \psi_2 \rangle}{i m \omega_{12}} \tag{12}$$

so that

$$\varepsilon(\omega) = 1 - \frac{e^2}{\varepsilon_0 m^2 \omega^2} \sum_{v,c} \int \frac{d^3k}{8\pi^3} |p_{cv}(\mathbf{k})|^2 \left[\frac{\mathscr{P}}{E_{cv}(\mathbf{k}) - \hbar\omega} - i\pi\delta(E_{cv}(\mathbf{k}) - \hbar\omega) \right]. \tag{13}$$

For semiconductors, the momentum matrix elements p_{cv} are not strongly dependent on $\mathbf{k}$, and the imaginary part depends mainly on the joint density of states $J(\omega)$:

$$\mathrm{Im}[\varepsilon(\omega)] \sim \frac{1}{\omega^2} J(\omega)$$

$$J(\omega) = \sum_{v,c} \int d^3k \, \delta(E_c(\mathbf{k}) - E_v(\mathbf{k}) - \hbar\omega). \tag{14}$$

In crystals other than semiconductors, the matrix elements play a role as well as the joint density of states. In any case, Eqs. (13) and (14) clearly show the link between band structure $E_{c,v}(\mathbf{k})$ and the linear optical response $\varepsilon(\omega)$.

Figure 3b displays the dielectric function of undoped GaAs at room temperature. The main features, labeled E_0, E_1, and E_2, are common to group IV and III–V semiconductors, although their locations and relative sizes vary from material to material. E_0 marks the fundamental band edge below which there is no interband absorption. The peaks E_1 and E_2 arise from regions in the band structure where valence and conduction bands are roughly parallel, leading to a high joint density of states for direct interband transitions. Transitions in the region around the L point and along the ΓL line produce the E_1 peak, while the stronger E_2 peak results from large regions on the square faces around the X points, around the edges of the hexagonal faces, and partway along the lines from these surface points toward the central Γ point (Waghmare and Kaxiras, 1998). The absorption peaks correspond to points of inflection in the real part of the dielectric function. E_2 is roughly coincident with the zero crossing of $Re[\varepsilon(\omega)]$, and its location approximately gives the average splitting between bonding and antibonding, or valence and conduction states, in GaAs.

5. THE DIELECTRIC FUNCTION OF CRYSTALLINE SOLIDS: INTRABAND CONTRIBUTIONS

In metals, the structure and/or the number of valence electrons per atom leads to occupied and unoccupied states that have no intervening energy gap. Figure 4 shows the band structure and dielectric function of Cu. Copper has an fcc lattice and hence the same Brillouin zone shape as c-GaAs, shown in Fig. 3a. States below the *Fermi level* E_F are occupied; states above it are unoccupied. The dielectric function is dominated by intraband contributions, where an electron is excited from an occupied to

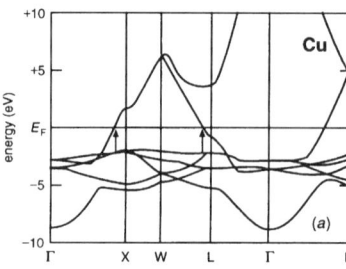
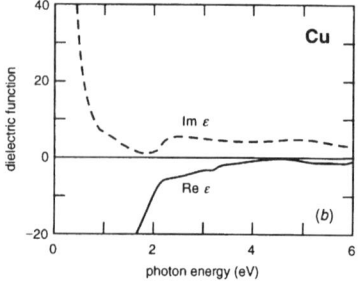

FIG. 4. (a) Band structure (Ashcroft and Mermin, 1976) and (b) dielectric function (Palik, 1985) of crystalline Cu. E_F is the Fermi level.

an unoccupied state in the same band. As there is a change in electron crystal momentum **k**, this transition requires a phonon or an impurity to conserve momentum.

For a Fermi sea of electrons, as exists in a metal or in a highly doped or excited semiconductor, the contribution to the dielectric function is described by the *Drude model*, namely (Ashcroft and Mermin, 1976; Siegal, 1994)

$$\varepsilon(\omega) = 1 + \omega_p^2 \left[\frac{i\tau}{\omega(1 - i\omega\tau)} \right], \qquad (15)$$

where N is the electron density, m is the effective mass of electrons near the Fermi level,

$$\omega_p = \left[\frac{Ne^2}{\varepsilon_0 m} \right]^{1/2}, \qquad (16)$$

is the *plasma frequency*, and τ is the mean scattering time (usually determined by electron–phonon interactions but sometimes by impurities or other factors). For an undoped semiconductor or insulator in which electrons have been excited, thermally or otherwise, from the valence to the conduction band, the holes left behind in the valence band also contribute to the dielectric function. The total free carrier contribution is

$$\varepsilon(\omega) = 1 + \frac{ie^2}{\varepsilon_0 \omega} \left[\frac{N_e \tau_e}{m_e(1 - i\omega\tau_e)} + \frac{N_h \tau_h}{m_h(1 - i\omega\tau_h)} \right] \qquad (17)$$

where the subscripts e and h refer to electrons and holes, respectively. The general shape of the dielectric function is still given by Eq. (15) if one uses a reduced effective mass m^* such that $1/m^* = 1/m_e + 1/m_h$ and takes $\tau_e = \tau_h = \tau$. The Drude dielectric function appears in Fig. 5. Note that the imaginary part of the Drude dielectric function is monotonically decreasing while the real part passes through zero near the plasma frequency.

The dielectric function of Cu, shown in Fig. 4b, has a large Drude contribution, as expected. However, at energies above 2 eV, $\text{Im}[\varepsilon(\omega)]$ increases because of contributions from interband transitions between the fully occupied d-bands and unoccupied states above the Fermi level. These interband transitions are indicated by the arrows in Fig. 4a and give copper its characteristic color. In nonmetals with significant free carrier densities, such as semimetals or excited semiconductors and insulators, the Drude dielectric function due to these carriers is usually smaller than the interband component caused by bound carriers.

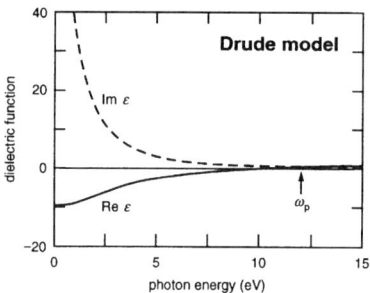

FIG. 5. Dielectric function of a free electron gas, described by a Drude model with plasma frequency $\omega_p = 12\,\text{eV}$ and relaxation time $\tau = 0.18\,\text{fs}$.

6. Lattice Structure Effects on the Dielectric Function

Links between band structure and optical properties are relevant for the study of crystalline materials, such as crystalline GaAs and Cu. However, crystal momentum **k** is not a well-defined concept in nonordered materials, and all that matters is the total joint density of states as a function of energy separation between occupied and unoccupied states. When a solid is disordered, this relaxation of crystal momentum conservation leads to less well-defined features in interband contributions to the dielectric function.

For example, compare the dielectric function of amorphous GaAs (a-GaAs) displayed in Fig. 6 with that for c-GaAs in Fig. 3. The peak in the imaginary part of the dielectric function is at roughly the same energy but is smoother and broader for the disordered material. The dielectric function of liquid Si, shown in Fig. 7, also does not display sharp absorption features. Many semiconductors become metals upon melting and have Drude-like dielectric functions. The Drude term is not affected by lattice structure.

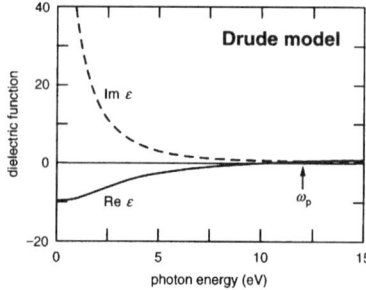

FIG. 6. Dielectric function of amorphous GaAs (Erman *et al.*, 1984).

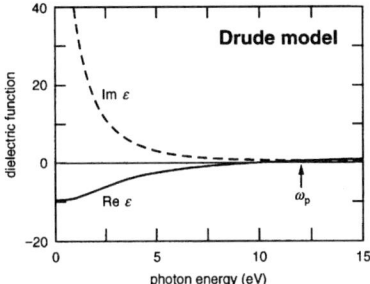

FIG. 7. Dielectric function of liquid Si (Jellison, 1997).

However, the disordered structure of the liquid material broadens interband contributions to the dielectric function, so that there are no sharp absorption features similar to that at 2 eV in crystalline copper.

7. ELECTRONIC CONFIGURATION EFFECTS ON THE DIELECTRIC FUNCTION

Structure affects the dielectric function indirectly, namely by determining the allowed quantum states for electrons. The occupation of these allowed states by electrons, which I will refer to as the *electronic configuration* of the material, is the second factor that determines the dielectric function. In Eq. (13), we assumed a fully occupied set of valence bands v separated by a gap from a fully unoccupied set of conduction bands c. However, excitation of a material — by heat, a laser beam, an electron beam, or whatever means — will excite electrons so that they do not occupy all the lowest energy states. This was alluded to in Subsection 5, where we wrote an expression for the intraband contribution of excited electrons and holes in semiconductors and insulators.

The presence of excited carriers requires us to modify the interband expression for the dielectric function to account for occupied conduction band states and unoccupied valence states. This is in addition to the intraband Drude contribution to $\varepsilon(\omega)$ due to these carriers. Define an *occupation factor*

$$f_s = \begin{cases} 1 & \text{if state } s \text{ is occupied} \\ 0 & \text{if state } s \text{ is unoccupied} \end{cases}. \tag{18}$$

Stimulated emission, and also absorption between different valence or different conduction bands, is now possible. The dielectric function for any solid can be written as (Bransden and Joachain, 1983)

$$\varepsilon(\omega) = 1 - \frac{e^2}{\varepsilon_0} \sum_{s',s} [f_s - f_{s'}] |x_{s's}|^2 \left(\frac{\mathscr{P}}{E_{s'} - E_s - \hbar\omega} - i\pi\delta(E_{s'} - E_s - \hbar\omega) \right). \tag{19}$$

Thus, the net absorption, described by the imaginary part of the dielectric function, depends on the difference in occupation between the two states. Indeed, if the higher state has a greater population, stimulated emission dominates. In a crystal, we can replace the state s with a band index b and crystal momentum $\mathbf{k}$, and

$$\varepsilon(\omega) = 1 + \frac{e^2}{\varepsilon_0 m^2 \omega^2} \sum_{b',b} \int \frac{d^3k}{8\pi^3} [f_b(\mathbf{k}) - f_{b'}(\mathbf{k})]$$

$$|p_{b'b}(\mathbf{k})|^2 \left(\frac{\mathscr{P}}{E_{b'b}(\mathbf{k}) - \hbar\omega} - i\pi\delta(E_{b'b}(\mathbf{k}) - \hbar\omega) \right) \tag{20}$$

where the conservation of crystal momentum is, as before, automatically included in the expressions.

In addition to the effect of electron occupation factors on the dielectric function, the electronic configuration also influences the energy eigenstates themselves. Electrons interact via the Coulomb potential. The interaction is stronger for free conduction electrons (or for holes) because they can come closer together than valence electrons, whose wave functions are concentrated near atomic sites. In principle, the electrons should be described by a single, multiparticle wave function. However, this degree of complexity is not generally necessary for practical purposes, and it is possible to describe the effect of electron–electron interactions as a perturbation on the single electon states, where the "electrons" are now actually quasiparticles (Ashcroft and Mermin, 1976; Mahan, 1990). One can ascribe physical pictures to some of the most important terms in this perturbation of the allowed energy levels (Kim et al., 1994). First, the ionic potential is screened. Second, each electron interacts with the average field of the other free electrons (Hartree term). Third, when we introduce Pauli exclusion and prevent electrons of the same energy and spin coming close together, the total electrostatic energy of the carriers reduces (exchange term). Fourth, the remaining corrections to the independent electron picture are called correlation terms.

In the ground state, each of these terms contributes to the effective one-electron Hamiltonian and hence to the allowed energy states. Screening, exchange and correlation effects all tend to broaden bands and close energy gaps. When additional carriers are excited, however, their relative contributions are different (Kim et al., 1994). For instance, when a small density of carriers is excited in a semiconductor, the third and fourth terms are most important and cause a decrease in the bandgap referred to as *bandgap renormalization* (Zimmermann, 1988). At excited carrier densities of 10^{22} carriers per cm^{-3}, such as those in the experiments in this chapter, changes in the first, third, and fourth terms can all be substantial (Kim *et al.*, 1994). The Hartree term is not much affected because the contributions of the electrons and holes cancel.

Thus, the optical properties give information about the electronic configuration of a material. However, we must be careful. Changes in the electronic configuration can change the dielectric function in many different ways. The intraband contribution depends just on the number of free carriers, while the interband part of $\varepsilon(\omega)$ is sensitive to the occupation factors for different energy levels and to changes in those energy levels caused by carrier–carrier interactions.

8. Why We Measure the Dielectric Function

Over the course of this section, we have seen how the optical properties we observe are related to the microscopic state of a material, that is, the structure and electronic configuration. Figure 8 illustrates the various steps in making the connection:

1. Using quantum mechanics, the structure and electronic configuration determine the allowed energy levels and the occupation of those levels.
2. The energy eigenstates and their occupation factors determine the linear or nonlinear response of the material to applied fields from a light beam, described by the macroscopic quantities ε and $\chi^{(n)}$, respectively.
3. Other optical properties, such as reflectivity or second-harmonic generation, depend on the fundamental material optical properties ε and $\chi^{(n)}$ combined with the experimental geometry, through Maxwell's equations and appropriate boundary conditions.

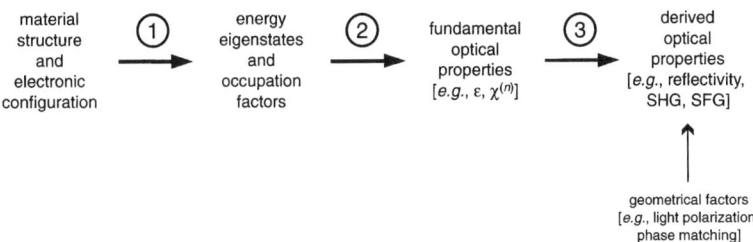

FIG. 8. Steps connecting material structure and electronic configuration with measurable optical properties.

Although one can derive an optical property such as reflectivity knowing the structure and electronic configuration, it is not possible to solve the inverse problem uniquely. A given value of reflectivity, or other optical property, can correspond to many different values of the dielectric function or nonlinear susceptibility. Even if we know the fundamental property, $\varepsilon(\omega)$ or $\chi^{(n)}$, we cannot determine the structure and electronic configuration uniquely. However, it is possible to make reasonable inferences about what happens to the state of a material when its optical properties change. These inferences are much more accurate if one does not have to reverse step 3 but instead measures $\varepsilon(\omega)$ or $\chi^{(n)}$ directly.

This is the first reason that we opt for measuring the dielectric function $\varepsilon(\omega)$ directly in our experiments. The dielectric function data are by far the simplest to interpret: $\varepsilon(\omega)$ is the fundamental material property and does not depend on any geometric factors. Furthermore, it only depends on single photon transitions in the material, whereas nonlinear properties depend on multiple transitions and are thus more difficult to understand. Our second motivation for determining the dielectric function is that it is an excellent indicator of the state of a material. The dielectric function can answer such questions as: Is the material a solid or a liquid? Is it crystalline or disordered? Is it metallic, semiconducting, or insulating? What is its temperature and pressure? Third, and finally, the broadband dielectric function is sensitive to electron dynamics across a large part of the band structure, which would not be possible with single-frequency measurements. This wealth of information that comes from the dielectric function justifies the additional effort required to find the dielectric function. But before describing how we measure the dielectric function and presenting our results, we spend some time reviewing the various processes that can contribute to ultrafast carrier and lattice dynamics.

III. Dynamics of Electrons and Atoms from Femtoseconds to Microseconds

The femtosecond and picosecond time scales are characteristic of many fundamental and interesting physical processes in condensed matter. When an intense, short laser pulse "hits" a solid, it will cause melting and/or ablation within a few nanoseconds. At lower pulse energies, the solid may just be heated by the energy it absorbs from the light. Such heating or melting can be observed using electrical measurements. But femtosecond laser pulses and the pump-probe technique reveal the sequence of steps by which the energy in the excitation pulse is transferred first to the electrons and then to the lattice. Several régimes of carrier excitation and relaxation are identifiable: (1) carrier excitation, (2) redistribution, thermalization, and cooling of the free carriers through scattering, (3) carrier-lattice thermalization, (4) carrier density reduction through recombination and diffusion, and finally (5) thermal structural effects. These regimes do not occur in strict sequence, but rather may overlap in time. For example, in many materials the carriers form a thermal distribution among themselves at the same time as they cool by transferring energy to lattice phonons. In addition, as we shall see, some nonthermal structural effects can occur while the lattice is still cold. This chapter details the variety of ultrafast physical processes that comprise these different regimes and specify what is known about their typical time scales under different excitation conditions.

Information on ultrafast electron and phonon dynamics comes, as we have said, from pump-probe experiments on laser-excited semiconductors. Typically, the pump excitation is weak, in that it generates a plasma density of 10^{15}–10^{19} carriers per cm^3, which is about 10^{-8}–10^{-4} of the valence electrons. Each experiment is cleverly designed to pick out an individual physical mechanism and determine its characteristic time. Experiments can be made sensitive to a single process through, for example, selection of appropriate pump and probe wavelengths or using a material with a specific bandgap.

It is not possible to isolate individual processes for study when exciting a high carrier density, that is, one above 10^{20} cm^{-3} or 0.1% of the valence electrons. Indeed, this chapter presents the first observations of the dominant scattering and relaxation mechanisms for densities between 1% and 10% of the valence electrons. However, before describing what *does* happen at high carrier density, it behooves us to outline the various processes that *can* occur in solids on femtosecond and picosecond time scales. These descriptions draw both on experimental evidence at lower excited carrier densities and on theoretical predictions of how time scales should vary with carrier density.

1. MECHANISMS OF CARRIER EXCITATION

Figure 9 illustrates different mechanisms for the excitation of electrons in solids by visible light. Interband *single photon absorption* (SPA) is the most important mechanism in semiconductors when the photon energy is higher than the smallest direct gap, that is, the smallest gap between conduction and valence states having the same **k**-value. Figure 9a shows the absorption of a single photon and the consequent generation of an electron–hole pair. The figure also presents the effect of SPA on the energy E and number density N of excited carriers: Interband absorption of light increases both. The number of carriers generated is equal to the number of photons absorbed, which is the absorbed energy divided by the photon energy $\hbar\omega$. The excited electron (or hole) density is this number divided by the volume into which the energy is deposited. Thus, for a laser pulse of incident fluence

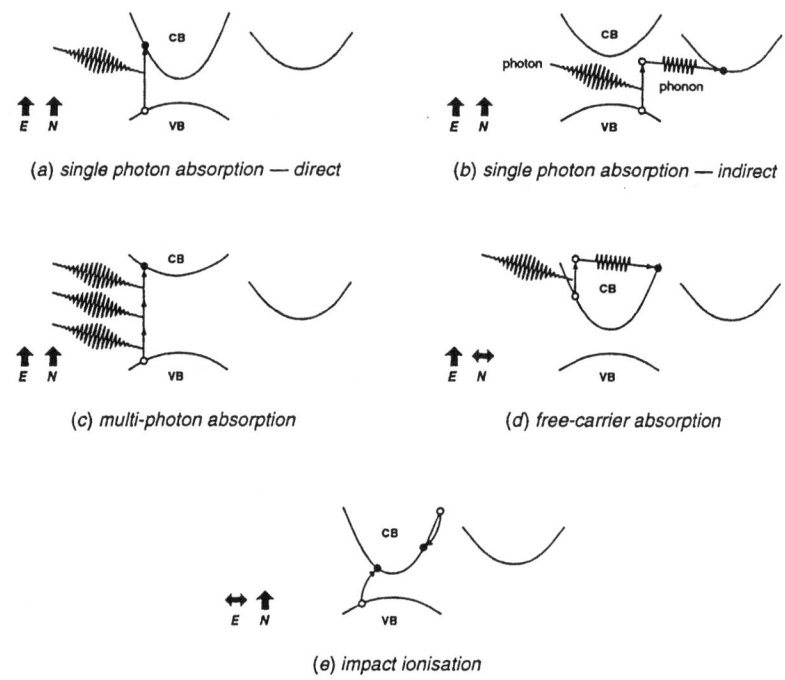

FIG. 9. Mechanisms for exciting carriers in a solid. The symbols for photons and phonons are labeled in panel (b).

(i.e., energy per unit area) F, the maximum free-carrier density is

$$N = \frac{(1-R)F\alpha}{\hbar\omega} \quad (21)$$

where the reflectivity R and absorption depth α are determined by dielectric function $\varepsilon(\omega)$. In reality, the absorbed energy density falls exponentially with distance into the material, and so does the excited carrier density. However, the exponential density gradient rapidly flattens out near the surface because of diffusion and, furthermore, the probe absorption depth is normally much less that of the pump pulse due to increased absorption by the carriers. These effects combine to make Eq. (21) a reasonable estimate of the carrier density when the pump induces only single photon absorption.

For indirect gap semiconductors, such as Si, single photon absorption can still occur with photons of energy greater than this gap, but only with the assistance of a phonon to conserve momentum. Figure 9b displays such a phonon-assisted, *indirect transition*. Direct, vertical transitions dominate indirect ones when both are allowed because direct transitions are lower order processes.

Multiphoton absorption (MPA), shown in Fig. 9c, can be important if the direct gap is greater than the photon energy or if single photon absorption is inhibited by filling of states in the conduction band. Since several photons must be absorbed "simultaneously," MPA is is not very probable compared with SPA. However, its rate increases as I^n and hence becomes important with very intense pulses. *Free carrier absorption* creates more energetic carriers from the excited electron–hole plasma or from initially free electrons in a metal. As Fig. 9d indicates, free carrier absorption increases the energy of the free carrier population but not the number of free carriers. Carriers thus excited well above the gap can generate additional excited carriers via impact ionization. Thus, impact ionization increases the number of free carriers but does not affect their total energy. Because the process is nonradiative, it can continue after the laser pulse is finished. All of these excitation processes take place in tandem, making it very difficult to estimate a carrier density. In some situations, such as those that are the subject of this chapter, Eq. (21) is valid. Often it is not, and the carrier density must be estimated by more complicated models or from experimental data.

2. Carrier Redistribution, Thermalization, and Cooling

A nearly monochromatic laser pulse deposits free electrons and holes at specific points in the band structure, because the only electrons excited are at **k**-vectors for which $E_c(\mathbf{k}) - E_v(\mathbf{k})$ matches the photon energy. The electrons and holes are redistributed throughout the conduction and valence bands via carrier–carrier scattering and carrier–phonon scattering, as illustrated schematically by Fig. 10. The electron and hole populations quickly approach Fermi–Dirac distributions. The number of carriers does not change. However, the energy E of the carriers does decrease, because of spontaneous phonon emission that transfers energy to the lattice.

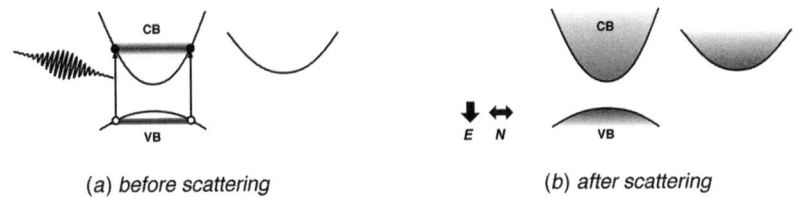

Fig. 10. Carrier redistribution and thermalization: (a) shows the carrier distribution following laser excitation but before scattering, while (b) illustrates the thermalized carrier distribution created by carrier–carrier and carrier–phonon scattering.

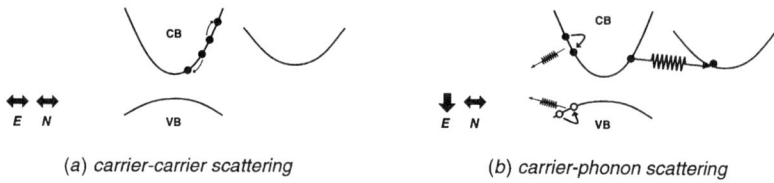

FIG. 11. Carrier scattering mechanisms in a semiconductor.

a. Carrier–Carrier Scattering

Carrier–carrier scattering is a two-body process in which two carriers collide. Figure 11a shows a single scattering event in **k**-space. The collisions are mediated by the Coulomb interaction between carriers. Carrier–carrier scattering does not change the total energy or the number of carriers, but individual carrier energies and momenta are randomized by multiple scattering events. Because the scattering involves two bodies, the scattering rate should increase with density. Experimental estimates of the scattering time vary. For example, Oudar et al. obtained 0.3 ± 0.1 ps for an excitation density of 10^{18} electrons per cm^3 using transient transmission spectra (Oudar et al., 1985), whereas Elsaesser and co-workers use time-resolved luminescence data to estimate that carrier–carrier scattering takes place in about 100 fs or less when the excited carrier density is in the range of $1.7–7.0 \times 10^{17}$ cm^{-3} (Elsaesser et al., 1991). Monte Carlo simulations predict a $N^{-1/3}$ dependence for the scattering rate (Lundstrom, 1990). However, at high densities exceeding 10^{20} cm^{-3}, screening should reduce the strength of the Coulomb interaction between carriers, and therefore decrease the scattering cross section. The scattering rate does not continue to increase at the same rate with increasing density and could conceivably decrease at higher densities.

b. Carrier–Phonon Scattering

Free carriers can lose (or gain) energy and momentum by emission (or absorption) of a phonon or lattice vibration. Carrier–phonon scattering in crystalline semiconductors is subdivided into two categories. In *intravalley scattering*, the carrier remains in the same conduction or valence band valley (i.e., the region around a local minimum in the conduction band or a local maximum in the valence band). Or it can transition to a different valley, which is called *intervalley scattering*. Figure 11b shows the two cases. Because of spontaneous phonon emission, more phonons are emitted than absorbed in carrier–phonon scattering events. Thus, energy is removed from the carrier system and transferred to the lattice. Two different interactions between carriers and phonons can result in scattering. The first is called the *deformation potential* and occurs because perturbations of the lattice by phonons change the forces seen by carriers. The second mechanism, *polar scattering*, occurs only in polar materials such as GaAs. These materials have local dipole moments between pairs or groups of atoms. Phonons perturb these dipoles, which in turn exerts a force on free carriers.

It can be shown that the scattering rate for deformation potential scattering is independent of the phonon wave vector **q** (Lundstrom, 1990). Hence, the deformation potential can cause both intravalley (which requires low-**q** phonons) and intervalley (which requires high-**q** phonons) scattering. On the other hand the scattering rate for polar scattering with optical phonons turns out to be inversely proportional to the square of the phonon wave vector (Lundstrom, 1990):

$$\frac{1}{\tau_{\text{e-ph, polar-optical}}} \propto \frac{1}{q^2}. \tag{22}$$

This dependence on phonon wave vector implies that polar scattering has a greater role in intravalley scattering than in intervalley scattering.

Irrespective of the nature of the physical interaction, or whether the scattering is intra- or intervalley, carrier–phonon scattering is always a two-body process. The number of collisions per second, or the scattering rate, roughly has a linear dependence on the number of carriers N and on the number of phonons. However, at very high carrier densities, this simple relation is no longer valid because of screening effects. Yoffa predicts that in Si, for which polar scattering is not possible, the critical densities at which screening becomes important are approximately 2.5×10^{19} cm^{-3} for intravalley scattering and $1-2 \times 10^{21}$ cm^{-3} for intervalley scattering (Yoffa, 1980). She predicts that the per carrier phonon emission rate is proportional to $1/N^2$ for the strong-scattering regime, that is, above the critical density.

Experimental work on carrier–phonon scattering in GaAs suggests that intervalley scattering from the central Γ valley to side valleys is very fast. Transient transmission measurements by Lin and colleagues, as interpreted by Bailey *et al.*, indicate $\Gamma \to L$ scattering times of $\tau_{\Gamma \to L} = 30$ fs at 10^{17} cm^{-3}, 17 fs at 3×10^{17} cm^{-3} and 13 fs at 10^{18} cm^{-3} (Lin *et al.*, 1987, 1988; Bailey *et al.*, 1989). Later transient transmission measurements by Becker *et al.*, using 6-fs, broadband pump and probe pulses suggest scattering times of 80 fs for $\Gamma \to L$ scattering and 55 fs for $\Gamma \to X$ scattering (Becker *et al.*, 1988).

Intravalley scattering was first studied in GaAs by Kash, Tsang, and Hvam using time-resolved anti-Stokes Raman spectroscopy. The anti-Stokes Raman scattering signal generated by the probe is proportional to the population of LO phonons spontaneously emitted by electrons losing energy to fall down the Γ valley (Kash *et al.*, 1985). The single LO phonon emission time is found to be 165 fs. Leitenstorfer and colleagues have roughly confirmed this estimate. They estimate the LO phonon emission time in GaAs to be about 100 fs, using transient transmission to probe the electron population evolution rather than that of the phonons (Leitenstorfer et al., 1996).

3. CARRIER–LATTICE THERMALIZATION

We have seen already that redistribution and thermalization of carrier populations happens in conjunction with the transfer of energy from the carriers to the lattice via electron–phonon interactions. Hence, the division between this section and the last is rather artificial. That said, phonon emission by carriers is only the first step in the process of heating the lattice. The phonons emitted by carriers undergoing intervalley scattering all have large wave vectors, whereas intravalley scattering in GaAs mainly produces LO phonons of low q. These phonons do not constitute an equilibrium thermal distribution, which preferentially populates the low energy, low-q acoustic modes. Only when phonons enter these acoustic modes will optical properties approach those of the material when it is heated conventionally. In their time-resolved Raman scattering measurement mentioned earlier, Kash, Tsang, and Hvam observe the decay of LO phonons to acoustic modes to take approximately 4 ps at room temperature (Kash et al., 1985). At 80 K, the decay time is about 7.5 ps because anharmonic coupling terms become less important at low temperatures.

4. CARRIER RECOMBINATION

Carrier–lattice thermalization brings the temperatures of the free carriers and the lattice to the same value. However, the free electrons and holes have Fermi–Dirac distributions about different Fermi levels. The distributions have the same temperature as that of the lattice, but there is still an excess of free carriers compared to the true thermodynamic equilibrium. These carriers can be removed in two ways: recombination of electrons and holes, or diffusion of carriers away from the photoexcited region. Of course, there is no reason why recombination and diffusion cannot occur before carrier–lattice thermalization. It is possible that carriers could recombine faster than they thermalize with the lattice.

When an electron and hole recombine, the excess energy can be removed through several means. First, a photon can be emitted (luminescence): This is *radiative recombination*. Second, another free electron or hole can take up the energy: This is *Auger recombination*. Third, the recombination can occur via defect or surface states. Figure 12 shows these three recombination

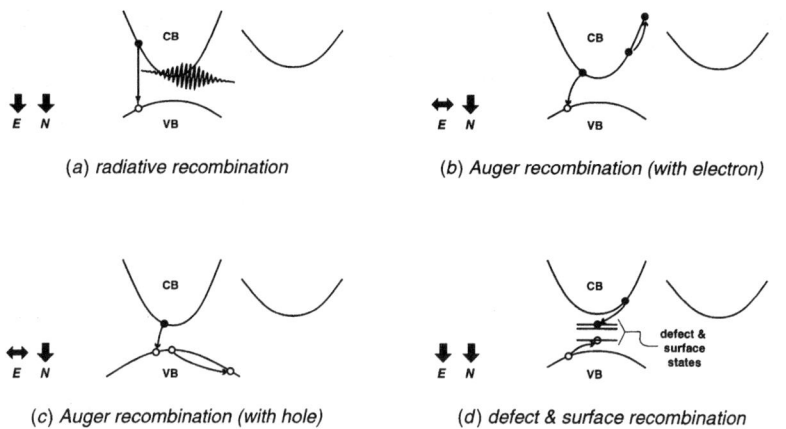

FIG. 12. Mechanisms for carrier recombination in a solid.

mechanisms. All three processes reduce the number of carriers. Radiative and defect recombination both reduce the energy in the carrier system. However, Auger recombination does not affect the total carrier energy; instead, the average energy per carrier increases as the density goes down.

Defect, radiative, and Auger recombination involve one, two, and three carriers, respectively. Hence, their rates depend on the density N, N^2, and N^3, respectively, and the combined recombination rate equation is

$$\frac{dN}{dt} = -C_{\text{recomb:defects}}N - C_{\text{recomb:rad}}N^2 - C_{\text{recomb:Auger}}N^3. \quad (23)$$

Strauss and co-workers have conducted the most careful measurements of recombination in GaAs to date (Strauss et al., 1993). They estimate that $C_{\text{recomb:defects}} < 5 \times 10^7 \text{ s}^{-1}$ (specific to their samples, which had minimal defects), $C_{\text{recomb:rad}} = (1.7 \pm 0.2) \times 10^{-10} \text{ cm}^3 \text{ s}^{-1}$, and $C_{\text{recomb:Auger}} = (7 \pm 4) \times 10^{-30} \text{ cm}^6 \text{ s}^{-1}$. This last number is an overall Auger coefficient averaging over different processes. These estimates imply that radiative recombination dominates for densities below $N \sim 2 \times 10^{19} \text{ cm}^{-3}$; for higher densities, Auger recombination is faster.

Above a density of about 10^{20} cm^{-3}, the recombination coefficients are all expected to decrease with increasing N because of the screening effects of mobile charge carriers. Screening diminishes the Coulomb interaction between free carriers, or between free carriers and defects, and thus decreases the probability of recombination events. Thus, at high carrier density, all three recombination rates do not continue to increase with density as much as predicted in Eq. (23). However, it is difficult to estimate which rate is affected most. One calculation based on static screening theory for Si suggests that the Auger lifetime saturates at about 6 ps for carrier densities above 10^{21} cm^{-3} (Yoffa, 1980). Nonetheless, it is reasonable to expect Auger recombination to be the dominant recombination mechanism at densities of 10^{20} cm^{-3} and higher.

5. Carrier Diffusion

In contrast to carrier recombination, diffusion does not actually decrease the number of free carriers in the material. It does, however, remove them in real space from the region of the sample in which they were originally excited, as shown schematically in Fig. 13. Thus, because the probe beam normally interrogates a subset of the photoexcited volume, diffusion reduces the observed carrier density and has a similar effect on optical properties as carrier recombination.

Diffusion is, by definition, governed by the equation

$$\frac{\delta}{\delta t} N(\mathbf{x}, t) = -D_N \nabla^2 N(\mathbf{x}, t) \qquad (24)$$

where D_N is the diffusion constant. For optically excited carriers, one uses an ambipolar diffusion constant that takes account of both electrons and holes (Seeger, 1991). The extent of diffusion after a given time is given roughly by

$$\text{diffusion distance} \sim \sqrt{D_N[\text{diffusion time}]}. \qquad (25)$$

Using this relation, the time for carriers to diffuse outside the photoexcited region in GaAs is about 7 ps for 635-nm light, whose absorption depth is about 270 nm (Palik, 1985). However, the rough formula can produce a poor estimate for diffusion over short distances if the carrier distribution is not gently sloped.

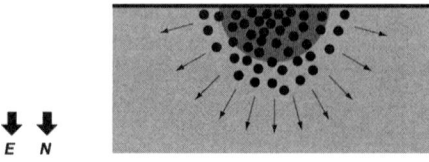

Fig. 13. Diffusion of excited carriers away from the laser-excited region (darkly shaded region).

As for other processes, the initial estimates for diffusion are only valid for weak laser excitation. Diffusion is affected by carrier temperature and by the band structure, both of which are altered greatly by strong excitations. First, $D_N = k_B T_e \mu(T_e)/e$ increases with electron temperature T_e, where $\mu(T_e)$ is a reduced mobility for electrons and holes (Seeger, 1991; Yoffa, 1980). Therefore, the diffusion time decreases as the T_e increases. On the other, the bandgap in a highly excited semiconductor is reduced by BGR close to the surface, which attracts carriers to the surface. This process, called *carrier confinement*, works against normal diffusion (van Driel, 1987).

6. STRUCTURAL EFFECTS

After the free carriers and the lattice come to an equilibrium temperature, and after the free carriers in excess of those excited by thermal effects have dissipated because of recombination, the material is essentially the same as if it had been heated in an oven. However, the laser pulse achieves this heating within a spectacularly short time — tens or hundreds of picoseconds. If the lattice temperature is higher than the melting (boiling) point, then *melting (vaporization)* can occur. However, this cannot happen instantaneously. The semiconductor is superheated, but remains a solid, until regions of liquid or gas nucleate. Starting from nucleation sites, at the surface, the liquid and/or gas phase expands into the material. Figure 14a schematically illustrates melting and evaporation, showing the expansion of the melt front that separates the liquid and solid phases.

The molten phase does not last indefinitely because of *thermal diffusion*. The laser only heats a region governed by the spot size and the absorption depth of the light. Surrounding parts of the material remain at room temperature, and a thermal gradient is formed. Heat diffuses from regions of high temperature to regions of low temperature, as depicted in Fig. 14b. For GaAs, the thermal diffusion constant is about $0.3\,\text{cm}^2\,\text{s}^{-1}$. For an absorption depth of 270 nm (635-nm pump pulse) thermal diffusion takes around 2 ns.

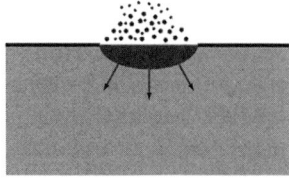
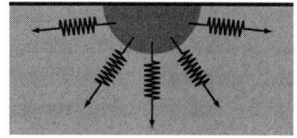

(a) melting, vaporization & ablation   (b) thermal diffusion

FIG. 14. Structural and thermal changes in a laser-excited solid.

Thermal diffusion cools the material in the photoexcited region. If no phase transition has occurred, this will simply result in a decrease of the temperature back to the ambient value. If liquification or vaporization has taken place, then *resolidification* or *condensation* will ensue as the temperature falls below the melting or boiling temperatures, respectively. Note, however, that resolidification does not guarantee a return to the structure that existed before the laser excitation. On a macroscopic scale, a crater forms if material is ablated or evaporated, because some of the ejected material does not recondense. In addition, on a microscopic scale, the solid formed on cooling may be in a different phase than the original material. For example, crystalline-to-amorphous transitions are common (Glezer *et al.*, 1995b; Solís *et al.*, 1994).

Observation of the onset of melting and evaporation requires femtosecond pulses. When nanosecond laser pulses are used to melt a solid, melting already starts during the laser pulse (Solís *et al.*, 1994). The best experimental observations of rapid melting and ablation utilize the technique of *femtosecond microscopy* (Downer *et al.*, 1985; von der Linde *et al.*, 1997; Sokolowski-Tinten *et al.*, 1998a). First introduced by Downer *et al.*, this technique involves taking images of the sample (including the pumped area) with the probe beam for different pump-probe delays. Essentially, one conducts a transient reflectivity measurement for a whole set of points across the pumped region. The series of pictures enables one to see how the whole pumped area responds to the excitation.

Several key observations are made with this technique. First, a high-reflectivity phase appears within about 1 ps after excitation of a semiconductor. Both Downer and Sokolowski-Tinten attribute this to a molten phase, expected to be metallic for most semiconductors. Furthermore, in GaAs, time-resolved second-harmonic generation shows that the high-reflectivity phase is disordered. The second-harmonic signal is non-zero in bulk GaAs because it lacks a center of symmetry. However, this signal drops to zero within a few hundred femtoseconds, indicating that the material has been disordered on the length scale of the light's wavelength (hundreds of nanometers) (Tom *et al.*, 1988; Saeta *et al.*, 1991; Glezer *et al.*, 1995b). Second, ablation starts within 5–10 ps of the excitation (Downer *et al.*, 1985; Saeta, 1991; von der Linde *et al.*, 1997). In semiconductors, the pump pulse energy must exceed one threshold for melting and a higher threshold for ablation. Finally, femtosecond microscopy experiments suggest that resolidification starts to take place starting about 5 ns after the excitation (von der Linde *et al.*, 1997). Above the ablation threshold, some material is lost and resolidification leaves a crater.

There is an ongoing debate as to whether the "melting" and ablation just described are the result of thermal or nonthermal processes. That is, has the material reached a well-defined thermodynamic state before the "melting" or ablation occurs? Are the "melting" and ablation transitions really thermodynamic phase changes, resulting in liquid and gas phases also recognizable as thermodynamic states? Indeed, there is debate as to whether the high-reflectivity, disordered phase observed in strongly excited semiconductors can be called a liquid phase. Certainly, structural changes do not have to await thermal equilibration of the carriers and lattice if the excitation is sufficiently intense. For instance, several workers propose a nonthermal mechanism for the high-reflectivity, disordered phase (sometimes called a "molten" phase) observed in GaAs a few picoseconds after excitation (van Vechten et al., 1979; Stampfli and Bennemann, 1990, 1992, 1994; Silvestrelli et al., 1996; Graves and Allen, 1998; Dumitrică and Allen, 1999). The contribution of Allen and co-workers to this volume describes in detail the proposed mechanism and their molecular dynamics simulations of nonthermal melting in GaAs. Briefly, the mechanism is this: The laser excites bonding, valence electrons to antibonding, conduction states, and thus weakens the bonds between atoms. This causes the atoms to move to new positions and the material disorders in the process.

Although the high-reflectivity, disordered phase is likely the result of a nonthermal transition, ablation at 5–10 ps may or may not be a thermal process. There is possibly enough time for equilibration of the carrier and lattice temperatures (via optical phonon emission by the carriers and subsequent decay of optical phonons to acoustic modes). Rapid Auger recombination could reduce the carrier density to the thermal level, but that is also not clear. Perhaps the biggest question is whether several picoseconds is sufficient for density changes to occur in a material. Such density changes are necessary if one is to describe ablation as a thermodynamic phase transformation of the material. Time-resolved X-ray measurements, early examples of which have recently been reported (Larsson et al., 1998; Rose-Petruck et al., 1999; Chin et al., 1999), can track the spacing between atoms and offer the hope of observing when density changes occur in a material.

Meanwhile, arguments over what is a thermal process and what must be nonthermal will no doubt continue. We have more to say about nonthermal phase transitions later. Such transformations are the most remarkable physical phenomena that take place for the extremely strong laser excitations considered in this work.

7. Summary

Figure 15 displays a range of characteristic times for the various processes, intended to reflect the likely range of times as the excitation strength varies. The excitation strength can affect characteristic times in a number of

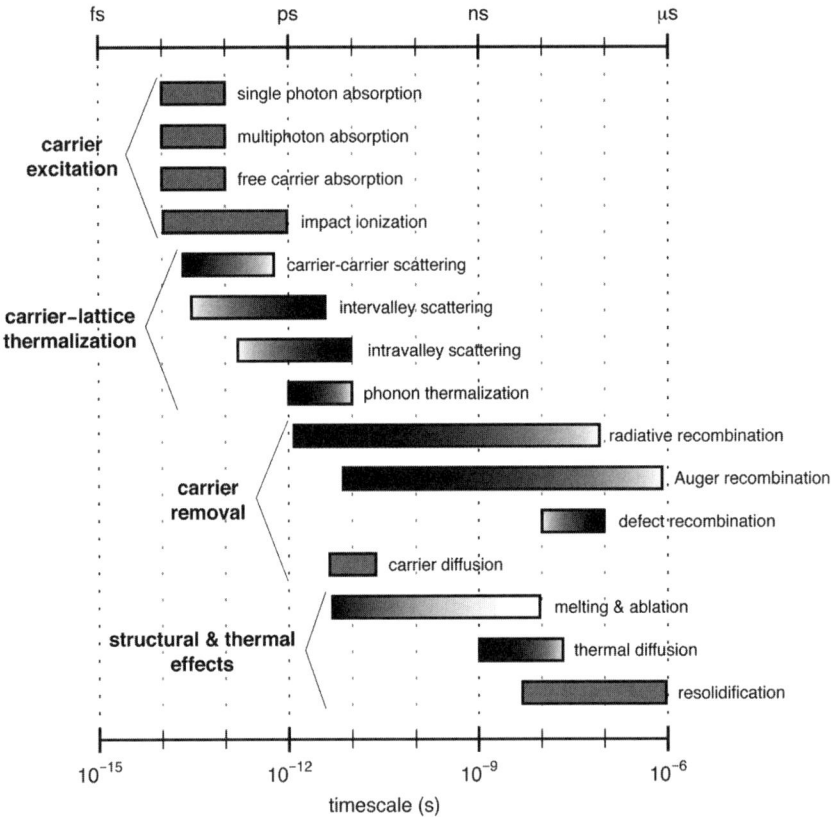

Fig. 15. Characteristic time scales for various electron and lattice dynamical processes in semiconductors, including mechanisms for excitation, carrier–lattice thermalization, carrier removal, and structural and thermal changes. Each bar represents a rough range of characteristic times that is expected over a carrier density range of about 10^{17} to 10^{22} cm^{-3}. Where there is a known dependence of the characteristic time of a process on carrier density, this is indicated by a gradient within the bar. The dark end indicates the characteristic time for the process at high densities while the light end represents low density. In some cases, there is no dependence on carrier density (e.g., excitation) or the dependence is not well known (e.g., diffusion). The bars for such processes are filled with a constant gray.

ways: through the carrier density, through the electronic and lattice temperature, and through changes in the band structure. We consider excitation strengths that generate carrier densities ranging from about 10^{17} to 10^{22} cm^{-3}. The experiments described in this section provide values for the time scales at the low end of this excited density range, and theoretical predictions are used to estimate lifetimes at higher carrier densities. Of course, one of the purposes of this contribution is to present new experimental evidence on the dominant processes at carrier densities in the range 10^{21}–10^{22} cm^{-3}.

In general, excitation takes places within or soon after the laser pulse, which is about 70 fs long in our experiments. Intervalley scattering rapidly removes carriers from the central valley if such scattering is energetically permitted, while carrier–carrier scattering helps to thermalize the electron populations in each valley. Intravalley scattering, which occurs simultaneously with intervalley and carrier–carrier scattering, takes energy away from the electron system on average, transferring it to LO phonons. Subsequently, these phonons relax to acoustic modes and the lattice moves to thermal equilibrium. Recombination and diffusion remove free carriers from the excited volume. At low densities, they typically occur after carrier redistribution; however, Auger recombination can in theory occur faster than carrier scattering at densities above about 10^{21} cm^{-3}. Similarly, structural effects are generally slow for weak excitation, but nonthermal structural changes such as disordering and ablation can take place within a few picoseconds for strong excitations.

We now know what processes contribute to electron and lattice dynamics in solids. We have some idea about their relative time scales, and rough predictions of lifetimes in our experimental density range of 10^{21} to 10^{22} cm^{-3}. Our experiments test some of these predictions, by figuring out the dominant mechanisms in the laser-induced dynamics and characteristic times for the dominant processes. Before presenting our results, we must describe our experimental methods.

IV. Measuring the Time-Resolved Dielectric Function

In order to measure the time-resolved dielectric function of a material, we measure the reflectivity at two angles of incidence for each time delay between the pump and probe pulses. Two reflectivities uniquely determine the real and imaginary parts of the dielectric function, by inversion of the Fresnel reflectivity formulas. In order to obtain the spectral dielectric function, we must measure reflectivity spectra with a broadband probe. Thus, this section describes the three key steps in our experimental technique:

1. Generate broadband (1.5–3.5 eV) probe pulses
2. Measure transient reflectivity spectra at two angles of incidence
3. Derive the transient dielectric function by numerically inverting the Fresnel formulas

1. WHITE-LIGHT PROBE GENERATION

The pump and probe pulses are both generated from a single pulse produced by a colliding pulse mode-locked (CPM) dye oscillator. A beamsplitter divides this pulse into two pulses. We amplify one pulse to generate a 635-nm, 250-μJ pump pulse. The other pulse is amplified to about 20 μJ, and then focused into a 2-mm-thick CaF_2 crystal. This causes white-light generation, which broadens the spectrum of the pulse to one ranging from 1.5 eV (850 nm) to 3.5 eV (350 nm). We choose CaF_2 because it gives the broadest continuum of any solid apart from LiF, which is hygroscopic and therefore difficult to work with (Brodeur and Chin, 1999).

Dispersion of the white-light probe in the CaF_2 crystal and other optical elements before the sample stretch the pulse to several hundreds of femtoseconds. Different frequencies in the broadband probe pulse arrive at the sample at different times. Rather than compressing the probe pulse and losing energy, we measure the temporal dispersion using sum-frequency generation of the probe with the pump pulse in a thin nonlinear BBO crystal at the sample location. The results we present in this chapter are corrected for the measured dispersion by temporally shifting the data at each frequency. More experimental details concerning our white-light probe are given elsewhere (Callan et al., 1998).

2. MEASURING TRANSIENT REFLECTIVITY SPECTRA

Figure 16 shows the experimental apparatus for measuring reflectivity spectra. Because reflectivity is the ratio between reflected and incident

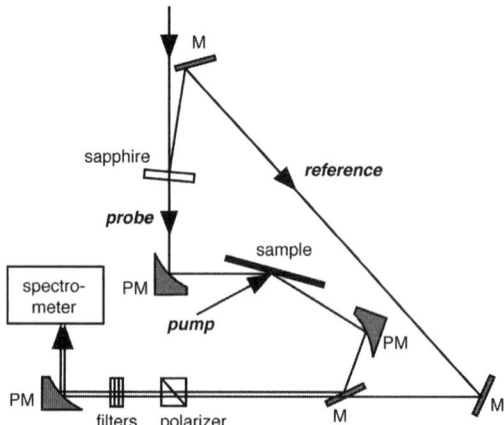

FIG. 16. Experimental setup for measuring transient reflectivity spectra. M, flat mirror; PM, off-axis paraboloid mirror.

beams, we take a reference spectrum by splitting off about 15% of the incident probe pulse with a 1-mm-thick, single crystal sapphire disk. Off-axis parabolic mirrors are used to focus the probe pulses to a spot within the pumped region of the sample, and to collimate the reflected light. Meanwhile, the reference beam is guided around the sample and directed nearly collinearly to, but slightly above, the probe beam reflected from the sample.

Both beams pass through a Glan–Taylor polarizer and a set of filters, namely two Schott glass filters (BG1 and BG24A) and a mixture of organic dyes. The polarizer selects s or p polarization for study. The filters flatten the probe spectrum. White-light generation produces a spectrum that is strongly peaked at the frequency of the seed pulse, but the filters reduce the dynamic range between high and low parts of the spectrum to accommodate the dynamic range of the CCD detector in the spectrometer. Note that dispersion due to elements such as these after the sample does not affect the temporal resolution of the experiment; only dispersion before the sample is important.

Finally, we focus the two beams onto the entrance slit of an imaging spectrometer, displaced slightly from each other in the vertical direction. A CCD camera is placed at the output to record the spectra of reflected and reference pulses. The spectrometer uses a prism rather than a grating because the white light spans more than a factor of two in wavelength.

The reflectivity spectrum is defined as the ratio of reflected spectrum to the incident spectrum. Because we measure a reference spectrum rather that the incident spectrum itself for each probe pulse, we must calibrate our system to find the reflectivity from the ratio of reflected to reference spectra.

3. Extracting the Transient Dielectric Function

In order to find the transient dielectric function $\varepsilon(\omega)$, we must conduct transient reflectivity experiments for two different angles of incidence. Thus, for each pump beam fluence, for each time delay between pump and probe, and for each frequency in the probe spectrum, we have values for the sample reflectivity at two known angles of incidence. This provides us with two Fresnel reflectivity equations [cf. Eqs. (6) and (3)]. We use a three-layer model to account for the oxide layer on GaAs, for which $\varepsilon = 4$ (Potter, 1969) and which we estimate to be about 3.5 nm thick. In our two reflectivity equations, there are two unknowns, namely the real and imaginary parts of the dielectric function at each frequency, pump fluence, and time delay. The solution cannot, however, be derived analytically. We use a numerical optimization algorithm based on the downhill simplex method (Nelder and Mead, 1965; Press et al., 1988). Figure 17 shows the measured reflectivity spectra of unexcited c-GaAs for p-polarized light at angles of incidence of 58° and 76°, together with the spectra for the real and imaginary parts of

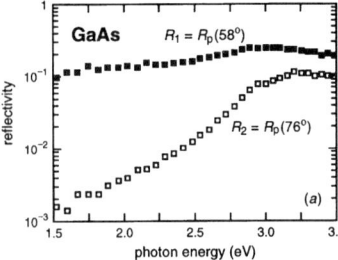 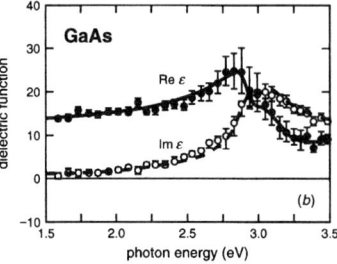

FIG. 17. Finding the dielectric function for reflectivity pairs. (a) Measured reflectivity spectra for unexcited c-GaAs for p-polarized light at angles of incidence of 58° (filled squares) and 76° (open squares). (b) Dielectric function spectrum for unexcited c-GaAs calculated from the reflectivity pairs using our numerical optimization algorithm to invert the Fresnel formulas. Filled circles, $\text{Re}[\varepsilon(\omega)]$; open circles, $\text{Im}[\varepsilon(\omega)]$. The curves show previous measurements of $\varepsilon(\omega)$ (solid line, $\text{Re}[\varepsilon(\omega)]$; dashed line, $\text{Im}[\varepsilon(\omega)]$) of c-GaAs at room temperature using ellisometry with continuous wave light (Palik, 1985).

the dielectric function calculated using the inversion algorithm. We used a (100) GaAs sample (Cr doped, $\rho > 7 \times 10^5, \Omega\,\text{m}$).

The error bars on the dielectric function values in Fig. 17b are calculated by assuming a 5% error in each of the reflectivities. For a given reflectivity pair, (R_1, R_2), the extrema of the error bars are given by the maximum and minimum values of $\text{Re}[\varepsilon]$ and $\text{Im}[\varepsilon]$ calculated from the four reflectivity pairs $(R_1-5\%, R_2-5\%)$, $(R_1-5\%, R_2+5\%)$, $(R_1+5\%, R_2-5\%)$ and $(R_1+5\%, R_2+5\%)$. As the figure indicates, the error varies greatly with photon energy. The reason for this is that the sensitivity of the reflectivity R to changes in the dielectric function ε varies as a function of ε itself. The sensitivity of the R to changes in ε is also dependent on the angle of incidence and the polarization of the light. At least one reflectivity must be for p-polarized light and an angle of incidence close to or above the Brewster angle in order to obtain a precise measurement of ε.

The curves in Fig. 17b are the real and imaginary parts of $\varepsilon(\omega)$ of GaAs derived from ellipsometry measurements with continuous light (Palik, 1985). The good agreement of the measured $\varepsilon(\omega)$ with previous results demonstrates that our technique gives an accurate measurement of the dielectric function.

V. Time-Resolved Dielectric Function of Highly Excited GaAs

We now present time-resolved measurements of the dielectric function of GaAs across the spectral range from the near infrared to the near ultraviolet after the sample is excited by an intense femtosecond laser pulse. The data, as promised, reveal much of the nature of the changes in the material following excitation. In the next section, we analyze the dielectric function data to learn more about electron and lattice dynamics. Parts of the data and analysis presented here has been previously published (Huang et al., 1998).

The response of GaAs to the femtosecond excitation depends strongly on the incident fluence, or energy per unit area, F. Permanent damage to the sample is observed under an optical microscope when the fluence exceeds a threshold fluence of $F_{th} = 1.0$ kJ/m^2. Below this threshold, changes in the optical properties of the material are completely reversible. However, to avoid any possible cumulative effects from repeated excitation of the same spot, we move the sample after every shot. Each shot excites and probes a fresh region of the sample. The behavior on picosecond timescales falls into three regimes depending on fluence.

1. Low Fluence Regime ($F < 0.5$ kJ/m^2)

Figure 18 shows the evolution of the dielectric function of GaAs after excitation by a pulse of 0.32 kJ/m^2, representative of the low fluence regime. In addition to the data from the broadband probe, the figures include earlier measurements of the dielectric function conducted at 4.4 eV (Glezer et al., 1995a). The solid and dashed curves in each of the panels represent the real and imaginary parts of the dielectric function of GaAs at room temperature (Palik, 1985). For the first 500 fs after excitation, the data show a decrease in Re[ε] while the peak in Im[ε] falls by about 15–20% in amplitude and broadens slightly. The dielectric function partly recovers after about 1 ps. But the data do not indicate a return to the original state of the material. Instead, the data at 4 and 16 ps after excitation agree well with ellipsometric measurements of the dielectric function of GaAs at 473 K and 732 K, respectively (Yao et al., 1991). See Figs. 19e and 19f. For 2 ps and earlier, there is no temperature for which the measurements of Yao et al. agree well with our experimental data. The dielectric function of laser-excited GaAs on this time scale still has contributions due to the excited free carriers and is not solely altered by lattice heating.

The lattice heating after 4 ps indicated by Fig. 19 is typical of the low fluence regime. Figure 20 shows the lattice temperature vs. time delay for three different excitation fluences, obtained from fits such as those in Figs. 19e and 19f. As might be expected, higher excitation fluences lead to higher lattice temperatures. (We cannot estimate the temperature at 8 and 16 ps after an excitation of 0.45 F_{th} because measurements of the dielectric function of GaAs have only been reported up to 884 K.) Exponential fits to the data at 0.20 F_{th} and 0.32 F_{th} yield a rise time of about 7 ps. These exponential fits can be used to predict values for the lattice

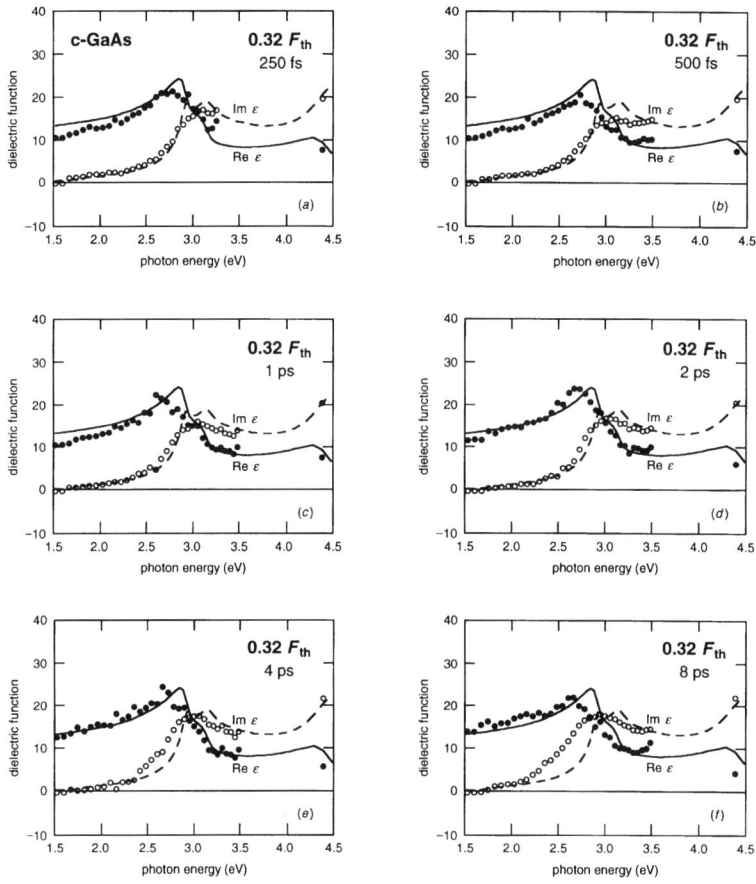

FIG. 18. Evolution of the dielectric function of crystalline GaAs (filled circles, $\text{Re}[\varepsilon(\omega)]$; open circles, $\text{Im}[\varepsilon(\omega)]$) after excitation by pulses of about $0.32\,F_{\text{th}}$. The dielectric function of c-GaAs at room temperature (Palik, 1985) is shown for comparison in all panels (solid curves, $\text{Re}[\varepsilon(\omega)]$; dashed curves, $\text{Im}[\varepsilon(\omega)]$).

temperature at times less than 4 ps, when electronic effects also contribute to the dielectric function. In each of the panels (a)–(d) of Fig. 19, the curves show the published dielectric function from the literature (Yao et al., 1991) for GaAs at a lattice temperature close to that predicted by the exponential fit for $0.32\,F_{\text{th}}$ in Fig. 20. The difference between the data and the curves gives the change in dielectric function due to electronic effects. We make use of these difference spectra later to elucidate more details on the electron dynamics.

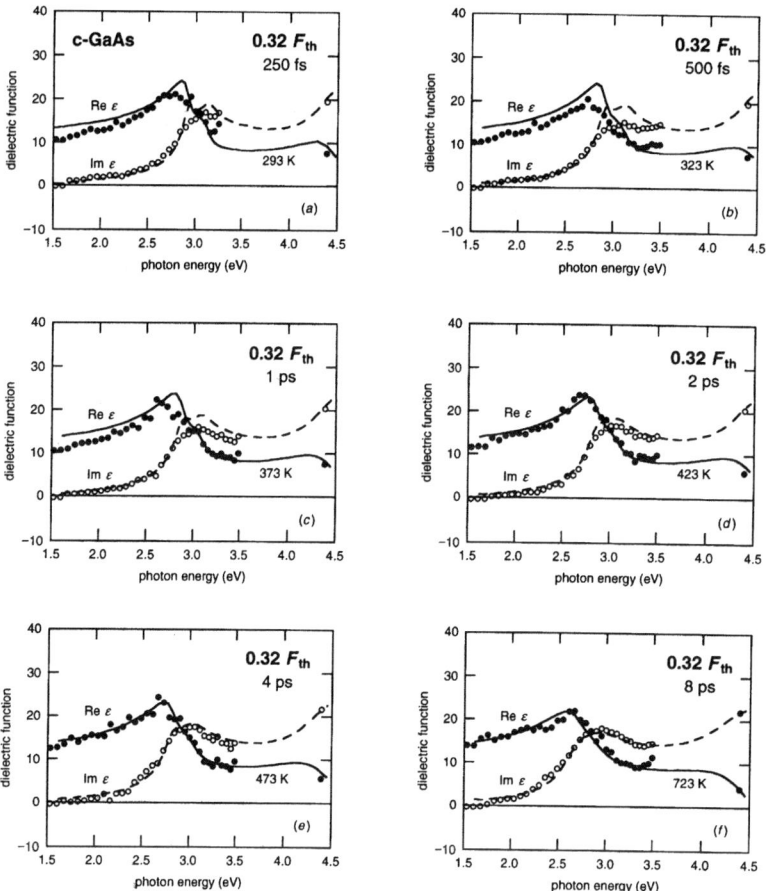

FIG. 19. Evolution of the dielectric function of crystalline GaAs (filled circles, Re[$\varepsilon(\omega)$]; open circles, Im[$\varepsilon(\omega)$]) after excitation by pulses of about $0.32 F_{th}$. The curves show Re[$\varepsilon(\omega)$] (solid line) and Im[$\varepsilon(\omega)$] (dashed line) for c-GaAs heated to various temperatures: (a) 293 K, (b) 323 K, (c) 373 K, (d) 423 K, (e) 473 K and (f) 723 K (Yao et al., 1991). The temperatures are selected by fitting curves to the data (for 4 ps and after) or derived from the exponential curves for lattice temperature versus time delay shown in Fig. 20.

2. MEDIUM FLUENCE REGIME ($F = 0.5$–0.8 kJ/m^2)

The changes in $\varepsilon(\omega)$ are more substantial at $0.70\ F_{th}$ than for excitations in the low fluence range, as shown in Fig. 21. The drop in the real part during the first few hundred femtoseconds is larger, and there is a dramatic increase in the imaginary part for photon energies up to 2.7 eV. Over several picoseconds, we observe the development of a single broad peak in the

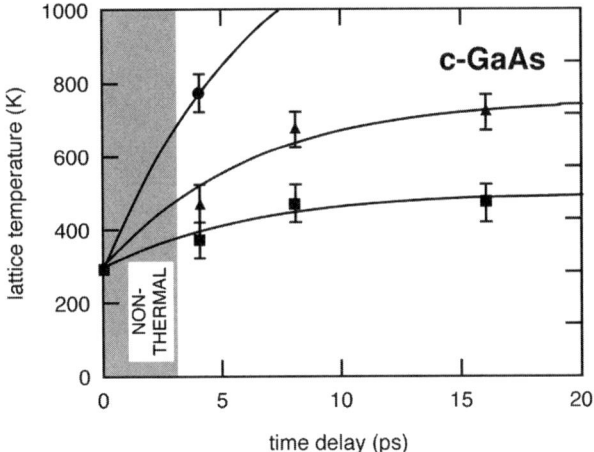

FIG. 20. Laser-induced lattice heating of c-GaAs for three different excitations in the low fluence regime: $0.20\,F_{th}$ (squares), $0.32\,F_{th}$ (triangles), and $0.45\,F_{th}$ (circles). The fitted exponential curves yield a common rise time of about 7 ps.

imaginary part and its center shifted to lower energy. After 4 ps, for instance, the peak is at about 2.8 eV (compared with 3.1 eV initially) and its full width at half maximum (FWHM) is over 2.8 eV (compared with about 0.7 eV initially). Meanwhile, the real part develops a smooth negative slope across most of the frequency range.

We cannot fit the data at 4 and 16 ps with measurements of $\varepsilon(\omega)$ for heated GaAs. However, such measurements have only been reported up to 884 K. As the temperature increases, the E_1 peak broadens and shifts to lower energy. Thus, it is possible that the material is superheated well above the melting point at 1510 K (Blakemore, 1982). Nonetheless, given that $\varepsilon(\omega)$ data for GaAs exist only for temperatures ranging from room temperature to 884 K, it is unreasonable to extrapolate to 1500 K or beyond and make comparisons with our measured $\varepsilon(\omega)$ following laser excitation.

On the other hand, the dielectric function at these times is similar to that for amorphous GaAs at room temperature (Erman et al., 1984), as demonstrated in Figs. 21d–f. The data is shifted to lower energies compared with the curves. However, this shift is to be expected because the laser-excited material is heated well above room temperature, and peaks in the dielectric-function move to lower energies as the temperature rises. Furthermore, not all amorphous phases are the same, and they may have slightly different dielectric functions. The curves show the dielectric function for amorphous GaAs prepared by ion implantation of As into crystalline GaAs. The phase generated by laser excitation of GaAs is not the same; for one thing, laser

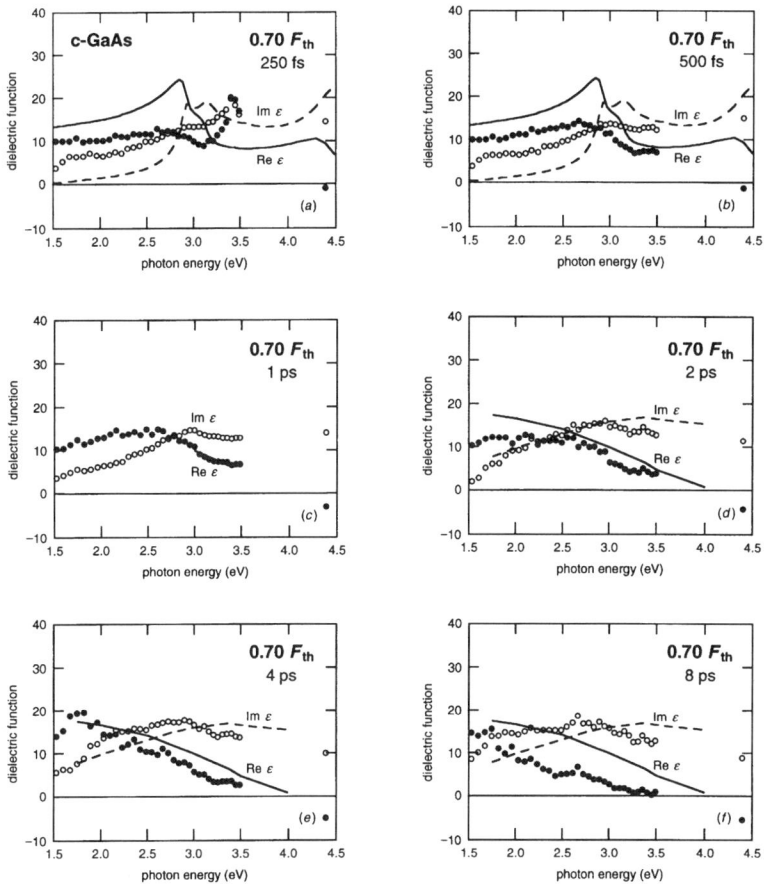

FIG. 21. Evolution of the dielectric function of crystalline GaAs (filled circles, $\text{Re}[\varepsilon(\omega)]$; open circles, $\text{Im}[\varepsilon(\omega)]$) after excitation by pulses of about $0.70\,F_{\text{th}}$. The curves in (a) and (b) show $\text{Re}[\varepsilon(\omega)]$ (solid line) and $\text{Im}[\varepsilon(\omega)]$ (dashed line) for c-GaAs at room temperature (Palik, 1985). The curves in (d), (e), and (f) represent the dielectric function of amorphous GaAs at room temperature (Erman et al., 1984).

excitation maintains the 1:1 stoichiometry of the sample, at least until the onset of ablation.

Sokolowski-Tinten et al. propose another model to explain the observed optical properties (Sokolowski-Tinten et al., 1991). Based on single-angle, single-frequency reflectivity measurements, they posit, like us, that there are three regimes of behavior with increasing fluence. However, they suggest that a thin liquid layer forms in the medium fluence range. They fit the reflectivity data with a model based on a thin layer of liquid GaAs (1-GaAs) formed at the surface above a substrate of bulk c-GaAs at room temperature. Assuming that liquid GaAs is metallic, as is the case for other liquid semiconductors such as silicon and germanium, they use a Drude model for its dielectric function. The best fit to the reflectivity data occurs for a plasma frequency of 16.4 eV, corresponding to a free carrier density of 1.95×10^{23} cm^{-3}, and a damping time that is slightly frequency dependent, varying from 0.1 fs at 1064 nm to 0.13 fs at 532 nm (Sokolowski-Tinten, 1997). However, no simple Drude model produces a very good fit to our spectral reflectivity data for two angles of incidence, as shown in Figure 22. The first panel of the figure shows the reflectivity spectra at 58° and 75° obtained 4 ps after an excitation of $0.70\,F_{th}$, together with the reflectivities predicted from the dielectric function of amorphous GaAs with a 3.5-nm-thick oxide layer. The shape of the reflectivity spectra for room-temperature a-GaAs match the data quite well but the spectra are shifted to higher energies than the data, in accordance with the dielectric function data.

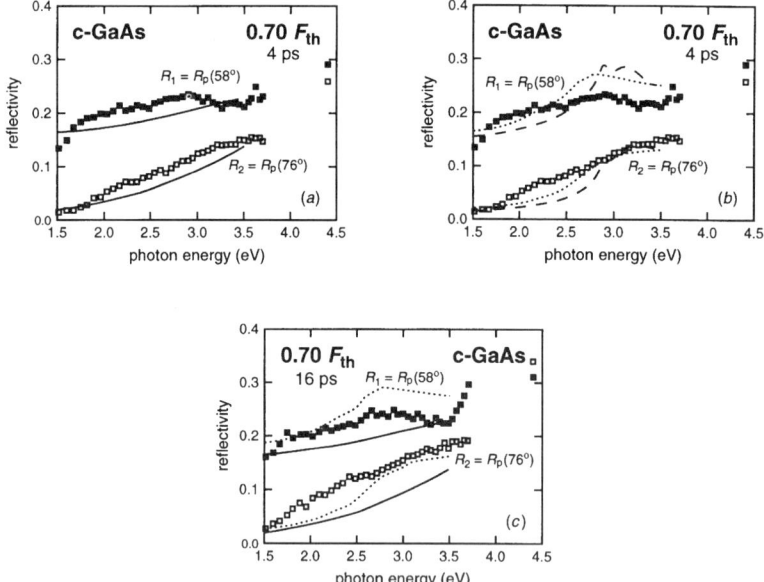

FIG. 22. Comparison with various models of the reflectivity spectra at 58° and 75° of crystalline GaAs following an excitation of $0.70\,F_{th}$. (a) Data at 4 ps compared with the calculated reflectivities for amorphous GaAs using the room temperature dielectric function data of Erman et al. (b) Data at 4 ps compared with models assuming a metallic liquid layer of thickness = 3.0 nm, $\omega_p = 12$ eV, and $\tau = 0.12$ fs. The dashed curves show the reflectivities at 58° and 75° assuming that the substrate is at room temperature. The dotted curves arise by considering a substrate heated to 884 K (Yao et al., 1991). (c) Data at 16 ps compared with amorphous GaAs (solid curves) and a model with a metallic liquid layer of thickness = 4.5 nm, $\omega_p = 12$ eV, and $\tau = 0.12$ fs (dashed curves).

Figure 22b compares the data with liquid-layer models. The first model considers a liquid layer on top of room-temperature GaAs, and the best fit is obtained for a 3-nm layer with $\omega_p = 12$ eV and $\tau = 0.12$ fs. The fits improve somewhat when the material underneath the liquid layer is heated: The dashed-dotted and simple dotted curves are for a 3-nm layer of liquid with $\omega_p = 12$ eV and $\tau = 0.12$ fs sitting on top of GaAs at 884 K. The dielectric function for GaAs at 884 K comes from Yao and co-workers (Yao et al., 1991). Finally, Fig. 22c shows the reflectivity data 16 ps after an excitation of $0.70 F_{th}$, together with reflectivities expected from room-temperature a-GaAs and from a liquid layer model with a layer of thickness = 4.5 nm, $\omega_p = 12$ eV, and $\tau = 0.12$ fs on a substrate of GaAs at 884 K. Neither model produces spectacular fits. As mentioned before, the amorphous data are at room temperature, and thus the fit could be improved by accounting for the higher temperatures in our experiment. On the other hand, the increase in thickness of the liquid layer from 4 to 16 ps tends to support the notion of Sokolowski-Tinten and co-workers that a liquid layer forms in this fluence regime and propagates into the material over time.

At this point, it is not possible to say which of the three models — superheating, amorphization, or formation of a liquid layer — best explains the observed behavior. Measurements of second-harmonic generation (Tom et al., 1988) and of the second-order susceptibility (Glezer et al., 1995b) indicate that GaAs becomes centrosymmetric (on the length scale of the light) a few picoseconds after excitations in this fluence range. There is no way to form a centrosymmetric crystalline phase, so this implies a loss of ordering on length scales up to 500–1000 nm. However, this result is also consistent with the three models. Even superheating is a possible interpretation, for the amplitude of lattice oscillations in a solid superheated well above the melting temperature could be so large as to make it "look" disordered to the incident light. Time-resolved X-ray diffraction offers the best hope for distinguishing between these three proposed structural phases.

3. HIGH FLUENCE REGIME ($F > 0.8$ kJ/m^2)

Above $0.80\ F_{th}$, the real part of the dielectric function becomes negative within the observed spectral range after a fluence-dependent time delay. The photon energy at which $Re[\varepsilon]$ crosses zero decreases with time delay, as shown in Fig. 23 for a fluence of 1.6 times the permanent damage threshold. Recall from Subsection 4 of Section II that the zero crossing of the real part of the dielectric function is an indicator of the bonding–antibonding gap in GaAs. Thus, Fig. 23a shows that, at $F = 1.60\ F_{th}$, most of the oscillator strength moves from an initial value of 4.75 eV (Palik, 1985) to below 3.2 eV after only 250 fs. By 4 ps, the zero crossing of the real part has shifted to well below 2 eV. The data below 1.8 eV lie in a region of dielectric

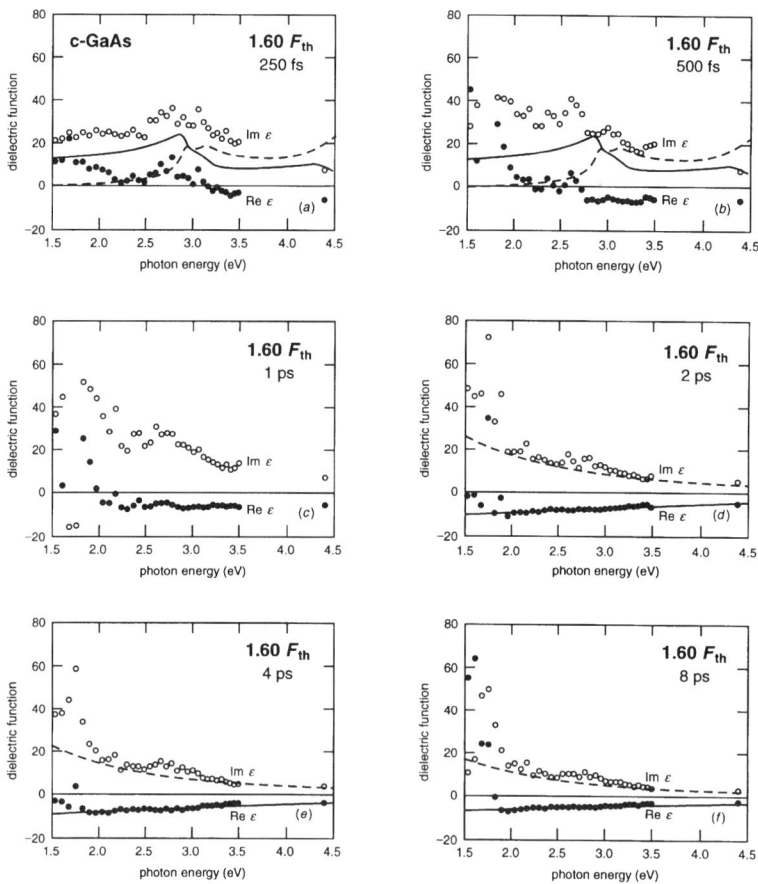

FIG. 23. Evolution of the dielectric function of crystalline GaAs (filled circles, Re[$\varepsilon(\omega)$]; open circles, Im[$\varepsilon(\omega)$]) after excitation by pulses of about $1.60\,F_{th}$. The curves in (a) and (b) show Re[$\varepsilon(\omega)$] (solid line) and Im[$\varepsilon(\omega)$] (dashed line) for c-GaAs at room temperature (Palik, 1985). The curves in (d), (e), and (f) show Drude-model dielectric functions with 0.18-fs relaxation time and plasma frequencies of 13.0, 12.0, and 10.5 eV, respectively.

function space where the reflectivity is not very sensitive to changes in $\varepsilon(\omega)$ and hence the uncertainty in $\varepsilon(\omega)$ is large. The curves in the panels for 2, 4, and 8 ps of Fig. 23 show fits to a Drude model. All have a relaxation time of 0.18 fs, while the plasma frequencies are 13.0, 12.0, and 10.5 eV, respectively. The large plasma frequencies imply, from Eq. (15), that nearly all the valence electrons in GaAs behave as free carriers. This indicates that the bandgap has closed completely and a *semiconductor-to-metal transition* has occurred.

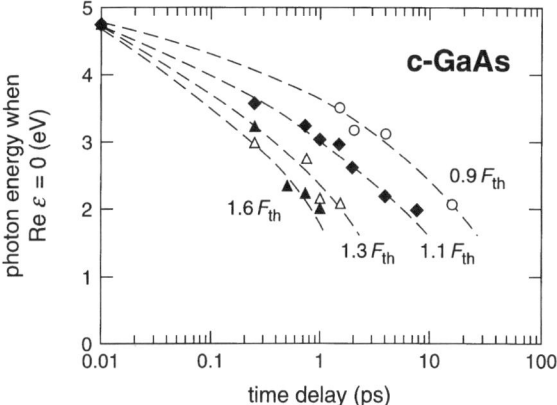

FIG. 24. Photon energy at which Re[ε(ω)] crosses zero as a function of time delay for different excitations in the high fluence regime: $0.9\,F_{th}$ (open circles), $1.1\,F_{th}$ (filled diamonds), $1.3\,F_{th}$ (open triangles), and $1.6\,F_{th}$ (filled triangles). The dashed curves are guides for the eye.

Note that, although the Drude fits are excellent, there is a discernible peak in Figs. 23d–23f centered at 2.75 eV. This peak implies a contribution due to interband transitions of between 2.5 and 3 eV in photon energy, even after the bandgap has collapsed. Recall that the same phenomenon occurs in copper, whose dielectric function is well described by a Drude model except for some interband transitions above 2 eV, transitions that are responsible for the reddish color of copper.

We observe the semiconductor-to-metal transition for fluences above $0.8\,F_{th}$. Figure 24 shows the progress of the zero-crossing of Re[ε] with time at different fluences, indicative of the collapse of the bonding–antibonding gap. Once the zero-crossing photon energy goes below 2 eV, it is difficult to determine where the crossing occurs because the data are often noisy at the edges of the probe spectral bandwidth. Figure 25 displays the plasma frequency as a function of time delay after the gap has collapsed sufficiently to describe the dielectric function well with a Drude model. The bandgap collapse and semiconductor-to-metal transition take place more quickly for higher fluences. Surprisingly, however, the plasma frequency is smaller for higher fluences and also decreases with time delay. We propose some explanations for these observations in Subsection 2 of Section VI.

VI. Carrier and Lattice Dynamics

By identifying the state of the material from its spectral dielectric function, we have described the general nature of the changes in GaAs following

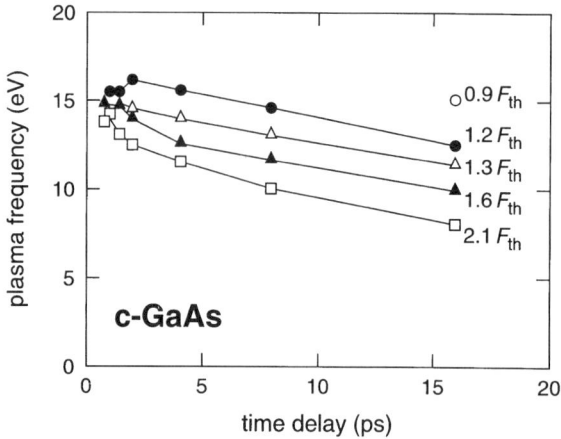

FIG. 25. Plasma frequency as a function of time delay for different excitations in the high fluence regime: $0.9 F_{th}$ (open circles), $1.2 F_{th}$ (filled circles), $1.3 F_{th}$ (open triangles), $1.6 F_{th}$ (filled triangles), and $2.1 F_{th}$ (open squares). The Drude model fits to the dielectric function data that produce these values of ω_p all take the relaxation time to be $\tau = 0.15$ fs. Good fits are possible for relaxation times in the range of 0.12–0.20 fs. The lines are guides for the eye.

excitation. For instance, we can say unambiguously that the lattice heats in the low fluence regime and that high fluences cause a semiconductor-to-metal transition. It is now time to look at the electronic and structural dynamics in more detail. Because the dielectric function is directly linked to the transitions between allowed energy levels in the material, and because the spectra provide data across a huge range of frequencies, we can extract a wealth of specific information on the dynamics. Carrier dynamics — excitation, scattering, and relaxation — dominate during the first picosecond after excitation at all fluences. Structural dynamics become more important as carriers relax, and dominate the changes in dielectric function after several picoseconds.

1. CARRIER DYNAMICS: EXCITATION, SCATTERING, AND RELAXATION

Excited free carriers affect both intra- and interband contributions to the dielectric function. First, the free carriers change the dielectric function by free carrier absorption of the probe, as described by the Drude model. Second, excited carriers fill states in the conduction band and block various interband transitions. Third, screening of the ionic potential due to the free carriers and electronic many-body effects change the band structure (Kim et al., 1994), which in turn affects the dielectric function. These free-carrier effects arise immediately after the excitation and dominate the changes in

$\varepsilon(\omega)$ for up to a picosecond after excitation, before changes in the lattice become important. Later, heating, disordering, and other thermal or nonthermal changes to the lattice alter the dielectric function through changes in the allowed energy states and/or in the effective masses of free carriers.

a. Carrier Dynamics at Low Fluences

Below 0.5 F_{th}, the most significant change in the dielectric function of GaAs during the first picosecond is a decrease in the E_1 peak around 3 eV. This decrease can arise from two sources. First, excited electrons or holes that fill states around the L valleys block absorption resulting from transitions between these states. This is called *band filling* or *Pauli blocking* (cf. Subsection 7 of Section II). Second, free carriers inhibit the formation of excitons by screening the Coulomb interaction between electrons and holes; such excitons make a substantial contribution to the absorption peaks in III–V semiconductors (Zollner et al., 1998; Yu and Cardona, 1996). Free carriers around the L points have the greatest screening effect on L-point excitons because of the $|\mathbf{k}_1 - \mathbf{k}_2|^{-2}$ dependence of the Coulomb potential. Thus, both mechanisms for decreasing the E_1 absorption peak require the presence of electrons (and/or holes) in the conduction (or valence) band states around the L point. There will be few holes at the L points because it is an energy maximum for holes in the valence band. Therefore, the decrease in absorption must be due to electrons in the conduction band L valleys. Because electrons are initally excited only to the Γ valley (via linear absorption of 1.9-eV photons), some of these electrons must scatter into the L valleys. The time scale for this scattering mechanism can be estimated by studying the transient changes in the absorption peak, using $\text{Im}[\varepsilon(3.0\,\text{eV})]$. To isolate the changes due to electrons in the L valleys, we remove the effects of lattice heating. We do this by taking, for each time delay, the difference in $\text{Im}[\varepsilon(3.0\,\text{eV})]$ between the data and the value for GaAs heated to the temperature expected at that time, that is, the data and the curves in Fig. 19. The resulting data for $\Delta\text{Im}[\varepsilon(3.0\,\text{eV})]$ are displayed as a function of time delay in Fig. 26. As expected, the changes are larger for higher fluences, which excite more free electrons. The time for the drop in $\Delta\text{Im}[\varepsilon(3.0\,\text{eV})]$, indicative of the Γ to L scattering time, is of the order of a few hundred femtoseconds for all three fluences.

The free carriers relax through a combination of phonon emission, Auger recombination, radiative recombination, and carrier diffusion, as discussed in Section III. The experimental observations allow us to determine the dominant relaxation mechanisms at carrier densities above $10^{21}\,\text{cm}^{-3}$. Recall that the effects of free carriers on $\varepsilon(\omega)$ subside within 4 ps, whereas the lattice heats in 7 ps (cf. Figs. 19 and 20). These two facts suggest the following model for the carrier relaxation dynamics: Auger recombination

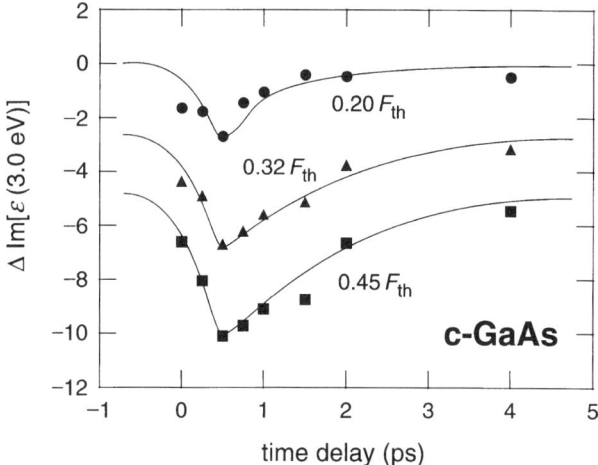

FIG. 26. Temporal evolution of the electronic contribution to Im[$\varepsilon(\omega)$] at 3.0 eV for different fluences in the low fluence range: $0.20\,F_{th}$ (circles), $0.32\,F_{th}$ (triangles), and $0.45\,F_{th}$ (squares). This electronic contribution is the difference between the measured value and the curves in Fig. 19, as described in the text. The curves are guides for the eye, and each data series is shifted down by 2.5 with respect to the previous one.

reduces the carrier density without affecting the total carrier energy. The average energy per carrier increases, so each remaining carrier must emit more phonons to reach the band edge, which increases the energy relaxation time. Thus, the carrier density decreases more rapidly than the total carrier energy, consistent with the observations.

There is an alternative explanation for the observation that the effects of free carriers disappear faster than the lattice heats. Emission of LO phonons by carriers as they relax in the Γ valley is only the first step in lattice heating; these nonequilibrium phonons must also relax to acoustic modes. Therefore, diffusion or some other mechanism could reduce the carrier density faster than the LO phonon decay, and produce the observed phenomena. However, a second piece of evidence casts doubt on this hypothesis and supports the initial explanation of delayed Auger heating.

Table I compares the energy density deposited by the laser in the excited volume with the energy density required for the observed lattice heating to occur throughout the excited region, for three excitations in the low fluence range. The deposited energy density is

$$E_{\text{absorbed}} = F(1 - R)\alpha \qquad (26)$$

where $R = 0.23$ is the pump pulse reflectivity and $\alpha = [270\,\text{nm}]^{-1}$, the

TABLE I

COMPARISON OF DEPOSITED LASER ENERGY WITH OBSERVED LATTICE HEATING IN THE LOW FLUENCE REGIME

Pump fluence	Deposited laser energy	Lattice temp. increase	Energy required for heating
0.20 F_{th}	0.48 kJ cm^{-3}	175 K	0.32 kJ cm^{-3}
0.32 F_{th}	0.80 kJ cm^{-3}	425 K	0.77 kJ cm^{-3}
0.45 F_{th}	1.13 kJ cm^{-3}	>475 K	>0.87 kJ cm^{-3}

inverse of the absorption depth. The heat energy density is

$$E_{\text{heat}} = \rho C_V \Delta T \tag{27}$$

where $\rho \approx 5.3 \text{ g cm}^{-3}$ is the density of GaAs (taken at room temperature), and $C_V \approx 250 \text{ mJ g}^{-1} \text{ K}^{-1}$ is the specific heat capacity at constant volume (Blakemore, 1992). Note that the final lattice temperature (values at 8 ps are used for the calculations in Table I) increases nonlinearly with fluence. Therefore, a greater portion of the absorbed energy goes to lattice heating at higher fluences. In addition, the fraction of absorbed energy going into lattice heating is surprisingly high, because excited electrons are initially only 0.5 eV above the bottom of the conduction band, corresponding only to about 25% of the laser photon energy. The fractions $E_{\text{heat}}/E_{\text{absorbed}}$ are probably overestimated by a factor of 2 or 3 since the laser deposits that much more energy density in the probed region close to the surface compared with the photoexcited region of depth α^{-1} as a whole. Although the energy deposited is higher in the probed region than specified in the table, the proportion of absorbed energy used for lattice heating and the increase in this proportion with fluence both still require explanation.

The model based on Auger recombination provides an explanation, whereas the alternative cannot. Because Auger recombination transfers the full 1.9-eV excitation energy of an electron–hole pair to promote a free electron or hole by 1.9 eV in the conduction or valence band, all of this energy can subsequently be transferred to the lattice. Thus, rapid Auger recombination makes it possible to transfer substantially more than 25% of the energy absorbed from the laser to the lattice. Furthermore, the Auger rate increases with carrier density, accounting for the increasing percentage of the deposited energy used to heat the lattice at higher fluences. If alternative carrier removal mechanisms — radiative recombination and diffusion — were faster than Auger recombination, the fraction of energy used for heating would decrease at higher fluences, as the radiative recombination or diffusion rates increased with carrier density.

The delayed Auger heating model of carrier relaxation is consistent with other experiments (Downer and Shank, 1986) and with theoretical understanding. As described in Section III, theory predicts that Auger recombination should dominate radiative recombination, and that diffusion should be limited by carrier confinement for carrier densities in our experimental range. Furthermore, the experiments of Kash et al. described in Subsection 2 of Section III show that the rate of decay of LO phonons to acoustic modes increases with increasing temperature (Kash et al., 1985). Thus, the decay time should be much shorter than 4 ps for our experimental conditions. Hence, a lattice heating time of 7 ps could not be explained by the delay caused by phonon decay. Our data suggest an Auger recombination rate of a few picoseconds at densities of $1-5 \times 10^{21}$ cm^{-3}. This is two orders of magnitude slower than the Auger coefficient at low carrier densities would predict, but is in reasonable agreement with Yoffa's predictions taking account of screening (Yoffa, 1980).

In summary, we can identify three dominant contributors to the scattering and relaxation of carriers. Intervalley scattering moves carriers to the side valleys within a few hundred femtoseconds after excitation. Energy relaxation and recombination cannot be separated in time. Instead, rapid Auger recombination decreases the carrier density faster than phonon emission removes energy and transfers it to the lattice.

b. Carrier Dynamics at Medium and High Fluences

As the fluence increases above $0.5 F_{th}$ and the excited carrier density exceeds 5×10^{21} cm^{-3}, the initial changes in the dielectric function do not match the strength and spectral shape of the Drude model (see Subsection 5 of Section II. The observed changes can, however, be explained if we consider many-body effects in addition to a Drude contribution to $\varepsilon(\omega)$. Kim et al. calculated the change in band structure in GaAs caused by a high density of free carriers through two many-body terms (see Subsection 7 of Section II): (1) screening of the ionic potential, and (2) self-energy corrections arising from exchange-correlation effects (Kim et al., 1994). Both of these terms broaden bands and reduce energy gaps, consistent with the broadening and slight downward shift of the E_1 peak observed within the first 500 fs after an excitation of $0.7 F_{th}$. The calculation of Kim and co-workers predicts large decreases in bandgap with increasing carrier density. At an excited carrier density of about 2×10^{22} cm^{-3}, or 10% of the valence electrons, they expect the L valley minimum to fall below the valence band maximum at Γ, closing the gap completely. The drastic decrease in the gaps at L and X they predict are consistent with our observations above $0.8 F_{th}$, which show the bonding–antibonding gap falling from 4.75 eV to within our spectral range of 1.5–4.5 eV.

Unfortunately, the subpicosecond alteration of the band structure at the highest fluences makes it impossible to deduce anything more about the electron and hole dynamics without detailed modeling. We cannot tell what changes in $\varepsilon(\omega)$ are due to carrier scattering or recombination or lattice heating when band structure change is also occuring. Of course, significant phonon emission and lattice heating do occur, and the Auger recombination rate is still high, although screening theory predicts saturation of the Auger rate (Yoffa, 1980).

2. STRUCTURAL DYNAMICS: LATTICE HEATING, DISORDERING, AND PHASE TRANSITIONS

After the first few picoseconds, structural changes dominate the changes in the dielectric function. At low fluences, the sample merely heats and cools. Structural changes are more dramatic at fluences above $0.5 F_{th}$, where there is evidence for disordering and melting. If the excitation is sufficiently strong, ablation occurs after 5–10 ps (Saeta, 1991; von der Linde et al., 1997), but that is at the outer limit of our time scale of interest. There are two key questions to answer about structural dynamics. First, does the material melt, evaporate, or undergo a solid-to-solid disordering transition? Second, are the structural changes thermal or nonthermal? That is, does the phase change occur under conditions of thermodynamic equilibrium, from one thermodynamic state to another? In an attempt to answer these questions, we look more closely at the behavior of $\varepsilon(\omega)$ for GaAs on many picosecond time scales for different fluences.

a. Structural Dynamics at Low Fluences

The structural dynamics at low fluences are simple. The lattice heats in a few picoseconds, and then cools because of thermal diffusion. However, one interesting question arises from these observations: If $\varepsilon(\omega)$ matches that of heated GaAs within 4 ps, does this imply that thermodynamic equilibration occurs on this time scale? Unfortunately, we cannot answer this question definitively. Conventional heating (in which a material remains in thermodynamic equilibrium at all times) causes both expansion of the lattice and increased vibrational motion. However, models of the dielectric function of heated semiconductors suggest that the main contribution of conventional heating comes from increased vibrations of atoms (Samara, 1983; Gopalan et al., 1987). In our experiments, the measured $\varepsilon(\omega)$ at 4 ps after excitation would still match that of conventionally heated GaAs even if no density changes happened within the 4 ps. Such density changes are necessary to say that the material is in thermodynamic equilibrium. Hence, the theoretical work of Samara and others means that we can reach no conclusion on this point.

Time-resolved X-ray diffraction experiments are already beginning to provide some information on lattice expansion following intense laser excitation (Larsson et al., 1998; Rose-Petruck et al., 1999; Chin et al., 1999). Chin et al. observe a delay of about 10 ps in the onset of lattice expansion in InSb. Lattice expansion, or strain, nucleates at the surface because the surface is free to move. Subsequently, the X-ray diffraction data indicate, a strain wave propagates into the material at the speed of sound, approximately $5.4\,\text{km}\,\text{s}^{-1}$ in GaAs and $5\,\text{km}\,\text{s}^{-1}$ in InSb (Rose-Petruck et al., 1999; Chin et al., 1999). This implies that it takes a further 20 ps for lattice expansion to occur within a 100-nm depth. The initial delay in density changes observed by Chin and co-workers is probably due to a combination of delayed Auger heating (which slows the rise time of the observed lattice temperature to about 7 ps) and the time for LO phonons to decay into the acoustic modes that can cause lattice expansion. Thus, the first X-ray diffraction experiments point to a time scale of about 10–30 ps for lattice density changes to take place following laser excitation in semiconductors. Their first results therefore suggest that thermal equilibrium cannot be reached within about 10 ps after the excitation.

b. *Structural Dynamics at Medium and High Fluences*

For excitations of more than $0.5\,F_{\text{th}}$, we observe disordering of GaAs within several picoseconds, based both on our dielectric function data and on earlier measurements of the second-order susceptibility (Glezer et al., 1995b). However, optical methods cannot probe the exact structure. In the medium fluence regime, the structural change may be amorphization, superheating, or the formation of a thin liquid layer at the surface. We could not distinguish between these options because of lack of data on the dielectric function for GaAs above 880 K or for amorphous GaAs at any temperature higher than room temperature. We also cannot tell clearly whether the transition is thermal or nonthermal. From time-resolved X-ray measurements on InSb, Chin et al. see subpicosecond disordering of the lattice in a surface layer for fluences below the single shot damage threshold (Chin et al., 1999). There is no lattice expansion on this time scale, and thus the authors conclude that the disordering is nonthermal.

In the high fluence regime, our data does provide evidence of nonthermal structural change. We observe the bandgap to collapse within a few picoseconds, becoming faster as the fluence increases. Although the calculations of Kim et al. predict that the gap in GaAs disappears completely when 10% of the valence electrons are excited (Kim et al., 1994), their model cannot explain the time dependence of our data. If the bandgap were solely changed by free carrier effects, the gap would be smallest immediately after excitation, when the carrier density is largest, and would subsequently return to its original value as the carriers relax. However, for fluences just above the threshold, we observe a gradual drop in the zero crossing of $Re[\varepsilon(\omega)]$ over several picoseconds. Band structure modifications caused by many-body effects do depend on carrier temperature as well as density, and the carrier temperature does change with time. However, theoretical work by Zimmermann and others (Zimmermann, 1988; Kalt and Rinker, 1992) indicates that bandgap renormalization is only weakly dependent on carrier temperature for the carrier densities in our experiments. Thus, we conclude that structural changes, rather than electronic effects, are responsible for the observed decrease in the gap over several picoseconds. These structural changes must be nonthermal, since the semiconductor-to-metal transition can occur in less than 2 ps, that is, before electrons have fully relaxed or dissipated, when the excitation fluence is about $1.2 F_{th}$ or above. As described in Subsection 6 of Section III, several theoretical papers (Stampfli and Bennemann, 1990, 1992, 1994; Silvestrelli et al., 1996; Graves and Allen, 1998; Dumitrică and Allen, 1999) propose a mechanism for femtosecond-laser-induced nonthermal structural changes in semiconductors. The laser excitation promotes electrons from bonding states in the valence bands to antibonding states in the conduction bands. This weakens the lattice bonding, and if the excitation is sufficiently strong, there is a lattice instability. This can lead to deformation of the zincblende structure and a metallic transition (Stampfli and Bennemann, 1994; Graves and Allen, 1998). The various theoretical approaches all predict that an excitation of about 10% of the carriers is sufficient to cause a semiconductor-to-metal transition in GaAs or Si, in reasonable agreement with our data. In the most recent work, Graves, Dumitrică, and Allen use molecular dynamics simulations to predict the dielectric function (Graves and Allen, 1998; Dumitrică and Allen, 1999). Figure 1 of the chapter in this volume by Allen and co-workers shows the calculated evolution of the imaginary part of the dielectric function following excitation of 20% of the valence electrons, that is, at twice the threshold fluence. The qualitative agreement with our data is excellent. Within a few hundred femtoseconds, the lattice disorders and the simulated dielectric function matches the Drude model quite well. Indeed, comparing the results in Fig. 23 with the simulated $Im[\varepsilon(\omega)]$, we see that the simulations predict two details observed in the data. First, there is a sharp rise in the imaginary part at the bottom end of the spectral range around 1.5 eV. Second, both experiment and simulation show a residual interband contribution around 2.5–3.0 eV after the semiconductor-to-metal transition has occurred. The simulations suggest that this contribution comes from some of the states in the valence and conduction bands that originally produced the E_2 in unexcited GaAs, but that are much closer together as a result of the excitation and consequent bandgap collapse.

The evolution of the metallic state in the high fluence regime is surprising in two respects. Recall from Fig. 25 that the plasma frequency, which is proportional to the square root of the carrier density, decreases with time after the semiconductor-to-metal transition. Furthermore, the plasma frequency at any time delay decreases with increasing fluence. The decrease in plasma frequency with time is probably due to diffusion of carriers into the material. Because the probe beam is only sensitive to the first 10 nm when in the metallic state, carriers can diffuse from the probed region very rapidly.

The second observation, namely that the plasma frequency decreases with increasing fluence, is more difficult to explain. Indeed, it is still a puzzle. Note that there is no reason to expect a higher carrier density with higher fluence, because the free carriers that generate the Drude dielectric function are produced mainly by the bandgap collapse and not the original excitation. That said, there is no reason to expect a lower carrier density with higher fluence either. Perhaps the resolution of this puzzle lies in noting that the plasma frequency depends not only on the carrier density but also on the effective masses of the free carriers, as expressed in Eq. (16). If the effective masses of carriers, electrons and holes, in the metallic state were to increase with fluence, then the plasma frequency would fall without any change in carrier density. Future simulations might be able to calculate effective masses and plasma frequencies at different fluences, and help us to understand the observations better.

VII. Conclusions

The time-resolved dielectric function provides a wealth of new information on the electron dynamics and the nature of structural changes in GaAs following femtosecond laser excitation of more than 10^{21} carriers per cm^3. Figure 27 summarizes what we now know. There are three distinct regimes of behavior in GaAs depending on the excitation fluence. For fluences below $0.5\,kJ/m^2$, the free carriers scatter quickly from the central Γ valley to the side valleys in about 500 fs. They relax by rapid Auger recombination and phonon emission, which causes the lattice to heat. For pump fluences above $0.5\,kJ/m^2$, the initial electronic effects on $\varepsilon(\omega)$ are stronger but less easy to interpret. The excited carriers influence the dielectric function through a combination of free carrier absorption, band filling, and changes in the band structure due to ionic screening and other many-body effects. Carrier dynamics are difficult to ascertain, but it is likely that intervalley scattering, Auger recombination, and phonon emission still dominate. After a few

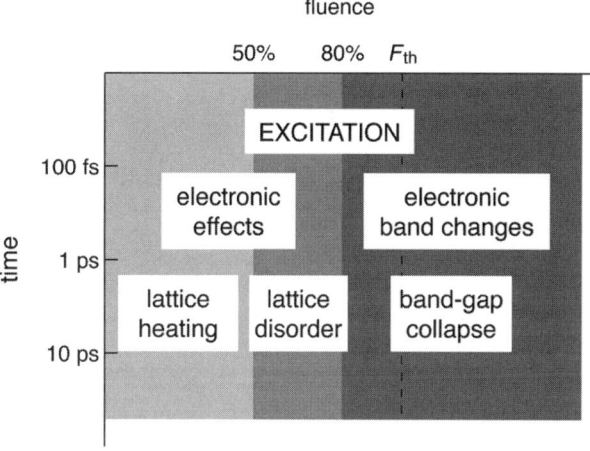

FIG. 27. Summary of the electronic and structural dynamics in GaAs following femtosecond laser excitations of 0.1–2.0 kJ/m².

picoseconds, the lattice undergoes disordering, either remaining in the solid phase or via a transformation to a liquid phase. These structural changes may be thermal or nonthermal below 0.8 kJ/m². Above this threshold, the threshold for the semiconductor-to-metal transition, the changes appear to be nonthermal because the structural changes start even before carrier relaxation is complete.

Despite the new insights gained through the present work, many open questions remain in the field of highly excited semiconductors. In the future, a combination of optical and X-ray experiments, together with a complementary theoretical effort, promise to resolve many of the key remaining debates.

Acknowledgments

Drs. Li Huang and Eli N. Glezer participated in the experiments described in this chapter. We thank Professors R. Allen, N. Bloembergen, H. Ehrenreich, E. Kaxiras, C. Klingshirn, K. Sokolowski-Tinten, and F. Spaepen for many valuable discussions. This work is supported by the National Science Foundation under contract no. NSF DMR-9807144 and the Army Research Office under contract no. ARO DAAG5S-98-1-0077. J. P. Callan gratefully acknowledges support from a Harvard Graduate School of Arts and Sciences Merit Fellowship, and A. M.-T. Kim gratefully acknowledges support from the Deutscher Akademischer Auslandsdienst (DAAD).

References

Ashcroft, N. W., and Mermin, N. D. (1976). *Solid State Physics* (Saunders College, Philadelphia).
Bailey, D. W., Stanton, C. J., Hess, K., LaGasse, M. J., Schloenlein, R. W., and Fujimoto, J. G. (1989). *Solid State Electron.* **32**, 1491.
Becker, P. C., Fragnito, H. L., Brito-Cruz, C. H., Shah, J., Fork, R. L., Cunningham, J. E. Henry, J. E. and Shank, C. V. (1988). *Appl. Phys. Lett.* **53**, 2089.
Blakemore, J. S. (1982). *J. Appl. Phys.* **53**, R123.
Born, M., and Wolf, E. (1980). *Principles of Optics* (Pergamon, Oxford).
Bransden, B. H., and Joachain, C. J. (1983). *Physics of Atoms and Molecules* (Longman, Harlow, England).
Brodeur, A. and Chin, S. L. (1999). *J. Opt. Soc. Am. B* **16**, 637.
Callan, J. P., Kim, A. M.-T., Huang, L., Glezer, E. N., and Mazur, E. (1998). *Mat. Res. Soc. Symp. Proc.* **481**, 395.
Chin, A. H., Schoenlein, R. W. Glover, T. E., Balling, P., Leemans, W. P., and Shank, C. V. (1999). *Phys. Rev. Lett.* **83**, 336.
Cohen, M. L., and Chelikowsky, J. (1989). *Electronic Structure and Optical Properties of Semiconductors*, 2nd ed. (Springer, Berlin).
Downer, M. C., and Shank, C. V. (1986). *Phys. Rev. Lett.* **56**, 761.
Downer, M. C., Fork, R. L., and Shank, C. V. (1985). *J. Opt. Soc. Am. B* **2**, 595.
Dumitricǎ, T., and Allen, R. E., (1999). To be published.
Electricity Supply Board of Ireland (1999). World Wide Web site http://www2.esb.ie/htm/ncc/index.htm.
Elsaesser, T., Shah, J., Rota, L., and Lugli, P. (1991). *Phys. Rev. Lett.* **66**, 1757.
Erman, M., Theeten, J. B., Chambon, P., Kelso, S. M., and Aspnes, D. E. (1984). *J. Appl. Phys.* **56**, 2664.
Glezer, E. N. Siegal, Y., Huang, L., and Mazur, E. (1995a). *Phys. Rev. B* **51**, 6959.
Glezer, E. N., Siegal, Y., Huang, L., and Mazur, E. (1995b). *Phys. Rev. B* **51** 9589.
Gopalan, S., Lautenschlager, P., and Cardona, M. (1987). *Phys. Rev. B* **35**, 5577.
Graves, J. S., and Allen, R. E., (1998). *Phys. Rev. B* **58** 13627.
Hecht, E. (1987). *Optics*, 2nd ed. (Addison-Wesley, Reading, MA).
Huang, L., Callan, J. P., Glezer, E. N., and Mazur, E. (1998). *Phys. Rev. Lett.* **80**, 185.
Jackson, J. D. (1975). *Classical Electrodynamics*, 2nd ed. (Wiley, New York).
Jellison, G. E. (1997). Private communication.
Kalt, H., and Rinker, M. (1992). *Phys. Rev. B* **45**, 1139.
Kash, J. A., Tsang, J. C., and Hvam, J. M. (1985). *Phys. Rev. Lett.* **54**, 2151.
Kim, D. H., Ehrenreich, H., and Runge, E. (1994). *Solid State Commun.* **89**, 119.
Larsson, J., Heimann, P. A., Lindenberg, A. M., Schuck, P. J., Bucksbaum, P. H. Lee, R. W., Padmore, H. A., Wark, J. S., and Falcone, R. W. (1998). *Appl. Phys. A* **66**, 587.
Leitenstorfer, A., Fürst, C., Laubereau, A., Kaiser, W., Tränkle, G., and Weimann, G. (1996). *Phys. Rev. Lett.* **76**, 1545.
Lin, W. Z., Fujimoto, J. G., Ippen, E. P., and Logan, R. A. (1987). *Appl. Phys. Lett.* **50**, 124.
Lin, W. Z., Schoenlein, R. W., and Fujimoto, J. G. (1988). *IEEE J. Quant. Elect.* **24**, 267.
Lundstrom, M. (1990). *Fundamentals of Carrier Transport*, Modular Series on Solid State Devices, Vol. X (Addison-Wesley, Reading, MA).
Mahan, G. D. (1990). *Many-Particle Physics*, 2nd ed. (Plenum Press, New York).
Nelder, J. A., and Mead, R. (1965). *Comp. J.* **7**, 308.
Oudar, J. L., Hulin, D., Migus, A., Antonetti, A., and Alexandre, F. (1985). *Phys. Rev. Lett.* **55**, 2074.
Palik, E. D. (1985). *Handbook of Optical Constants of Solids* (Academic Press, New York).

Potter, R. F. (1969). In *Optical Properties of Solids*, eds. Nudelman, S., and Mitra, S. S. (Plenum, New York).
Press, W. H., Flannery, B. P., Teukolsky, S. A., and Vetterling, W. T. (1988). *Numerical Recipes in C: The Art of Scientific Computing* (Cambridge University Press, Cambridge).
Rose-Petruck, C., Jimenez, R., Guo, T., Cavalleri, A., Siders, C. W., Raksi, F., Squier, J. A., Wlker, E. C., Wilson, K. R., and Barty, C. P. J. (1999). *Nature* **398**, 310.
Saeta, P. N. (1991). Ph.D. thesis, Harvard University.
Saeta, P., Wang, J.-K., Siegal, Y., Bloembergen, N., and Mazur, E. (1991). *Phys. Rev. Lett.* **67**, 1023.
Samara, G. A. (1983). *Phys. Rev. B* **27**, 3494.
Seeger, K. (1991). *Semiconductor Physics: An Introduction* (Springer-Verlag, Berlin).
Shah, J. (1996). *Ultrafast Spectroscopy of Semiconductors and Semiconductor Nanostructures* (Springer Verlag, Berlin), and references therein.
Siegal, Y. (1994). Ph.D. thesis, Harvard University.
Silvestrelli, P., Alavi, A., Parrinello, M., and Frenkel, D. (1996). *Phys. Rev. Lett.* **77**, 3149.
Sokolowski-Tinten, K. (1996). Private communication.
Sokolowski-Tinten, K., Schulz, H., Bialkowski, J., and von der Linde, D. (1991). *Appl. Phys. A* **53**, 227.
Sokolowski-Tinten, K., Bialkowski, J., Cavalleri, A., von der Linde, D., Oparin, A., Meyer-ter-Vehn, J., and Anisimov, S. I. (1998a). *Phys. Rev. Lett.* **81**, 224.
Sokolowski-Tinten, K., Solís, J., Bialkowski, J., Siegel, J., Afonso, C. N., and von der Linde, D. (1998b). *Phys. Rev. Lett.* **81**, 3679.
Solís, J., Afonso, C. N., Trull, J. F., and Morilla, M. C. (1994). *J. Appl. Phys* **75**, 7788.
Stampfli, P., and Bennemann, K. H. (1990). *Phys. Rev. B* **42**, 7163.
Stampfli, P., and Bennemann, K. H. (1992). *Phys. Rev. B* **46**, 10686.
Stampfli, P., and Bennemann, K. H. (1994). *Phys. Rev. B* **49**, 7299.
Strauss, U., Rühle, W. W., and Köhler, K. (1993). *Appl. Phys. Lett.* **62**, 55.
Sutter, D. H., Steinmeyer, G., Gallmann, L., Matuschek, N., Morier-Genoud, F., Keller, U., Scheuer, V., Angelow, G., and Tschudi, T. (1999). *Opt. Lett.* **24**, 631.
Tom, H. W. K., Aumiller, G. D. and Brito-Cruz, C. H. (1988). *Phys. Rev. Lett.* **60**, 1438.
van Driel, H. M. (1987). *Phys. Rev. B* **35**, 8166.
van Vechten, J. A., Tsu, R., and Saris, F. W. (1979). *Physics Letters A* **74a**, 422.
von der Linde, D., Sokolowski-Tinten, K., and Bialkowski, J. (1997). *Applied Surf. Sci.* **109/110**, 1.
Waghmare, U., and Kaxiras, E. (1998). Private communication.
Yao, H., Snyder, P. G., and Woollam, J. A. (1991). *J. Appl. Phys.*, **70**, 3261.
Yoffa, E. J. (1980). *Phys. Rev. B* **21**, 2415.
Yu, P. Y., and Cardona, M. (1996). *Fundamentals of Semiconductors*, (Springer-Verlag, Berlin).
Zewail, A. H. (1994). *Femtochemistry: Ultrafast Dynamics of the Chemical Bond* (World Scientific, Singapore), and references therein.
Zimmermann, R. (1988). *Many Particle Theory of Highly Excited Semiconductors* (Teubner, Leipzig).
Zollner, S., Myers, K. D., Dolan, J. M., Bailey, D. W. and Stanton, C. J. (1998). *Thin Solid Films* **313**, 578.

CHAPTER 5

Quantum Kinetics for Femtosecond Spectroscopy in Semiconductors

Hartmut Haug

INSTITUT FÜR THEORETISCHE PHYSIK
J. W. GOETHE UNIVERSITÄT FRANKFURT
FRANKFURT, GERMANY

I. INTRODUCTION 205
II. SEMICONDUCTOR BLOCH EQUATIONS FOR PULSE EXCITATION WITH QUANTUM
   KINETIC SCATTERING INTEGRALS 209
III. LOW-EXCITATION FEMTOSECOND SPECTROSCOPY 215
   1. *Femtosecond FWM with LO-Phonon Quantum Kinetic Scattering* 217
   2. *Femtosecond DTS with LO-Phonon Quantum Kinetic Scattering* 219
IV. FEMTOSECOND DTS FOR SCREENED COULOMB AND LO-PHONON QUANTUM
   KINETIC SCATTERING 220
V. RESONANT FWM WITH SCREENED COULOMB AND LO-PHONON QUANTUM
   KINETIC SCATTERING 223
   REFERENCES 228

I. Introduction

Two-beam femtosecond spectroscopy (Shah, 1996) allows to prepare and detect the nonequilibrium time evolution of the photoexcited carriers in a regime that can no longer be described by semiclassical, Markovian kinetics. In this ultrashort time regime, the kinetics is influenced by the quantum mechanical coherence of the photoexcited carriers. Because of the energy–time uncertainty relation, the scattering rates are no longer governed by the Golden Rule, but are expressed by memory integrals over the past of the system. The resulting relaxation and dephasing kinetics of the carriers is delayed in comparison with the instantaneous semiclassical kinetics. The general theory of quantum kinetics, which is not only relevant for femtosecond optical experiments but also for quantum transport in nanostructures, has been described in a textbook (Haug and Jauho, 1996). The quantum kinetics of the optically excited carriers due to their interaction with longitudinal optical phonons is the fastest scattering process for low electron densities. Because the screening of the interaction by the excited carriers plays only a minor role at these low densities, the quantum kinetics due to longitudinal optical (LO) phonon scattering is relatively simple; therefore, it is relatively well developed and has been tested successfully by corresponding femtosecond four-wave mixing (FWM) experiments (Bányai et al., 1995), more recently with coherent control (Wehner et al., 1998), and by femtosecond pump and probe experiments, also called differential transmission spectroscopy (DTS) (Fürst et al., 1997; Schmenkel et al., 1998; Haug, 1998). These experiments have all been performed with GaAs, which is only a weakly polar semiconductor. The key assumptions of the quantum kinetic theory due to LO-phonon scattering are as follows: (a) The use of only the lowest self-energy diagram, which is, however, calculated by using fully dressed carrier and phonon nonequilibrium Keldysh Green functions. In order to simplify the problem, the phonons are most often taken as a thermal bath. For this case the phonon Green functions reduce to free-particle Green functions with a given bath temperature. Particularly, one neglects vertex corrections in this approximation. (b) The nonequilibrium Green functions that depend on two time arguments are expressed in terms of the one-time dependent density matrix by using the so-called generalized Kadanoff–Baym approximation (GKBA) (Lipavsky et al., 1986; Haug and Jauho, 1996). In this relation, the off-diagonal time development is described, for example, for $t > t'$, by the retarded Green function, while the density matrix is taken at the earlier time t'. For the retarded Green function, further simplified forms are used. Both assumptions (a) and (b) are only valid in the weak coupling regime. Therefore, this scheme was quite successful for GaAs for which the dimensionless polaron coupling constant $\alpha = 0.069$, while improved concepts have to be used for more polar materials such as the II–VI compounds in which α becomes comparable to one.

In order to overcome the approximation (b), several groups calculated directly for this scattering mechanism the nonequilibrium particle propagator, which depends on two time arguments (Hartmann and Schäfer, 1992; Bonitz et al., 1996; Gartner et al., 1999). However, because these complex calculations have not yet been used to describe actual experiments and usually still use approximation (a), we do not include these attempts in the present review. These more complex approaches are, however, important to test the validity of the just-mentioned approximations, which enter into a quantum kinetic description in terms of density matrices.

The quantum kinetics due to Coulomb carrier–carrier scattering that has to include the self-consistent time-dependent screening has been developed only recently (Bányai et al., 1996). The buildup of screening of the Coulomb interaction by intraband scattering needs about an inverse plasma frequency

(El Sayed *et al.*, 1994). For times much shorter than a plasma oscillation period, the Coulomb potential between the pulse-excited carriers is essentially unscreened.

When screening sets in, the Coulomb potential becomes a function of the space and time coordinates of the two interacting particles. The two-time-dependent screened Coulomb potential obeys an integral equation in which the integral kernel is in RPA given by the intraband polarization bubble, also called the GG approximation. It is important to note that the GG-approximation is not a conserving approximation. The charge neutrality of the system requires that the polarization function vanishes at zero momentum transfer, that is, at $q = 0$. This requirement compensates the $1/q^2$ singularity of the Coulomb potential. In the two-time formalism the GG polarization without vertex corrections does not fulfil this requirement. However, if one combines the GG approximation with the GKBA, the singularity at $q = 0$ is cancelled, a fact that is well known from RPA equilibrium theory.

By memory saving linear programming techniques, it has become possible to include a self-consistent solution of this integral equation into the numerical treatment of the semiconductor optical Bloch equations combined with the quantum kinetic relaxation and dephasing rates for the carrier–carrier scattering (Bányai *et al.*, 1998). In this theory the integral equation for the two-time-dependent screened Coulomb potential is solved by expressing all two-time-dependent nonequilibrium particle propagators via the GKBA in terms of the two-time-dependent spectral functions and the reduced density matrix elements that depend on one time only. Whereas the spectral functions are taken in the free-particle approximation, the density matrix elements are determined self-consistently by the semiconductor Bloch equations. In order to include also the energy relaxation of the excited electron–hole (e-h) gas, the scattering by LO phonons is taken additionally into account. The semiconductor Bloch equations for the density matrix elements with these quantum kinetic scattering integrals are solved, yielding the time development of the e-h distributions and of the interband polarization components in the presence of the strong resonant laser pulses.

In Section II we present the theoretical model for the considered two-band (if necessary extended up to four bands) semiconductor. In particular, the semiconductor Bloch equations for a laser pulse excited crystal are given with quantum kinetic integrals for both carrier–phonon and carrier–carrier scattering including the time-dependent screened Coulomb potential.

In Section III we review the application of the quantum kinetic relaxation and dephasing kinetics in the low-density regime with carrier–LO-phonon scattering only, for femtosecond DTS by the group of Leitenstorfer (Fürst *et al.*, 1997) and femtosecond FWM experiments by Wegener and coworkers (Bányai *et al.*, 1995). For both experiments essential features can only be explained using the quantum kinetics with memory effects. In the case of the DTS experiment, the sharpening of the one-phonon replica in time and the delayed buildup of the second phonon replica are only obtained in the framework of quantum kinetics, but not with the semiclassical Boltzmann-type kinetics. In the case of the FWM experiment, the clearly observed oscillations in the time-integrated FWM signal due to an interference of interband-polarization components coupled by the coherent exchange of a phonon are only obtained with quantum kinetics and are completely absent in a semiclassical kinetics with energy-conserving transition rates. In both cases a very good agreement of the quantum kinetic calculations and the corresponding experimental measurements has been obtained.

In Section IV the application of the theory with carrier–carrier and carrier–LO-phonon scattering for resonant femtosecond differential transmission (DTS) on thin layer of GaAs at 15 K is reviewed. The results are compared with corresponding experiments on a wide range of delay times between the pump and probe pulse and excitation densities (Vu *et al.*, 1999). In this experiment the pump pulse excites electrons in the conduction band due to transitions from the heavy- and light-hole valence band, respectively, while the test pulse is tuned to transitions between the spin-orbit split-off valence band and the conduction band (see Fig. 1). This setup allows observation of the relaxation of the electron distribution in the conduction band directly. The parameter-free quantum kinetics gives a good qualitative description of the electron relaxation kinetics for all considered parameters. In particular the quantum kinetics yields the correct time scales for the whole relaxation process.

In Section V the application of the theory to the calculation of femtosecond four-wave mixing is described (Hügel *et al.*, 1999; Vu, 1999). The resulting time-resolved and time-integrated FWM signals are compared with corresponding experiments (Hügel *et al.*, 1999). In particular, the time-resolved FWM signals are photon-echo-like at short delay times, but appear much earlier at long delay times in full agreement with the experiment. For a simple Markovian dephasing in terms of a phenomenological T_2 transverse relaxation time, the FWM signal would always be a photon echo. From the decay of the time-integrated FWM signal an inverse dephasing time is deduced that is shown to vary with the excitation density n as $n^{1/3}$ in quantitative agreement with the experiment. In all these calculations of the last two sections, the quantum kinetic collision integrals due to screened carrier–carrier and carrier–phonon scattering are included.

II. Semiconductor Bloch Equations for Pulse Excitation with Quantum Kinetic Scattering Integrals

In the following we treat direct-gap semiconductors with dipole-allowed optical transitions between the valence bands and the conduction band. All explicit applications of the presented quantum kinetic theory will be given for GaAs. The band structure around the Γ-point is shown in Fig. 1. The optical interband transitions are indicated by arrows. If necessary all three valence bands (light- and heavy-hole band, as well as the spin-orbit split-off band) will be included; otherwise, only a two-band model is treated. For simplicity we use this simpler two-band model in the following description and refer for the extended formulations to the original literature. The quantum kinetics of the optically excited electrons will be given in terms of particle propagators $G^<(\vec{r}_1, t_1, \vec{r}_2, t_2) = i \langle \psi^\dagger(\vec{r}_2, t_2) \psi(\vec{r}_1, t_1) \rangle$, where $\psi^\dagger(\vec{r}, t)$ and $\psi(\vec{r}, t)$ are the electron creation and annihilation operators. For the conventions and relations used; we refer to Haug and Jauho (1996). We assume that the electrons are distributed homogeneously in space, so that the particle propagator depends only on the relative space coordinate $\vec{r}_1 - \vec{r}_2$. By means of an expansion into Bloch wave functions, the propagator depends then on the wave vector $\vec{q}$ and on the band index v. In all applications we limit ourselves to isotropic situations in momentum space. Under the excitation with short coherent femtosecond pulses, the

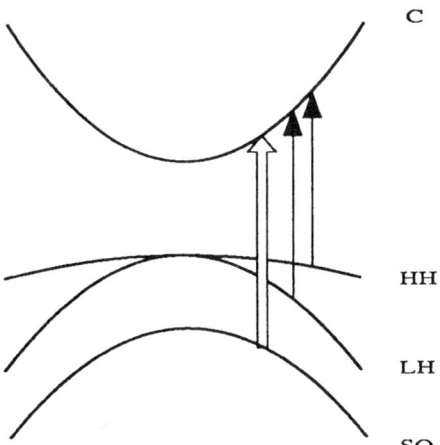

FIG. 1. Schematic band structure of cubic III–V semiconductor compounds like GaAs near the Γ point.

propagator depends explicitely on two time arguments. We do not introduce a separation in central and relative time, nor do we use Fourier transformation with respect to the relative time. Thus, we use the particle propagator in the form (we use in this paper $\hbar = 1$, $V = 1$)

$$G^<_{v,\mu,\vec{q}}(t_1, t_2) = i\langle a^+_{\mu,\vec{q}}(t_2) a_{v,\vec{q}}(t_1)\rangle. \tag{1}$$

For the calculation of the optical response one needs only the equal-time limit of the particle propagator, namely the density matrix elements $\rho_{v,\mu,\vec{q}}(t) = -iG^<_{v,\mu,\vec{q}}(t, t)$. Obviously the diagonal density matrix elements describe the electron distributions in the various bands. For the valence bands, one eventually replaces the electron distributions by the distributions of the missing electrons, that is, the holes. For the heavy-hole valence band (hhv), for example, one uses the relation $\rho_{hhv,hhv,\vec{q}}(t) = 1 - \rho_{hh,hh,-\vec{q}}(t)$, where $\rho_{hh,hh,\vec{q}}(t)$ is the distribution of the heavy holes in the top valence band. The off-diagonal elements of the density matrix describe the interband polarization components, which are induced by the excitation with the coherent laser light. The coherent light pulses excite a polarization between the conduction band and one or more valence bands, depending on the selection rules for a given light polarization. For a two-band model, for example, with only one valence band, the macroscopic interband polarization parallel to the applied field is given by

$$P(t) = \sum_{\vec{q}} (d_{c,v,\vec{q}} \rho_{v,c,\vec{q}}(t) + c.c.), \tag{2}$$

where $d_{c,v,\vec{q}}$ is the dipole matrix element for the optical transition. Close to the band edge, the wave-vector dependence can be neglected; only if contributions deeper in the band have to be taken into account, must one include the q-dependence, such as in the framework of Kane's theory (Haug and Koch, 1993). In all applications discussed here, we assume isotropy in momentum space, so that the density matrix and the dipole matrix element depend both only on the wave number but not on the direction of the wave vector.

The mean-field evolution of density matrix under the influence of a coherent optical laser pulse and the Coulomb interaction is given by the so-called semiconductor Bloch equations

$$\left(\frac{\partial t}{\partial t} + i(e_{v,\vec{k}} - e_{\mu,\vec{k}})\right) \rho_{v,\mu,\vec{k}}(t)$$

$$= i\sum_\rho (\Sigma_{v,\rho,\vec{k}}(t)\rho_{\rho,\mu,\vec{k}}(t) - \rho_{v,\rho,\vec{k}}(t)\Sigma_{\rho,\mu,\vec{k}}(t)) + \left.\frac{\partial \rho_{v,\mu,\vec{k}}(t)}{\partial t}\right|_{coll}. \tag{3}$$

Here, $e_{v,\vec{k}}$ are the quadratic free-electron energies in band $v = c, v$ expressed by the corresponding effective masses. The mean-field self-energies that depend only on one time argument are given by

$$\Sigma_{v,\mu,\vec{k}}(t) = d_{v,\mu,k} E(t) + \sum_{\vec{q}} V_q (\rho_{v,\mu,\vec{k}-\vec{q}}(t) - \delta_{v,\mu}\delta_{v,v}). \quad (4)$$

Here, $d_{v,\mu,k}$ is the previously discussed dipole matrix element for a transition between the bands v and μ (one of them has to be the conduction band, while the other has to be one of the valence bands), and $E(t) = E_0(t)e^{-i\omega t}$ is the classically treated electric field of the laser pulse. The pulse amplitudes $E_0(t)$ are taken to be in the form of a Gaussian or in that of a hyperbolic secant, depending on the forms used in the experiments. The Fourier transform of the bare Coulomb potential is $V_q = 4\pi e^2/\varepsilon_0 q^2$ with the background dielectric constant of the unexcited crystal ε_0. In the diagonal valence band element of the Hartree–Fock exchange energies, the contribution from the interaction with the full valence band is subtracted, because this term has been included already in the band structure calculation. Without the Coulomb exchange terms, these equations are equivalent to the Bloch equations of an inhomogeneous system of $\frac{1}{2}$ spins under the influence of a dc and rotating magnetic field, respectively, together with phenomenological damping terms. Here we replace these phenomenological damping terms by quantum kinetic collision integrals given by (Haug and Jauho, 1996)

$$\left.\frac{\partial \rho_{\vec{k}}(t)}{\partial t}\right|_{\text{coll}} = -\int_{-\infty}^{t} dt' [\Sigma_{\vec{k}}^{\gtrless}(t,t') G_{\vec{k}}^{\lessgtr}(t',t) - \Sigma_{\vec{k}}^{\lessgtr}(t,t') G_{\vec{k}}^{\gtrless}(t',t) \\ - G_{\vec{k}}^{\gtrless}(t,t')\Sigma_{\vec{k}}^{\lessgtr}(t',t) + G_{\vec{k}}^{\lessgtr}(t,t')\Sigma_{\vec{k}}^{\gtrless}(t',t)]. \quad (5)$$

Note that all Green functions and self-energies are matrices in the band index. In the following we describe briefly the scattering self-energies for the carrier–phonon scattering and for the carrier–carrier scattering, also called Coulomb scattering. In both self-energies vertex corrections are not included, partly because they contain further time integrals that give only a small contribution for the ultrashort time regime and partly because we limit ourselves to weak interactions. For a more extended presentation we refer to a textbook on quantum kinetics (Haug and Jauho, 1996). The scattering self-energy due to scattering with a thermal bath of LO phonons with a frequency ω_0 is given by

$$\Sigma_{\vec{k}}^{\gtrless}(t,t') = i \sum_{\vec{q}} g_q^2 G_{\vec{k}-\vec{q}}^{\gtrless}(t,t') D_q^{0,\gtrless}(t,t'), \quad (6)$$

where $D_q^{0,\gtrless}$ are the free (unrenormalized) propagators of the thermal phonons,

$$D_q^{0,<}(t,t') = -i \sum_{\pm} N_q^{\pm} e^{\pm i\omega_0(t-t')}, \quad D_q^{0,>}(t,t') = -D_q^{0,<}(t,t')^*, \tag{7}$$

with

$$N_q^{\pm} = N + \frac{1}{2} \pm \frac{1}{2} \quad \text{and} \quad N = \frac{1}{e^{\omega_0\beta} - 1}, \tag{8}$$

where $\beta = 1/(KT)$ is the inverse thermal energy. For a very strong excitation of hot carriers where many LO phonons are generated in the relaxation process (Shah, 1989), one has to go beyond this approximation and treat, together with the electron kinetics, the kinetics of the LO phonons (Schilp et al., 1995; Haug and Jauho, 1996). Under high excitation conditions, the time-dependent screening of the carrier-LO-phonon interaction should also be included, as is done for Coulomb scattering later (Vu, 1999).

We take the Coulomb scattering self-energies again in the self-consistent GW-approximation without vertex correction:

$$\Sigma_k^{\gtrless}(t,t') = i \sum_{\vec{q}} G_{\vec{k}-\vec{q}}^{\gtrless}(t,t') V_q^{\gtrless}(t,t'). \tag{9}$$

The effective interaction W, here called $V_q(t,t')$, is the two-time-dependent screened Coulomb potential, which will be determined self-consistently. Again it is important to note that the screened Coulomb potential is not only a function of the time difference. The reason for this fact is that the carriers excited by the femtosecond pulse need some time to screen the bare potential V_q. The equation for the screened potential is

$$V_q(t_1,t_2) = V_q \delta(t_1 - t_2) + V_q L_q(t_1,t_3) V_q(t_3,t_2), \tag{10}$$

where the time arguments lie on the Keldysh time contour and an integration convention over repeated time arguments is implied. $L_q(t_1,t_2)$ is the intraband polarization function. We will use the nonequilibrium self-consistent RPA intraband polarization

$$L_q^{\lessgtr}(t,t') = -2i \sum_{\vec{p}} G_{\vec{p}+\vec{q}}^{\lessgtr}(t,t') G_{\vec{p}}^{\gtrless}(t',t). \tag{11}$$

For equilibrium situations one regains the usual RPA polarization from this formulation immediately. The equation for the scattering potential $V_q^>(t, t')$ that follows from the general potential equation is (Bányai et al., 1998)

$$V_q^>(t, t') = V_q \left\{ L_q^>(t, t')V_q + \int_{-\infty}^{t} dt'' L_q^r(t, t'')V_q^>(t'', t') \right.$$

$$\left. + \int_{-\infty}^{t'} dt'' L_q^>(t, t'')[V_q^>(t', t'') - V_q^>(t'', t')] \right\}, \quad (12)$$

where symmetry relations such as

$$V_q^{(\lessgtr)}(t, t')^* = -V_q^{(\lessgtr)}(t', t) \quad \text{and} \quad V_q^>(t, t') = V_q^<(t', t) \quad (13)$$

have been used. It is advantageous to use this formulation, which contains only single-time integrals for the numerical determination of the screened potential.

Alternatively, one could first solve the equation for the retarded Coulomb potential $V_q^r = V_q + V_q L_q^r V_q^r$ and use the relation $V_q^< = V_q^r L_q^< V_q^a$ to calculate the scattering potential. The latter relation includes two time integrations (the implied matrix notation for the time arguments is not shown explicitly) that need a great deal of computer memory space.

If one calculates only the one-time density matrix by the semiconductor Bloch equations, one has to make a connection between the two-time particle propagators in the intraband polarization and the one-time density matrices. We follow Lipavský et al. (1986) and use generalized Kadanoff–Baym approximation (GKBA), here extended to matrices in the band index (Haug and Jauho, 1996):

$$G_{\vec{k}}^\lessgtr(t, t') = -G_{\vec{k}}^r(t, t')\rho_{\vec{k}}(t') + \rho_{\vec{k}}(t)G_{\vec{k}}^a(t, t'). \quad (14)$$

The idea is to use for the spectral two-time functions in this relation, that is, the retarded and advanced functions, relatively simple approximations. The simplest approximation is to use a free-particle Wigner–Weisskopf form with a suitably chosen damping constant $\gamma_{i,k}$:

$$G_{i,j,\vec{k}}^r(t, t') = G_{j,i,\vec{k}}^a(t', t)^* = -i\theta(t - t')\delta_{i,j}e^{(-ie_{i,k} - \gamma_{i,k})(t - t')}. \quad (15)$$

Better approximations are to calculate the spectral functions self-consistently with the mean-field terms contained in the Bloch equation. Such functions contain density-dependent Hartree–Fock–Coulomb renormalizations, excitonic effects, and the band mixing due to the coherent laser pulses, that is, the optical Stark effect (Haug and Jauho, 1996; Tran Thoai *et al.*, 1995). It may be also important to use a nonexponential decay. It has been shown (Haug and Bányai, 1996) that in the quantum kinetic regime a damping in the form of a hyperbolic secant can be derived as an analytical approximation from the two-time formulation of the Dyson equation for the retarded Green function. This damping law takes into account that the damping also is not instantaneous on a femtosecond time scale. However, limits in the computer time and memory do not allow in all applications the use of all these more refined approximations for the spectral functions.

Before we demonstrate that the just-described quantum kinetics of Coulomb scattering allows analysis of recent femtosecond experiments, we illustrate the buildup of screening after a single 15-fs pulse tuned 50 meV above the bandgap of a two-band semiconductor. The time-dependent screening can best be illustrated by using an incomplete Fourier transformation of the retarded Coulomb potential (Bányai *et al.*, 1998):

$$V_q^r(\omega, t) = \int_0^\infty d\tau e^{i\omega\tau} V_q^r(t, t - \tau). \tag{16}$$

The 15 fs-pulse peaks at $t = 0$ and has a pulse strength of $\pi/4$, which corresponds to an excited e-h density of 5.2×10^{17} cm^{-3}. In Fig. 2, Im $\varepsilon_q^{-1}(\omega, t)$ shows the buildup of the plasmon pole within ca. 200 fs for $qa_0 = 1$. One sees that in the time domain of femtosecond spectroscopy the potential is

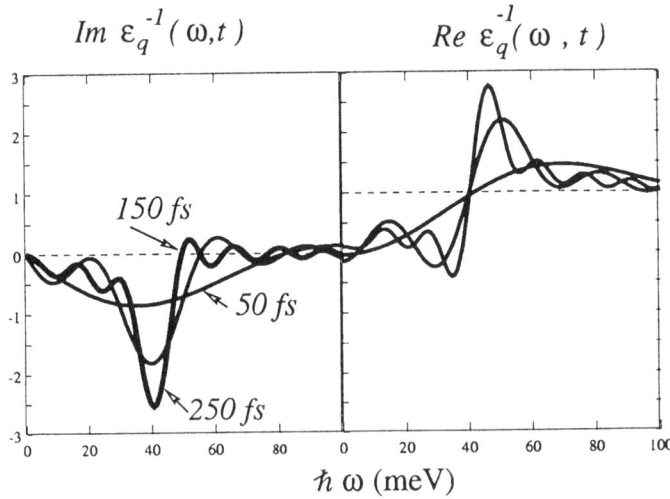

FIG. 2. Real and imaginary part of $\varepsilon_{qa_0=1}^{-1}(\omega, t)$ for three different times.

neither a bare one nor a fully screened one; thus, the two-time-dependent Coulomb potential has to be used in these applications.

Finally we want to note that it is possible also to include the screening for both the carrier–carrier interaction and the carrier–LO-phonon interaction self-consistently (Vu, 1999) by introducing one effective interaction matrix element containing both processes. Such a treatment is relatively straightforward in the equilibrium many-body theory (Mahan, 1981), but because of the different time structure of the bare Coulomb interaction (which is singular in time as $\delta(t_1 - t_2)$) and the free phonon propagator (which oscillates with the time difference $t_1 - t_2$) a joint treatment in the nonequilibrium many-body theory is considerably more difficult (Vu, 1999).

III. Low-Excitation Femtosecond Spectroscopy

At low excitation densities the quantum kinetics of the LO-phonon scattering can be studied by two-pulse femtosecond spectroscopy. The typical experimental scheme is shown in Fig. 3.

One can measure the transmission of the probe beam with and without pump pulse. From these measurements one can obtain the differential transmission spectrum (DTS) as a function of the delay time and, for example, the intensity of the pump pulse. In this type of experiment the pulse

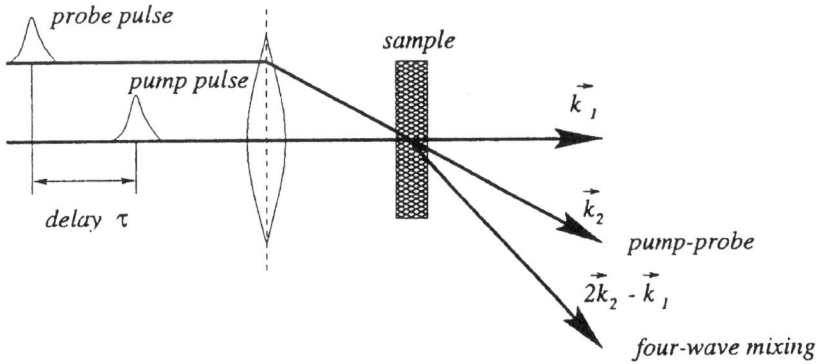

FIG. 3. Schematics of two-pulse experiments: Two successive pulses called pump and probe pulse with a delay time τ propagate in the directions of $\vec{k}_1$ and $\vec{k}_2$ through the sample. The DTS is measured in the direction $\vec{k}_2$ of the test pulse; the FWM signal is measured in the direction of the beam diffracted from the lattice induced by the two pulses $2\vec{k}_2 - \vec{k}_1$.

width of the two pulses is usually different: One uses a relatively broad pump pulse for good spectral resolution and a short, very weak test pulse with which a broad spectrum can be measured. This type of experiment yields mainly information on the kinetics of the carriers excited by the stronger pump pulse. We will review the analysis of an experiment (Fürst et al., 1997) in which the two pulses have been obtained from a two-color laser, so that the two pulses had a fixed phase relation. It will be shown that this coherence between pump and probe pulse determines the measured DTS spectra in addition to the quantum kinetics of the carrier relaxation. Alternatively, one can measure the signal diffracted from the lattice induced by the two noncopropagating pulses. The lattice vector is $\vec{k}_2 - \vec{k}_1$. Adding a lattice vector to the wave vector $\vec{k}_2$ of the probe pulse, one gets a first-order diffracted signal in the direction of $2\vec{k}_2 - \vec{k}_1$. This is the four-wave mixing signal, which can be measured either time-resolved or time-integrated. In degenerate FWM the central frequencies (as well as the width) of the two pulses are equal. We will review a pioneering femtosecond FWM experiment by Wegener et al. and its quantum kinetic analysis (Bányai et al., 1995), which historically gave the first clear experimental observation of an essentially new prediction of quantum kinetics, namely phonon-related beats on top of the decaying time-integrated FWM signal.

Following the history of quantum kinetics and its experimental verification, we start by discussing first the degenerate FWM experiment by Wegener and co-workers (Bányai et al., 1995; Wegener et al., 1996), in which 14-fs pulses tuned to the lowest exciton resonance of GaAs have been applied. Using the form of quantum kinetic collision rate (5) together with the GKBA (14), one gets for the carrier–LO phonon scattering alone the following form of the collision rates for the density matrix element $\rho_{\mu,\nu,\vec{k}}(t)$ (Haug, 1992; Haug and Jauho, 1996):

$$\left.\frac{\partial \rho_{\mu,\nu,\vec{k}}}{\partial t}\right|_{\text{scatt}} = \sum_{\vec{k}',\sigma,\tau,\zeta=\pm 1} \int_{-\infty}^{t} dt' \left[G^r_{\mu,\sigma,\vec{k}}(t,t') G^a_{\mu,\sigma,\vec{k}}(t',t) \right.$$
$$\times \left[e^{i\zeta\omega_0(t-t')} (N_{-\zeta,\vec{k}-\vec{k}'} \rho_{\sigma,\tau,\vec{k}}(t') - N_{\zeta,\vec{k}-\vec{k}'} \rho_{\sigma,\tau,\vec{k}'}(t') \right.$$
$$\left. \left. + \zeta \sum_{\lambda} \rho_{\sigma,\lambda,\vec{k}'}(t) \rho_{\lambda,\tau,\vec{k}}(t')) \right] - (\vec{k} \Leftrightarrow \vec{k}') \right]. \quad (17)$$

Here we use again the thermal phonon distributions $N_{\zeta,\vec{q}} = \frac{1}{2} + \zeta\frac{1}{2} + N$ with $\zeta = \pm 1$ and $N = (\exp(\omega_0/kT) - 1)^{-1}$. Notice that because of the band mixing by the coherent light field, both the density matrix and the spectral functions have elements that are off-diagonal in the band index. Because in the considered experiment the pulses have been tuned to the 1s exciton resonance, we determine the spectral functions self-consistently by solving, for example, the Dyson equation of the retarded Green function with the mean-field self-energy (4), which depends on the density matrix elements. The advanced Green function can be obtained from the retarded Green function by the relation $G^a_{v,\mu,\vec{q}}(t, t') = G^r_{\mu,v,\vec{q}}(t', t)^*$. Thus, the semiconductor Bloch equation with the quantum kinetic collision rate (17) and the Dyson equation of the retarded Green function have to be solved numerically together. If we do this, excitonic effects, the optical Stark effect, and energy renormalization are included in the single-particle spectra of the collision rate. For more details we refer to the original literature (Tran Thoai et al., 1995; Haug and Jauho, 1996).

The main difference of the quantum kinetic collision integrals and the corresponding semiclassical Boltzmann collision rates are the memory structure or non-Markovian form of the former one. The Boltzmann scattering rates at time t depend only on the values of the density matrix elements at the same time t. Furthermore, the energy in each collision is conserved exactly in the Boltzmann kinetics, while the quantum kinetics that is also valid for short time intervals Δt yields no energy conservation, which is clear already from the energy–time uncertainty relation $\Delta E \Delta t \geqslant \hbar$. A deeper-lying reason for the memory structure of the quantum kinetics is the quantum coherence of the excited carriers on short time intervals after their excitation. In this regime the carriers exhibit their quantum mechanical wave nature rather than their particle aspect, and coherence means that the memory to the past is not lost. We will see in the results of the numerical calculations specified to the femtosecond FWM experiment of Wegener *et al.* that the quantum coherence and memory structure of the collision integrals result in coherent LO-phonon related quantum beats, which are completely absent if one treats the collision rates by means of a semiclassical, that is, Markovian Boltzmann-like, kinetics.

1. Femtosecond FWM with LO-Phonon Quantum Kinetic Scattering

In a four-wave mixing configuration, two light pulses with the same intensity, carrier frequency, and pulse duration are considered that propagate in the directions $\vec{k}_1$ and $\vec{k}_2$, respectively. The delay between pulse 1 and 2 is $t_{12} = \tau$. The electric field of the light pulses is thus

$$E(t) = e^{i\vec{k}_1 \cdot \vec{r}}(E_1(t) + E_2(t - t_{12})e^{(i\vec{k}_2 - \vec{k}_1) \cdot \vec{r}}), \tag{18}$$

where $E_1(t) = E_2(t) = E_0(t)e^{-i\omega t}$. The pulses have been taken in accordance with the experiment as hyperbolic secants,

$$E_0(t) = E_0 \frac{1}{\cosh(t/\Delta t)}. \tag{19}$$

The relative phase factor is $e^{i\phi} = e^{(i\vec{k}_2 - \vec{k}_1) \cdot \vec{r}}$. In thin samples the dependence on this phase can be treated adiabatically (Bányai et al., 1995). Therefore, we calculate the induced interband polarization for various values of ϕ project out of this quantity the FWM signal in the direction $2\vec{k}_2 - \vec{k}_1$. Because a residual Coulomb scattering is also present at relatively low excitation intensities, we used in addition to the dephasing by phonon scattering, for the off-diagonal density matrix elements, that is, for interband polarization, a phenomenological damping in the form $\gamma = \gamma_0 + \gamma_1 n(t)$. For 14-fs pulses applied to high-quality bulk GaAs samples with a carrier frequency tuned to the exciton resonance, we get a nearly perfect agreement with corresponding experiments by Wegener and co-workers (Bányai et al., 1995; Wegener et al., 1996), as can be seen from Fig. 4 (Reitsamer et al., 1996). The integrated FWM signal shows on the exponential decay an oscillation with a frequency $(1 + m_e/m_h)\omega_0$, which can be understood as the

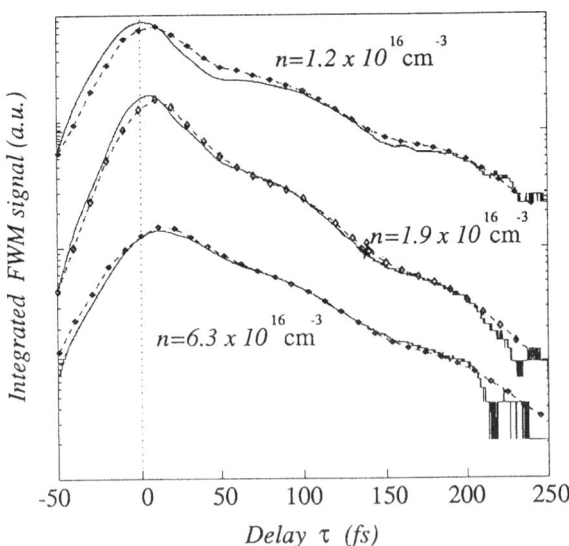

FIG. 4. Measured (solid lines) and calculated (dashed lines) time-integrated FWM signal for three different excitation densities with LO-phonon scattering.

beating frequency between two interband polarization components coupled by a coherent quantum kinetic LO-phonon scattering process in the conduction band. A Boltzmann kinetics would only give an exponential decay but cannot explain the superimposed oscillations. These observed oscillations are therefore a clear manifestation of delayed quantum kinetics. The calculated FWM spectrum shows no sideband structure, because the oscillations take place only with respect to the delay time τ, but do not occur in the dependence of the interband polarization on the real time t. Also, these theoretical predictions for the FWM spectrum are in complete agreement with the measurements (Bányai *et al.*, 1995). Recently, the experimental measurements have been refined by splitting the first pulse into a double pulse with varying delays between this doublet (Wehner *et al.*, 1998). With this coherent control technique, the still reversible scattering quantum kinetics can be influenced strongly, that is, it can be partly reversed with the second pulse of the doublet. An analytically solvable simplified two-level model with diagonal coupling to LO phonons (Axt *et al.*, 1999; Steinbach *et al.*, 1999) allows us to understand most of the observed features of these coherent control experiments.

2. Femtosecond DTS with LO-Phonon Quantum Kinetic Scattering

Leitenstorfer and co-workers (Fürst *et al.*, 1999) have performed resonant low-intensity femtosecond pump-and-probe experiments in bulk GaAs in which the relaxation of the excited electrons by successive LO-phonon emission has been studied in detail. In this experiment a 120-fs pump pulse propagating in the direction $\vec{k}_1$ tuned 150 meV above the band gap and a 25-fs test pulse propagating in the direction $\vec{k}_2$ tuned 120 meV above the bandgap have been used in order to get a spectrally sufficient narrow range of excited carriers. Both beams excite transitions between the heavy-hole band and the conduction band. The carrier concentration excited by the pump pulse has been only about 8×10^{14} cm^{-3}, so that the electron–LO-phonon interaction dominates the relaxation kinetics. The experimentally measured differential transmission spectra (DTS) are shown at the top of Fig. 5 according to Schmenkel *et al.* (1998) and Haug (1998) for various delay times τ. In the second section of Fig. 5 we show the DTS calculated (Schmenkel *et al.*, 1998; Haug, 1998) by a standard linearization of the polarization for the test beam and using semiclassical Boltzmann kinetics. The excited particles block the transitions and give rise to the main positive differential transmission signal.

The negative DTS above the main line is due to the Hartree–Fock exchange effects, that is, mainly due to excitonic enhancement and to a minor degree due to bandgap renormalization. We see that at negative and at small positive delay times, the structures induced by the pump pulse are spectrally much broader than in the experiment. Actually, the time–energy uncertainty relation does not allow the sharp structures seen experimentally. The solution to this dilemma is that at negative delay times (test before probe) a linearization in the test beam is not valid. A general nonlinear coherent calculation (Schmenkel et al., 1998; Haug, 1998) of the interband polarization in the direction $\bar{k}_2$ of the test beam by the same projection technique that has been used for the calculation of the FWM signal is shown in the third section of Fig. 5. Now one sees that the changes of the transmission by the pump pulse at early and negative delay times are similar to those seen experimentally. However, the use of the instantaneous Boltzmann scattering results in a buildup of the first phonon replica, which is too fast in comparison with the experiment. If one replaces the Boltzmann kinetics by the delayed quantum kinetic scattering integrals (Schmenkel et al., 1998; Haug, 1998), the resulting buildup time for the first LO-phonon cascade is in agreement with the experiment, as shown in the lowest section of Fig. 5.

IV. Femtosecond DTS for Screened Coulomb and LO-Phonon Quantum Kinetic Scattering

In order to study the buildup of Coulomb correlations, renormalizations, and particularly the relaxation of the hot carriers, the technique of differential transmission spectroscopy is appropriate. In these experiments a strong resonant pump pulse excites carriers. One measures with a second delayed probe (also called test) pulse how its transmission spectrum is changed by the preceding pump pulse excitations. The measured spectra need a theoretical analysis, because often several many-body effects contribute to the observed changes.

A particularly well-designed experiment has been performed by Camescasse and co-workers (Vu et al., 1999) in which a relatively long, spectrally narrow pump pulse of 150 fs excites electrons into the conduction band from the heavy- and from the light-hole valence band, respectively. The spectral resolution of this pulse is high enough to generate an electron distribution with two peaks. In order to study only the relaxation of this nonequilibrium electron distribution in the conduction band, the following 30-fs, spectrally wide test pulse is tuned to transitions between the completely filled lower-lying spin-orbit split-off valence band to the conduction band. The DTS spectrum is thus not influenced by the hole relaxation in the two upper

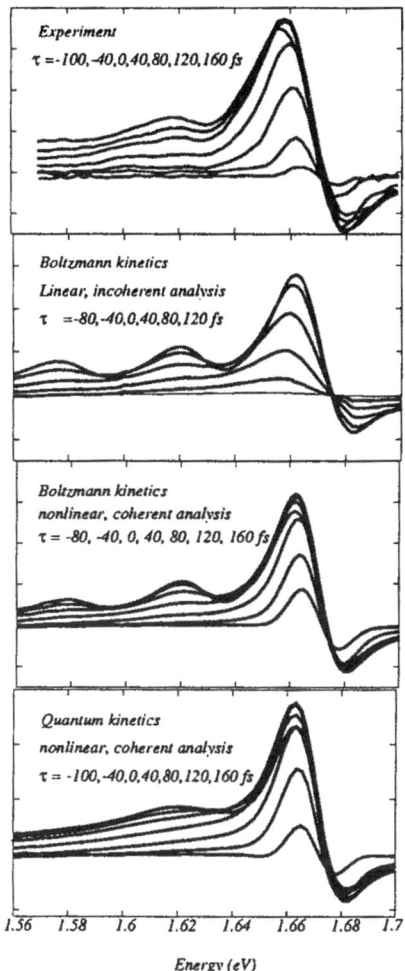

FIG. 5. Measured (top) and calculated DTS spectra in various approximations for delay times ranging from −100 fs to 160 fs.

valence bands, but is only influenced by the distribution of electrons in the conduction band. A sketch of the excitation scheme is shown in Fig. 1.

We solve the just-described Bloch equations in a three-band model for the pump pulse, including both carrier–carrier scattering with the time-dependent screening and carrier–phonon scattering. The resulting electron nonequilibrium distribution is inserted in the linearized Bloch equation of the test pulse for the off-diagonal density matrix elements $\rho_{c,so,k}(t)$. From these interband polarization components, one can calculate the absorption (or transmission) spectrum of the test pulse. In order to reduce the numerical complexity, the dephasing kinetics of $\rho_{c,so,k}(t)$ is simply described using a phenomenological dephasing with a transverse relaxation time T_2. Obviously, the disadvantage of this approximation is that the pump-induced line broadening is not included accurately. However, the very long time integration that is needed in order to calculate the spectrum of the test beam polarization is with quantum kinetic dephasing integrals presently not possible. The measured and calculated DTS spectra are shown for two excitation densities and for various delay times between the pump and test pulses in Figs. 6 and 7 (Vu *et al.*, 1999).

Because of bandgap renormalization and induced line broadening, one sees induced absorption below the exciton resonance at the original $E_{so,g}$. Similarly, the excitonic enhancement causes a weak induced absorption above the populated states related to the so-called Mahan exciton. The double peak structure can be seen in the calculated spectra up to delay times of about $+80$ fs, while for larger delays, particularly with $\tau = 160$ and 200 fs, already monotonically decreasing distributions are reached.

All the complicated features seen in the experiment are extremely well reproduced by the theory.

In particular, the double-peak structure at early delay times, the induced absorption regions below and above the populated states, and the speed of the evolution of the spectra are correctly given by the theory. Because of the energy relaxation by LO-phonon emission, the present calculations predict correctly that for larger delay times there is no longer a dip in the DTS spectrum above the exciton resonance, which one would get without the phonon scattering. However, because of the oversimplified, phenomenological description of the dephasing of the test pulse polarization, the broadening of the spectra — particularly around the exciton resonance — is insufficient, as expected. Because the pump-pulse-induced broadening is too small in comparison to the gap shift, one gets at early negative delay times below the exciton resonance a small spectral range with positive DTS signal that is not observed in the experiment.

A comparison of Fig. 6 with Fig. 7 shows that at higher density the spectra at later delay times change from a triangular shape to a more square shape because of the more degenerate e distribution. The calculated spectra follow these general trends quite well. In particular, both the measured and the calculated spectra show the filling of states with increasing delay. One can follow this effect by watching the high-energy crossover to induced absorption above the filled state due to excitonic enhancement.

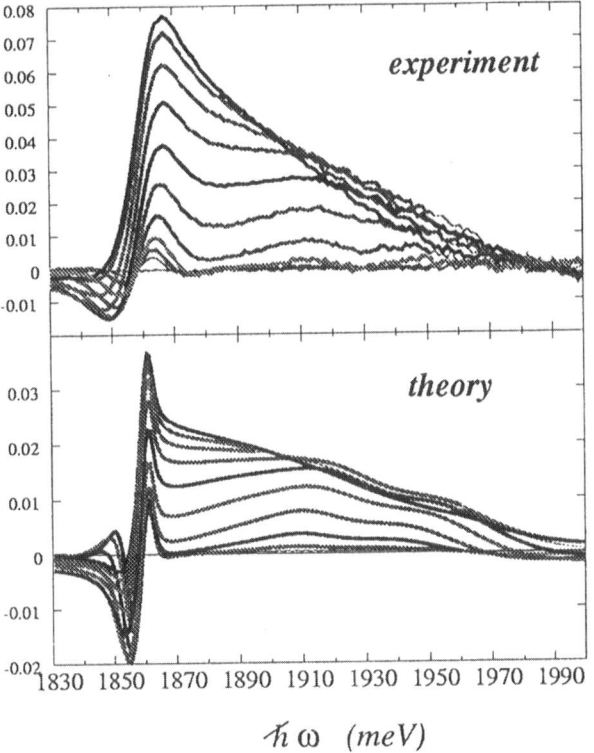

FIG. 6. Measured (top) and calculated (bottom) DTS spectra in GaAs at 15 K for various delay times in steps of 40 fs ranging from $\tau = -160$ fs to 200 fs for a pump pulse with a strength of $\chi = 0.72$, which generates the excitation density $n = 4.5 \times 10^{17}$ cm^{-3}. The central pump frequency has been 1.589 eV, corresponding to a detuning with respect to the unrenormalized lh-hh bandgap of $\Delta = 70$ meV. For the calculation of the test spectra, we used $T_2 = 170$ fs and $\gamma = 0.6 \times 10^{-20}$ cm^3 fs^{-1}.

V. Resonant FWM with Screened Coulomb and LO-Phonon Quantum Kinetic Scattering

In a four-wave mixing experiment, a laser-induced lattice is formed in the sample by two nonparallel beams. The beams themselves are diffracted from this lattice giving rise to FWM signals in various order.

Because the lattice is formed by the induced polarizations, the FWM signal decays with increasing delay between the pulsed beams. If the two

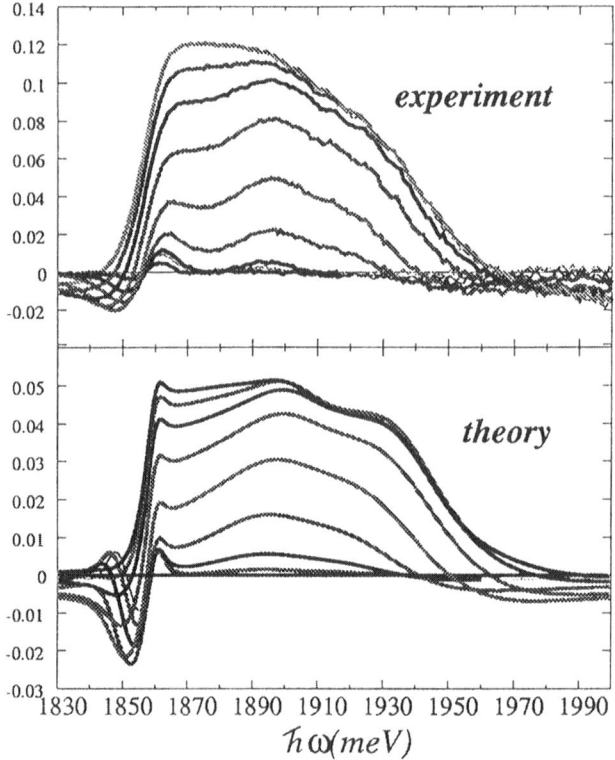

FIG. 7. Measured and calculated DTS spectra for GaAs at 15 K for various delay times in steps of 40 fs ranging from $\tau = -200$ fs to 1600 fs for a pump pulse that generates an excitation density of $n = 8.5 \times 10^{17}$ cm^{-3}. The central pump frequency has been taken as 1.569 eV, which corresponds to a detuning with respect to the unrenormalized lh-hh bandgap of $\Delta = 50$ meV. For the calculation of the test spectra we used $T_2 = 150$ fs and $\gamma = 0.4 \times 10^{-20}$ cm^3 fs^{-1}.

beams propagate in the directions of $\vec{k}_1$ and $\vec{k}_2$, the lowest FWM signal is observed in the direction of $2\vec{k}_2 - \vec{k}_1$. This signal can be calculated from the semiconductor Bloch equations by considering an excitation with two 11-fs hyperbolic secant pulses with equal central frequency and intensities according to the experiment, but propagating in different directions

$$e^{i(\vec{k}_1 \cdot \vec{r} - \omega t)}(E_0(t) + E_1(t - \tau)e^{i\phi}),$$

where the relative phase $\phi = (\vec{k}_2 - \vec{k}_1) \cdot \vec{r}$. For thin samples one can calculate the FWM signal by calculating the interband polarization in the direction corresponding to $e^{i(\vec{k}_1 \cdot \vec{r} - 2\phi)} = e^{(2\vec{k}_2 - \vec{k}_1) \cdot \vec{r}}$. For this purpose we use again the projection technique (Haug and Jauho, 1996) described previously. Because the FWM signal is generated by the interband polarization in the FWM direction, here one studies directly an off-diagonal element of the density matrix, that is, one directly observes the quantum coherence of the system. This quantum coherence can show up in quantum beats between various contributions to the FWM signal. The decay of the signal allows one to measure the dephasing processes directly.

In the calculations (Hügel et al., 1999), carrier–carrier scattering with time-dependent screening and carrier–phonon scattering are again included as described earlier. In Fig. 8 we show calculated time-resolved FWM for three densities and for various delay times according to Hügel et al. (1999) and compare them with a typical measured signal of Fig. 9. Both the calculated and measured time-resolved FWM signals are photon-echo-like at small delay times, while the signals shift to earlier times at large delays. One sees that these characteristic changes of the time-resolved signals are well reproduced within the non-Markovian screened Coulomb scattering, while a Markovian description of the scattering kinetics does not reproduce this behavior. If one increases the strength of the Coulomb effects by confinement of the carriers to a quasi-2D quantum well, the breakup of the time-resolved signals becomes more pronounced again both in theory and experiment (Mieck et al., 2000), showing that this feature is indeed a Coulomb effect.

The time-integrated FWM signal decays exponentially in time as shown in Fig. 10. For higher excitation intensity the decay becomes faster. The resulting dephasing time $\tau(n)$, which is obtained from exponential fits to the decaying time-integrated FWM signal of Fig. 10, is shown in Fig. 11. The dephasing time that is calculated quantum kinetically without adjustable parameters agrees astonishingly well with the measured ones. Both sets of data can be fitted by the power law

$$\frac{1}{\tau(n)} = \frac{1}{\tau_0} + cn^{1/3}, \tag{20}$$

where the first term describes the dephasing by room-temperature longitudinal optical phonons. The simplest excitation induced dephasing would be linear in the e-h density, but screening and Pauli blocking reduce the exponent of n to $\frac{1}{3}$.

Although Shank and co-workers (Becker et al., 1988) found this exponent in early photon echo experiments, they tried to understand the exponent in

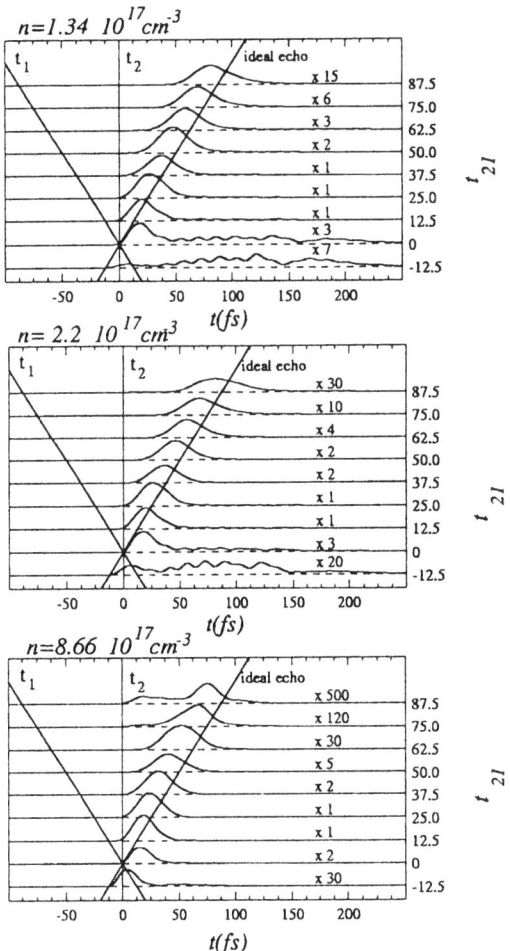

FIG. 8. Calculated time-resolved FWM signals for GaAs at 300 K for various delay times t_{21} and three excitation densities. The left and center lines mark the centers of the two pulses; the right line is the ideal echo line.

terms of the dimension D, that is, as $1/D$ supported by a hard sphere scattering model. They believed they had confirmed this conjecture with photon echo experiments on quantum wells (Bigot et al., 1991). Recent 11-fs experiments and corresponding quantum kinetic calculation in the just-described scheme (Mieck et al., 2000) showed, however, that in quasi-two-dimensional quantum well structures the dephasing also follows Eq. (20), again with the exponent 1/3.

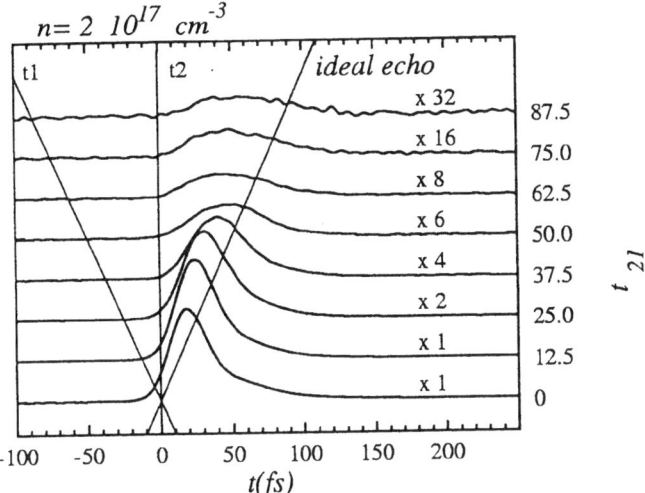

FIG. 9. Measured time-resolved FWM signals for GaAs at 300 K for various delay times t_{21} and an excitation density of $n = 2 \times 10^{17}\,\text{cm}^{-3}$. The left and center line mark the centers of the two pulses, the right line is the ideal echo line.

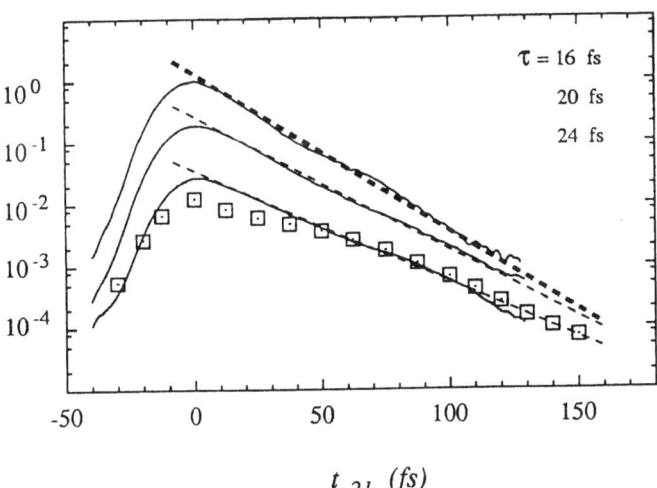

FIG. 10. Calculated (symbols) and measured (lines) time-integrated FWM signals for GaAs at 300 K as a function of the delay time t_{21} for three excitation densities together with exponential fits.

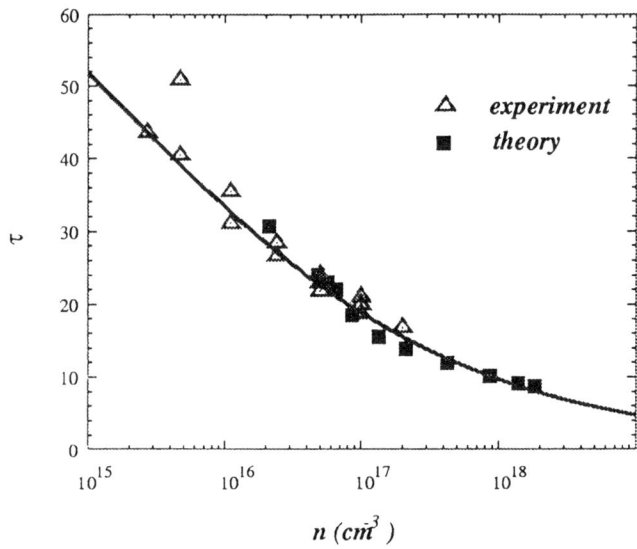

FIG. 11. Calculated and measured dephasing times as a function of the excitation density.

ACKNOWLEDGMENTS

This work has been supported by the DFG-Schwerpunktprogramm *Quantenkohärenz in Halbleitern*. I appreciate the fruitful cooperation within our research group, in particular with L. Bányai, C. Ell, K. El Sayed, P. Gartner, B. Mieck, E. Reitsamer, C. Remling, A. Schmenkel, P. Tamborenea, D. B. Tran Thoai, and Q. T. Vu. Furthermore, I acknowledge the fruitful cooperation with our experimental colleagues M. Wegener, A. Leitenstorfer, A. Alexandrou, and F. X. Camescasse.

REFERENCES

V. M. Axt, M. Herbst, and T. Kuhn (1999). *Superlattices and Microstructures* **26**, 117.
L. Bányai, D. B. Tran Thoai, E. Reitsamer, H. Haug, D. Steibach, M. U. Wehner, M. Wegener, T. Marschner, and W. Stolz (1995). *Phys. Rev. Lett.* **75**, 2188.
L. Bányai, Q. T. Vu, B. Mieck, and H. Haug (1998). *Phys. Rev. Lett.* **81**, 882.
P. C. Becker, H. L. Fragnito, C. H. Brito Cruz, R. L. Fork, J. E. Cunningham, J. E. Henry, and C. V. Shank (1988). *Phys. Rev. Lett.* **61**, 1647.
J. Y. Bigot, M. T. Portella, R. W. Schoenlein, J. E. Cunningham, and C. V. Shank (1991). *Phys. Rev. Lett.* **67**, 636.

M. Bonitz, D. Kremp, D. C. Scott, R. Binder, W. D. Kraeft, and H. S. Köhler (1996). *J. Phys. Cond. Matter* **8**, 6057.

K. El Sayed, S. Schuster, H. Haug, F. Herzel, and K. Henneberger (1994). *Phys. Rev. B* **49**, 7337.

C. Fürst, A. Leitenstorfer, A. Laubereau, and R. Zimmermann (1997). *Phys. Rev. Lett.* **78**, 3733.

P. Gartner, L. Bányai, and H. Haug (1999). *Phys. Rev. B* **60**, 14234.

M. Hartmann and W. Schäfer (1992). *Phys. Stat. Sol. B* **173**, 149.

H. Haug (1992). *Phys. Stat. Sol. B* **173**, 139.

H. Haug (1998). *J. Nonl. Opt. Physics and Materials* **7**, 227.

H. Haug and L. Bányai (1996). *Sol. State Commun.* **100**, 303.

H. Haug and A. P. Jauho (1996). *Quantum Kinetics in Transport and Optics of Semiconductors.* Springer, Berlin.

H. Haug and S. W. Koch (1993). *Quantum Theory of the Optical and Electronic Properties of Semiconductors*, 3d ed. World Scientific, Singapore.

W. A. Hügel, M. F. Heinrich, M. Wegener, Q. T. Vu, L. Bányai, and H. Haug (1999). *Phys. Rev. Lett.* **83**, 3373.

P. Lipavský, V. Špička, and B. Velicky (1986). *Phys. Rev. B* **34**, 6933.

G. Mahan (1981). *Many-Particle Physics.* Plenum, New York.

B. Mieck, H. Haug, W. A. Hügel, M. F. Heinrich, and M. Wegener (2000). *Phys. Rev. B*, in press.

E. Reitsamer, L. Bányai, D. B. Tran Thoai, P. I. Tamborenea and H. Haug (1996). In *Physics of Semiconductors* (M. Scheffler and R. Zimmermann, Eds.), p. 685. World Scientific, Singapore.

J. Schilp, T. Kuhn, and G. Mahler (1995). *Phys. Stat. Sol. B* **188**, 417.

A. Schmenkel, L. Bányai, and H. Haug (1998). *J. Luminesc.* **75/76**, 134.

J. Shah (1989). *Solid State Electron.* **32**, 1051.

J. Shah (1996). *Ultrafast Spectroscopy of Semiconductors and Semiconductor Microstructures.* Springer, Berlin.

D. Steinbach, G. Kocherscheidt, M. U. Wehner, H. Kalt, M. Wegener, K. Ohkawa, D. Hommel, and V. M. Axt (1999). *Phys. Rev. B* **60**, 12079.

D. B. Tran Thoai, L. Bányai, E. Reitsamer, and H. Haug (1995). *Phys. Stat. Sol. B* **188**, 1387.

Q. T. Vu (1999). Ph.D. Thesis (Dissertation), Universität Frankfurt am Main.

Q. T. Vu, L. Bányai, H. Haug, F. X. Camescasse, J.P. Lickforman, A. Alexandrou (1999). *Phys. Rev. B* **213**, 397.

M. Wegener, M. U. Wehner, D. Steinbach, L. Bányai, D. B. Tran Thoai, E. Reitsamer, H. Haug, and W. Stolz (1996). In *Physics of Semiconductors* (M. Scheffler and R. Zimmermann, Eds.), p. 633. World Scientific, Singapore.

M. U. Wehner, M. Ulm, D. S. Chemla, and M. Wegener (1998). *Phys. Rev. Lett.* **80**, 1992.

CHAPTER 6

Coulomb Correlation Signatures in the Excitonic Optical Nonlinearities of Semiconductors

T. Meier and S. W. Koch

DEPARTMENT OF PHYSICS AND MATERIAL SCIENCES CENTER
PHILIPPS UNIVERSITY
MARBURG, GERMANY

I.	INTRODUCTION	231
II.	THEORETICAL APPROACH AND MODEL	242
	1. *The Coherent $\chi^{(3)}$ Limit*	242
	2. *Nonlinear Optical Signals*	248
	3. *One-Dimensional Model System*	250
III.	APPLICATIONS TO PUMP-PROBE SPECTROSCOPY	254
	1. *Resonant Absorption Changes for Low Intensities*	254
	2. *Off-Resonant Absorption Changes for Low Intensities*	266
	3. *Higher Intensities up to the Coherent $\chi^{(5)}$ Limit*	274
IV.	ABSORPTION CHANGES INDUCED BY INCOHERENT OCCUPATIONS	280
V.	APPLICATIONS TO FOUR-WAVE-MIXING SPECTROSCOPY	290
	1. *Biexcitonic Beats*	290
	2. *Disorder-Induced Dephasing*	297
VI.	SUMMARY	304
VII.	OUTLOOK	307
	REFERENCES	309

I. Introduction

The analysis of optical nonlinearities in semiconductors and semiconductor heterostructures is an active field of current research. The near bandgap semiconductor response is determined by the optical interaction with the resonant or near resonant material polarization that may result in the excitation of carriers (electron–hole pairs) and/or transient coherent nonlinearities. The microscopic analysis of these effects requires the investigation of the relevant quasiparticles and their interactions. In this chapter we focus on the carrier interaction effects, that is, the many-body Coulomb correlations and their influence on bandgap semiconductor optical nonlinearities.

For low carrier densities the Coulomb interaction may lead to the formation of bound electron–hole complexes, that is, one electron–hole pair states (excitons), two electron–hole pairs states (excitonic molecules, biexcitons), etc. With increasing density Fermionic phase space filling, many-body screening effects, as well as collective excitations such as plasmons gradually become important; see Klingshirn and Haug (1981) for an early review of these topics and also the textbooks Zimmermann (1988), Haug and Koch (1994), and Haug and Jauho (1996). To consistently account for such many-body effects, the Coulomb interaction among the carriers needs to be included in the theory on a microscopic basis. This, however, introduces the typical many-body problem of an infinite equation hierarchy for the quantum mechanical correlation functions, which can be solved exactly only for very special excitation conditions. Generally, however, consistent approximation schemes are needed in order to reliably model the nonlinear optical response of semiconductors and semiconductor nanostructures.

As background for the work presented in this chapter we briefly survey the key developments that led to our current understanding. The remarkable general interest in the topic of semiconductor nonlinearities is reflected in an overwhelming number of publications. Since it is not feasible to refer to all the relevant work, we mostly focus on articles that are directly relevant in the context of the topics discussed in this chapter and apologize for all omissions.

One subject area of particular interest, that has already been studied intensively in the 1980s and is still receiving considerable current attention are the nonlinear absorption changes induced by the presence an electron–hole plasma. In the theoretical description of such nonlinearities many-body effects were included microscopically mostly by using Green's function techniques (Ivanov and Keldysh, 1983; Haug and Schmitt-Rink, 1984; Schäfer and Treusch, 1986; Henneberger and Haug, 1988; Hartmann et al., 1988). This approach provides a good basic understanding of the plasma induced absorption changes. As summarized, for example, in Haug and Schmitt-Rink (1984), the weakening of the exciton absorption with increasing plasma density, that is, exciton saturation, as well as the appearance of optical gain for high plasma densities is qualitatively well described by this theory. These nonlinear absorption changes are induced by Pauli blocking (also called phase-space filling), as well as genuine many-body effects such as bandgap renormalization and plasma screening, both of which are a consequence of the Coulomb interaction.

The role of Pauli blocking, which is the only nonlinearity that is present in the hypothetical absence of the Coulomb interaction (Allen and Eberly, 1975; Sipe and Ghahramani, 1993), is relatively easy to understand qualitatively. The blocking reflects the fact that the phase space available for the optical excitation is reduced if the system is already (partially) excited. It has its origin in the Fermionic nature of electrons and holes. More formally, Pauli blocking results in a factor $1 - f_e - f_h$ multiplying the optical field E, where f_e and f_h are the electron and hole occupation probabilities, which can assume values between zero and one (Haug and Koch, 1994). As a consequence of this prefactor a finite population of electrons and holes with $f_e + f_h < 1$ reduces the strength of the optical excitation. Furthermore, for $f_e + f_h > 1$ a change of sign occurs, which corresponds to the transition from absorption to optical gain, that is, light amplification or "negative absorption."

Whereas the role of Pauli blocking for absorption changes is rather simple to understand intuitively, the Coulomb-induced screening and band-gap renormalization are somewhat more complicated (Haug and Schmitt-Rink, 1984; Schmitt-Rink et al., 1989; Haug and Koch, 1994). Screening refers to the change of the effective Coulomb interaction potential between a pair of charge carriers in the presence of other carriers in an excited system. Typically, in the presence of an electron–hole plasma the effective real space interaction potential decays more rapidly than in vacuum, indicating the reduction of the overall interaction strength. Simultaneously, the single-particle energies of electrons and holes also depend on the level of excitation. This energy renormalization results in a modification of the single-particle energies, which can often be regarded as a shrinkage of the bandgap with increasing excitation density. Both these Coulomb effects strongly modify the optical semiconductor response resulting in many-body induced absorption and refractive index changes. The strength of the different contributions depends furthermore sensitively on the plasma density, the plasma temperature, and in particular also on the dimensionality of the physical system (Schmitt-Rink et al., 1989; Haug and Koch, 1994).

Besides the more rigorous Green's function theory, a simplified plasma theory (Bányai and Koch, 1986) has also been developed to model the nonlinear semiconductor response. In particular, optical absorption in the presence of an electron–hole plasma can be described by a generalized Elliott formula (Bányai and Koch, 1986; Koch et al., 1988b). This theory has the advantage of being much easier to evaluate than the Green's function results, while it still includes the most important nonlinearities. The plasma theory therefore was heavily used as input to study further effects, for example, dispersive optical bistability and increasing absorption induced optical bistability (Koch et al., 1988a).

Typically the many-particle scattering processes in semiconductors lead to dephasing of optical excitations in the picosecond to subpicosecond range. Since the experimental investigation of coherent effects in semiconductors requires laser pulses shorter than the dephasing time, it was not before the development of femtosecond laser sources in the late 1980s that detailed coherent nonlinear optical semiconductor measurements became possible. Such ultrafast optical experiments are ideally suited to study details of the microscopic processes governing the nonlinear optical response. Two well-established techniques in particular, yielding information about various interaction processes of the material excitations, are pump and probe (PP) and four-wave mixing (FWM) spectroscopy; see, for example, Fröhlich et al. (1985), Mysyrowicz et al. (1986), Von Lehmen et al. (1986), Schultheis et al. (1986a, 1986b), Schmitt-Rink and Chemla (1986), Fluegel et al. (1990), Göbel et al. (1990), Noll et al. (1990), Wegener et al. (1990), Leo et al. (1990), Webb et al. (1991), Cundiff et al. (1992), Kim et al. (1992a), and Lindberg et al. (1992).

Among the first coherent PP semiconductor experiments were measurements of the dynamical optical Stark effect of excitons (Fröhlich et al., 1985; Mysyrowicz et al., 1986; Von Lehmen et al., 1986). To investigate the Stark effect, typically a strong pump pulse is tuned below the lowest excitonic transition and the pump induced absorption changes are monitored by a weak probe pulse (Haug and Koch, 1994). The experiments showed that for large detuning the pump pulses lead to a blue shift of the excitonic resonance (Fröhlich et al., 1985; Mysyrowicz et al., 1986; Von Lehmen et al., 1986). This behavior is qualitatively similar to that of atomic systems, which can be understood in the "dressed atom picture" where the Stark shift corresponds to level repulsion induced by the coupling between the optical transition and the light field.

Whereas the basic occurrence of a blue shift can be motivated already on the basis of a simple two-level model, a more detailed understanding of the optical Stark effect in semiconductors requires a microscopic modeling of the relevant bands as well as the inclusion of the many-body Coulomb interaction. In ideal two-band systems and for large detuning below the exciton, the light-induced blue shift can be well described on the basis of the semiconductor Bloch equations (SBE) in the time-dependent Hartree–Fock approximation (TDHF) (Schmitt-Rink et al., 1988; Lindberg and Koch, 1988; Koch et al., 1988b; Binder et al., 1991; Haug and Koch, 1994).

Also, FWM experiments provide clear evidence for the importance of many-body Coulomb interactions. In this context, for example, the observation of time-integrated (TI-FWM) signals for negative delays as well as the long temporal rise of the time-resolved (TR-FWM) signal after the second pulse can only be understood if the Coulomb-induced optical nonlinearities are consistently included in the analysis (Wegener et al., 1990; Leo et al., 1990; Lindberg et al., 1992; Weiss et al., 1992; Kim et al., 1992b). It soon became apparent that the relative importance of the various nonlinear contributions resulting from a systematic microscopic analysis depends on various experimental and sample conditions, and especially also on the degree of disorder in the sample under investigation. For example, the just-mentioned Coulomb-induced effects are particularly strong in good quality samples with narrow inhomogeneous linewidths. Thus, sample inhomogeneities may sometimes mask interaction effects. However, as

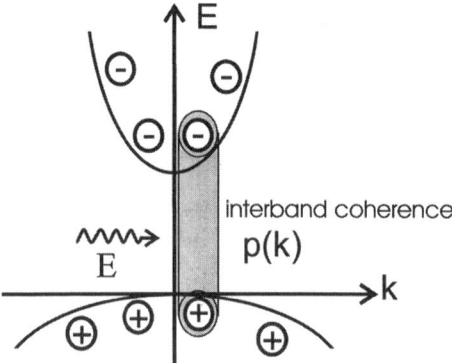

FIG. 1. Schematic illustration of the interband coherence p between holes in the valence and electrons in the conduction band created by a light field E.

already shown in Jahnke *et al.* (1994), even in disordered quantum wells there is still significant influence of many-body interactions requiring a simultaneous treatment of Coulomb and disorder effects, which is a major challenge for current theories and computer resources.

As already mentioned, the consistent theoretical analysis of Coulomb interaction effects leads to the hierarchy problem characteristic for many-body physics. This problem appears already when we set up the equation of motion for the interband coherence or single-exciton amplitude p (Haug and Koch, 1994). As shown in Fig. 1, the quantity p describes the optically induced coherence between a hole in the valence and an electron in the conduction band. In dipole approximation the total optical polarization P is given by a sum over all p's times the optical matrix element. Because of the many-body Coulomb interaction, the expectation value for the two-point function p couples to four-point functions. Evaluating the equations of motions for the four-point functions one finds that they couple to six-point function, etc., that is, the infinite hierarchy of equations of motion (Haug and Koch, 1994). Such a many-body problem can be solved exactly only for very small systems with just a few sites (Soos *et al.*, 1993; Guo *et al.*, 1995; Sauter-Fischer *et al.*, 1998). If one is interested in the optical response of extended semiconductors and semiconductor nanostructures, approximations are needed to derive closed systems of equations that can be treated numerically.

One well-known approximation scheme is to limit the dynamical quantities entering the theory to the two-point level. This is what is done in the TDHF approximation, where all four-point functions are factorized into products of two-point functions (Lindberg and Koch, 1988; Haug and Koch, 1994). This procedure results in coupled equations for the interband polarization and the electron and hole populations. Within the TDHF the Coulomb interaction leads to excitonic resonances and manifests itself in renormalizations of the effective field and the single-particle energies, which are linear in the Coulomb interaction. The resulting coupled equations of motion for the interband coherence and the electron and hole populations are often referred to as the SBE in TDHF approximation (Haug and Koch, 1994). These equations have been used extensively to describe various nonlinear coherent optical effects in semiconductors. The SBE are a direct extension of the atomic optical Bloch equations to which they formally reduce in the limit of vanishing Coulomb interaction.

The Hartree–Fock SBE (HF-SBE) properly describe interaction effects that lead to FWM signals for negative time delays as well as the temporal rise of the TR-FWM after the second pulse (Wegener et al., 1990; Leo et al., 1990; Lindberg et al., 1992; Weiss et al., 1992; Kim et al., 1992b; Schäfer et al., 1993). Furthermore, these equations provide a detailed understanding of the transient optical Stark effect for excitation of a two-band semiconductor below the excitonic resonance (Schmitt-Rink and Chemla, 1986; Koch et al., 1988b; Binder et al., 1991; Haug and Koch, 1994). Within the HF-SBE, many-body effects are included via the just-mentioned Coulomb-induced field and energy renormalization terms, which appear as optical nonlinearities besides the Pauli blocking. Especially in the limit of resonant excitation of excitons with weak light intensities, these Coulomb-interaction-induced nonlinearities are typically much stronger than the Pauli blocking and dominate the nonlinear optical response (Wegener et al., 1990; Leo et al., 1990; Lindberg et al., 1992; Weiss et al., 1992; Kim et al., 1992b; Schäfer et al., 1993). Under resonant excitation conditions it may be allowed to consider only the energetically lowest $1s$ exciton. If this is the case, one can project the HF-SBE onto this single exciton state. The resulting equations describe the exciton as an effective two-level system, which, because of the Coulomb-induced field and energy renormalizations, is driven not only by the external field, but by a so-called local field. This local field is the sum of the external field plus an induced field that is proportional to the exciton polarization (Wegener et al., 1990; Lindberg et al., 1992; Meier et al., 1994).

To avoid the complexities of a fully microscopic analysis, the nonlinear optical response of semiconductors has often been modeled in terms of few-level systems; see for example Göbel et al. (1990), Leo et al. (1992), Koch et al. (1992, 1993), Feldmann et al. (1993), and Erland and Balslev (1993). In these approaches the optical nonlinearities are purely induced by Pauli blocking. Even though this may seem somewhat surprising, such simple approaches with properly adjusted parameters were often quite successful in describing many basic features of experiments. Examples are quantum beats in TI-FWM and photon echoes in TR-FWM (Göbel et al., 1990; Leo et al., 1992; Koch et al., 1992, 1993; Feldmann et al., 1993; Erland and Balslev, 1993).

For some time the HF-SBE were basically sufficient to analyze experiments on transient semiconductor nonlinearities. With improved sample quality, experimental techniques, and theoretical analysis, it became, however, clear that it is necessary to go beyond the Hartree–Fock approximation and to include Coulomb correlation effects. Some of the important experiments indicating the relevance of correlations used pulses with different polarization directions. Such experiments gave evidence for excitation-induced dephasing processes (Wang et al., 1993, 1994; Hu et al., 1994; Rappen et al., 1994) and clearly revealed that two-exciton states, especially the bound biexciton, may lead to significant contributions (Hulin and Joffre, 1990; Feuerbacher et al., 1991; Bar-Ad and Bar-Joseph, 1992; Lovering et al., 1992; Bott et al., 1993, 1996; Bigot et al., 1993; Mayer et al., 1994, 1995; Smith et al., 1994, 1995; Bartels et al., 1995; Albrecht et al., 1996; Schäfer et al., 1996; Langbein et al., 1997; Kner et al., 1997, 1998; Sieh et al., 1999a, 1999b).

Concerning the inclusion of correlations different schemes have been developed. One can, for example, extend the HF-SBE to include Coulomb correlations on the level of two-point functions (Lindberg and Koch, 1988; Haug and Koch, 1994). To that end one can derive equations of motion for the correlation contributions and then factorize the expressions appearing as driving terms into products of two-point functions. Using, for example, the second Born approximation (SBA) (Binder et al., 1992; Rappen et al., 1994; Schäfer et al., 1996; Jahnke et al., 1996, 1997), these equations are then formally solved and the result is inserted into the equations for the two-point functions. The correlation contributions within the SBA describe excitation induced scattering and dephasing, i.e., they lead to terms in the equations of motion that are quadratic in the Coulomb interaction potential. This type of theory has been successfully used to describe the density dependent nonlinear optical response of quantum-well excitons and of excitons in microcavities (Jahnke et al., 1996, 1997). Furthermore, good agreement with experiments could be obtained for quasiequilibrium gain spectra of excited quantum wells (Chow et al., 1997; Girndt et al., 1998). Hence, under many conditions the SBA seems to be sufficient to describe the most important aspects of correlations and their influence on optical nonlinearities in semiconductors in the presence of an electron–hole plasma.

One physical process that has been proven to contribute strongly to the nonlinear optical response for low densities and resonant exciton excitation is the formation of a bound two-exciton state (Hulin and Joffre, 1990; Feuerbacher et al., 1991; Bar-Ad and Bar-Joseph, 1992; Lovering et al., 1992; Bott et al., 1993, 1996; Bigot et al., 1993; Mayer et al., 1994, 1995; Smith et al., 1994, 1995; Bartels et al., 1995; Albrecht et al., 1996; Schäfer et al., 1996; Langbein et al., 1997; Sieh et al., 1999a, 1999b). In spectrally resolved FWM and PP experiments performed on high-quality samples, the exciton to biexciton transition appears below the exciton resonance with an energy difference corresponding to the biexciton binding energy. To theoretically describe biexcitons at the dynamical level one has to deal with the full quantum mechanical four-body problem, including, the Coulomb interaction among the particles (Lindberg et al., 1994b; Axt and Stahl, 1994a, 1994b). The SBA does not describe bound biexciton resonances, since the Coulombic interactions appearing in the homogeneous part of the equation of motion of the four-point function describing biexcitons are neglected (Schäfer et al., 1996).

To be able to calculate the nonlinear optical response of semiconductors including two-exciton resonances, it is thus necessary to include not only two-point functions, but also higher-order correlation functions in the theoretical description. Again, there are many possible ways to construct closed sets of equations involving, for example, nonequilibrium Green's functions techniques or equations of motion methods. One approach that is particularly well suited for coherent excitation conditions is to limit the theoretical analysis to a finite order in the optical field. In finite order in the light field, the Coulomb correlations also contribute only up to a finite order, and thus only a finite number of electron–hole correlation functions need to be included (Lindberg et al., 1994b; Axt and Stahl, 1994a, 1994b; Östreich et al., 1995). To be able to describe FWM and PP experiments one needs to consider at least processes up to third order in the light field ($\chi^{(3)}$).

For a complete description of the optical response up to third order within the coherent limit, where interactions with other quasiparticles, for example phonons, are neglected, it has been shown that it is sufficient to consider just two density matrices. These are the single-exciton amplitude (polarization amplitude) p and the two-exciton amplitude B, which describes the evolution of two interacting electron–hole pairs. Numerical solutions of the coherent $\chi^{(3)}$ equations are rather demanding since a four-body problem has to be solved. Thus, even nowadays the full coherent $\chi^{(3)}$ equations are mostly solved only for one-dimensional systems (Axt et al., 1996, 1998; Östreich et al., 1998; Meier and Koch, 1999; Sieh et al., 1999b; Weiser et al., 2000). Two-dimensional calculations are possible; however, they require elaborate programming and computer resources (Kner et al., 1997, 1998; Sieh et al., 1999a). In the coherent $\chi^{(3)}$ limit for excitation at and below the exciton resonance, the one- and two-dimensional results show qualitatively the same absorption changes around the exciton resonance, which are in agreement with experiment (Sieh et al., 1999a).

As for the description of excitons, for the analysis of biexcitonic contributions phenomenological discrete-level systems have also been proposed and applied. Again, by choosing suitable system parameters, that is, the relevant exciton and two-exciton states, and their energies, as well as the optical matrix elements between them, these schemes were able to qualitatively model a variety of experimental results (Bott et al., 1993, 1996; Finkelstein et al., 1993; Mayer et al., 1994, 1995; Smith et al., 1995; Albrecht et al., 1996;

Langbein et al., 1997; Wagner et al., 1999). Such an analysis is well suited for a qualitative understanding of the relevant states and symmetries arising as a consequence of the optical selection rules.

If one leaves the coherent limit, additional occupation-type dynamical variables need to be considered. Up to third order it has been shown that four density matrices have to be taken into account. Besides the coherences p and B, the pair (exciton) occupation N and the exciton to two-exciton transition Z also have to be included (Axt et al., 1996). Since Z is a six-point function, solutions of the coupled equations for these four density matrices are numerically extremely demanding (Axt et al., 1996; Meier and Koch, 1999). So far, few such calculations have been reported. In Meier and Koch (1999) it has been shown that if low-density incoherent occupations N are present, the six-point function Z does contribute strongly to the optical response. It thus needs to be considered for a proper description of excitonic bleaching and for transitions from incoherent single- to two-exciton states.

It should be noted here that for strongly bound Frenkel excitons, which are often used to describe optical excitations in molecular and organic systems, a similar many-body problem exists as for Wannier excitons (Mukamel, 1995). To describe the third-order response of Frenkel exciton systems, one needs to consider exactly the same four density matrices that have been introduced (Mukamel, 1995; Meier et al., 1997a, 1997b; Zhang et al., 1998; Chernyak et al., 1998a). Since a Frenkel exciton consists of an electron and a hole that are strongly bound and are assumed to stay together in the same unit cell all the time, the resulting equations for Frenkel excitons are much simpler to solve numerically than the Wannier exciton problem. For example, in Frenkel exciton systems the two-exciton amplitude B is only a correlation function with two (exciton) indices, and Z has only three indices, whereas for Wannier systems the internal dynamics of the electrons and holes has to be considered explicitly and these two quantities have twice as many indices. Therefore, for Frenkel exciton systems it has been possible to perform calculations of the nonlinear optical response using the full $\chi^{(3)}$ equations, including simultaneously the influence of disorder as well as exciton–phonon coupling (Meier et al., 1997a, 1997b; Zhang et al., 1998). So far such calculations have not been reported for Wannier systems.

If one is interested in the optical response for excitation with increasing intensity levels, the $\chi^{(3)}$ limit is no longer valid and higher order contributions of the light field need to be considered (Bartels et al., 1995; Albrecht et al., 1996). As outlined above, because of the many-body hierarchy this requires the inclusion of additional correlation functions. In the coherent limit it is possible to formulate the optical response up to $\chi^{(n)}$ using $(n + 1)/2$ transition-type correlation functions (Victor et al., 1995). For example, in the coherent $\chi^{(5)}$ limit, besides p and B, the three-exciton amplitude W contributes to the optical response (Victor et al., 1995). Some evidence for the existence of three- and four-exciton complexes has been found in size and intensity dependent photoluminescence studies of quantum dots (Bayer et al., 1998). Since calculations of the three-exciton amplitude W require the solution of a full six-body problem, so far they have not been possible for extended semiconductor systems. In our results, we give some hints on how the contributions induced by W may show up in PP spectroscopy.

Concerning the expansion in orders of the optical field, a few remarks are in order. The $\chi^{(n)}$ expansion is typically done with respect to the laser field amplitude E or similarly with respect to the interaction energy, which is the laser field amplitude times the dipole transition matrix element μE. Since these are not dimensionless quantities, it is clear that they cannot be the proper quantities that should be used in a systematic expansion. A proper dimensionless expansion quantity could be $\mu E t/\hbar$, which means that a $\chi^{(n)}$ expansion is only valid up to a maximum, interaction-determined time t. The appearance of an upper time limit can be motivated by considering high density effects, arising from infinite summations of the many-body hierarchy, for example, screening. Such effects that are missed in a truncation of the equation hierarchy may introduce additional energetic shifts $\delta\varepsilon$ of the single- and multiple-exciton resonances. Therefore the applicability of a $\chi^{(n)}$ expansion is naturally restricted to a time scale $t < \hbar/\delta\varepsilon$, where t decreases with increasing excitation intensity. A systematic investigation of the limited validity of this expansion scheme is still an open problem of current research.

In this chapter we review our recent many-body analysis of resonant semiconductor nonlinearities starting from the coherent low intensity $\chi^{(3)}$ limit. Next, we include higher excitation intensities up to $\chi^{(5)}$ and analyze the influence of incoherent occupations on the optical response (Sieh et al., 1999a, 1999b; Meier and Koch, 1999; Koch et al., 1999; Weiser et al., 2000). Mostly for numerical reasons we present in this chapter detailed results only for a one-dimensional model system that allows us to investigate signatures of correlations in different situations.

We start in Section II with a presentation of the theoretical approach. The equations of motion describing the coherent $\chi^{(3)}$ limit are derived and discussed in Subsection 1. Extensions of the theory toward higher orders in the light field ($\chi^{(5)}$) and the inclusion of incoherent contributions are given in Subsection 3 of Section III and Section IV, respectively. Applications to calculate nonlinear optical signals induced by optical excitation with two laser pulses are discussed in Subsection 2 of Section II. The one-dimensional model system used throughout this chapter is introduced in Subsection 3 of Section II.

In Section III we focus on various correlation-induced effects as seen in PP spectroscopy. We start by analyzing absorption changes following resonant excitation close to the exciton within the coherent $\chi^{(3)}$-limit in Subsection 1. It is shown that the inclusion of correlations leads to characteristic signatures in the process of exciton bleaching and introduces excited state absorption in the PP spectra at energetic positions corresponding to single- to two-exciton transitions. For excitation with opposite circularly and colinearly polarized pump and probe pulses, a bound biexciton appears energetically below the exciton in the PP spectra. If both heavy- and light-hole transitions are excited, one may also see a bound two-exciton state for cocircularly polarized excitation. This is a mixed biexciton that is made of a heavy- and a light-hole exciton. In Subsection 2, results on absorption changes induced by off-resonant pumping below the exciton resonance, that is, in Stark effect configuration, are presented. Here, we especially find that instead of the usual blue shift of the exciton resonance, it is possible under certain conditions to see a correlation-induced red shift. Furthermore, it is demonstrated that the influence of correlations generally decreases with detuning from the exciton resonance. Thus, for large detuning the differential absorption is well described within the TDHF approximation. The results discussed in Subsections 1 and 2 are extended toward higher pump intensities in Subsection 3, where the coherent $\chi^{(5)}$ limit is briefly introduced.

Absorption changes induced by incoherent low-density occupations are analyzed in Section IV. It is shown that in the presence of low-density incoherent occupations N, the six-point correlation function Z that describes exciton to two-exciton transitions contributes strongly to the optical response. This six-point function is important for a proper description of exciton bleaching and is necessary to describe resonances corresponding to single- to two-exciton transitions.

In Section V we analyze signatures of correlations in FWM spectroscopy. Results on two-exciton- and especially biexciton-induced beats in TR- and TI-FWM within $\chi^{(3)}$ and $\chi^{(5)}$ are presented in Subsection 1. In agreement with experimental observations, we find that no biexciton-induced beats are present in the TI-FWM of a homogeneous system within $\chi^{(3)}$. Such beats do, however, appear if higher orders ($\chi^{(5)}$) contribute significantly to the response. We also discuss the influence of disorder, which is treated phenomenologically as a simple correlated inhomogeneous broadening in Subsection 1, on the biexciton-induced beating in TI-FWM. In disordered systems polarization-dependent exciton–biexciton beats appear in the TI-FWM already at the $\chi^{(3)}$ level. For excitation with perpendicular polarized pulses, we furthermore find that in agreement with experimental observations, half-period beats may appear in the TI-FWM of disordered systems for higher excitation intensities. Results including both correlations and disorder on a microscopic basis are presented in Subsection 2. It is shown that weak diagonal disorder may introduce a polarization-dependent decay of the TI-FWM signal. This disorder-induced dephasing has been observed experimentally and can be understood only if correlations are included in the theoretical description.

Our main results are summarized in Section VI and an outlook is presented in Section VII.

II. Theoretical Approach and Model

In this section we review the theoretical analysis of coherent optical nonlinearities of semiconductors using the expansion in orders of the external optical field. In Subsection 1, we describe the derivation and explicitly list the equations that define the coherent $\chi^{(3)}$ limit. Extensions of this approach toward the inclusion of higher-order ($\chi^{(5)}$) terms and incoherent contributions are given in Subsection III.3 and Section IV, respectively. The theoretical analysis of nonlinear optical experiments involving two-pulse excitation like PP and FWM is outlined in Subsection 2 of the present section. The one-dimensional tight-binding model used for the numerical evaluation of the equations is introduced and described in Subsection 3.

1. The Coherent $\chi^{(3)}$ Limit

To theoretically describe the dynamics of optical excitations in semiconductors we use the general Hamiltonian (Haug and Koch, 1994):

$$H = H_0 + H_C + H_I. \tag{1}$$

Here, H_0 is the single-particle Hamiltonian, H_C describes the Coulomb interaction, and H_I the interaction with a classical electric field.

In a localized real-space basis the single-particle Hamiltonian H_0 reads

$$H_0 = \sum_{ijc} T_{ij}^c c_i^{c+} c_j^c + \sum_{ijv} T_{ij}^v d_i^{v+} d_j^v, \tag{2}$$

where i and j label real-space sites and $c(v)$ the conduction (valence) bands, respectively. $c_i^{c+}(c_j^c)$ creates (destroys) an electron at site $i(j)$ in band c, and $d_i^{v+}(d_j^v)$ creates (destroys) a hole at site $i(j)$ in band v. The diagonal terms of the matrices T contain the electronic site energies, whereas the couplings between the sites are defined by the off-diagonal matrix elements. Since we do not consider band-mixing effects in this paper, T is assumed to be band diagonal and thus carries only a single band index.

The Coulomb Hamiltonian H_C is used in the form

$$H_C = \frac{1}{2} \sum_{ijvv'} (c_i^{v'+} c_i^{v'} - d_i^{v'+} d_i^{v'}) V_{ij} (c_j^{v+} c_j^v - d_j^{v+} d_j^v), \tag{3}$$

where V_{ij} describes the monopole–monopole Coulomb interaction (Huhn and Stahl, 1984) between particles at sites i and j. H_C includes the repulsion between electrons and between holes, as well as the attraction between electrons and holes, which gives rise to the formation of bound electron–hole complexes, that is, excitons, biexcitons etc. As usual in descriptions of inorganic semiconductors with bandgap energies in the optical region (Haug and Koch, 1994), Coulombic interactions that interchange particles between the valence and conduction bands are assumed to be small and thus neglected in H_C.

The dipole interaction of the electronic system with a classical electric field is given by

$$H_I = -\mathbf{E}(t) \cdot \mathbf{P} = -\mathbf{E}(t) \cdot \sum_{ijvc} (\boldsymbol{\mu}_{ij}^{vc} d_i^v c_j^c + (\boldsymbol{\mu}_{ij}^{vc})^* c_j^{c+} d_i^{v+}), \qquad (4)$$

where $\boldsymbol{\mu}$ is the matrix element for the optical transition between the valence and conduction bands and $\mathbf{P}$ is the total optical interband polarization, which is given by summing over all microscopic polarizations $\boldsymbol{\mu}_{ij}^{vc} d_i^v c_j^c$. Since we are interested in the dependence of the nonlinear optical response on the polarization direction of the incident pulses, the vector character of $\mathbf{E}$, $\boldsymbol{\mu}$, and $\mathbf{P}$ is important and cannot be neglected.

Using the total Hamiltonian H and the Heisenberg equation, the equation of motion of the interband coherence $p_{12}^{v_1 c_2} = \langle d_1^{v_1} c_2^{c_2} \rangle$ is obtained as

$$-i\partial_t p_{12}^{v_1 c_2} = -\sum_j T_{2j}^c p_{1j}^{v_1 c_2} - \sum_i T_{i1}^v p_{i2}^{v_1 c_2} + V_{12} p_{12}^{v_1 c_2}$$

$$- \mathbf{E}(t) \cdot [(\boldsymbol{\mu}_{12}^{v_1 c_2})^* - \sum_{jc} (\boldsymbol{\mu}_{1j}^{vc})^* f_{j2}^{cc_2} - \sum_{iv} (\boldsymbol{\mu}_{i2}^{vc_2})^* f_{i1}^{vv_1}]$$

$$+ \sum_{av_a} V_{a1} \langle d_a^{v_a} d_1^{v_1} d_a^{v_a+} c_2^{c_2} \rangle - \sum_{av_a} V_{2a} \langle d_1^{v_1} d_a^{v_a} c_2^{c_2} d_a^{v_a+} \rangle$$

$$+ \sum_{ac_a} V_{a1} \langle c_a^{c_a+} d_1^{v_1} c_a^{c_a} c_2^{c_2} \rangle - \sum_{ac_a} V_{2a} \langle d_1^{v_1} c_a^{c_a+} c_2^{c_2} c_a^{c_a} \rangle. \qquad (5)$$

Here, the generalized electron and hole populations (intraband coherences) are defined as $f_{jj'}^{cc'} = \langle c_j^{c+} c_{j'}^{c'} \rangle$ and $f_{ii'}^{vv'} = \langle d_i^{v+} d_{i'}^{v'} \rangle$. The four-point correlation functions appearing in Eq. (5) represent the first step of the infinite hierarchy of many-particle correlations induced by the Coulomb interaction.

In the remainder of this section, we now derive dynamic equations assuming the coherent limit in which dephasing processes due to scattering with other quasiparticles are neglected. Furthermore, we consider optical excitation with weak intensities, which allows us to restrict the theoretical description to third order in the optical field. These two assumptions define the coherent $\chi^{(3)}$ limit. In this limit the electron and hole populations and coherences do not have to be treated as independent variables but can be expressed via the interband coherences p as (Haug and Koch, 1994;

Lindberg et al., 1994b; Axt and Stahl, 1994a, 1994b)

$$f_{12}^{c_1c_2} = \sum_{av_a} p_{a2}^{v_ac_2}(p_{a1}^{v_ac_1})^* \qquad (6)$$

and

$$f_{12}^{v_1v_2} = \sum_{ac_a} p_{1a}^{v_1c_a}(p_{2a}^{v_2c_a})^*. \qquad (7)$$

Similarly, the four-point terms appearing in Eq. (5) can be written as (Lindberg et al., 1994b; Axt and Stahl, 1994a)

$$\langle d_a^{v_a+} d_1^{v_1} d_a^{v_a} c_2^{c_2} \rangle = \sum_{bc_b} \langle d_1^{v_1} c_b^{c_b} d_a^{v_a} c_2^{c_2} \rangle (p_{ab}^{v_ac_b})^*. \qquad (8)$$

Applying this decoupling scheme to all terms in Eq. (5), the optical response in the coherent $\chi^{(3)}$-limit can be expressed using just two transition-type quantities that describe the coherences between a single and two interacting electron–hole pairs. These quantities are the interband coherences or single-exciton amplitudes $p_{12}^{v_1c_2}$ and the two-exciton amplitudes $B_{1234}^{v_1c_1v_c_2} = \langle d_1^{v_1} c_2^{c} d_3^{v} c_4^{c_2} \rangle$, respectively (Lindberg et al., 1994b; Axt and Stahl, 1994a, 1994b; Schäfer et al., 1996). Figure 2 schematically illustrates the physical meaning of p and B. The quantity p describes the coherence between the ground and a single-exciton state, whereas B describes the coherence between the ground and a two-exciton state, respectively. Note that Eqs. (6)–(8) remain valid in the presence of disorder (Weiser et al., 2000).

In order to analyze pure correlation effects, which go beyond the TDHF approximation (Haug and Koch, 1994), and to distinguish between coherent and incoherent contributions (Axt et al., 1996), it is advantageous to remove the uncorrelated parts from B, that is, to define (Schäfer et al., 1996; Sieh et al., 1999b)

$$\bar{B}_{1234}^{v_1cv_c_2} = B_{1234}^{v_1cv_c_2} - p_{12}^{v_1c} p_{34}^{v_c_2} + p_{14}^{v_1c_2} p_{32}^{vc}. \qquad (9)$$

This procedure results in closed equations of motion for the single-exciton amplitude p_{12}^{vc} and the two-exciton amplitude $\bar{B}_{1234}^{v_1cv_c_2}$. The equation of motion for the single-exciton amplitude p is

$$-i\partial_t p_{12}^{vc} = -\sum_j T_{2j}^c p_{1j}^{vc} - \sum_i T_{i1}^v p_{i2}^{vc} + V_{12} p_{12}^{vc}$$
$$+ \mathbf{E}(t) \cdot [(\boldsymbol{\mu}_{12}^{vc})^* - \sum_{abv'c'} ((\boldsymbol{\mu}_{1b}^{vc'})^*(p_{ab}^{v'c'})^* p_{a2}^{v'c} + (\boldsymbol{\mu}_{b2}^{v'c})^*(p_{ba}^{v'c'})^* p_{1a}^{vc'})]$$
$$+ \sum_{abv'c'} (V_{a2} - V_{a1} - V_{b2} + V_{b1})[(p_{ba}^{v'c'})^* p_{b2}^{v'c} p_{1a}^{vc'}$$
$$- (p_{ba}^{v'c'})^* p_{ba}^{v'c'} p_{12}^{vc} - (p_{ba}^{v'c'})^* \bar{B}_{ba12}^{v'c'vc}]. \qquad (10)$$

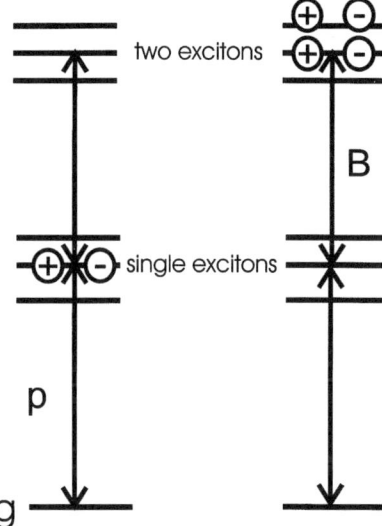

FIG. 2. Schematic illustration of the physical meaning of the two quantities p and B, which are relevant in the coherent $\chi^{(3)}$ limit in the basis of many-exciton eigenstates. $p = \langle dc \rangle$ describes single exciton, that is, one electron–hole pair excitations, and $B = \langle dcdc \rangle$ describes two-exciton coherences, that is, excitations of two electron–hole pairs. The destruction operators appearing in the definitions of p and B are symbolized by the circles with $-$ for the negatively charged electrons (c) and $+$ for the positively charged holes (d).

Here, the first line defines the homogeneous part, which includes the electronic energies and couplings (T) and the electron–hole Coulomb attraction (V_{12}). The next lines in Eq. (10) describe different types of inhomogeneities. In addition to the linear source term given by the external field times the dipole transition matrix element, $\mathbf{E}(t) \cdot \boldsymbol{\mu}^*$, in the coherent $\chi^{(3)}$ limit there are the optical nonlinearities arising from three processes. The first one is the Pauli blocking, $\mathbf{E}(t) \cdot \boldsymbol{\mu}^* p^* p$, which is the only nonlinearity present in the optical Bloch equations where Coulomb interactions are neglected (Allen and Eberly, 1975; Sipe and Ghahramani, 1993; Haug and Koch, 1994). The other two nonlinearities are induced by the many-body Coulomb interaction. They include the first-order Coulomb contribution (Vp^*pp) and the correlation contribution ($Vp^*\bar{B}$) (Schäfer et al. 1996; Sieh et al., 1999b), which involves two-exciton resonances. If one wants to analyze the origin of a particular nonlinear effect, it can become interesting to separately study the respective contributions of these three types of optical nonlinearities (Östreich et al., 1995; Schäfer et al., 1996; Axt et al., 1996; Sieh et al., 1999a, 1999b). In the TDHF approximation correlations are neglected and the optical response is determined by Pauli blocking and the first-order Coulomb contributions (Haug and Koch, 1994).

The first-order Coulomb term $\propto p_{ba}^{v'c'*} p_{ba}^{v'c'} p_{12}^{vc}$ looks like an energy renormalization, which for the typical selection rules for heavy-hole to conduction electron transitions (Bott et al., 1996; Schäfer et al., 1996), introduces a coupling between the spin subspaces. For a homogeneous system this term vanishes. This can be seen easily by recalling that the relation $p_{ba}^{vc} = p_{ab}^{vc}$ holds in a homogeneous system. Therefore, $p_{ba}^{v'c'*} p_{ba}^{v'c'} p_{12}^{vc}$ is symmetric with respect to interchanging a and b; the prefactor $V_{a2} - V_{a1} - V_{b2} + V_{b1}$, however, changes sign if a is replaced by b and vice versa, and thus the sum over a and b appearing in Eq. (10) vanishes. In inhomogeneous, for example, disordered systems, this first-order Coulomb term is, however, finite and does contribute to the nonlinear optical response; see Subsection 2 of Section V.

The equation of motion for the two-exciton amplitude $\bar{B}$ is

$$-i\partial_t \bar{B}_{ba12}^{v'c'vc} = -\sum_i (T_{2i}^c \bar{B}_{ba1i}^{v'c'vc} + T_{i1}^v \bar{B}_{bai2}^{v'c'vc} + T_{ai}^c \bar{B}_{bi12}^{v'c'vc} + T_{ib}^v \bar{B}_{ia12}^{v'c'vc})$$
$$+ (V_{ba} + V_{b2} + V_{1a} + V_{12} - V_{b1} - V_{a2}) \bar{B}_{ba12}^{v'c'vc}$$
$$- (V_{ba} + V_{12} - V_{b1} - V_{a2}) p_{1a}^{vc'} p_{b2}^{v'c}$$
$$+ (V_{1a} + V_{b2} - V_{b1} - V_{a2}) p_{ba}^{v'c'} p_{12}^{vc}. \tag{11}$$

The first two lines in Eq. (11) constitute the homogeneous part of the equation of motion, which includes the electronic energies and couplings (T) as well as the six possible Coulomb interactions between the two electrons and the two holes. The eigenmodes of $\bar{B}$ correspond to correlated complexes of two electrons and two holes, i.e., two-exciton states, including both bound biexcitonic and unbound continuum states. The last two lines in Eq. (11) represent the inhomogeneities. Since the uncorrelated contributions have been removed from $\bar{B}$, it is purely driven by sources that are proportional to the many-body interaction V (Schäfer et al., 1996; Axt et al., 1996; Sieh et al., 1999b), that is, by terms proportional to Vpp.

The total interband polarization $\mathbf{P}$ is obtained from the sum

$$\mathbf{P} = \sum_{ijvc} \boldsymbol{\mu}_{ij}^{vc} p_{ij}^{vc}. \tag{12}$$

Equations (10) and (11) fully determine the interband polarization $\mathbf{P}$ within the coherent $\chi^{(3)}$ limit (Lindberg et al., 1994b; Axt and Stahl, 1994a, 1994b; Östreich et al., 1995; Schäfer et al., 1996; Sieh et al., 1999b). For a spatially homogeneous system the complexity of solving the coupled equations can be reduced, since in this case the center of mass motion is irrelevant. Therefore, p and $\bar{B}$ depend only on the relative motion of the particles and it is advantageous to use the k-space version of the equations as presented in Schäfer et al. (1996). However, in spatially inhomogeneous (for example, disordered) systems, the complete Eqs. (10) and (11) have to be solved. These equations can easily be extended to include propagation effects (Stroucken *et al.*, 1996; Hübner *et al.*, 1996).

Instead of removing the uncorrelated parts from the four-point functions, one can also derive equations that describe the optical response up to $\chi^{(3)}$ using the original four-point functions $B_{1234}^{v_1 c v c_2}$ (Axt and Stahl, 1994a). This results in the following equations for p and B:

$$-i\partial_t p_{12}^{vc} = -\sum_j T_{2j}^c p_{1j}^{vc} - \sum_i T_{i1}^v p_{i2}^{vc} + V_{12} p_{12}^{vc}$$

$$+ \mathbf{E}(t) \cdot [(\boldsymbol{\mu}_{12}^{vc})^* - \sum_{abv'c'} ((\boldsymbol{\mu}_{1b}^{vc'})^*(p_{ab}^{v'c'})^* p_{a2}^{v'c} + (\boldsymbol{\mu}_{b2}^{v'c})^*(p_{ba}^{v'c'})^* p_{1a}^{vc'})]$$

$$- \sum_{abv'c'} (V_{a2} - V_{a1} - V_{b2} + V_{b1})[(p_{ba}^{v'c'})^* B_{ba12}^{v'c'vc}] \quad (13)$$

and

$$-i\partial_t B_{ba12}^{v'c'vc} = -\sum_i (T_{2i}^c B_{ba1i}^{v'c'vc} + T_{i1}^v B_{bai2}^{v'c'vc} + T_{ai}^c B_{bi12}^{v'c'vc} + T_{ib}^v B_{ia12}^{v'c'vc})$$

$$+ (V_{ba} + V_{b2} + V_{1a} + V_{12} - V_{b1} - V_{a2}) B_{ba12}^{v'c'vc}$$

$$- \mathbf{E}(t) \cdot [(\boldsymbol{\mu}_{12}^{vc})^* p_{ba}^{v'c'} + (\boldsymbol{\mu}_{ba}^{v'c'})^* p_{12}^{vc} - (\boldsymbol{\mu}_{1a}^{vc'})^* p_{b2}^{v'c} - (\boldsymbol{\mu}_{b2}^{v'c})^* p_{1a}^{vc'}]. \quad (14)$$

Equations (13) and (14) are totally equivalent to Eqs. (10) and (11), that is, solving either one of these equation pairs gives exactly the same result. What is, however, different is the appearance of the many-body Coulomb-induced optical nonlinearities in the equation of p. Whereas in Eq. (10) the Coulomb terms are split into a first-order ($Vp*pp$) and a correlation part ($Vp*\bar{B}$), no such distinction appears in Eq. (13), but all many-body nonlinearities are proportional to $Vp*B$.

The comparison of Eqs. (10) and (11) with Eqs. (13) and (14) indicates that strong compensations between the first-order and the higher-order Coulomb correlations can be expected. Examples of such cancellations have been reported in Jahnke *et al.* (1996, 1997) and Sieh *et al.* (1999a, 1999b). Because of these cancellations occurring in Eqs. (10) and (11), it may be numerically advantageous to solve Eqs. (13) and (14), which avoids the subtraction of two large numbers (Weiser *et al.*, 2000). Another difference between Eqs. (11) and (14) is that $\bar{B}$ is driven by terms proportional to Vpp (Schäfer *et al.*, 1996), whereas B is driven by μEp (Axt and Stahl, 1994a).

At this point it is instructive to compare the discussed approach with the SBA for the Coulomb interaction (Binder *et al.*, 1992; Rappen *et al.*, 1994; Schäfer *et al.*, 1996; Jahnke *et al.*, 1996, 1997). Generally, the SBA is not restricted to the $\chi^{(3)}$-regime and includes also the effect of screening of the Coulomb interaction. The SBA has been successfully applied for excitation conditions where a correlated electron–hole plasma dominates the nonlinear response. For a consistent comparison with the treatment of excitonic and biexcitonic correlations discussed in this chapter, we have to restrict the SBA to include only contributions up to $\chi^{(3)}$. At this level, the equations within the SBA are obtained if one neglects the six Coulomb terms in the homogeneous part of the equation for $\bar{B}$, that is, the third line in Eq. (11). The SBA in $\chi^{(3)}$ thus does not include bound two-exciton states and treats the unbound two-exciton continuum approximately. Since the driving terms in the $\bar{B}$ equation are proportional to V, and $\bar{B}$ entering in Eq. (10) is multiplied with V, the resulting correlation contributions in the polarization equation within the SBA are of second order in the Coulomb potential. If a dynamical equation for $\bar{B}$ is solved, Coulomb-memory effects are included in the SBA. Under appropriate conditions it is often sufficient to solve the equation for $\bar{B}$ within the Markov approximation and insert the result into the equation for p (Schäfer et al., 1996; Jahnke et al., 1996, 1997). Clearly, such a treatment eliminates Coulomb-memory effects, and its validity has to be checked for the respective situation under investigation.

In the theoretical treatment discussed here, contributions of two-exciton states are fully included on the $\chi^{(3)}$-level. Furthermore, the Markov approximation is not used, since in addition to the equation for p, an independent differential equation is solved for $\bar{B}$. Thus, on the $\chi^{(3)}$-level Coulomb-memory effects are fully included in the description.

2. Nonlinear Optical Signals

In the following we present the main aspects of the theoretical analysis of nonlinear optical experiments such as PP and FWM that are performed with two laser pulses. Within the rotating wave approximation the laser field is given by

$$\mathbf{E}(t) = \mathbf{e}_1 E_1(t) e^{i(\mathbf{k}_1 \cdot \mathbf{r} - \omega_1 t)} + \mathbf{e}_2 E_2(t) e^{i(\mathbf{k}_2 \cdot \mathbf{r} - \omega_2 t)}. \tag{15}$$

Here, $E_1(t) \propto e^{-((t+\tau)/\bar{t}_1)^2}$ and $E_2(t) \propto e^{-(t/\bar{t}_2)^2}$ denote the temporal envelopes of the Gaussian pulses that are centered at $t = -\tau$ and $t = 0$, respectively. A positive time delay τ corresponds to pulse 1 arriving before pulse 2. Furthermore, $\mathbf{e}_1$ and $\mathbf{e}_2$ define the polarization directions, ω_1 and ω_2 the central frequencies, $\bar{t}_1$ and $\bar{t}_2$ the pulse durations, and $\mathbf{k}_1$ and $\mathbf{k}_2$ the propagation directions of the two pulses; see Fig. 3.

In our numerical calculations, Eqs. (10) and (11) or equivalently Eqs. (13) and (14) are solved in the time domain up to third order in the laser field, Eq. (15). The single- and two-exciton amplitudes are decomposed into their different kinematic directions induced by the pulses (Lindberg et al., 1992). In first order in the light field one has to consider two linear interband

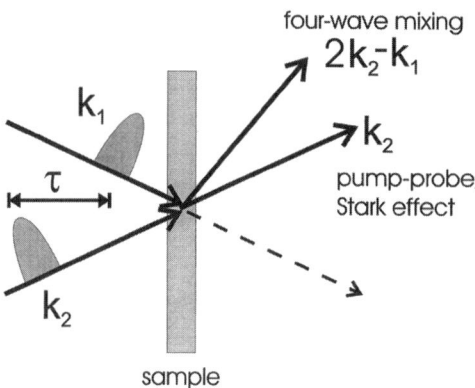

FIG. 3. Schematic illustration of nonlinear optical four-wave mixing and pump-probe experiments performed by excitation with two optical laser pulses having a time delay τ.

coherences resulting from the two pulses. These are denoted by $p^{(1|0)}$ and $p^{(0|1)}$, where the superscript $(n|m)$ labels the propagation direction $n\mathbf{k}_1 + m\mathbf{k}_2$. In second order, in principle, three two-exciton amplitudes $\bar{B}^{(2|0)}$, $\bar{B}^{(1|1)}$, and $\bar{B}^{(0|2)}$ appear. As outlined later, however, for the lowest order description of PP only $\bar{B}^{(1|1)}$ needs to be considered, whereas to FWM only $\bar{B}^{(0|2)}$ contributes.

In FWM performed in two-pulse self-diffraction geometry the field emitted in direction $2\mathbf{k}_2 - \mathbf{k}_1$ is detected; see Fig. 3. We thus need to calculate the interband coherence $p^{(-1|2)}$ associated with this direction. Since within $\chi^{(3)}$ pulse 1 is conjugated and included in first order, it is clear that only pulse 2, which is included in second order, can excite a two-exciton coherence. Thus, only $\bar{B}^{(0|2)}$ needs to be calculated in this case.

The TR-FWM *amplitude* depending on real time t and time delay τ is given by

$$P_{FWM}(t, \tau) = \sum_{ijvc} \mu_{ij}^{vc} (p_{ij}^{vc})^{(-1|2)}(t, \tau). \tag{16}$$

As outlined earlier, this amplitude is induced by optical nonlinearities corresponding to Pauli blocking and first- and higher-order Coulomb terms. It can be split into the sum over the individual amplitudes (Weiser et al., 2000)

$$P_{FWM}(t, \tau) = P_{FWM}^{pb}(t, \tau) + P_{FWM}^{CI,first}(t, \tau) + P_{FWM}^{CI,corr}(t, \tau), \tag{17}$$

where the superscript pb denotes the optical nonlinearity induced by Pauli blocking and the terms denoted with *CI* are due to Coulomb-interaction-induced nonlinearities CI, *first* is the first-order (Hartree–Fock) term, and CI, *corr* is the higher-order correlation contribution. In a TDHF calculation the correlations are neglected and the optical response is given by the sum over the Pauli blocking and the first-order Coulomb terms (Haug and Koch, 1994). For optically thin samples, the experimentally measured TR-FWM signal intensity is proportional to

$$S_{FWM}(t, \tau) = |P_{FWM}(t, \tau)|^2, \qquad (18)$$

and the TI-FWM signal is

$$I_{FWM}(\tau) = \int_{-\infty}^{\infty} |P_{FWM}(t, \tau)|^2 dt. \qquad (19)$$

In our analysis of PP signals we consider pulse 1 as the pump and pulse 2 as the probe pulse. A PP experiment then measures pump-induced changes of the field emitted into the probe direction (0|1); see Fig. 3. These changes are determined by the differential polarization $\delta \mathbf{P}$ (Haug and Koch, 1994), which is obtained by considering all contributions which (i) propagate in the direction of the probe pulse ($\mathbf{E}_2$), (ii) include two interactions with the pump pulse ($\mathbf{E}_1$), and (iii) are linear in the probe pulse. In this case a two-exciton coherence can only be induced by the combined action of pump and probe. Thus, only $\bar{B}^{(1|1)}$ needs to be calculated. If the duration of the probe pulse is short, that is, if the probe spectrum is broad, the differential absorption is determined by the Fourier transform of the differential time-domain polarization (Haug and Koch, 1994; Sieh *et al.*, 1999b),

$$\delta\alpha(\omega, \tau) \propto Im\left[\int dt (\mathbf{e}_2)^* \cdot \delta\mathbf{P}(t, \tau) e^{i\omega t}\right], \qquad (20)$$

where $\mathbf{e}_2$ denotes the polarization direction of the probe pulse. The total differential absorption can be written as the sum over the three nonlinearities (Sieh *et al.*, 1999b):

$$\delta\alpha(\omega, \tau) = \delta\alpha_{pb}(\omega, \tau) + \delta\alpha_{CI,first}(\omega, \tau) + \delta\alpha_{CI,corr}(\omega, \tau). \qquad (21)$$

3. One-Dimensional Model System

To keep the numerical complexity within reasonable limits, we present in this chapter numerical results for a one-dimensional model system. However, we have verified for several configurations that for appropriately chosen parameter combinations, the results are also qualitatively representative for realistic quantum-well systems; see, for example, Sieh et al. (1999a).

The electronic coupling in our model system is used in the tight-binding approximation, that is, $T_{ij}^{c,v} = 0$ for $|i - j| > 1$. The tight-binding coupling is described by the elements $T_{i,i+1}^{c,v} = T_{i+1,i}^{c,v} = J_{c,v}$. The diagonal parts $T_{ii}^{c,v} = \varepsilon_i^{c,v}$ are the energies of the electron and hole sites, respectively. Without disorder these energies are taken to be site independent and are chosen such that the energy of the 1s heavy-hole exciton coincides with the zero of the energy scale. In our studies of disorder effects we introduce Gaussian diagonal disorder where the site energies are chosen from Gaussian distributions.

To avoid the divergence of the exciton binding energy in one dimension, we use the regularized Coulomb-interaction potential

$$V_{ij} = U_0 \frac{d}{|i-j|d + a_0}. \qquad (22)$$

Here d is the distance between the sites and U_0 and a_0 are parameters characterizing the strength of the interaction and its spatial variation, respectively.

Figure 4(a) schematically visualizes the real-space tight-binding model used throughout this chapter. For small system sizes the equations for p and $\bar{B}$ can be solved numerically without further approximations. Considering an excitation with two optical pulses and using the usual Fourier expansion with respect to the pulse directions (Lindberg et al., 1992), PP and FWM signals are obtained as discussed in Subsection 2 of Section II.

The numerical calculations are performed for a system consisting of 10 sites using periodic boundary conditions. For a homogeneous system within the coherent $\chi^{(3)}$ limit it has been checked that the features present in the numerical results do not change for larger systems. Except for Subsection 2 of Section V, where disorder-induced dephasing is analyzed, we phenomenologically insert decay times $1/\gamma_p$ and $1/\gamma_{\bar{B}}$ into the equations of motion for p and $\bar{B}$. This yields a homogeneous broadening that is taken to be smaller than the energy differences between the relevant states. Consistent with the decomposition of B into $\bar{B}$ and terms proportional to pp (see Eq. (9)), we always use $\gamma_{\bar{B}} = 2\gamma_p$.

Concerning the band structure we consider four valence and two conduction bands. For the valence bands we include the two energetically degenerate heavy-hole bands and the two degenerate light-hole bands. These are characterized by the states $|-3/2h\rangle$ and $|3/2h\rangle$ and $|1/2l\rangle$ and $|-1/2l\rangle$, respectively. The two energetically degenerate conduction bands are $|-1/2e\rangle$ and $|1/2e\rangle$. The light fields are assumed to propagate perpendicular to the system extension in the z direction. We use the usual dipole matrix

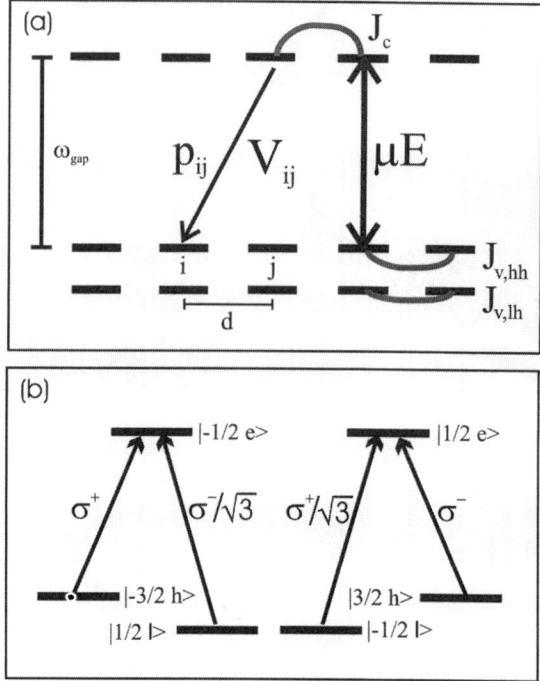

FIG. 4. (a) Schematic drawing of the one-dimensional tight-binding model. ω_{gap} is the bandgap frequency, p_{ij} describes the interband coherence between a hole at site i and an electron at site j, V_{ij} is the Coulomb interaction between sites i and j, and d is the distance between adjacent sites. J_c, $J_{v,hh}$, and $J_{v,lh}$ denote the electronic couplings between nearest neighbor sites in the different bands. μE depicts the optical interband excitation induced by the electric field E, which is proportional to the dipole matrix element μ. (b) Schematic drawing of the optical heavy- and light-hole to electron transitions, including selection rules.

elements (Bott *et al.*, 1993; Mayer *et al.*, 1994; Bott *et al.*, 1996; Schäfer *et al.*, 1996):

$$\boldsymbol{\mu}_{ij}^{11} = \delta_{ij}\mu_0\boldsymbol{\sigma}^+ = \delta_{ij}\frac{\mu_0}{\sqrt{2}}\begin{pmatrix}1\\i\end{pmatrix}$$

$$\boldsymbol{\mu}_{ij}^{12} = \boldsymbol{\mu}_{ij}^{21} = 0$$

$$\boldsymbol{\mu}_{ij}^{22} = \delta_{ij}\mu_0\boldsymbol{\sigma}^- = \delta_{ij}\frac{\mu_0}{\sqrt{2}}\begin{pmatrix}1\\-i\end{pmatrix}$$

$$\boldsymbol{\mu}_{ij}^{31} = \delta_{ij}\frac{\mu_0}{\sqrt{6}}\boldsymbol{\sigma}^- = \delta_{ij}\frac{\mu_0}{\sqrt{6}}\begin{pmatrix}1\\-i\end{pmatrix}$$

$$\mu_{ij}^{32} = \mu_{ij}^{41} = 0$$

$$\mu_{ij}^{42} = \delta_{ij}\frac{\mu_0}{\sqrt{3}}\sigma^+ = \delta_{ij}\frac{\mu_0}{\sqrt{6}}\begin{pmatrix}1\\i\end{pmatrix}, \qquad (23)$$

where μ_0 is the modulus of the matrix element for the heavy-hole transition Consistent with the tight-binding description, the optical transitions are taken as diagonal in the site index. Additionally, we assume the typical selection rules for zincblende semiconductors, that is, heavy-hole and light-hole to electron transitions that are allowed for circularly polarized light described by $|-3/2h\rangle \to |-1/2e\rangle(\mu_{ij}^{11})$, $|3/2h\rangle \to |1/2e\rangle(\mu_{ij}^{22})$, $|1/2l\rangle \to |-1/2e\rangle(\mu_{ij}^{31})$, and $|-1/2l\rangle \to |1/2e\rangle(\mu_{ij}^{42})$. Consistent with the assumption of a GaAs-type structure we use for the ratio between the oscillator strengths of the heavy-hole and light-hole transition $|\mu_{hh}|^2/|\mu_{lh}|^2 = 3$. Because of these selection rules (Fig. 4b), we have two separate subspaces of states, which are optically isolated but coupled by the many-body Coulomb interaction. For simplicity, in some of the numerical calculations the light-hole transitions have been neglected. This approximation is reasonable if the spectral pulse widths and the detuning of the optical pulses with respect to the heavy-hole exciton are smaller than the energetic splitting between the heavy- and light-hole excitons.

If only heavy-hole transitions need to be considered, in first order in the light field only p^{11} and p^{22} are excited. This is due to the fact that the optical matrix elements for p^{12} and p^{21} vanish and that in the linear regime no coupling between the subspaces of different spin states exists within our model. In second order in the light field, depending on the polarization directions of the incident laser pulses, different types of two-exciton states can be excited. If we consider only heavy-hole transitions, then after two interactions with cocircularly polarized pulses, $\bar{B}^{1111}$ or $\bar{B}^{2222}$ are relevant. The quantity $\bar{B}_{abcd}^{v'c'vc}$ is antisymmetric with respect to interchanging the band and real-space indices of the two electrons and the two holes, respectively, that is, $\bar{B}_{cbad}^{vc'v'c} = -\bar{B}_{abcd}^{v'c'vc} = \bar{B}_{adcb}^{v'cvc'}$. Since the contributions of the two electrons and the two holes that enter into $\bar{B}^{1111}$ and $\bar{B}^{2222}$ come from the same bands, the real-space part of the corresponding $\bar{B}$'s has to be antisymmetric. Such two-exciton states typically do not include bound states (Bott et al., 1993; Schäfer et al., 1996). If we consider two interactions with opposite circularly polarized pulses, then $\bar{B}^{1122}$ and $\bar{B}^{2211}$ are created in second order.

If both heavy- and light-hole transitions contribute, one additionally needs to include pure light-hole and mixed two-exciton states consisting of one light and one heavy hole. According to our numerical results discussed later, these mixed two-exciton states can form bound biexcitons, which have actually been observed recently in spectrally resolved FWM experiments (Wagner et al., 1999).

Because of the selection rules, in third order only p^{11}, p^{31}, p^{22}, and p^{42} contribute to the optical response. Therefore, within the coherent $\chi^{(3)}$ limit the response is fully determined by these quantities.

III. Applications to Pump-Probe Spectroscopy

In this section numerical results for correlation-induced signatures in PP spectroscopy are presented. We start in Subsection 1 by considering the low excitation limit and discuss differential absorption spectra after resonant excitation of the exciton resonance. Absorption changes in the Stark-effect configuration, where the pump pulse is detuned below the exciton resonance, are presented in Subsection 2. In Subsection 3 the formalism is extended up to the coherent $\chi^{(5)}$ limit and results on absorption changes induced by pump fields of higher intensities are analyzed.

1. Resonant Absorption Changes for Low Intensities

Here we discuss examples of numerical results on absorption changes induced by a pump pulse that resonantly excites the exciton transition. For simplicity, we start by assuming that the energetic splitting between the heavy-hole and light-hole excitons is much larger than the spectral width of the pump pulse, which allows us to neglect the light-hole transitions (Sieh et al., 1999a, 1999b). For the system parameters we use model HH-I of Table I, which results in a heavy-hole exciton binding energy of 8 meV. Homogeneous broadening is introduced phenomenologically through decay rates $\gamma_p = 1/3$ ps and $\gamma_{\bar{B}} = 1/1.5$ ps, respectively. We assume a pump pulse of duration $\bar{t}_1 = 1$ ps that is tuned to the heavy-hole exciton resonance. To obtain good spectral resolution the probe pulse is taken to be very short, $\bar{t}_2 = 10$ fs, that is, it is spectrally white in the region of interest. These dephasing and pulse parameters are used in most of the calculations.

The differential absorption spectra for resonant excitation at the exciton resonance with cocircular polarized pump and probe pulses having a time delay of $\tau = 2$ ps are shown in Fig. 5a. The differential absorption $\delta\alpha(\omega)$ is negative in the vicinity of the exciton resonance corresponding to a pump-pulse-induced bleaching of the exciton (note that the zero of the energy scale in Fig. 5 and all following figures coincides with the energy of the 1s heavy-hole exciton). For energies larger than the exciton energy we see in Fig. 5a positive contributions to $\delta\alpha(\omega)$, which are explained later. Besides the total signal in the upper part of Fig. 5a, the three individual contributions to $\delta\alpha(\omega)$ (see Eq. (21)), arising from Pauli blocking, and

TABLE I
Model Parameters[a]

Model	J_c (meV)	$J_{v,hh}$ (meV)	$J_{v,lh}$ (meV)	U_0 (meV)	a_0/d	⇒	hh-EX (meV)	lh-EX (meV)	hh-BX (meV)	lh-BX (meV)	hh-lh-BX (meV)
HH-I	8.0	0.8	—	8.0	0.5		8.0	—	1.4	—	—
HH-II	15.0	1.5	—	15.0	0.5		15.0	—	2.6	—	—
HH-LH	15.0	8.9	4.7	15.0	0.5		13.0	14.0	2.2	2.3	2.2

[a]Tight-binding couplings for the conduction band J_c and for the heavy-hole and light-hole valence bands $J_{v,hh}$ and $J_{v,lh}$, respectively. Strength of Coulomb interaction U_0. a_0/d determines its regularization and spatial variation according to Eq. (22). Resulting binding energies for heavy-hole (hh-EX) and light-hole (lh-EX) excitons. Binding energies of heavy-hole (hh-BX), light-hole (lh-BX), and mixed heavy-light-hole (hh-lh-BX) biexcitons.

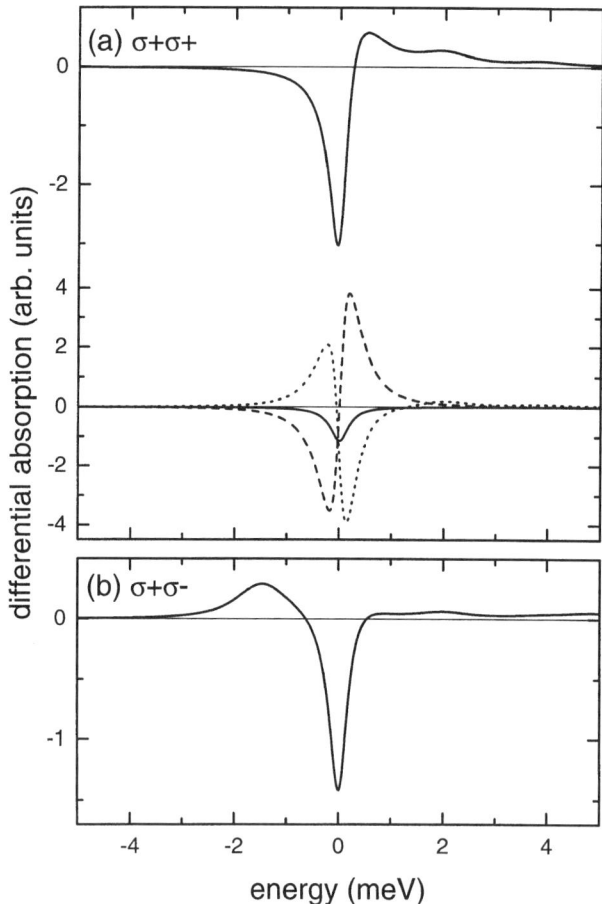

FIG. 5. Total differential absorption spectra $\delta\alpha(\omega)$ for resonant excitation at the exciton resonance and a time delay of $\tau = 2$ ps. (a) Cocircular polarized pump (σ^+) and probe (σ^+) pulses. Also displayed in the lower part are $\delta\alpha_{pb}$ (solid), $\delta\alpha_{CI,first}$ (dashed), and $\delta\alpha_{CI,corr}$ (dotted). (b) Opposite circular polarized pump (σ^+) and probe (σ^-) pulses. System parameters of model HH-I in Table I are used. Homogeneous broadening is introduced phenomenologically through decay rates $\gamma_p = 1/3$ ps and $\gamma_{\bar{B}} = 1/1.5$ ps, respectively. The pump pulse of duration $\bar{t}_1 = 1$ ps and is tuned in resonance with heavy-hole exciton. The probe pulse is very short, $\bar{t}_2 = 10$ fs, that is, it is white in the spectral region of interest.

first- and higher-order Coulomb contributions are displayed separately in the lower part. It is shown that $\delta\alpha_{pb}$ is weak and corresponds to a pure bleaching of the exciton resonance induced by the pump pulse. This is what is to be expected, since the Pauli blocking contribution corresponds to approximating the exciton as a simple two-level system. Formally one can show this by projecting Eq. (10) onto the 1s-exciton resonance and considering only the Pauli blocking nonlinearity $\mu E p p^*$. The resulting differential absorption can be calculated analytically in the limit of ultrashort pulses for positive time delays, as $\delta\alpha_{pb}(\omega) \propto -\gamma_p/((\varepsilon_p - \omega)^2 + \gamma_p^2)$ (Sieh et al., 1999b). This approximation predicts that the absorption change has a negative shape centered at the exciton resonance ε_p, which is exactly the line shape plotted as the solid line in the lower part of Fig. 5a.

As shown by the dashed line in the lower part of Fig. 5a, $\delta\alpha_{CI,first}$ is very strong and antisymmetric around the exciton resonance. It has a dispersive shape corresponding to a blue shift of the exciton. As for the Pauli blocking, this line shape of the first-order Coulomb term can be understood on the basis of a simple calculation. If one projects Eq. (10) onto the 1s-exciton resonance and considers only the first-order Coulomb term $Vppp^*$, one finds that compared to the Pauli blocking, the external field μE is replaced by an exciton coherence Vp. Adding the Pauli blocking and the first-order Coulomb terms, one arrives at the so-called local-field model, which corresponds to projecting the HF-SBE onto the 1s-exciton resonance (Wegener et al., 1990; Lindberg et al., 1992). Here, the effective field is given by $\mu E + Vp$, that is, the sum of the external and the internal fields. Apart from the prefactor, the main difference between μE and Vp is a change of phase that is responsible for the different line shapes of the Pauli blocking and the first-order Coulomb terms in the differential absorption spectra. This can be rationalized by solving the equation of motion governing the linear excitonic response $\partial_t p = i\omega_p p + i\mu E$ for excitation with an ultrashort pulse, that is, $E(t) = \delta(t)E_0$. One obtains $p(t) = i\mu E_0 \Theta(t) \exp(i\omega_p t)$ and thus a phase shift of $\pi/2$ between $p(t)$ and $E(t)$ at $t = 0$. It is basically this phase shift that induces the change from an absorptive differential absorption of the Pauli blocking to a dispersive absorption change of the first-order Coulomb term (Sieh et al., 1999b).

$\delta\alpha_{CI,corr}$ is shown as the dotted line in the lower part of Fig. 5a. Its shape is also mainly dispersive around the exciton resonance, but with opposite sign compared to $\delta\alpha_{CI,first}$, that is, this term yields a red shift. Besides contributions at the exciton energy, $\delta\alpha_{CI,corr}$ also includes terms with resonances at the energies corresponding to transitions from excitons to unbound two-exciton states. These transitions induce some positive differential absorption, so called excited-state absorption at energies above the exciton resonance. The strong compensation between $\delta\alpha_{CI,first}$ and $\delta\alpha_{CI,corr}$, which could already be suspected from a comparison of Eqs. (10) and (11) to Eqs. (13) and (14), can also be reproduced by analytical calculations in the limit of ultrashort pulses (Sieh et al., 1999b). Such calculations further show that $\delta\alpha_{CI,corr}$ includes positive absorptive resonances if the frequency coincides with the energy difference between the exciton ε_p and a two-exciton energy $\varepsilon_{\bar{B}}$, that is, at $\omega = \varepsilon_{\bar{B}} - \varepsilon_p$.

When adding up the three contributions via Eq. (21), the resulting differential absorption exhibits a predominantly absorptive spectral shape around the exciton resonance. Figure 5a shows that in fact the bleaching at the exciton resonance is dominated by Coulomb-interaction-induced nonlinearities and only weakly enhanced by $\delta\alpha_{pb}$ (Sieh et al., 1999a, 1999b).

We now consider excitation with opposite circularly polarized pulses. For this polarization geometry, both $\delta\alpha_{pb}$ and $\delta\alpha_{CI,first}$ vanish as long as the system is spatially homogeneous (Sieh et al., 1999a, 1999b). This is due to the fact that none of these contributions introduces any coupling between the subspaces of different spin states. Therefore, for this polarization geometry, in the $\chi^{(3)}$ limit the total signal is given purely by the correlation contribution, that is, $\delta\alpha = \delta\alpha_{CI,corr}$. The corresponding spectrum for $\tau = 2$ ps is displayed in Fig. 5b. As for cocircular excitation, for opposite circular excitation we find bleaching at the exciton resonance and excited-state absorption due to transitions to unbound two-exciton states appearing energetically above the exciton resonance. Whereas for cocircularly polarized excitation only contributions from unbound two-exciton states are present, now there is a clear signature of a bound biexciton in the differential absorption spectra, appearing about 1.4 meV below the excitonic resonance (Sieh et al., 1999a, 1999b). As shown in Sieh et al. (1999a), the differential absorption spectra in Fig. 5 are in qualitative agreement with two-dimensional calculations and experiments performed on high-quality quantum wells.

The signatures of transitions to two-exciton states in Fig. 5 can be understood qualitatively on the basis of the level scheme shown in Fig. 6. For $\sigma^+\sigma^+$ excitation two-exciton states $B++$ formed by two σ^+ excitons are generated. As discussed in Subsection 3 of Section II, two-exciton states formed from two excitons with the same spin typically do not include bound states. $B+-$, however, is formed by two excitons with opposite spin and thus includes both bound biexcitons and unbound two-exciton states (Schäfer et al., 1996).

Figures 7a and 7b display $\delta\alpha$ and the individual contributions to it for linear parallel (xx) and linear perpendicular (xy) excitation configurations, respectively. The comparison with Fig. 5 reveals that the signal for xy geometry, Fig. 7b, is very similar to the signal for cocircular polarized excitation. In both cases strong compensations among the first- and higher-order Coulomb nonlinearities occur, and no bound biexcitons show up. The fact that for linearly polarized pulses a bound biexciton cannot be excited by two pulses with perpendicular polarization directions has already been noted in Bott et al. (1993) on the basis of a phenomenological few-level system and is in agreement with analytical results based on the present model that are presented in the Appendix of Sieh et al. (1999b). The origin of this selection rule is that for xy excitation the two pathways to the biexciton involving successive excitations of σ^+ and σ^- excitons and vice versa interfere destructively and cancel exactly. For xx excitation (Fig. 7a), a bound biexciton shows up, and cancellations between $\delta\alpha_{CI,first}$ and $\delta\alpha_{CI,corr}$

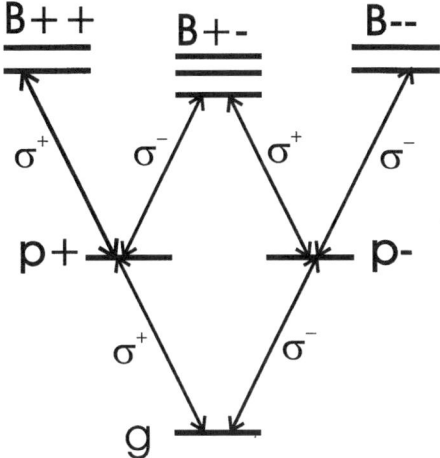

FIG. 6. Schematic level scheme including the heavy-hole transitions relevant in the coherent $\chi^{(3)}$ limit. $p+$ and $p-$ denote the spin-degenerate single-exciton states that can be excited with σ^+ and σ^- polarized light, respectively. $B++$ and $B--$ are the two-exciton states formed by two σ^+ and σ^- single excitons. These two excitons typically include only unbound states. $B+-$ denotes the two-exciton states formed by a σ^+ and a σ^- exciton. These mixed two excitons usually include bound biexcitons as well as unbound states.

are also present, resulting in bleaching at the exciton resonance. Thus, the spectrum for xx configuration looks like the sum of the spectra obtained for cocircular and opposite-circular excitation.

For a simple understanding of the selection rules governing the appearance of the biexciton contributions in the polarization-dependent differential absorption spectra in Figs. 5 and 7, we show a reduced level scheme in Fig. 8, where only the two degenerate 1s excitons and the bound biexciton are considered. As was discussed in Bott et al. (1993) the circular selection rules can be transformed into linear selection rules. According to the resulting level scheme, linearly polarized pump and probe pulses excite a bound biexciton only if the pulses are not orthogonally polarized. Thus, the bound biexciton should appear in the differential absorption spectra for $\sigma^+\sigma^-$ and xx excitation, but should not contribute for $\sigma^+\sigma^+$ and xy, which is in agreement with Figs. 5 and 7.

A systematic comparison between the differential absorption spectra for the different excitation configurations is given in Fig. 9, where a time delay $\tau = 0$ ps between the pump and probe pulses has been assumed. Except for small features, such as positive $\delta\alpha$ below the exciton resonance in Figs. 9a and 9d, which are induced by the overlapping pump and probe pulses, the spectra for $\tau = 0$ ps are very similar to the spectra for $\tau = 2$ ps. As in Figs. 5 and 7, we find in Fig. 9 that the line shape for cocircular is close to the

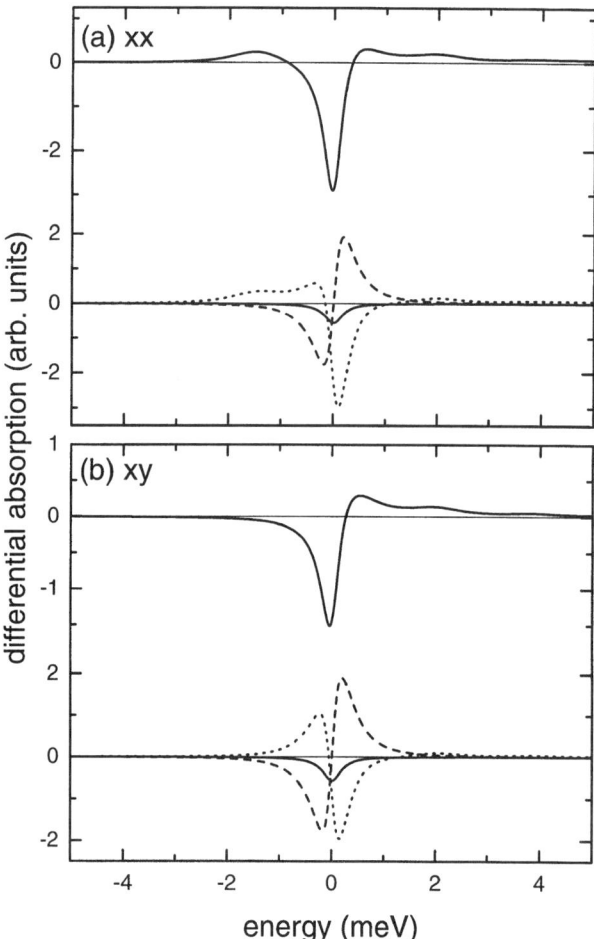

FIG. 7. Total differential absorption spectra $\delta\alpha(\omega)$ for resonant excitation at the exciton resonance and a time delay of $\tau = 2$ ps. (a) Linear-parallel polarized pump (x) and probe (x) pulses. (b) Linear-perpendicular polarized pump (x) and probe (y) pulses. Also displayed in the lower parts are $\delta\alpha_{pb}$ (solid), $\delta\alpha_{CI,first}$ (dashed), and $\delta\alpha_{CI,corr}$ (dotted). System parameters of model HH-I in Table I are used.

one for linear-perpendicular, and that opposite-circular is close to linear-parallel.

Figure 10 displays the differential absorption spectra for the various excitation configurations assuming a negative time delay $\tau = -2$ ps corresponding to the probe pulse arriving before the pump pulse. In this case all spectra are dominated by coherent oscillations that originate from the scattering of the pump pulse off the grating created by both pump and

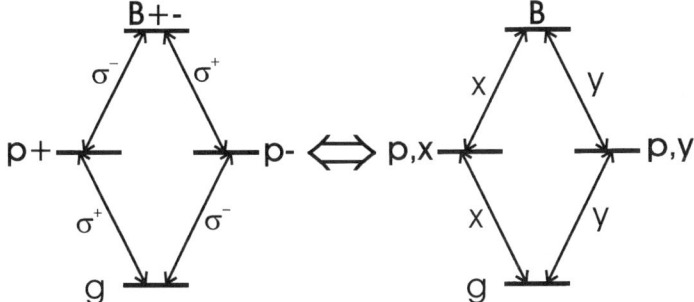

FIG. 8. Reduced level scheme including the two spin-degenerate 1s heavy-hole excitons $p+$ and $p-$ as well as the bound biexciton $B+-$. The model with circular polarized transition matrix elements is equivalent to the model with linear polarized selection rules.

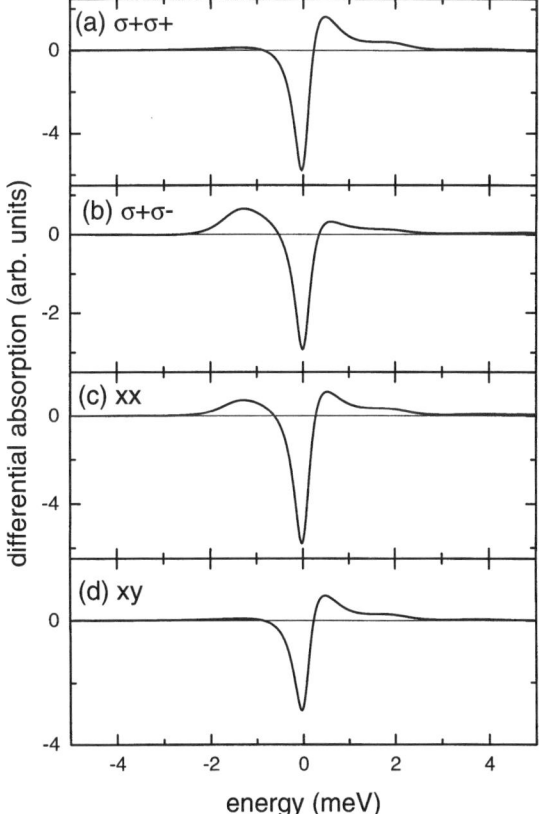

FIG. 9. Total differential absorption spectra $\delta\alpha(\omega)$ for resonant excitation at the exciton resonance and a time delay of $\tau = 0$ ps. (a) Cocircular ($\sigma^+\sigma^+$), (b) opposite-circular ($\sigma^+\sigma^-$), (c) linear-parallel (xx), and (d) linear-perpendicular (xy) polarized pump and probe pulses, respectively. System parameters of model HH-I in Table I are used.

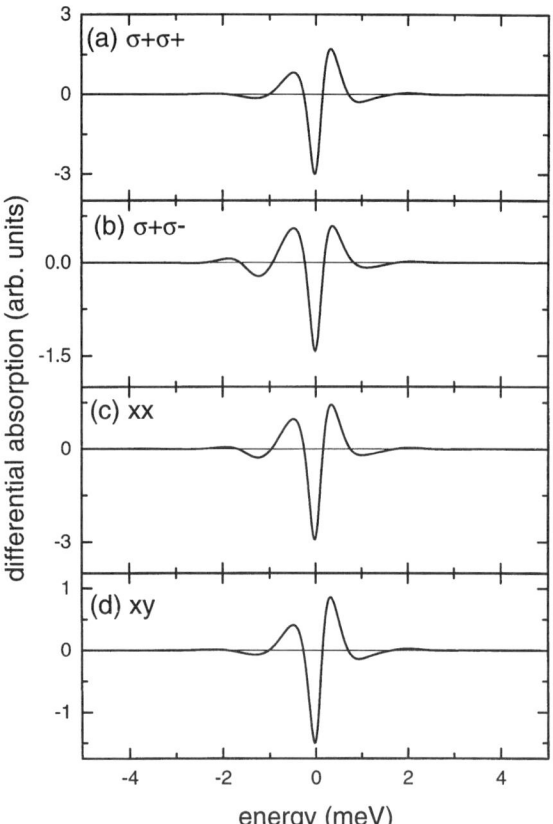

FIG. 10. Total differential absorption spectra $\delta\alpha(\omega)$ for resonant excitation at the exciton resonance and a time delay of $\tau = -2$ ps. (a) Cocircular ($\sigma^+\sigma^+$), (b) opposite-circular ($\sigma^+\sigma^-$), (c) linear-parallel (xx), and (d) linear-perpendicular (xy) polarized pump and probe pulses, respectively. System parameters of model HH-I in Table I are used.

probe pulses into the probe direction (Koch et al., 1988b). The energetic period of these oscillations is inversely proportional to $|\tau|$ and develops into the (nonoscillating) bleaching for $|\tau| \to 0$. These spectral oscillations are present in all three individual contributions to the differential absorption (Sieh et al., 1999b). Their appearance can be reproduced by analytical calculations performed on the basis of the present model in the limit of ultrashort pulses. As is shown in the Appendix of Sieh et al. (1999b), if the probe arrives before the pump pulse the frequency-domain differential polarization $\delta P(\omega, \tau)$ is multiplied by the phase factor $\exp(i(\omega_p - \omega)\tau)$, where ω_p is the exciton frequency. Since the differential absorption $\delta\alpha(\omega, \tau)$ is determined by the imaginary part of $\delta P(\omega, \tau)$, the phase factor introduces

$\sin((\omega_p - \omega)\tau)/(\omega - \omega_p)$-like spectral oscillations of $\delta\alpha(\omega, \tau)$ around ω_p (Sieh et al., 1999b).

To add to the complexity of our model system and to make the results more applicable to a wider variety of realistic systems, we now consider a situation where both heavy- and light-hole transitions contribute. To be able to concentrate on general excitonic signatures, we change the parameters and increase the exciton binding energies. We model the effective masses considering that the in-plane dispersion of the valence bands in quantum wells can be obtained using the Luttinger-Hamiltonian. For GaAs parameters we get $m_{hh} = m_0/(\gamma_1 + \gamma_2) = 0.112 m_0$ and $m_{lh} = m_0/(\gamma_1 - \gamma_2) = 0.211 m_0$, where we use the Luttinger parameters $\gamma_1 = 6.85$ and $\gamma_2 = 2.1$ Thus, the mass reversal between the heavy- and light-hole dispersions is included in our model, but any further band-mixing effects are neglected. For the conduction band electrons we use $m_e = 0.0665 m_0$. Taking the tight-binding coupling of the electrons as $J_c = 15$ meV, the valence band masses then enter into the model by considering the nearest neighbor coupling to be inversely proportional to the mass, that is, $J_{v,hh} = J_c m_e/m_{hh}$ and $J_{v,lh} = J_c m_e/m_{lh}$, respectively. For the list of system parameters and resulting binding energies, see model HH-LH in Table I. The site energies for the light holes are chosen relative to the heavy holes, allowing for an adjustable splitting between the heavy- and light-hole excitons, which is taken initially as 5 meV. As discussed in Subsection 3 of Section II, we assume the typical ratio of $\frac{1}{3}$ between the oscillator strengths of the light- and heavy-hole transitions.

Figures 11a and 11b show differential absorption spectra for the opposite-circular and cocircular excitation geometries, respectively, assuming a rather short pump pulse with $\bar{t}_1 = 200$ fs, which is tuned spectrally 2 meV below the light-hole exciton resonance, and a pump-probe delay of $\tau = 1$ ps. The dashed and dotted lines in Figs. 11a and 11b show the resulting differential absorption spectra obtained by considering either heavy-hole or light-hole transitions alone, whereas the solid lines present results where both transitions are considered simultaneously. As before, if only heavy-hole transitions (or similarly only light-hole transitions) are taken into account, we find bleaching at the exciton and excited state absorption induced by transitions from the exciton to unbound two-exciton states appearing energetically above the exciton for both $\sigma^+\sigma^+$ and $\sigma^+\sigma^-$. Additionally, for $\sigma^+\sigma^-$, excited-state absorption induced by exciton to biexciton transitions is present in the spectral region below the exciton.

The spectra including both heavy- and light-hole transitions are displayed by the solid lines in Figs. 11a and 11b. One can clearly see that these are not simply the sum of the results for heavy and light holes calculated separately. Thus, there is clear evidence for coupling between the heavy- and light-hole transitions, which within the present model is induced by two mechanisms: (i) the sharing of a common electronic state (Bott et al., 1996),

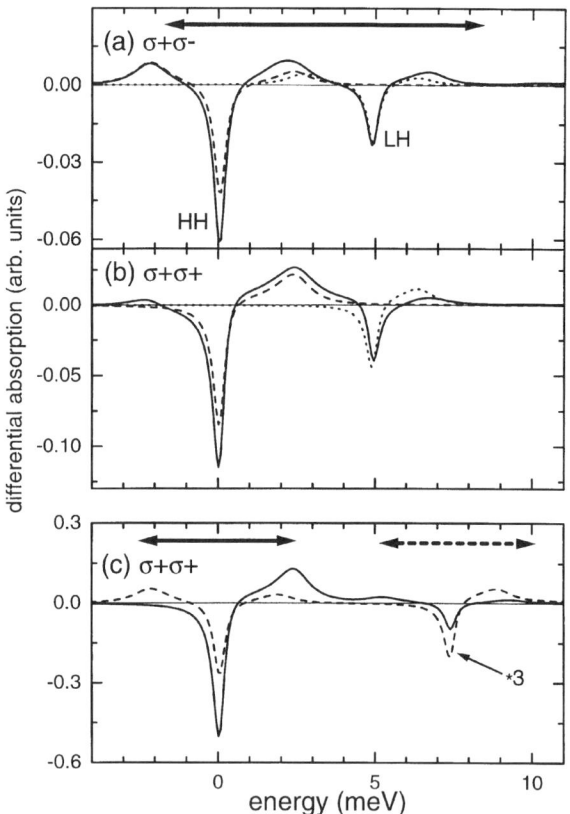

FIG. 11. Total differential absorption spectra $\delta\alpha(\omega)$ including heavy- and light-hole excitons. (a) Opposite-circular ($\sigma^+\sigma^-$) and (b) cocircular ($\sigma^+\sigma^+$) excitation using a pump pulse with $\bar{t}_1 = 200\,\text{fs}$ that is spectrally centered 2 meV below the light-hole exciton resonance and a time delay of $\tau = 1$ ps. Total signal including heavy and light holes (solid), only heavy holes considered (dashed), and only light holes considered (dotted). (c) Cocircular ($\sigma^+\sigma^+$) excitation for a larger heavy–light splitting and a longer ($\bar{t}_1 = 400\,\text{fs}$), that is, spectrally narrower, pump pulse. Excitation resonant at heavy-hole exciton (solid) and resonant at light-hole exciton (dashed). The arrows indicate the spectral width of the pump pulses. System parameters of model HH-LH in Table I are used. Homogeneous broadening is introduced phenomenologically through decay rates of $\gamma_p = 1/3$ ps and $\gamma_B = 1/1.5$ ps.

see Fig. 4b, and (ii) via the many-body Coulomb interaction. As a result of this coupling, the strong excitation of light-hole excitons for the present excitation configuration induces additional bleaching of the heavy-hole exciton; see Figs. 11a and 11b. Furthermore, one can see in Fig. 11b that if both heavy and light holes are also included for cocircular excitation, we obtain excited state absorption below the 1s heavy-hole exciton. This positive differential absorption below the lowest exciton can be ascribed to a mixed biexciton, that is, a bound two-exciton state, formed from a heavy- and a light-hole exciton. Clear evidence for the existence of such mixed biexcitons in quantum wells has been given recently using spectrally resolved FWM (Wagner et al., 1999). The mixed biexciton also appears as excited state absorption energetically below the light-hole exciton. For the present choice of parameters this contribution is, however, overshadowed by the induced absorption originating from transitions to unbound heavy-hole two-exciton states.

To verify that the low energy resonance appearing for $\sigma^+\sigma^+$ excitation below the heavy-hole exciton indeed involves both a heavy- and a light-hole exciton, we present in Fig. 11c additional results obtained for an increased splitting of 7.5 meV between the heavy- and light-hole excitons. In order to selectively excite each one of the two excitons we assume a longer pump pulse of $\bar{t}_1 = 400$ fs, which has a reduced spectral width indicated by the arrows in Fig. 11. The results for cocircular excitation, of the heavy-hole (solid) and light-hole exciton (dashed) are displayed in Fig. 11c. If the pump resonantly excites the heavy-hole exciton, the mixed biexciton appears below the light-hole exciton, and vice versa, confirming that both types of excitons participate in forming this mixed two-exciton state.

The bound biexciton states contributing for coupled heavy- and light-hole transitions can be depicted in the level scheme shown in Fig. 12, where in addition to the heavy- and light-hole biexcitons, the mixed biexciton states

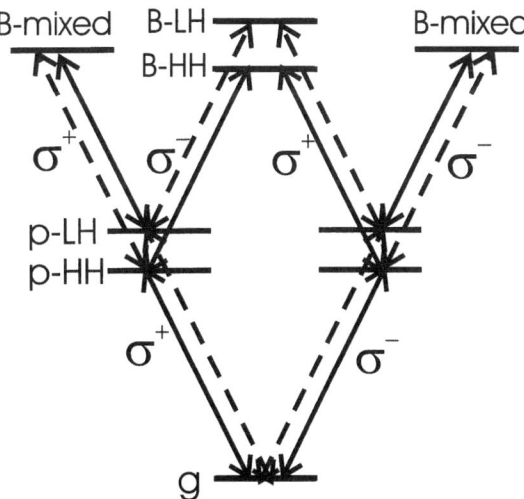

FIG. 12. Schematic level scheme including heavy- and light-hole excitons and biexcitons as well as the mixed biexcitons formed by a heavy- and a light-hole exciton. The solid lines represent optical transitions involving heavy holes and the dashed lines represent transitions involving light holes.

as introduced in Wagner *et al.* (1999) are included. This level scheme can be seen as a refinement of the simpler schemes used in Bott *et al.* (1993), Mayer *et al.* (1994, 1995), Smith *et al.* (1995), and Bott *et al.* (1996), where only bound heavy-hole biexcitons were considered.

Besides the numerical results on resonant absorption changes of a homogeneous system discussed here, our approach has been generalized to include effects of energetic disorder together with correlations on a microscopic basis and has also been applied to excitons in microcavities (Sieh *et al.*, 1999b). The numerical results on reflection changes of excitons in a microcavity are in good qualitative agreement with recent experiments (Fan *et al.*, 1998).

2. Off-Resonant Absorption Changes for Low Intensities

In this subsection we present numerical results on absorption changes induced by a pump pulse that is detuned below the heavy-hole exciton resonance. As in the previous subsection, we start by assuming that the splitting between the heavy-hole and light-hole excitons is large and initially neglect the light holes. The system parameters and the pulse widths are identical to those used in Figs. 5, 7, 9, and 10.

We first discuss the influence of detuning the pump pulse below the exciton resonance on the differential absorption for cocircular polarized excitation. Since for detuned excitation the pump pulse creates mainly off-resonant excitations (adiabatically following polarizations) that decay with the pulse envelope, we use a time delay of $\tau = 0$ ps throughout this subsection. Figure 13 shows the resulting differential absorption spectra for the detunings of 1, 3, 5, and 7.5 meV of the pump-pulse energy below the exciton energy. The higher detunings (5 and 7.5 meV) exceed both the spectral width of the pump pulse and the homogeneous width of the exciton resonance. The solid lines in Fig. 13 are the full $\delta\alpha$, whereas the dashed lines are the results within the TDHF approximation where correlations are neglected, that is, $\delta\alpha_{HF} = \delta\alpha_{pb} + \delta\alpha_{CI,first}$. We see that with increasing detuning the bleaching at the exciton resonance develops into a dispersive shape corresponding to a pure blue shift (Koch *et al.*, 1988b). Furthermore, the relative importance of Coulomb correlations diminishes rapidly with increasing detuning. We thus conclude here that at least for cocircular excitation the TDHF approximation gives a good description of the differential absorption as long as off-resonant excitation is considered.

Next we fix the detuning of the pump pulse to 4.5 meV below the exciton resonance and investigate the absorption changes for the various polarization configurations. Figures 14a–d displays the theoretical results for $\sigma^+\sigma^+$, $\sigma^+\sigma^-$, xx, and xy excitation. Whereas for the three cases (a), (c), and (d) the differential absorption corresponds to a blue shift, for $\sigma^+\sigma^-$ clearly

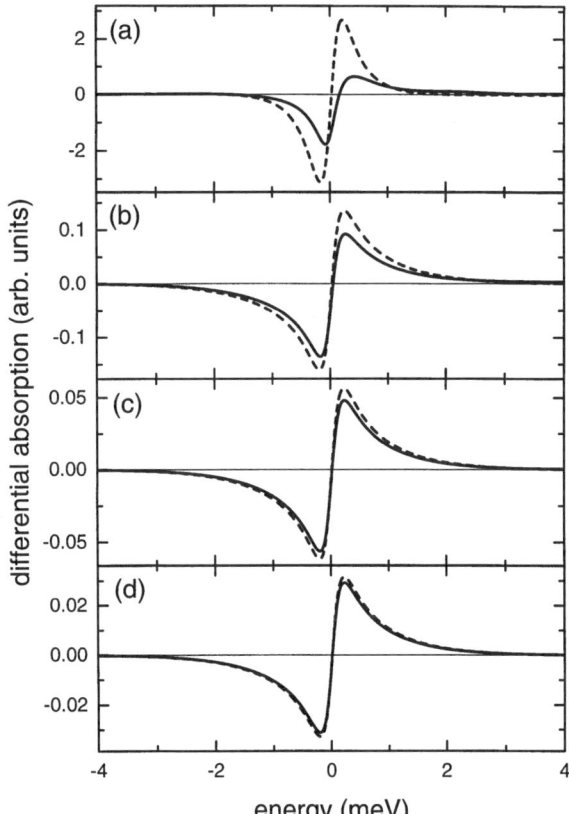

FIG. 13. Differential absorption spectra for various detunings of the pump pulse below the exciton resonance with cocircular polarized pump and probe pulses for time delay $\tau = 0$ ps. The central frequency of the pump pulse is chosen as (a) 1 meV, (b) 3 meV, (c) 5 meV, and (d) 7.5 meV below the energy of the 1s-exciton. The solid line gives the full $\delta\alpha$, and the dashed line the result of a Hartree–Fock calculation ($\delta\alpha_{pb} + \delta\alpha_{CI,first}$) neglecting correlations ($\delta\alpha_{CI,corr}$). System parameters of model HH-I in Table I are used.

a red shift appears. Note that for $\sigma^+\sigma^-$ excitation both the Pauli blocking and the first-order Coulomb-induced nonlinearities, that is, the Hartree–Fock contributions, vanish identically (Sieh et al., 1999a, 1999b). Thus, in this case the red shift of the differential absorption is purely induced by Coulomb correlations.

The physical origin of the red shift can be analyzed in more detail by looking at the individual contributions to the signal. For $\sigma^+\sigma^+$, xx, and xy polarization, both the Pauli blocking and the first-order Coulomb terms always induce a blue shift, whereas the Coulomb correlations always

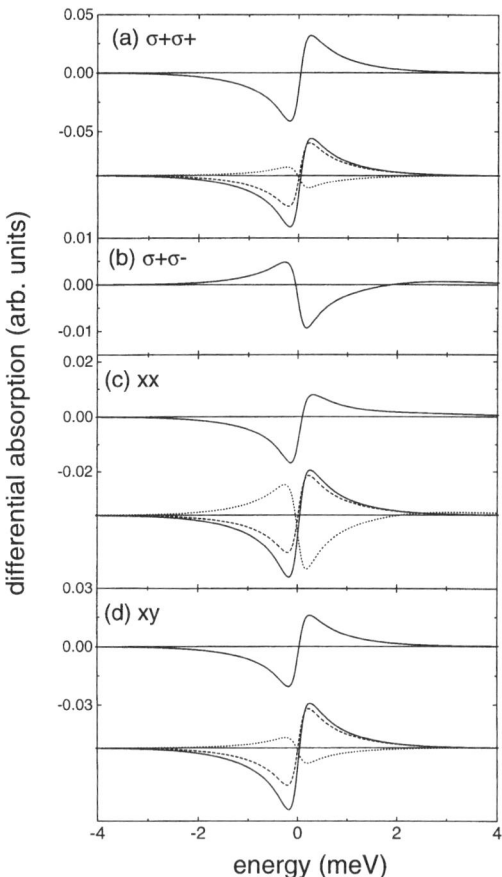

FIG. 14. Differential absorption spectra for excitation 4.5 meV below the exciton and time delay $\tau = 0$ ps. (a) Cocircularly polarized pump and probe pulses ($\sigma^+\sigma^+$); the lower panel shows the three contributions to the signal (solid: Pauli blocking; dashed: first-order Coulomb; dotted: Coulomb correlation). (b) Opposite-circularly polarized pump and probe pulses ($\sigma^+\sigma^-$). (c) and (d) Same for linear parallel (xx) and linear perpendicular (xy) polarizations, respectively. System parameters of model HH-I in Table I are used.

correspond to a red shift; see Figs. 14a–d). The fact that for $\sigma^+\sigma^+$ and xy excitation, where no bound biexcitons are excited, the correlation term alone still corresponds to a red shift clearly demonstrates that the correlation-induced reversal of the shift direction is not directly related to the existence of a *bound* biexciton. For $\sigma^+\sigma^+$, xx, and xy polarization, however, the correlation contribution is rather small and its red shift is always overcompensated by the Pauli blocking and the first-order Coulomb terms, resulting in a net blue shift.

To substantiate the claim that the existence of a bound biexciton is not necessary to obtain a correlation-induced red shift, we have performed additional calculations of the differential absorption spectra for the $\sigma^+\sigma^-$ configuration. To eliminate the bound biexciton contribution also for $\sigma^+\sigma^-$ excitation, we artificially dropped the six terms containing the attractive and repulsive Coulomb terms between the two electrons and two holes appearing in the homogeneous part of the equation of motion for the two-exciton amplitude $\bar{B}$; see Eq. (11). In this case, for $\sigma^+\sigma^-$ excitation as well, no bound biexcitons exist. The approximation of neglecting the Coulomb terms that lead to the formation of bound two-exciton states is identical to the SBA in $\chi^{(3)}$ approximation, since only terms up to second order in the Coulomb interaction are retained in the signal. By solving the dynamic equation for $\bar{B}$, Coulomb memory effects are still included on this level. The resulting spectrum is displayed by the dashed line in Fig. 15a and shows that the signal amplitude is somewhat reduced; however, the red shift clearly persists. We therefore conclude so far that it is clearly the dynamics of Coulomb correlations that is responsible for the presence of a red shift, but that the exact structure of the two-exciton states seems to be important only for the quantitative result.

To further investigate the origin of the red shift we performed additional calculations where we switched off memory effects in the SBA. This is done by solving Eq. (11) without the six Coulomb terms in the homogeneous part within the Markov limit (Schäfer *et al.*, 1996; Jahnke *et al.*, 1996, 1997). The resulting spectrum is displayed by the dotted line in Fig. 15a. It clearly does not correspond to a red shift, but instead shows increased absorption at the exciton that can be attributed to the Coulomb-induced coupling among the excitons with different spin. Because of this coupling, after excitation of a σ^+ exciton, there is an increased possibility for the creation of a σ^- exciton. Hence, the red shift is a genuine signature of Coulomb memory effects, which are missed if a Markov approximation is performed. These memory effects are different from those reported in Bányai *et al.* (1998), which involved excitation high above the bandgap, and the buildup of free-carrier Coulomb screening.

Since we included the decay rate of two-exciton excitations $\gamma_{\bar{B}}$ as the broadening in the energy denominators, the result obtained within the SBA in the Markov limit depends strongly on the choice of the decay rates γ_p and $\gamma_{\bar{B}}$. If one uses $\gamma_p = 1/0.6$ ps and $\gamma_{\bar{B}} = 1/0.3$ ps (see Fig. 15b), then the SBA without memory may also look like a red shift. However, for these parameters the full results (solid line in Fig. 15b), correspond to exciton broadening rather than a red shift. Hence, the red shift obtained within the SBA without memory depends on the exact choice of the phenomenological dephasing rates. It disappears for slower dephasing as shown in Fig. 15a and changes into exciton broadening for even more rapid dephasing modeled by $\gamma_p = 1/100$ fs and $\gamma_{\bar{B}} = 1/50$ fs; see Fig. 15c. In this case, the equation of

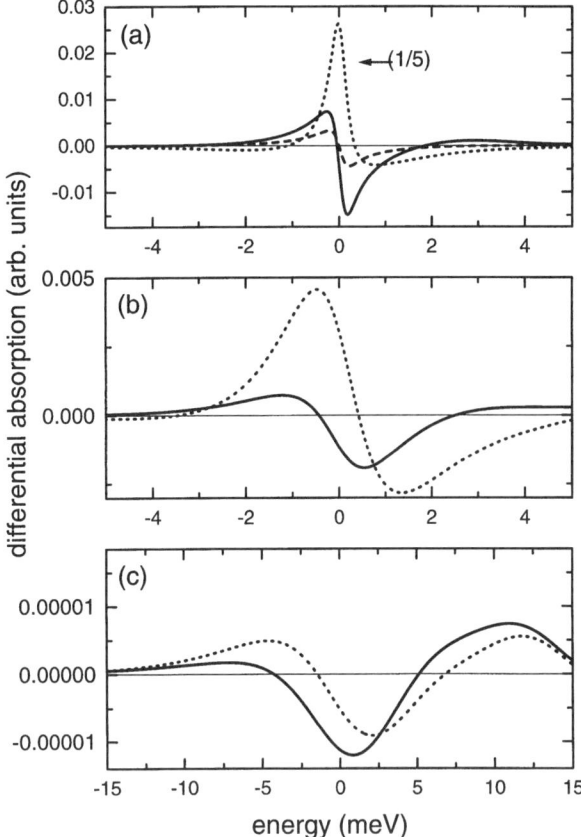

FIG. 15. Differential absorption spectra for excitation 4.5 meV below the exciton and time delay $\tau = 0$ ps using opposite-circularly polarized pump and probe pulses ($\sigma^+\sigma^-$). (a) Full calculation (solid) and results within the second Born approximation with (dashed) and without (dotted, divided by 5) Coulomb memory using $\gamma_p = 1/3$ ps and $\gamma_{\bar{B}} = 1/1.5$ ps. (b) Full calculation (solid) and result within the second Born approximation without Coulomb memory (dotted) for stronger dephasing $\gamma_p = 1/0.6$ ps and $\gamma_{\bar{B}} = 1/0.3$ ps. (c) Same as (b) for $\gamma_p = 1/100$ fs and $\gamma_{\bar{B}} = 1/50$ fs. System parameters of model HH-I in Table I are used.

motion for $\bar{B}$ is damped out rapidly with a time constant of 50 fs, which corresponds to a very short Coulomb memory depth. Thus, the result shown in Fig. 15c actually approaches the limit of vanishing Coulomb memory in which the full result and the SBA without memory become similar. Both look like exciton broadening, which dominates for larger dephasing rates.

The origin of the discussed red shift is different from that observed by Hulin and Joffre (1990). There, CuCl was investigated, which has a very large biexciton binding energy. A red shift of the differential absorption was reported to occur in a small spectral window when the pump pulse is tuned slightly below the exciton to biexciton transition (Hulin and Joffre, 1990). This red shift appeared only for linearly parallel polarized pump and probe pulses (Hulin and Joffre, 1990) and was therefore related to the existence of a bound biexciton (Combescot and Combescot, 1988).

Since there have been numerous experimental investigations on the optical Stark effect in GaAs/AlGaAs-type quantum wells in the past, one might wonder why the red shift as reported in Sieh *et al.* (1999a) was not observed before. In the following we address this point and show that the observability of the red shift crucially depends on the energetic separation between heavy- and light-hole excitons, as well as on the choice of the pump detuning. For this purpose we extend the calculations to include both heavy- and light-hole transitions using the typical selection rules and the three-to-one ratio of the oscillator strengths; see Subsection 3 of Section II.

As explained in Subsection 1 of Section III, the heavy- and light-hole transitions are coupled by two mechanisms: (i) via sharing a common electronic state (Bott *et al.*, 1996) and (ii) via Coulomb correlations. Figures 16a and 16b show calculated differential absorption spectra for cocircular and opposite circular excitation considering both heavy and light holes for various splittings between the heavy-and light-hole excitons. Here we use the same parameters to model heavy- and light-hole transitions as introduced in Subsection 1 of Section III and used in Fig. 11. The site energies for the light holes are chosen relative to the heavy holes to have an adjustable splitting between the heavy- and light-hole excitons. As in Figs. 14 and 15, the pump pulse is detuned 4.5 meV below the 1s heavy-hole exciton and a time delay of $\tau = 0$ ps is used. For a rather large splitting of 15 meV between the heavy- and light-hole excitons (which is the splitting present in the InGaAs quantum well sample investigated in Sieh *et al.*, 1999a), we reproduce the blue shift for $\sigma^+\sigma^+$ and the red shift for $\sigma^+\sigma^-$ excitation; see Figs. 16a and 16b. For reduced heavy–light splitting, the blue shift present for $\sigma^+\sigma^+$ excitation survives, and even its amplitude remains almost unchanged. As shown in Fig. 16b, for $\sigma^+\sigma^-$ excitation, however, the amplitude of the red shift strongly decreases with decreasing splitting. For a splitting of less than 1 meV, the red shift even changes into a blue shift. We thus conclude that the correlation-induced red shift should be pronounced only in samples with a significant splitting between the heavy- and light-hole excitons, as were used in Sieh *et al.* (1999a). For small splittings the coupling between heavy- and light-hole excitons induces a blue shift that overcompensates the red shift that is present if only the heavy-hole exciton is relevant (Koch *et al.*, 1999).

For a fixed heavy–light splitting taken as 15 meV in Figs. 17a and 17b, the differential absorption spectra strongly depend on the choice of the detuning. For cocircular excitation the blue shift present for a detuning of 4.5 meV also survives for larger detunings and is only reduced in amplitude

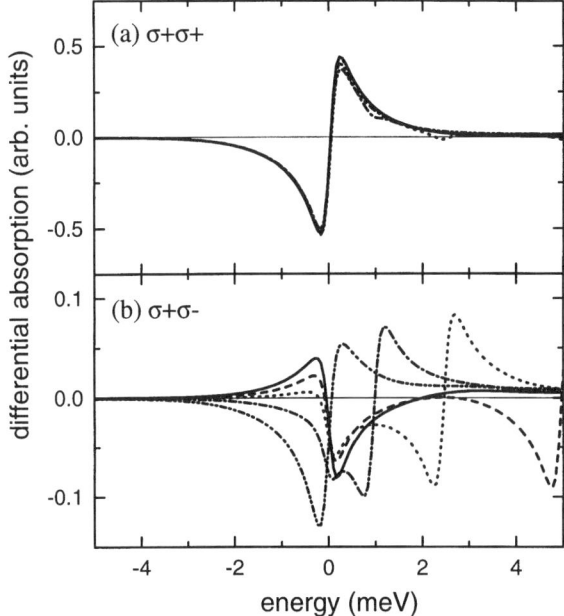

FIG. 16. Differential absorption spectra including heavy and light holes for time delay $\tau = 0$ ps. (a) Cocircularly and (b) opposite-circularly polarized pump and probe pulses, respectively, for a fixed detuning of 4.5 meV below the 1s heavy-hole exciton resonance and various splittings between the heavy- and light-hole excitons. Solid: 15 meV; dashed: 5 meV; dotted: 2.5 meV; dash-dot: 1 meV; and dash-dot-dot: 0 meV. System parameters of model HH-LH in Table I are used.

with increased detuning. Thus, the behavior displayed in Fig. 17a is very similar to Fig. 13, where only heavy holes were considered. For opposite circular excitation, however, besides a strong reduction in amplitude, the direction of the shift changes with increasing detuning. For very large detuning we again obtain a blue instead of a red shift. The reason for this behavior is that for very large detuning, heavy- and light-hole excitons are both excited completely off-resonantly. In this adiabatic following regime the frequency-dependent optical response depends rather weakly on the detuning between the pump pulse and the resonance frequency (Binder *et al.*, 1990). Therefore, although with increasing detuning the excitation of both heavy- and light-hole excitons decreases, the relative weight of the light-hole exciton increases (Joffre *et al.*, 1989; Koch *et al.*, 1999). Because of the heavy-hole–light-hole coupling, this relatively strong excitation of the light-hole transitions induces the blue shift at the heavy-hole exciton resonance.

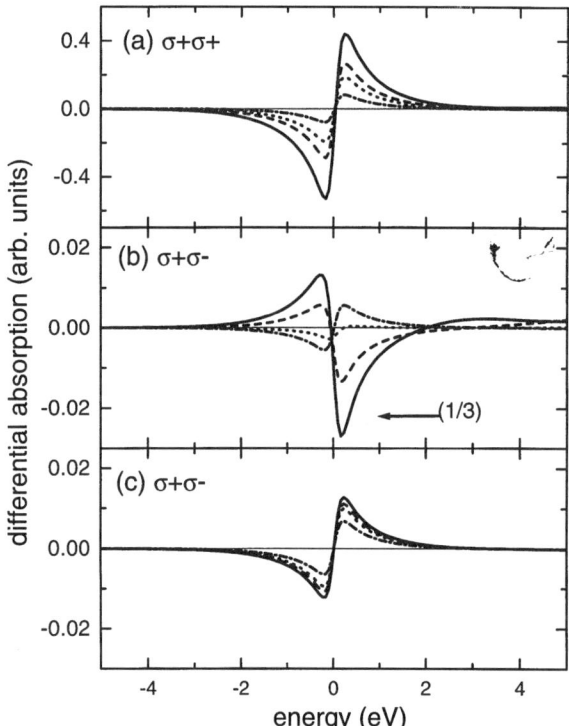

FIG. 17. Differential absorption spectra including heavy and light holes for time delay $\tau = 0$ ps. (a) Cocircularly and (b) opposite-circularly polarized pump and probe pulses, respectively, for a fixed splitting between the heavy- and light-hole excitons of 15 meV for various detunings below the heavy-hole exciton resonance. Solid: 4.5 meV; dashed: 6.75 meV; dotted: 9 meV; and dash-dot: 18 meV. (c) Same as (b) in Hartree–Fock approximation, neglecting correlations. System parameters of model HH-LH in Table I are used.

The numerical results presented in Figs. 16 and 17 clearly demonstrate that the Coulomb-memory-induced red shift for oppositely circular excitation is very sensitive to both the heavy-hole–light-hole splitting and the pump detuning. Thus, it should only be observable in samples with sufficiently large exciton splittings in a certain detuning range that depends on the heavy- and light-hole masses and on the ratio of their oscillator strengths.

As for cocircular excitation (Fig. 13), for opposite circular excitation for very large detuning below the exciton, the differential absorption lineshapes are also well described within the TDHF approximation. This can be verified by comparing the results of the full calculation displayed in Fig. 17b with the TDHF results displayed in Fig. 17c. Within the TDHF we obtain a blue shift for all detunings (see Fig. 17c). Note that without the light holes the TDHF result vanishes for opposite circular excitation; thus, the blue shift in Fig. 17c is solely induced by heavy-hole–light-hole coupling. Since the energy difference of 15 meV between the heavy-hole and light-hole excitons and thus the detuning between the pump and the light-hole exciton is rather large, the amplitude of the blue shift decreases only weakly with increasing detuning below the heavy-hole exciton (see Fig. 17c). For the largest detuning of 18 meV, both the full and the TDHF result (dashed-dotted lines in Figs. 17b and 17c correspond to a blue shift with about the same amplitude. Thus, as for cocircular excitation, for opposite circular excitation the importance of correlations decreases rapidly with increasing detuning. It should be noted that this result can only be obtained if in addition to the heavy holes further valence bands are considered. Without light holes and their coupling to the heavy holes, the red shift at the heavy-hole exciton obtained for opposite circular excitation would remain for extremely large detuning. It is the coupling between the heavy-hole and light-hole transitions that introduces TDHF terms, which overcompensate the heavy-hole correlation-induced red shift for detunings larger than the heavy–light splitting and turn the differential absorption into a blue shift.

3. Higher Intensities up to the Coherent $\chi^{(5)}$ Limit

So far we have discussed resonant and off-resonant absorption changes induced by a relatively weak pump pulse within the coherent $\chi^{(3)}$ limit. To find out how the correlation-induced polarization dependencies of the differential absorption change with increasing pump intensity, we now extend the theoretical description up to the coherent $\chi^{(5)}$ limit. As long as the coherent limit is assumed, the nonlinear optical response can be fully described using transition-type correlation functions only (Victor *et al.*, 1995).

Omitting the various sums and indices, the basic structure of the coherent $\chi^{(3)}$ limit, Eqs. (13) and (14), is given by

$$\partial_t p = i[\omega_p p + \mu E(1 - bp^*p) + Vp^*B] \qquad (24)$$

$$\partial_t B = i[\omega_B B + \mu E p]. \qquad (25)$$

Here, ω_p and ω_B are energies of a single- and a two-exciton state, respectively. μ is the dipole matrix element and b and V denote the strength of the Pauli blocking and the Coulomb-interaction-induced nonlinearities. Linear in the light-field E, a single-exciton transition p can be excited through the source term μE in Eq. (24). In second order, the light field generates transitions to two excitons B via the source term $\mu E p$ in Eq. (25). Because of the Coulomb interaction, B couples back to the equation for single excitons through the third-order nonlinearity $Vp*B$. Within the coherent $\chi^{(3)}$ limit, the nonlinear optical response is solely induced by the Pauli blocking ($b\mu E p^* p$) and the Coulomb-interaction-induced contributions ($Vp*B$), which involve single- and two-exciton excitation to lowest orders.

In the coherent $\chi^{(5)}$ limit, besides terms containing p and B in higher orders, the three-exciton amplitude W also contributes to the optical response. The derivation of the coherent $\chi^{(5)}$ limit can be performed applying similar procedures as outlined in Section II. Omitting the various indices, the resulting equations can be written schematically as (Victor et al., 1995)

$$\partial_t p = i[\omega_p p + \mu E(1 - b[p^*p + p^*p^*pp + B^*pp + Bp^*p^* + B^*B])$$
$$+ V[p^*B + p^*p^*pB + B^*pB + B^* W + p^*p^*W]] \quad (26)$$

$$\partial_t B = i[\omega_B B + \mu E[p + p^*B] + Vp^*W] \quad (27)$$

$$\partial_t W = i[\omega_W W + \mu EB]. \quad (28)$$

As in Eqs. (24) and (25) one can distinguish in Eqs. (26)–(28) between source terms directly related to optical excitation involving μE and source terms arising because of the presence of the many-body Coulomb interaction V. In comparison to the $\chi^{(3)}$ equations, in $\chi^{(5)}$ a number of additional terms appear. These terms include one- and two-exciton amplitudes in higher orders, as well as the three-exciton amplitude W. Whereas in $\chi^{(3)}$ B is driven purely by sources proportional to the field E, in $\chi^{(5)}$ it also has Coulombic sources involving the three-exciton amplitude ($Vp*W$). The source term μEB in Eq. (28) shows that W is induced by an additional optical transition on top of a two-exciton amplitude B. B is of at least second order in the light field and therefore W is of at least third order. Since the optical polarization is determined by p and since those source terms in Eq. (26) that include W are also proportional to p^*p^* or B^*, which both are of at least second order in the field, the three-exciton amplitude W only contributes in fifth or higher order. The structure of Eqs. (26)–(28) thus shows that to lowest order the optical excitation is responsible for the *creation* of multiple excitons described by B and W. The Coulomb interaction, on the other hand, introduces a coupling of these multiple excitons *back* to p, which determines the optical polarization.

It should be noted that the equation of motion for the three-exciton transition W represents a full six-body problem, whose solution requires a tremendous numerical effort. So far, no such solutions have been reported for extended semiconductors. In Bartels et al. (1995) W has been decomposed into genuine six-particle correlations and lower factorized correlations according to $W = \bar{W} + ppp + pB$, and for the analysis of $\chi^{(5)}$ in FWM $\bar{W}$ was neglected.

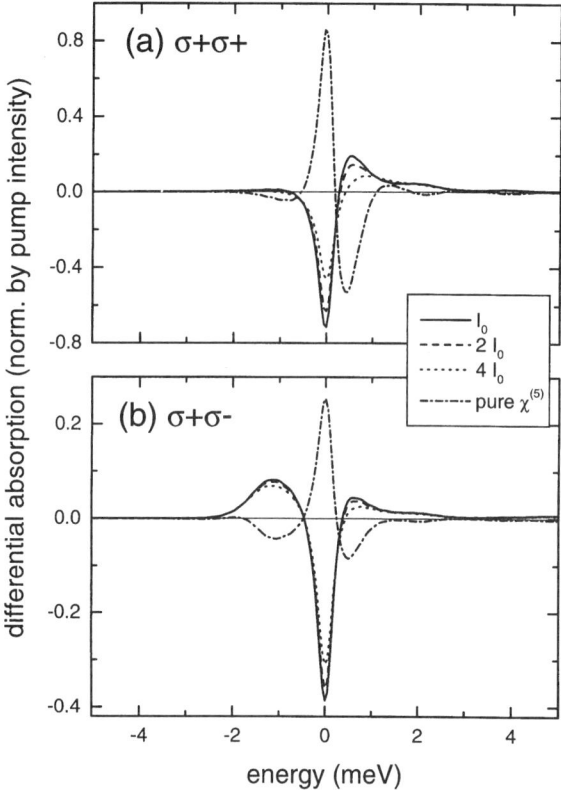

FIG. 18. Normalized differential absorption spectra for resonant excitation at the exciton resonance and time delay $\tau = 0$ ps. (a) Cocircularly polarized pump and probe pulses ($\sigma^+\sigma^+$) and (b) opposite-circularly polarized pump and probe pulses ($\sigma^+\sigma^-$) for various pump pulse intensities. At I_0 (solid) the spectra are dominated by $\chi^{(3)}$. With increasing intensity $2I_0$ (dashed) and $4I_0$ (dotted) $\chi^{(5)}$ contributions become stronger. The pure $\chi^{(5)}$ result (dash-dot) is also shown. System parameters of model HH-I in Table I are used.

To simplify our analysis of $\chi^{(5)}$ absorption changes we start by first considering only contributions from p and B up to fifth order and neglecting the three-exciton amplitude W in Eqs. (26)–(28). We furthermore restrict our investigations to heavy-hole transitions only and use the same parameters as in Subsections 1 and 2 of this section.

Numerical results of resonant absorption changes assuming a time delay of $\tau = 0$ ps and different pump pulse intensities are displayed in Fig. 18, where all spectra have been normalized by the intensity of the pump pulse. By comparing the low intensity results (solid lines) and the pure $\chi^{(5)}$ results (dashed-dotted lines) of Figs. 18a and 18b we notice that basically the spectral structure of the absorption changes induced by $\chi^{(5)}$ are the negative of the $\chi^{(3)}$ contributions. Thus, instead of bleaching at the exciton, $\chi^{(5)}$ corresponds to induced absorption. Furthermore, spectral oscillations show up in $\chi^{(5)}$ that do not appear in $\chi^{(3)}$. These features disappear with increasing time delay. With increasing intensity we see that in the normalized spectra displayed in Figs. 18a and 18b, both the bleaching at the exciton and the excited state absorption at the exciton to two-exciton transitions are weakened. Close inspection of Figs. 18a and 18b reveals that the excited state absorption decreases more rapidly with increasing pump intensity than the bleaching. Furthermore, the intensity-induced changes are somewhat stronger for cocircular than for oppositely circular excitation.

The computed absorption changes for excitation 4.5 meV below the exciton are displayed in Fig. 19. Since the detuned pump pulse creates less excitation than a resonant pulse, one has to strongly increase the pump pulse intensity to see significant intensity-dependent changes of the spectra. For oppositely circular excitation it is shown in Fig. 19b that in contrast to $\chi^{(3)}$, which corresponds to a red shift, the pure $\chi^{(5)}$ results in a blue shift of the exciton. Consequently, the amplitude of the red shift in the total spectra decreases with increasing pump intensity. For cocircular excitation, however, we find that the pure $\chi^{(5)}$ results corresponds to a broadening of the exciton, whereas a blue shift is obtained within $\chi^{(3)}$ (see Fig. 19a). Thus, with increasing pump intensity the spectra in this configuration gradually change from a shift toward a broadening.

The results in Figs. 18 and 19 clearly reveal that for increased pump intensities the differential absorption spectra still depend sensitively on the polarization of the exciting pulses. This can be expected to remain valid as long as scattering events, which become increasingly rapid with increased intensity, do not destroy this polarization sensitivity. If scattering becomes too fast, the coherent limit assumed here is no longer realistic and additional incoherent contributions need to be included; see, for example, Section IV.

Whereas the $\chi^{(3)}$ limit corresponds to excitation with very low intensities, which have to be low enough that a further reduction in intensity results in no further change of the lineshapes, the range of validity of the $\chi^{(5)}$ limit is not so well defined. The range of intensities where $\chi^{(5)}$ dominates over higher orders $\chi^{(n)}$ with $n > 5$ could be quite small. Thus, the results presented here should only serve as an illustration of the lowest order intensity-dependent changes. It should be noted, however, that the $\chi^{(5)}$ signatures presented in Figs. 18 and 19 are in qualitative agreement with recent experiments (Meier et al., 2000a). This agreement can be taken as an indication that under certain circumstances the approximations made so far, that is, neglecting the three-exciton resonance W and using the coherent $\chi^{(5)}$ contributions of p and B, may provide a reasonable description of the experiment.

To have at least some feeling about the relevance and the signatures induced by three-exciton resonances W in PP spectroscopy, we have

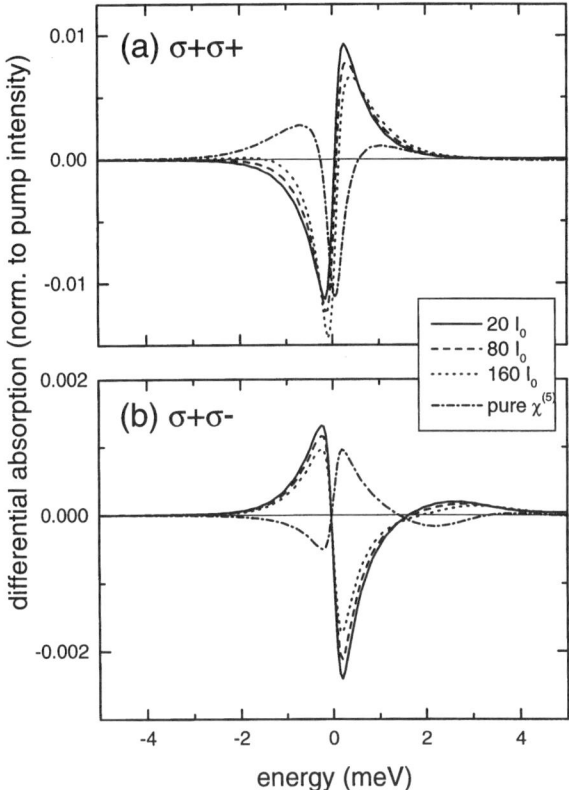

FIG. 19. Normalized differential absorption spectra for excitation 4.5 meV below the exciton resonance and time delay $\tau = 0$ ps. (a) Cocircularly polarized pump and probe pulses ($\sigma^+\sigma^+$) and (b) opposite-circularly polarized pump and probe pulses ($\sigma^+\sigma^-$) for various pump pulse intensities. At $20I_0$ (solid) the spectra are dominated by $\chi^{(3)}$. With increasing intensity $80I_0$ (dashed) and $160I_0$ (dotted) $\chi^{(5)}$ contributions become stronger. The pure $\chi^{(5)}$ result (dash-dot) is also shown. System parameters of model HH-I in Table I are used.

performed calculations solving the full set of Eqs. (26)–(28). Because of constraints in the numerical resources, these calculations were only possible for a reduced system size of $N = 8$ sites. Since for the present system parameters already $N = 10$ sites are needed to ensure a converged energetic position of the biexciton, it is thus clear that the even more extended three-exciton states are not converged for such a small system size. With these precautions in mind, we show in Fig. 20 the computed resonant absorption changes assuming a time delay of $\tau = 0$ ps and different pump pulse intensities. The $\chi^{(3)}$ (solid) and $\chi^{(5)}$ results without (dashed) and with (dotted) the three-exciton contributions W are displayed separately. The selection rules and the relevant single-, two-, and three-exciton states are

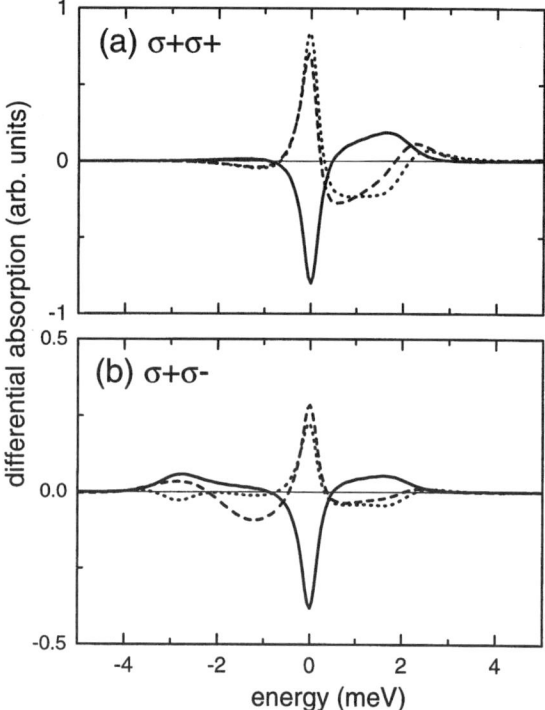

FIG. 20. Differential absorption spectra for resonant excitation at the exciton resonance and time delay $\tau = 0$ ps. (a) Cocircularly polarized pump and probe pulses ($\sigma^+\sigma^+$) and (b) opposite-circularly polarized pump and probe pulses ($\sigma^+\sigma^-$). The $\chi^{(3)}$ (solid) and the $\chi^{(5)}$ contributions without (dashed) and with (dotted) three-exciton resonance W are displayed separately. Here the system size was reduced to 8 sites. System parameters of model HH-I in Table I are used. In the solution of the equation of motion for W, a phenomenological decay rate of $\gamma_W = 1/1$ ps is used.

shown schematically in the level scheme of Fig. 21. Differences between the dashed and the dotted lines in Figs. 20a and 20b are induced by W. The spectral region where such differences appear corresponds to a transition from two single- to a three-exciton state or to a transition from a two- to a three-exciton state, which are described by the inhomogeneities Vp^*p^*W and VB^*W in Eq. (26), respectively. Generally we see that the three-exciton states have only very small influence on the differential absorption if the pump pulse is detuned far below the exciton (not shown in the figure). We thus expect that the intensity dependence of the absorption changes displayed in Fig. 19 for excitation below the exciton remain basically unchanged if W is included in the calculation.

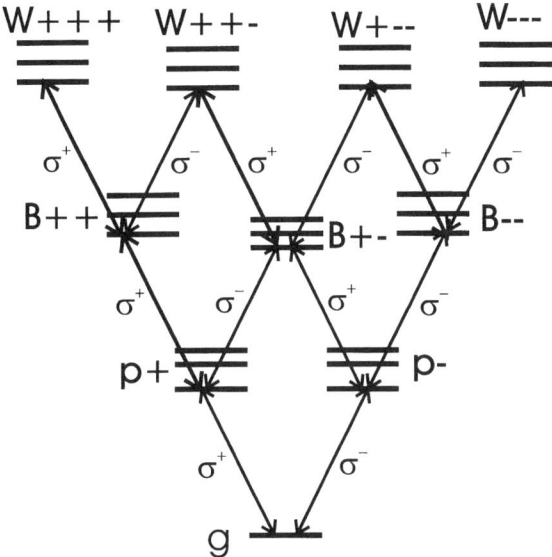

FIG. 21. Schematic level scheme including the heavy-hole transitions relevant in the coherent $\chi^{(5)}$ limit. As in Fig. 6, $p+$ and $p-$ denote the spin-degenerate single-exciton states that can be excited with σ^+ and σ^- polarized light. $B++$ and $B--$ are the two-exciton states formed by two single excitons with the same spin, whereas $B+-$ is formed by two excitons with opposite spin. $W+++$, $W++-$, $W+--$, and $W---$ represent the different types of three-exciton states that may contribute to the nonlinear optical response.

IV. Absorption Changes Induced by Incoherent Occupations

In the previous section, excitonic absorption changes induced by coherent pump fields have been analyzed within the coherent $\chi^{(3)}$ and $\chi^{(5)}$ limits. In this section we extend the theoretical analysis by including also incoherent processes, that is, we compute absorption changes caused by incoherent occupations of excitons and/or electron–hole pairs.

To describe the third-order optical response of semiconductors including incoherent processes, we have to evaluate four density matrices (Axt and Stahl, 1994a; Axt et al., 1996; Bartels et al., 1997; Meier and Koch, 1999):

$$p^{\eta}_{\lambda} = \langle \hat{d}_{\eta} \hat{c}_{\lambda} \rangle, \quad \text{single-exciton transitions} \qquad (29)$$

$$B^{\eta\eta'}_{\lambda\lambda'} = \langle \hat{d}_{\eta} \hat{c}_{\lambda} \hat{d}_{\eta'} \hat{c}_{\lambda'} \rangle, \quad \text{two-exciton transitions} \qquad (30)$$

$$N^{\eta\eta'}_{\lambda\lambda'} = \langle \hat{c}^+_{\lambda} \hat{d}^+_{\eta} \hat{d}_{\eta'} \hat{c}_{\lambda'} \rangle, \quad \text{pair occupations} \qquad (31)$$

$$Z^{\eta\eta'\eta''}_{\lambda\lambda'\lambda''} = \langle \hat{c}^+_{\lambda} \hat{d}^+_{\eta} \hat{d}_{\eta'} \hat{c}_{\lambda'} \hat{d}_{\eta''} \hat{c}_{\lambda''} \rangle, \quad \text{exciton–two-exciton transitions.} \qquad (32)$$

Here, $\hat{c}_\lambda^+(\hat{d}_\eta^+)$ creates and $\hat{c}_\lambda(\hat{d}_\eta)$ destroys an electron (a hole) in state $\lambda(\eta)$. To improve readability of the equations, the notation has been simplified and in the following the indices λ and η include both real-space site and subband (spin) information. In addition to p and B, which already show up in the coherent limit, the density matrices N and Z, which are (partly) occupation-type, that is, involve *both* creation and destruction operators, need to be considered in incoherent situations. A simple physical interpretation of p and B in the basis of single- and two-exciton eigenstates is indicated in Fig. 2, and N and Z are illustrated in Fig. 22.

In order to be able to separate not only coherent and incoherent contributions but also first- and higher-order contributions induced by the many-body Coulomb interaction it is advantageous to use a cumulant expansion and to define the following quantities (Schäfer *et al.*, 1996; Axt *et al.*, 1996; Bartels *et al.*, 1997; Meier and Koch, 1999):

$$\bar{B}_{\lambda\lambda'}^{\eta\eta'} = B_{\lambda\lambda'}^{\eta\eta'} - p_\lambda^\eta p_{\lambda'}^{\eta'} + p_{\lambda'}^\eta p_\lambda^{\eta'} \tag{33}$$

$$\bar{N}_{\lambda\lambda'}^{\eta\eta'} = N_{\lambda\lambda'}^{\eta\eta'} - p_\lambda^{*\eta} p_{\lambda'}^{\eta'} \tag{34}$$

$$\bar{Z}_{\lambda\lambda'\lambda''}^{\eta\eta'\eta''} = Z_{\lambda\lambda'\lambda''}^{\eta\eta'\eta''} - p_\lambda^{*\eta}\bar{B}_{\lambda'\lambda''}^{\eta'\eta''} - p_{\lambda'}^{\eta'}\bar{N}_{\lambda\lambda''}^{\eta\eta''} + p_{\lambda''}^{\eta'}\bar{N}_{\lambda\lambda'}^{\eta\eta''} - p_{\lambda''}^{\eta''}\bar{N}_{\lambda\lambda'}^{\eta\eta'} + p_{\lambda'}^{\eta''}\bar{N}_{\lambda\lambda''}^{\eta\eta'}$$
$$- p_\lambda^{*\eta}(p_{\lambda'}^{\eta'} p_{\lambda''}^{\eta''} - p_{\lambda''}^{\eta'} p_{\lambda'}^{\eta''}). \tag{35}$$

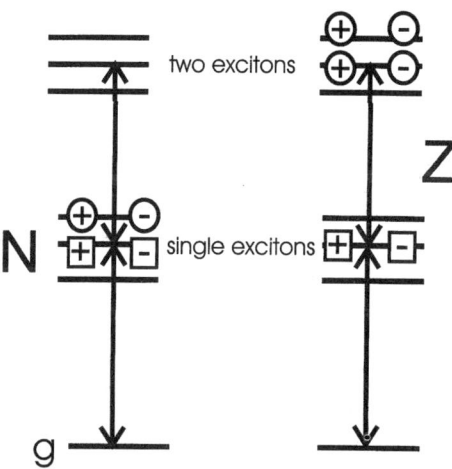

FIG. 22. Schematic illustration of the physical meaning of the two quantities N and Z, which contribute to the optical response within $\chi^{(3)}$ if incoherent processes are relevant in the basis of many-exciton eigenstates. The four-point function $N = \langle c^+ d^+ dc \rangle$ describes on electron–hole pair occupation, and the six-point function $Z = \langle c^+ d^+ dcdc \rangle$ describes single- to two-exciton transitions. As in Fig. 2, the destruction operators appearing in the definitions of N and Z are symbolized by the circles with $-$ for the negatively charged electrons (c) and $+$ for the positively charged holes (d). The squares with $+$ and $-$ denote the corresponding creation operators.

Without dephasing, that is, in the coherent $\chi^{(3)}$ limit, the pair occupations $\bar{N}$ and the exciton–two-exciton transitions $\bar{Z}$ (which is of mixed transition and occupation type) vanish (Lindberg et al., 1994b; Axt et al., 1996; Bartels et al., 1997); see Subsection 1 of Section II.

In the following we are interested in the absorption changes induced by an incoherent occupation $\bar{N}$. As shown in Meier and Koch (1999), the linear optical response to a test pulse $\mathbf{E}(t)$ in the presence of such an incoherent occupation $\bar{N}$ can be analyzed by solving the coupled equations of motion for p and $\bar{Z}$. Starting from the Heisenberg equations with the many-body Hamiltonian, the resulting equations are (Axt et al., 1996; Bartels et al., 1997; Meier and Koch, 1999)

$$\left[-i(\partial_t + \gamma_{p_2^1}) p_2^1 + \sum_{1'2'} \Omega_{p_{22'}^{11'}} p_{2'}^{1'} \right]$$

$$= \mathbf{E}(t) \cdot (\boldsymbol{\mu}_2^1 - \sum_{ji} [\boldsymbol{\mu}_i^1 \bar{N}_{i2}^{jj} + \boldsymbol{\mu}_2^j \bar{N}_{ii}^{j1}])$$

$$- \sum_{ji} V(_2^1|_i^j)(p_i^1 \bar{N}_{i1}^{ij} + p_2^j \bar{N}_{ii}^{j1} - p_2^j \bar{N}_{ii}^{ij} - p_i^j \bar{N}_{i2}^{j1}) + \sum_{ji} V(_2^1|_i^j) \bar{Z}_{jj2}^{ii1} \quad (36)$$

and

$$\left[-i(\partial_t + \gamma_{\bar{Z}_{146}^{235}}) \bar{Z}_{146}^{235} + \sum_{1'2'3'4'5'6'} \Omega_{\bar{Z}_{11'44'66'}^{22'33'55'}} \bar{Z}_{1'4'6'}^{2'3'5'} \right]$$

$$= V(_6^3|_4^5)(\bar{N}_{16}^{23} p_4^5 + \bar{N}_{14}^{25} p_6^3) - V(_4^3|_6^5)(\bar{N}_{14}^{23} p_6^5 + \bar{N}_{16}^{25} p_4^3). \quad (37)$$

In Eqs. (36) and (37) the upper (lower) indices refer to holes (electrons). It should be noted here that in Eqs. (36) and (37) all relative and center-of-mass coordinates are included, and these equations are also valid in inhomogeneous situations. In Eq. (36) $\boldsymbol{\mu}_2^1$ is the interband dipole matrix element. $\gamma_{p_2^1}$ and $\gamma_{\bar{Z}_{146}^{235}}$ are phenomenologically introduced dephasing rates. The propagator appearing in the homogeneous part of Eq. (36) is

$$\Omega_{p_{22'}^{11'}} = \delta_{22'} T_{11'}^v + \delta_{11'} T_{22'}^c - \delta_{11'} \delta_{22'} V_{12}, \quad (38)$$

which is identical to the homogeneous part of Eq. (10). As in Eq. (10), the matrix $T^c(T^v)$ contains the electron (hole) site energies in their diagonal part, and the couplings between the sites in their off-diagonal elements, and V_{12} is the electron–hole Coulomb attraction (Sieh et al., 1999a, 1999b). The propagator appearing in the homogeneous part of Eq. (37) is given by

$$\Omega_{Z_{11'44'66'}^{22'33'55'}} = -\Omega_{p_{11'}^{22'}} \delta_{33'} \delta_{44'} \delta_{55'} \delta_{66'} + \Omega_{p_{44'}^{33'}} \delta_{11'} \delta_{22'} \delta_{55'} \delta_{66'}$$

$$+ \Omega_{p_{66'}^{55'}} \delta_{11'} \delta_{22'} \delta_{33'} \delta_{44'} + V(_4^3|_6^5) \delta_{11'} \delta_{22'} \delta_{33'} \delta_{44'} \delta_{55'} \delta_{66'}$$

$$= -\Omega_{p_{11'}^{22'}} \delta_{33'} \delta_{44'} \delta_{55'} \delta_{66'} + \Omega_{\bar{B}_{44'66'}^{33'55'}} \delta_{11'} \delta_{22'} \quad (39)$$

Here, the abbreviation $V(^a_b|^c_d) = V_{ac} - V_{ad} - V_{bc} + V_{bd}$ has been used. According to the last line of Eq. (39) the six point function $\bar{Z}$ corresponds to the simultaneous creation of a two-exciton amplitude $\bar{B}$ and destruction of a single-exciton amplitude p. Thus, $\bar{Z}$ describes exciton to two-exciton transitions. As in Eq. (14), the propagator for the two-exciton transition $\bar{B}$ is

$$\Omega_{\bar{B}^{33'55'}_{44'66'}} = \Omega_{p^{33}_{44}}\delta_{55'}\delta_{66'} + \Omega_{p^{55}_{66}}\delta_{44'}\delta_{33'} + V(^3_4|^5_6)\delta_{33'}\delta_{44'}\delta_{55'}\delta_{66'}. \quad (40)$$

Note that for the calculation of the differential absorption induced by $\bar{N}$ linear in the test pulse $\mathbf{E}(t)$, no two-exciton amplitude ($\bar{B}$) needs to be considered since it can only be excited by two coherent fields; see Subsection 2 of Section II.

The driving term $\mathbf{E}(t) \cdot \boldsymbol{\mu}^1_2$ in Eq. (36) describes the linear optical excitation of an interband coherence. All other inhomogeneous terms in Eq. (36) are proportional to $\bar{N}$ and represent absorption changes due to three types of processes (Meier and Koch, 1999): (i) contributions directly involving the light field $\propto \mathbf{E}(t) \cdot \boldsymbol{\mu}\bar{N}$ that are due to Pauli blocking, (ii) the terms $\propto Vp\bar{N}$ that represent first-order Coulomb-induced nonlinearities, and (iii) the term $\propto V\bar{Z}$ that represents higher-order Coulomb-induced nonlinearities. $\bar{Z}$ is driven by terms proportional to Coulomb matrix elements V times $p\bar{N}$.

In our numerical studies the coupled Eqs. (36) and (37) are solved directly in the time domain assuming an ultrashort, that is, spectrally white, test pulse $\mathbf{E}(t)$. As in the previous section, the differential absorption is given by the imaginary part of the frequency domain differential interband polarization $\delta \mathbf{P}(\omega)$ (see Eq. (20)), which is calculated by solving Eqs. (36) and (37) neglecting the direct linear optical source that does not involve $\bar{N}$. Since we start from an incoherent occupation $\bar{N}$ and investigate the differential absorption linear in the test pulse $\mathbf{E}$, there is no particular polarization dependence in the signals (Meier and Koch, 1999).

Since two-dimensional calculations including the sixpoint correlation $\bar{Z}$ could not be done so far, we again use the one-dimensional tight-binding model (Sieh et al., 1999a; Meier and Koch, 1999; Sieh et al., 1999b) with 10 sites, that is, model HH-I of Table I. Concerning the subbands we consider two spin-degenerate electron and heavy-holes bands, respectively. The phenomenologically introduced dephasing rates $\gamma_{p_1^2}$ and $\gamma_{\bar{Z}^{235}_{146}}$ are both set to 1/2 ps.

The homogeneous part of the equation of motion of $\bar{N}^{23}_{14}$ contains $\Omega_{p_4^3} - \Omega_{p_1^2}$, that is, the difference of the energies of two single-exciton amplitudes; thus, $\bar{N}^{23}_{14}$ describes pair occupations and coherences (Axt et al., 1996; Bartels et al., 1997). This pair occupation can be expanded (Meier and Koch, 1999) using the complete set of excitonic eigenstates $\Psi_\alpha(m, n)$, which are the eigenstates of the homogeneous part of the equation of p, via

$$\bar{N}^{23}_{14} = \sum_{\alpha\beta} a_{\alpha\beta}\Psi^*_\alpha(1', 2')\Psi_\beta(4', 3'). \quad (41)$$

As in the previous equations, here also the indices 1, 2, 3, 4 run over both real-space sites and subbands, whereas on the right-hand side of Eq. (41), 1', 2', 3', 4' include real-space information only and the subband information is contained in α and β. In Eq. (41) all relative and center-of-mass coordinates are included. The total density n is given by

$$n = \frac{1}{L}\sum_\alpha a_{\alpha\alpha}, \qquad (42)$$

where L is the length of the system.

For low densities (nondegenerate limit) and in thermal equilibrium, we assume that $\bar{N}$ is given by a summation over thermally populated exciton states,

$$\bar{N}^{23}_{14} = \sum_{\alpha' v_e v_h} \frac{\exp(-\varepsilon_{\alpha'}/k_B T)}{A} \delta_{v_e v_h} \Psi^*_{\alpha'}(1', 2') \Psi_{\alpha'}(4', 3'), \qquad (43)$$

with

$$n = \frac{1}{L}\sum_{\alpha'} 2\frac{\exp(-\varepsilon_{\alpha'}/k_B T)}{A}, \qquad (44)$$

where the prefactor 2 comes from the summation over the two optically coupled pairs of subbands, that is, $v_e = v_h = 1$ and $v_e = v_h = 2$ (Schäfer et al., 1996; Sieh et al., 1999b) and A is a constant determining the total density. In Eq. (43) we have assumed that no coherences between different exciton states are present and that excitons of both optically coupled pairs of degenerate electron and heavy-hole subbands are equally populated.

Besides assuming that initially thermally populated excitons are present, we can also consider the situation where unbound electron–hole states are populated. To model this situation we simply replace the exciton wave functions $\Psi_\alpha(m, n)$ in Eq. (41) by the complete set of single-particle eigenstates $\Phi_\alpha(m, n)$ with energies $\tilde{\varepsilon}_\alpha$, which are the eigenstates of the homogeneous part of the equation of p if the electron–hole Coulomb attraction V is neglected. For the thermal pair-occupation $\bar{N}$ we then use Eq. (43), where Ψ_α is replaced by Φ_α and ε_α by $\tilde{\varepsilon}_\alpha$.

We start our numerical analysis with the low-temperature limit, where we assume that only the energetically lowest exciton state, that is, the optically active 1s exciton, is populated. For this situation we calculate the differential absorption induced by the incoherent occupation $\bar{N}$ as seen by a σ^+ polarized probe pulse. As an illustration of what can be expected in this situation, we schematically display in Fig. 23 the quantities induced linearly in the probe pulse. Starting from the incoherent occupation $\bar{N}$, the σ^+

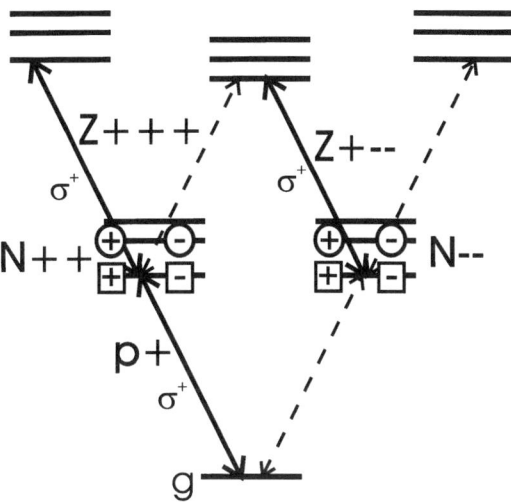

FIG. 23. Schematic level scheme illustrating absorption changes induced by an incoherent occupation N as viewed by a σ^+ polarized probe pulse. As indicated by the solid transition lines, the probe field can either create a single-exciton coherence $p+$ or create coherences between single- and two-exciton states that are described by $Z+++$ and $Z+--$.

polarized probe field can either create a single-exciton coherence $p+$ or can induce coherences between single- and two-exciton states, which are described by $Z+++$ and $Z+--$.

Figure 24 shows the numerically calculated differential absorption spectra at $T = 0\,\text{K}$. Besides the total signal displayed in Fig. 24a, the three contributions due to Pauli blocking and first- and higher-order Coulomb interactions are displayed in Fig. 24b. We see that Pauli blocking induces a pure and rather small bleaching of the exciton, whereas the first-order Coulomb terms have a dispersive shape at the exciton corresponding to the well-known blue shift (Peyghambarian *et al.*, 1984). As for the resonant coherent excitation discussed in Subsection 1 of Section III, this blue shift is strongly compensated by the higher-order Coulomb term ($\propto \bar{Z}$), which besides excited-state absorption due to single- to two-exciton transitions also yields a red shift at the exciton resonance (Meier and Koch, 1999); see Fig. 24b. Adding the contributions we find that the total signal shows no net shift and thus that only bleaching of the exciton resonance remains; see Fig. 24a. This bleaching of the exciton is mainly a consequence of the strong Coulomb-interaction-induced terms, whereas the Pauli blocking introduces only a weak additional contribution. Additionally, we obtain excited state absorption above and below the exciton that is due to the excitation of unbound and bound two-exciton states. Similar correlation-induced line shapes in the differential absorption spectra due to thermalized excitons,

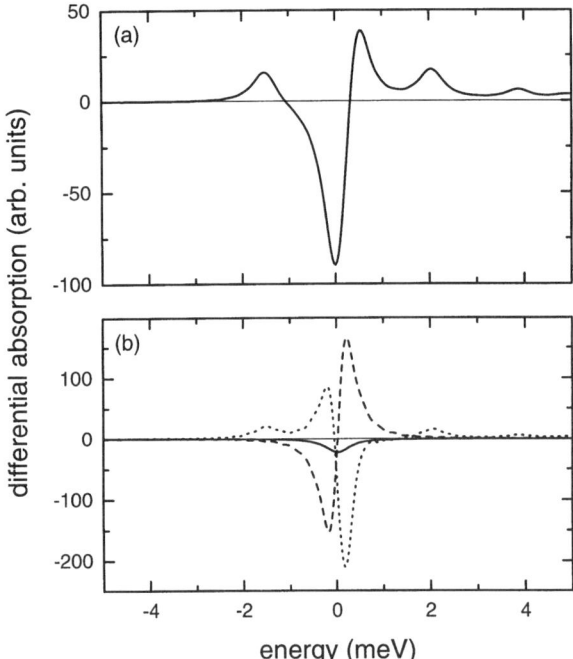

FIG. 24. Differential absorption spectra induced by an incoherent occupation of the lowest exciton state. The zero of the energy scale is chosen to coincide with the 1s-exciton resonance. (a) Total signal and (b) Pauli blocking (solid), first-order Coulomb-terms (dashed), and higher-order Coulomb-correlations (dotted). System parameters of model HH-I in Table I are used.

that is, bleaching at the exciton resonance and excited-state absorption due to the presence of two excitons were also calculated recently for aggregates of Frenkel excitons; see, for example, Meier et al. (1997a). The results displayed in Fig. 24 clearly demonstrate that for a proper description of both exciton bleaching and excited-state absorption, higher-order Coulomb correlations represented by the six-particle density matrix $\bar{Z}$ become important.

Figure 25 compares the differential absorption caused by (a) thermalized excitons and (b) thermalized electron–hole pairs for the three temperatures $T = 2, 25$, and 100 K. Figure 25b shows that by raising the temperature of an occupation of unbound electron–hole pairs, both the exciton bleaching and the amplitude of the excited state absorption decrease. The result obtained for thermalized exciton occupations (Fig. 25a) superficially looks rather similar; however, the reduction of the bleaching is nonmonotonous as function temperature. This indicates that the Coulomb-induced coupling

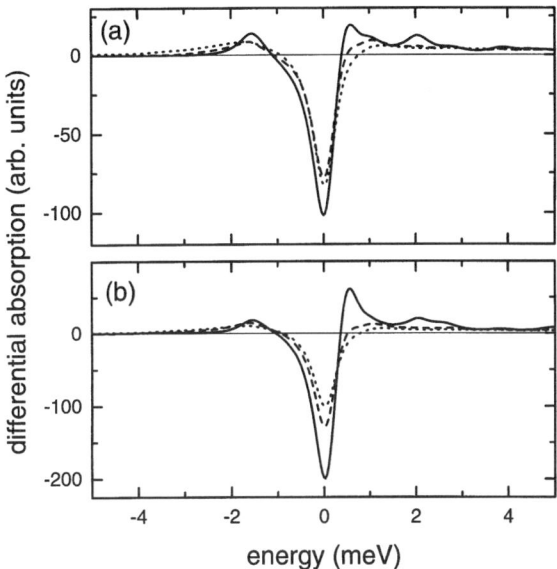

FIG. 25. (a) Differential absorption spectra for $T = 2\,\mathrm{K}$ (solid), 25 K (dashed), and 100 K (dotted) induced by thermally occupied excitons. (b) Same as (a) for thermally occupied electron–hole states. System parameters of model HH-I in Table I are used.

of energetically higher excitons to the 1s exciton has a nontrivial energy dependence. At very low temperatures, where only the lowest exciton is populated, one sees clear signatures of the individual two-exciton states in Fig. 25. These positive contributions become less structured at elevated temperatures because for increased temperatures energetically higher excitons are also populated.

Overall, the computed results for excitons and electron-hole pairs are rather similar. However, for a thermalized electron–hole pair population, the bleaching of the exciton resonance is stronger than for an exciton population. This stronger bleaching induced by electron–hole pairs compared to excitons implies a stronger nonlinear coupling of low-energy electron–hole pairs to the 1s exciton as compared to the nonlinear coupling among the excitons themselves. For the Pauli blocking contribution, these differences can be analyzed simply by using the exciton and electron–hole wave functions, as described in Schmitt-Rink *et al.* (1985). If Coulomb correlations are included in the description, no such simple explanation is possible; however, the overall conclusions remain valid.

Figure 26 summarizes our results on the temperature dependence of the exciton bleaching. It is shown that for low temperatures the bleaching induced by thermalized electron–hole pairs is about twice as strong as the

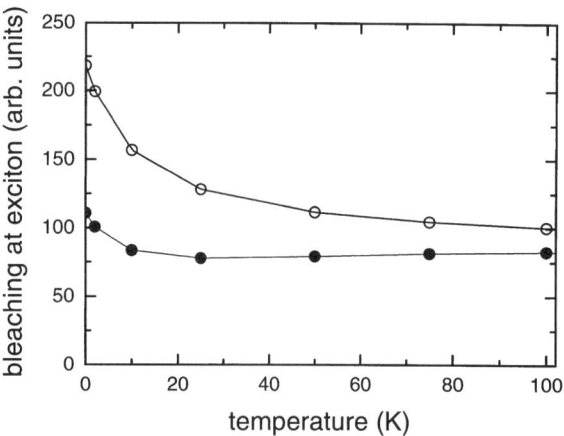

FIG. 26. Amplitudes of exciton bleaching as a function of temperature induced by thermally populated excitons (solid circles) and thermally populated electron–hole pairs (open circles). The lines are guides for the eye. System parameters of model HH-I in Table I are used.

bleaching induced by thermalized excitons. Concerning the temperature dependence of the bleaching induced by electron–hole pairs, we find that it decreases with increasing temperature, which is in qualitative agreement with the predictions of Schmitt-Rink et al. (1985). For the excitons we obtain a slightly nonmonotonous behavior as function of temperature. For very high temperatures (not shown in figure) within our finite bandwidth tight-binding model, the differential absorption induced by excitons and electron–hole pairs becomes exactly identical and approaches a nonvanishing value. This is due to the fact that in the high temperature limit, where $k_B T$ becomes much larger than the bandwidth Δ, thermally populated electron–hole pairs and excitons correspond to the same $\bar{N}$. For an effective mass model with an infinite bandwidth, this limit would be different (Schmitt-Rink et al., 1985).

The dominance of the unbound electron–hole over the exciton contributions to the resonance bleaching is most likely the reason for the success of electron–hole plasma theories explaining experimentally observed exciton saturation at low temperatures (Schmitt-Rink et al., 1985; Lee et al., 1986; Haug and Koch, 1989, 1994). This trend is expected to be even more pronounced in systems with dimensionalities greater than one because of the overall reduction of the relative importance of Coulomb-correlation effects with increasing dimensionality.

To analyze experimental findings where at low temperatures a blue shift of the exciton line in the presence of exciton and electron–hole pair populations was observed (Peyghambarian et al., 1984), we evaluate Eqs.

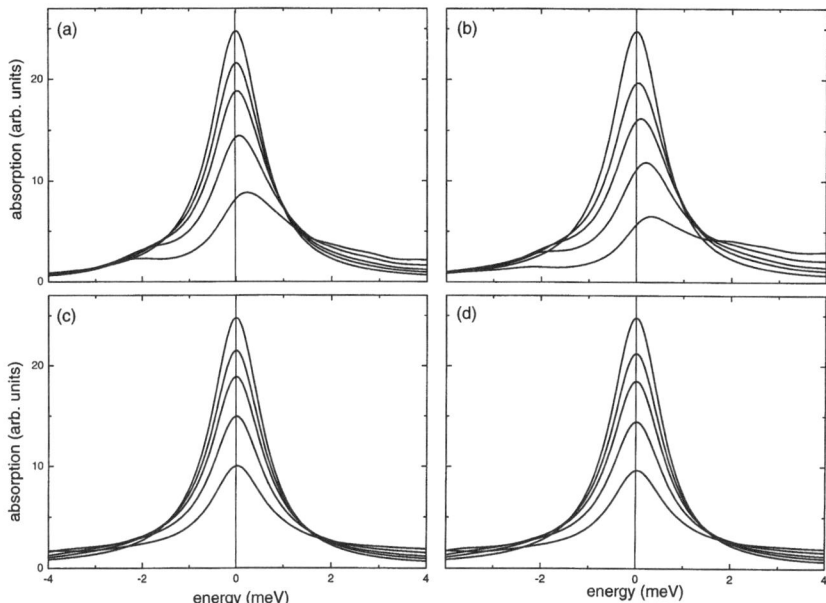

FIG. 27. Absorption spectra (a) and (c) for thermally populated excitons and (b) and (d) for thermally populated electron–hole pairs at temperatures of (a) and (b) $T = 10$ K, and (c) and (d) $T = 200$ K. The highest line depicts the linear absorption; for the lower ones the density is increased by a factor of 2 between each line from $n = 0.02/d$, $0.04/d$, $0.08/d$, to $0.16/d$. The spectral position of the exciton in the linear absorption is depicted by the vertical lines.

(36) and (37) for incoherent populations at higher densities. We now assume that in thermal equilibrium the exciton population is described by a Bose function with a self-consistently determined density- and temperature-dependent chemical potential. For the case of electron–hole pairs the Bose function is replaced by a Fermi function.

The resulting absorption spectra are displayed in Fig. 27. We find a blue shift of the exciton line with increasing density for low temperatures ($T = 10$ K). The blue shift is larger for a fixed density of electron–hole pairs than for excitons. With increased temperature ($T = 200$ K), the difference between electron–hole pairs and excitons is very small, because $k_B T$ is larger than the exciton binding energy. This is also in agreement with the results presented earlier (see Fig. 26). The spectra displayed in Fig. 27 clearly show that the blue shift of the exciton is present only at low temperatures, whereas at elevated temperatures a pure bleaching is found, which is in agreement with the experiments of Peyghambarian et al. (1984). Thus, even for higher density populations, Eqs. (36) and (37) qualitatively describe absorption changes experimentally observed in quantum wells.

V. Applications to Four-Wave-Mixing Spectroscopy

In this section we analyze the influence of correlations on FWM spectroscopy. As outlined for the case of PP, FWM signals also depend sensitively on the polarization directions of the incident pulses. Because of the propagation directions of the signals monitored in PP and FWM, $\mathbf{k}_2$ versus $2\mathbf{k}_2 - \mathbf{k}_1$, the two-exciton manifold that represents carrier correlations is probed in different ways in both experiments (compare Subsections 2 and 3 of Section II). Whereas PP performed with opposite circularly polarized pulses allows one to concentrate on effects induced by bound biexcitons, for this geometry the lowest order FWM signal vanishes. On the other hand, for both linear-parallel and -perpendicular polarized excitation, the biexciton does contribute to FWM, but is seen in PP only if pump and probe pulses are not orthogonally polarized. As discussed in more detail later, in FWM performed with linear-perpendicular polarized pulses the spectral components at the exciton resonance are strongly reduced and the signal is dominated by transitions at frequencies between the exciton and two-exciton states. In PP such a selective reduction of the contributions due to exciton bleaching is not possible. Hence, analyzing the polarization dependencies of FWM and PP makes it possible to investigate the dynamics of different correlation-induced effects.

In Subsection 1 of this section we focus on beats in TI and TR-FWM induced by biexcitons. Whereas in TR-FWM such exciton–biexciton beats are already present within $\chi^{(3)}$ (Axt et al., 1998), for a homogeneous system biexciton-induced beats do not show up for positive time delays in TI-FWM (Bott et al., 1993; Bartels et al., 1995). For higher intensities (Bartels et al., 1995) and in disordered systems (Albrecht et al., 1996), however, these beats also appear in TI-FWM.

In Subsection 2 it is shown that disorder may induce a polarization-dependent decay of the TI-FWM. The analysis of this polarization-dependent disorder-induced dephasing requires the consistent treatment of both disorder and correlations on a microscopic basis.

1. BIEXCITONIC BEATS

The importance of Coulomb-correlation-induced contributions became apparent in the attempts to analyze experiments exploring the dependence of the optical response on the polarization directions of the exciting pulses (Hulin and Joffre, 1990; Feuerbacher et al., 1991; Bar-Ad and Bar-Joseph, 1992; Lovering et al., 1992; Bott et al., 1993, 1996; Bigot et al., 1993; Wang et al., 1993, 1994; Hu et al., 1994; Rappen et al., 1994; Mayer et al., 1994, 1995; Smith et al., 1994, 1995; Bartels et al., 1995; Albrecht et al., 1996; Schäfer et al., 1996; Langbein et al., 1997; Kner et al., 1997, 1998; Sieh et al.,

1999a, 1999b). Clear evidence for correlations in such experiments is provided by the observation of polarization-dependent beats induced by the excitation of the bound biexciton. In agreement with the selection rules governing the biexciton excitation (see, for example, Fig. 8), such beats are observed in FWM for linear-parallel and linear-perpendicular polarized pulses, but are absent for cocircular polarized pulses. Note that PP and FWM involve different pulse sequences and signal directions (compare Subsection 2 of Section II) leading to different polarization dependencies for the excitation of biexcitons. For example, as shown in Subsection 1 of Section III, in PP performed with linear-perpendicular polarized pump and probe pulses, no biexciton is excited, whereas in FWM performed with the same polarization configuration of the incident pulses the biexciton contributes strongly; see Bott et al. (1993, 1966), Mayer et al. (1994, 1995), Smith et al. (1995), Albrecht et al. (1996), Schäfer et al. (1996), and later discussion. Furthermore, it should be noted that in FWM, unlike in PP, in lowest order the response vanishes identically for excitation with opposite circularly polarized pulses (Lindberg et al., 1994a).

In the following we analyze exciton–biexciton beats in TR- and TI-FWM within the coherent $\chi^{(5)}$ limit using the theoretical model introduced in Subsection 3 of Section III. For this purpose we compare the FWM signals obtained for linear-parallel and linear-perpendicular polarized incident pulses. Our results show that in agreement with experiment, for a homogeneous system in TI-FWM biexciton-induced beats are absent for positive delays in the $\chi^{(3)}$ limit. They are, however, induced by either higher intensities ($\chi^{(5)}$) (Bartels et al., 1995) or by disorder (Bigot et al., 1993; Albrecht et al., 1996).

In this subsection we use model HH-II of Table I. All wave functions obtained for the parameter set HH-II are identical to HH-I, since only the energy has been rescaled by 15/8 (see Table I) in order to have a larger heavy-hole exciton binding energy of 15 meV (Sieh et al., 1999b), which allows us more easily to concentrate on effects in the spectral vicinity of the exciton. In our numerical analysis we consider all terms arising in the coherent limit $\chi^{(5)}$ limit, that is, we include one-, two-, and three-exciton amplitudes represented by p, B, and W, respectively, by solving the full Eqs. (26)–(28). For numerical convenience we reduce the system size to $N = 6$, which introduces some well-known finite size effects, for example, an enhanced biexciton binding energy (3.3 meV instead of 2.6 meV for a larger system; see Table I). Since, however, the features of the FWM response reported here are in agreement with calculations for larger systems and with experimental observations, we assume for numerical convenience that even this small system size is sufficient for a qualitative description.

Figure 28 displays numerical results on the TR-FWM intensities for various time delays between $\tau = 0.8$ and 2.0 ps after excitation with two linear-parallel polarized pulses (xx). In Fig. 28a the $\chi^{(3)}$ result is displayed

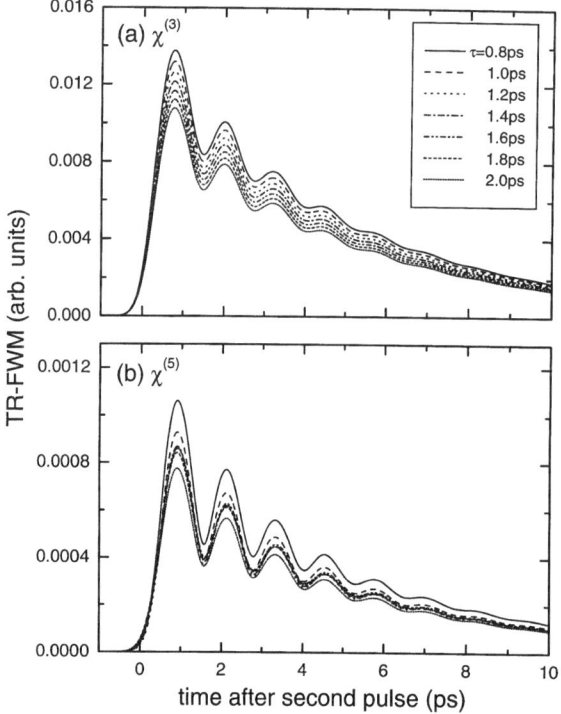

FIG. 28. Time-resolved four-wave mixing versus real time for delays from $\tau = 0.8$ ps to $\tau = 2.0$ ps as indicated in graph for excitation with linear-parallel polarized pulses. (a) Pure $\chi^{(3)}$ and (b) pure $\chi^{(5)}$. The temporal width of the exciting pulses was chosen as $\bar{t}_1 = \bar{t}_2 = 500$ fs, and both pulses are tuned 1 meV below the exciton resonance. The phenomenologically introduced dephasing rates are $\gamma_p = 1/5$ ps, $\gamma_B = 2\gamma_p = 1/2.5$ ps, and $\gamma_W = 3\gamma_p = 1/1.667$ ps. System parameters of model HH-II in Table I are used.

and it shows temporal beats with a period of 1.2 ps. This beat period is just the inverse of the biexciton binding energy of about 3.3 meV, which has also been obtained independently by calculating the differential absorption for excitation with opposite circular polarized pump and probe pulses (not shown in figure).

The appearance of the exciton–biexciton beats in the TR-FWM in Fig. 28a can be understood by analyzing the schematic Eqs. (24) and (25) that describe the nonlinear response within the coherent $\chi^{(3)}$ limit. Considering only a short-lived source term, such as $b\mu E p^* p$ in Eq. (24), the exciton amplitude evolves freely and oscillates with the exciton frequency ω_p, that is, $p(t) \propto \exp(i\omega_p t)$. However, because of the presence of another source term that itself is decaying slowly, as is the case for the Coulomb source Vp^*B in Eq. (24), one obtains an oscillation with a frequency that is given by the difference between ω_B and ω_p, that is, $p*B \propto \exp(i(\omega_B - \omega_p)t)$. Solving the equation of motion for the exciton coherence p driven by such a source term, one finds that in this case p includes oscillations with both ω_p as well as $\omega_B - \omega_p$, that is,

$$p(t) \propto \alpha \exp(i\omega_p t) + \beta \exp(i(\omega_B - \omega_p)t) = \exp(i\omega_p t)[\alpha + \beta \exp(i(\omega_B - 2\omega_p)t)],$$

where α and β determine the weight of the two spectral components. Consequently, the TR intensity $|p(t)|^2$ shows temporal beating. If the excitation is chosen such that mainly the biexciton is excited, as is the case for the results displayed in Fig. 28, one thus expects beats in the TR-FWM with a temporal period inversely proportional to the biexciton-binding energy $\omega_B - 2\omega_p$.

Since in Fig. 28 the time axis is chosen such that the last pulse, which is pulse 2 for positive time delay, always arrives at $t = 0$ (see also the definition of the pulse envelopes given in Subsection 2 of Section II), the phase of the beating is independent of the time delay between the two pulses. This is a consequence of the fact that in FWM a two-exciton amplitude B is only excited by pulse 2, which contributes in second order to the $\chi^{(3)}$-FWM signal monitored in direction $2\mathbf{k}_2 - \mathbf{k}_1$. The arrival of pulse 2 thus defines the onset of the biexciton-induced beats. With increasing time delay τ the TR-FWM intensity displayed in Fig. 28a simply shows a decrease determined by the decay rate γ_p, since the polarization induced by pulse 1 decays during the time interval τ until pulse 2 arrives and the FWM response is generated. The fact that for positive time delays the phase of the biexciton beats are independent of τ automatically implies the absence of biexciton beating in the TI-FWM. The TI-FWM is the integral over the TR-FWM intensity (see Eq. (19)), and thus up to $\chi^{(3)}$ it will show only a simple decay but no modulation with increasing τ.

In comparison to the $\chi^{(3)}$ results, the τ dependence is different for the pure $\chi^{(5)}$ contribution to the TR-FWM (see Fig. 28b). Although this contribution alone cannot be observed directly in experiments, it is instructive to consider it separately here. In the pure $\chi^{(5)}$ contribution, exciton–biexciton beats are present in TR-FWM; however, unlike in $\chi^{(3)}$, we find a nontrivial dependence of the amplitude of the TR-FWM intensity on the time delay τ. Whereas within $\chi^{(3)}$ the amplitude simply decays monotonously (exponentially) with increasing τ, in $\chi^{(5)}$ the amplitude decay is modulated. Thus, for excitation with higher intensities one has an additional dependence of the TR-FWM response on time delay τ. As discussed later, this τ-dependent modulation of the TR-FWM amplitude in $\chi^{(5)}$ introduces biexcitonic beats for higher intensities in TI-FWM.

Figure 29 displays numerical results for TR-FWM intensities at various time delays between $\tau = 0.8$ and 2.0 ps after excitation with two linear-perpendicular polarized pulses (xy). As for xx excitation, for xy excitation beats appear in TR-FWM. In the case of xy excitation the signal intensity

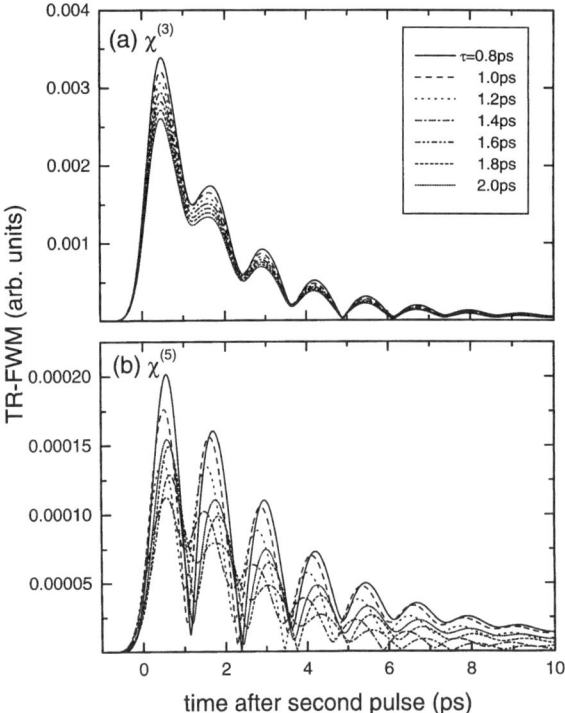

FIG. 29. Time-resolved four-wave mixing versus real time for delays from $\tau = 0.8$ ps to $\tau = 2.0$ ps as indicated in graph for excitation with linear-perpendicular polarized pulses. (a) Pure $\chi^{(3)}$ and (b) pure $\chi^{(5)}$. System parameters of model HH-II in Table I are used.

is reduced and the modulation depth is stronger than for xx. Both these effects are because of a strong suppression of spectral FWM components at the exciton frequency (see Axt et al., 1998, and discussion in the following subsection). The modulations seen in Fig. 29a correspond to the inverse of the difference between the biexciton and the lowest unbound two-exciton state, which is energetically situated slightly above twice the energy of the 1s exciton (compare Sieh et al., 1999a and 1999b). As discussed in more detail in the following subsection, within the TDHF approximation there is no difference between the FWM signal for xx and xy excitation; thus, the polarization dependence of the $\chi^{(3)}$ response (compare Figs. 28a and 29a) is solely induced by Coulomb correlations. As for xx excitation also for xy excitation within $\chi^{(3)}$ the phase of the beats is independent of the time delay, and the amplitude decays monotonously with increasing time delay; see Fig. 29a.

If we look to the pure $\chi^{(5)}$ contribution to the TR-FWM for xy, see Fig. 29b, we find that both the amplitude of the exciton–biexciton beats and their phase change with time delay. This is different compared to xx

excitation (see Fig. 28b), where only the amplitude of the TR-FWM was modulated with τ. We can thus expect some nontrivial behavior of the TI-FWM, which is analyzed later.

Figure 30 shows numerical results on the intensity dependence of TI-FWM for xx and xy excitations. As already explained on the basis of TR-FWM, in the coherent $\chi^{(3)}$ limit no exciton–biexciton beats are present for positive time delays in TI-FWM for either polarization configuration (see Fig. 30a). For higher excitation intensities, however, $\chi^{(5)}$ contributions become relevant and consequently for both xx and xy configurations beats appear (see Figs. 30b and 30c). These beats are a consequence of the amplitude modulations of the biexciton-induced beats appearing in TR-FWM in $\chi^{(5)}$ for the case of xx and also of the phase shifts in $\chi^{(5)}$ for xy excitation. In agreement with experiments (Mayer *et al.*, 1994; Bartels *et al.*, 1995), the biexciton-induced beats in TI-FWM become stronger with increasing intensity. Furthermore, one can see a weak polarization dependence of the beat phase in Fig. 30c.

Besides the excitation with higher intensities, disorder may also lead to exciton–biexciton beats in TI-FWM (Bigot *et al.*, 1993; Albrecht *et al.*,

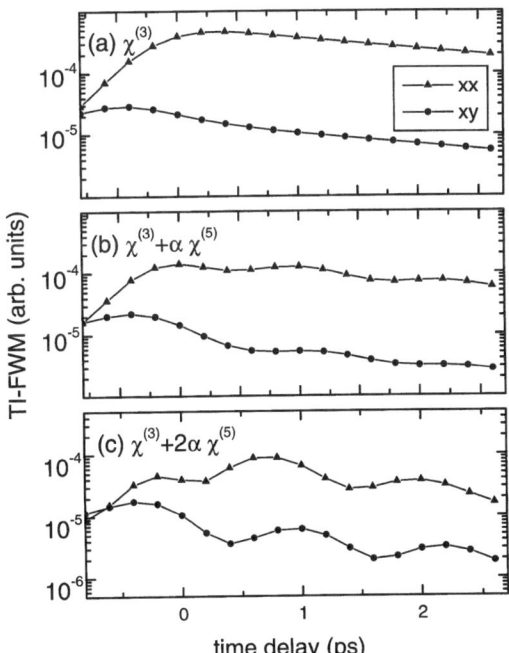

FIG. 30. Time-integrated four-wave mixing versus time delay for linear parallel (triangles) and linear perpendicular (circles) excitation. (a) Pure $\chi^{(3)}$, (b) $\chi^{(3)} + \alpha\chi^{(5)}$, and (c) $\chi^{(3)} + 2\alpha\chi^{(5)}$. The lines are guides for the eye. System parameters of model HH-II in Table I are used.

1996). To explain this effect, it is sufficient to consider a phenomenological disorder-induced inhomogeneous broadening. If this inhomogeneous broadening is assumed to be correlated for the exciton and the exciton to two-exciton transitions, it will lead to a photon-echo-like TR-FWM signal. Assuming a Gaussian inhomogeneous broadening thus introduces a Gaussian temporal envelope $\propto \exp(-((t-\tau)/\bar{t})^2)$ of the TR-FWM response, where lower limits for the temporal widths of the photon echo $\bar{t}$ are given by the inverse of the width of the inhomogeneous distribution and by the temporal width of the exciting pulses.

To phenomenologically include disorder effects into our calculations, we simply assume that such a correlated Gaussian inhomogeneous broadening of the exciton and two-exciton energies is present. The spectral width of the inhomogeneous broadening is assumed to be larger than the spectral pulse widths; thus the inhomogeneous broadening simply introduces a photon-echo emission of the TR-FWM response with a width determined by the incident pulses ($\bar{t}_1$ and $\bar{t}_2$). Therefore, to compute polarization-dependent TI-FWM signals in the presence of inhomogeneous broadening, we simply multiply the TR-FWM amplitude obtained for the homogeneous system (see Eq. (17)) by $\exp(-((t-\tau)/\bar{t})^2)$ using $\bar{t} = \bar{t}_1 = \bar{t}_2$ and then integrate over time to obtain the TI-FWM.

The TI-FWM signals including inhomogeneous broadening are displayed in Fig. 31. As shown in Fig. 31a, already in $\chi^{(3)}$, inhomogeneous broadening induces exciton–biexciton beats for both xx and xy. Since in our calculations we have used rather long pulses of $\bar{t}_1 = \bar{t}_2 = 500$ fs that are not much shorter than the period of the beats (≈ 1.2 ps), the disorder-induced beating and the polarization dependence of the beat phase are not very pronounced (Albrecht et al., 1996). Another result of the inhomogeneous broadening is the strong suppression of FWM signals for negative delays, which can be noticed by comparing Figs. 30a and 31a (note the different time delay axis in these figures). This suppression is a consequence of the fact that for negative time delays τ the maximum of the echo amplitude $\propto \exp(-((t-\tau)/\bar{t})^2)$ occurs at $t = \tau < 0$, that is, at a temporal position before the pulses have excited the system.

With increased intensity we find in Fig. 31b that the beats for xx become slightly stronger. For xy, however, one can see in Fig. 31b, and also in Fig. 31c where the pure $\chi^{(5)}$ contribution is shown, that half-period beats appear. Such half-period beats in TI-FWM have actually been observed experimentally on disordered samples for xy excitation and increasing excitation intensity; see Albrecht et al. (1996). For an explanation of these half-period beats, which only appear for the xy and not for the xx configuration, one should consider the $\chi^{(5)}$ contributions to the TR-FWM, as displayed in Figs. 28b and 29b. The main difference between xx and xy is that for xy the $\chi^{(5)}$ contributions introduce an additional time-delay-dependent phase shift of the TR-FWM, which is not present for xx. Thus,

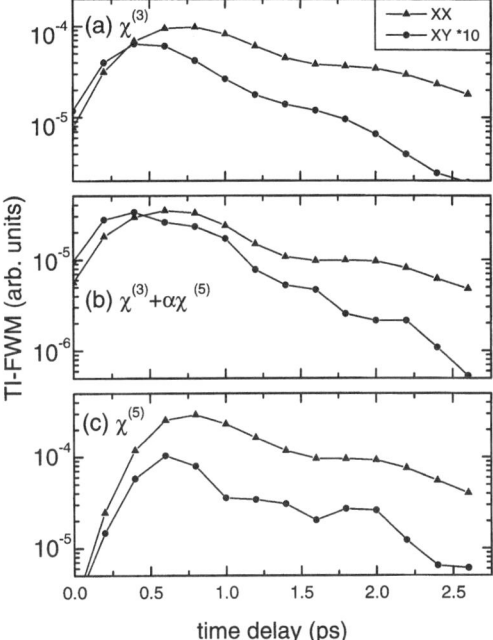

FIG. 31. Time-integrated four-wave mixing versus time delay for linear parallel (triangles) and linear perpendicular (circles) excitation. Disorder is included as correlated inhomogeneous broadening as discussed in the text. (a) Pure $\chi^{(3)}$, (b) $\chi^{(3)} + \alpha\chi^{(5)}$, and (c) pure $\chi^{(5)}$. The lines are guides for the eye. System parameters of model HH-II in Table I are used.

in disordered systems because of the emission of the TR-FWM as a photon echo, the TI-FWM may be strongly influenced by the dependence of the phase on the time delay that introduces the half-period beats.

2. Disorder-Induced Dephasing

The calculations presented in this subsection are performed by microscopically including both disorder and many-body correlations on the coherent $\chi^{(3)}$ level. Since we want to analyze dephasing, that is, the decay of the TI-FWM induced by disorder, no phenomenological decay rates for the single- and two-exciton amplitudes p and B are used here. The parameters for the model system are listed as model HH-II in Table I. In order to concentrate on predominantly excitonic excitation conditions, the duration of the incident pulses is taken as $\bar{t}_1 = \bar{t}_2 = 500$ fs and the central frequencies are tuned 1 meV below the energy of the 1s exciton in the ordered system. Here we use a system size of $N = 10$ sites, which has been shown to ensure converged energetic positions for the 1s exciton and the biexciton for the present parameters (Sieh et al., 1999b).

To set the stage for the analysis of disorder-induced effects, we start by discussing the polarization dependent TR-FWM in the absence of disorder. Figure 32 displays the calculated TR-FWM signal for (a) cocircular ($\sigma^+\sigma^+$), (b) linear-parallel (xx), and (c) linear-perpendicular (xy) excitation on a logarithmic scale. To get insight into signatures of the total signal as well as the three different types of optical nonlinearities induced by Pauli blocking and the first- and higher-order Coulomb terms, we separate all those terms in Fig. 32. Because the 500 fs pulse only excites the 1s exciton and no energetically higher single-exciton states, the Pauli blocking and the first-

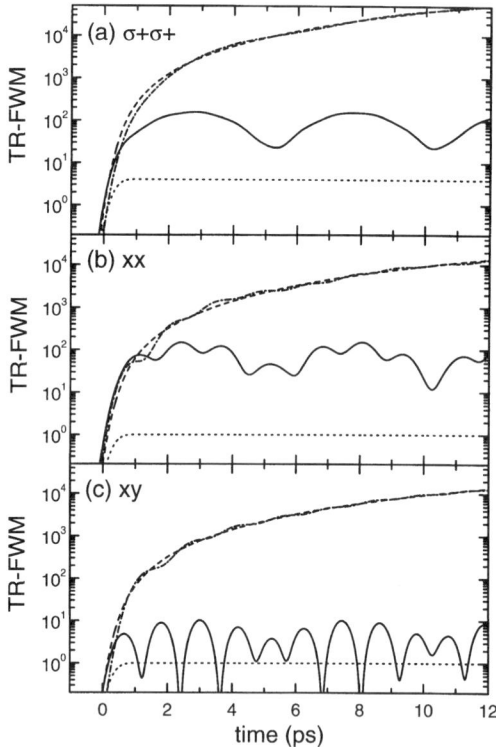

FIG. 32. Time-resolved four-wave mixing for a homogeneous system on a logarithmic scale for delay $\tau = 2$ ps using no phenomenological decay rates. (a) Cocircular $\sigma^+\sigma^+$, (b) linear parallel xx, and (c) linear perpendicular xy excitation. Besides the total signal (solid) the results for the three nonlinearities corresponding to Pauli blocking (dotted), first-order Coulomb (dashed), and higher-order Coulomb correlations (dashed-dotted) are displayed separately. The duration of the incident pulses is taken as $\bar{t}_1 = \bar{t}_2 = 500$ fs, and the central frequencies are tuned 1 meV below the energy of the 1s exciton in the ordered system. System parameters of model HH-II in Table I are used.

order Coulomb terms show no temporal modulations. Furthermore, both components are identical for xx and xy excitation. Also, the $\sigma^+\sigma^+$ TR-FWM has the same temporal line shape, but because of the selection rules the intensity is higher by a factor of 4. As shown by the dotted lines in Figs. 32a–c, the Pauli blocking simply rises with the integral over the second pulse and then remains constant. It yields only a small contribution to the total FWM signal. The TR-FWM intensity induced by first-order Coulomb terms (see dashed lines in Figs. 32a–c) increases quadratically with time. As long as no dephasing is included in the calculation, this unphysical increase, which is a consequence of the local-field-like nonlinearity of the $1s$ exciton, continues infinitely (Lindberg et al., 1992). To have finite signals as well as a nontrivial polarization dependence of the temporal line shapes, it is again essential to treat Coulomb correlations. As is shown by the dashed-dotted lines in Fig. 32a–c, the TR-FWM intensity of the correlation term alone basically increases quadratically with time. On top of this increase one can see polarization-dependent beats of the correlation term, which are caused by the excitation of bound and unbound two-exciton states.

Constructing the total FWM polarization according to Eq. (17), solid lines in Figs. 32a–c, the quadratic increase vanishes because of strong compensations among the Coulomb terms. The signal does not decay since no dephasing is included here. For all excitation configurations the total TR-FWM signal is modulated as function of time. For $\sigma^+\sigma^+$ excitation no bound biexciton can be excited, and the resulting slow modulations are caused by the excitation of unbound two-exciton states that are energetically situated slightly above twice the energy of the $1s$ exciton (compare Sieh et al., 1999a and 1999b). The rapid modulations for xx, (Fig. 32b) correspond to exciton–biexciton beats and have a frequency determined by the inverse of the biexciton binding energy. The signatures of unbound two-exciton states survive as additional weak slow modulations of the rapid exciton–biexciton beats for xx excitation. For xy excitation (Fig. 32c) the rapid modulation period corresponds to the inverse of the energy difference between bound and unbound two-exciton states. As pointed out in the previous subsection, this is because of the fact that in this case the FWM components at exactly the exciton frequency cancel because of the selection rules (Axt et al., 1998). We furthermore see, by comparing Figs. 32b and 32c, that the signal for xy is weaker and more strongly modulated than that for xx. It is noteworthy that all this strong polarization dependence of the FWM response is solely induced by Coulomb correlations.

In the following we focus on xx and xy excitation since there have been a number of experiments that reported different decay times of TI-FWM, comparing excitation with linear-parallel, and linear-perpendicular polarized pulses (Schmitt-Rink et al., 1992; Cundiff et al., 1992; Bennhardt et al., 1993; Carmel and Bar-Joseph, 1993; Albrecht et al., 1996). Often, the TI-FWM decays more rapidly for linear perpendicular excitation compared to linear parallel excitation. As already suggested in Bennhardt et al. (1993), this difference in the decay times seems to be related to the inhomogeneous line width of the exciton, that is, to disorder. Whereas one has the same decay times in perfectly homogeneous samples (Mayer et al., 1994), a difference between linear-parallel and linear-perpendicular configurations is present in inhomogeneous, that is, disordered samples, and this difference increases with the inhomogeneous linewidth (Schmitt-Rink et al., 1992; Bennhardt et al., 1993; Carmel and Bar-Joseph, 1993; Albrecht et al., 1996).

To explain the origin of disorder-induced dephasing, microscopic simulations of FWM in the presence of disorder are needed. As an example, we consider here diagonal disorder in the conduction band characterized by a Gaussian distribution of the site energy via $g(\delta\varepsilon_i^c) \propto \exp(-(\delta\varepsilon_i^c)^2/(2\sigma_c^2))$. The width of the distribution function is taken as $\sigma_c = 2.5$ meV, that is, σ_c is only about one-sixth of the exciton binding energy. The disorder is assumed to be uncorrelated in real space such that the energies for all electron sites are taken from independent distribution functions. Figure 33 shows a schematic drawing of the one-dimensional tight-binding model including diagonal disorder.

Numerical results for TR-FWM at three time delays are displayed in Fig. 34. To obtain the FWM signals with disorder we have averaged over 100 disorder realizations. It has been checked for some representative

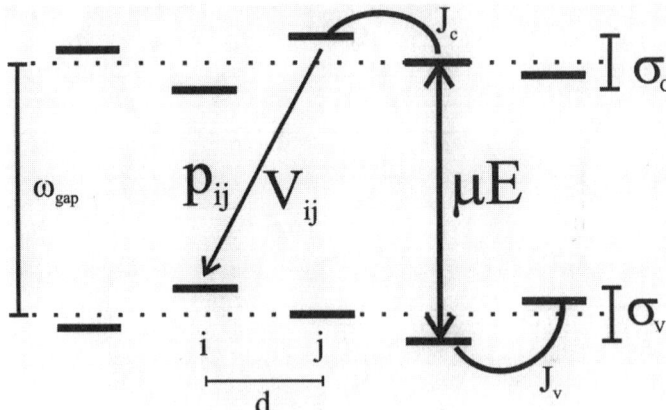

FIG. 33. Schematic drawing of the one-dimensional tight-binding model including diagonal disorder. ω_{gap} is the bandgap frequency, p_{ij} describes the interband coherence between a hole at site i and an electron at site j, V_{ij} is the Coulomb interaction between sites i and j, and d is the distance between adjacent sites. J_c and J_v denote the electronic couplings between nearest neighbor sites in the different bands. μE denotes the optical interband excitation induced by the light field E, which is proportional to the dipole matrix element μ. Energetic disorder is included via taking the site energies from Gaussian distribution functions with widths σ_c and σ_v, respectively. System parameters of model HH-II in Table I are used.

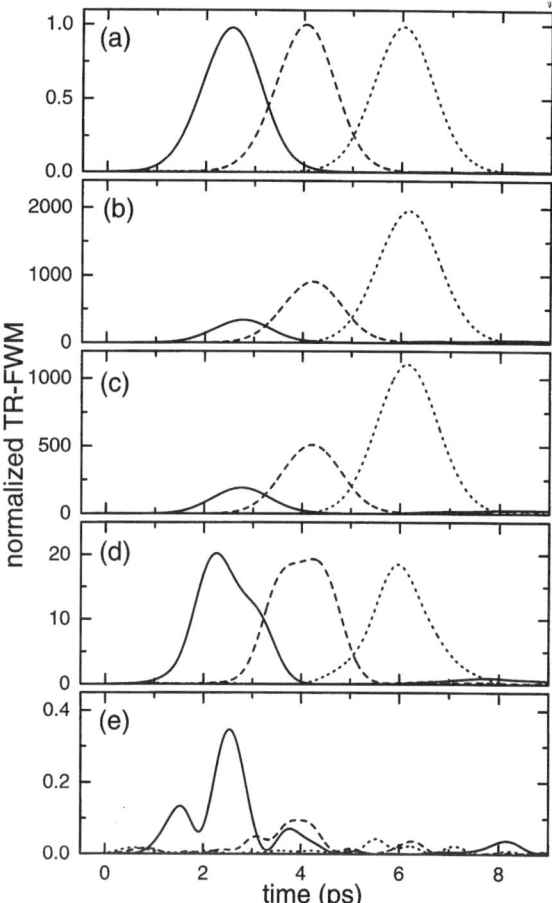

FIG. 34. Time-resolved four-wave mixing for time delays $\tau = 2.5$ ps (solid), 4 ps (dashed), and 6 ps (dotted) for conduction-band disorder of strength $\sigma_c = 2.5$ meV. (a) Only Pauli blocking ($xx = xy$), (b) Hartree–Fock calculation given by Pauli blocking plus first-order Coulomb-terms for xx, (c) Hartree–Fock calculation given by Pauli blocking plus first-order Coulomb-terms for xy, (d) total signal including correlations for xx, and (e) total signal including correlations for xy excitation. Pauli blocking (a) is normalized to 1; relative heights are comparable. System parameters of model HH-II in Table I are used.

calculations that by adding more realizations no qualitative change of the signals occurs. In Fig. 34a only Pauli blocking is considered, which gives identical results for xx and xy excitation. It is shown that the disorder-induced inhomogeneous broadening of the 1s-exciton energy leads to the emission of the signal as a photon echo. Apart from the shift of the echo maximum with τ, no obvious dependence of the TR-FWM intensity on the time delay is present. Figures 34b and 34c display the TR-FWM signals for *xx* and *xy* excitation obtained on the TDHF level. Except for a different overall amplitude that for both geometries is much larger than for Pauli blocking, the signals in Figs. 34b and 34c are very similar. As for the Pauli blocking, also within the TDHF calculations the disorder-induced inhomogeneous broadening results in a photon echo emission of the TR-FWM. However, because of the local-field-like excitonic nonlinearity, the amplitude of the echoes now *increases* with time delay. Thus, for resonant excitation of the 1s-exciton and short-range disorder on an energetic scale much smaller than the exciton binding energy, the TDHF result does not show disorder-induced dephasing, but is characterized by an unrealistic increase of the signal with increasing time delay. Hence, as for the ideal system, in the presence of disorder Coulomb correlations are needed to recover TR-FWM signals that do not increase with time delay (see Figs. 34d and 34e). The modulations of the TR-FWM that are present in the ordered case (see Fig. 32), partly survive in the presence of disorder and induce some polarization-dependent structure on top of the echoes in Figs. 34d and 34e. As in the ordered case, with disorder the modulations are stronger and the TR-FWM intensity is weaker for *xy* than for *xx* excitation. Furthermore, a disorder-induced decrease of the FWM intensity with increasing delay is obvious for *xy*, but absent for *xx*. Thus, Fig. 34 gives a first hint toward the existence of polarization-dependent disorder-induced dephasing.

The increase of the photon echo amplitude with increasing time delay for excitation at the exciton resonance and weak disorder has been verified by various numerical calculations using the present model in TDHF. This phenomenon has actually obscured theoretical investigations on disorder-induced dephasing in semiconductors for quite a while. In this context, it should furthermore be noted that the final Eq. (30) of Zimmermann (1992), which suggests a Gaussian-type dephasing of the excitonic FWM signal in the presence of disorder, is actually not correct. Later, it was discovered that the difference between two energies appearing in the exponent of the Gaussian is zero (Zimmermann, 1999), which implies that the infinite rise of the excitonic FWM signal is recovered.

Numerical results on TI-FWM with diagonal disorder of two strengths $\sigma_c = 1.5$ and 2.5 meV in the conduction band are displayed in Figs. 35a for *xx* and 35b for *xy* excitation. For comparison, the results for the diagonal valence band disorder of $\sigma_v = 1.5$ meV are also included. In all cases, apart from some numerical noise induced by the finite number of disorder realizations, there is no obvious decay of the TI-FWM with time delay for *xx*. For *xy*, however, the TI-FWM intensity clearly decreases with increasing time delay. Furthermore, it can be seen in Fig. 35b that the decay becomes more rapid with increasing disorder. To obtain an estimate of the strength of the disorder-induced dephasing, we fit the curves in Fig. 35b with a single exponential decay corresponding to $\alpha_0 \exp(-\tau/(T_2/4))$, where α_0 is

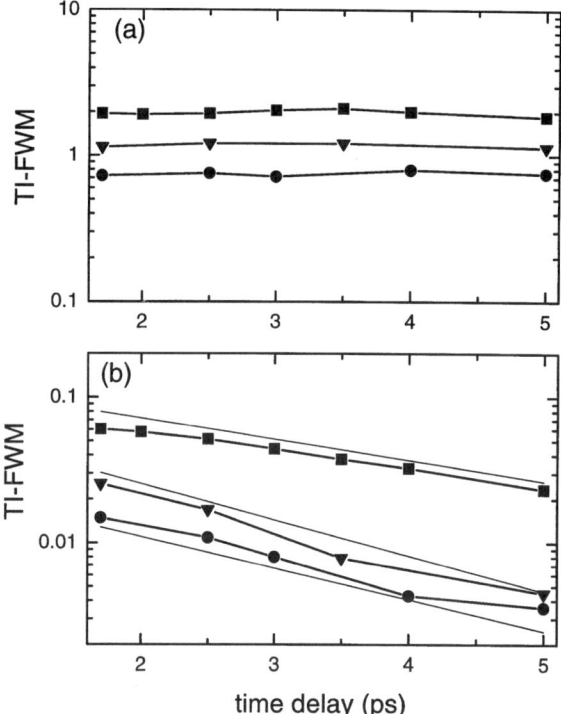

FIG. 35. Time-integrated four-wave mixing versus time delay for three conduction-band disorder strengths of $\sigma_c = 1.5$ meV (squares) and 2.5 meV (circles), as well as a valence-band disorder strength of $\sigma_v = 1.5$ meV (triangle down) (a) for xx and (b) for xy excitation. The symbols indicate the time delays considered in the numerical calculations and the lines are guides for the eye. The thin lines appearing in (b) describe exponential decays $\propto \exp(-\tau/(T_2/4))$, with $T_2 = 12$, 7, and 8 ps, from top to bottom, respectively. System parameters of model HH-II in Table I are used.

a prefactor determining the overall strength (thin lines in Fig. 35b). Note that $T_2/4$ would be the decay constant of the TI-FWM of an inhomogeneously broadened two-level system, if a dephasing time T_2 is introduced phenomenologically into the equation of motion for the interband coherence p. For the case of conduction band disorder we obtain $T_2 = 12$ ps for $\sigma_c = 1.5$ meV and $T_2 = 8$ ps for $\sigma_c = 2.5$ meV, respectively. For the valence band disorder of $\sigma_v = 1.5$ meV, we obtain $T_2 = 7$ ps. Because of the larger mass of the heavy holes, the same value of disorder considered in the valence band has a stronger effect and thus induces a more rapid dephasing in comparison to disorder in the conduction band.

In Weiser et al. (2000) the TI-FWM for conduction band disorder and xy excitation induced by Pauli blocking and Coulomb interaction has been analyzed separately. As expected, for the present disorder strength much smaller than the exciton binding energy, Pauli blocking alone shows no dephasing (Bennhardt et al., 1991). If the dephasing of the total TI-FWM would be induced by an inhomogeneous distribution of the biexciton binding energy (Langbein et al., 1997), one would expect that the Coulomb interaction term alone is sufficient to obtain the dephasing. In Weiser et al. (2000) it has, however, been shown that this is not the case. Therefore, we conclude that for the present model with short-range disorder the dephasing for xy excitation is not because of an inhomogeneous distribution of the biexciton binding energy, but is the consequence of an *interference effect* between the Pauli blocking and the Coulomb interaction induced non-linearities. For the xx-configuration such a strong interference does not happen, because in this case the Pauli blocking is much smaller than the Coulomb interaction term, which thus dominates the FWM response (compare Fig. 34).

More details on the phenomenon of disorder-induced dephasing can be found in Weiser et al. (2000), where besides the two extremes of one ordered and one disordered band, results of calculations considering simultaneously disordered valence and conduction bands are also reported. Furthermore, it is shown in that within our model a magnetic field applied perpendicular to the system reduces the disorder-induced dephasing, which is in agreement with experiment (Carmel and Bar-Joseph, 1993). Also, the theoretical estimates for the disorder-induced dephasing as function of the inhomogeneous contribution to the linewidth are in good agreement with the experimental results obtained on quantum wells (Mayer, 1995; Weiser et al., 2000). This agreement may seem rather surprising, considering the simple one-dimensional model used in the calculations, but on the other hand, this agreement substantiates that the results obtained within the present model are also experimentally relevant.

VI. Summary

In this chapter we present and numerically evaluate a theoretical approach that is well suited to describing nonlinear optical processes in semiconductors, including many-body Coulomb correlations for excitation with low-intensity pulses. Whereas plasma theories and the HF-SBE have been able to successfully explain a number of experimentally observed phenomena, they often do not yield the full result. As is shown here, to adequately describe the nonlinear optical response in the low-density limit, the explicit inclusion of higher-order correlation functions is essential. As relevant examples we discuss particular experimental configurations that analyze the dependence of PP and FWM signals on the polarization directions of the incident pulses. Besides coherent excitation, the presence of incoherent carriers has also been studied, and even in this case higher-order many-body correlations turn out to be very important.

In the weak excitation limit Coulomb correlations manifest themselves by two-exciton contributions, including both unbound and bound biexcitonic states. Because of the optical selection rules, biexcitons are only excited for certain polarization geometries. The excitation of different types of two-exciton states introduces polarization dependencies of the nonlinear optical signals. Generally, both PP and FWM include spectral components corresponding to single- to two-exciton transitions. In the PP spectra they lead to induced absorption and in FWM they may induce temporal beats.

Within the coherent $\chi^{(3)}$ limit we find that many-body correlations are important for the proper description of excitonic nonlinearities. For resonant excitation they strongly compensate the blue shift at the exciton induced by the first-order Coulomb terms. The resulting spectra at the exciton predominantly show bleaching and no net shift. Furthermore, the correlations represented by two-exciton excitations introduce excited state absorption in the PP spectra at energetic positions corresponding to single- to two-exciton transitions. For excitation with opposite circularly and colinearly polarized pulses, the optical excitation of a bound biexciton is possible. The exciton to biexciton transition induces excited state absorption energetically below the exciton in the differential absorption spectra. If both heavy- and light-hole transitions are excited, a bound two-exciton state shows up for cocircularly polarized excitation. This mixed biexciton is formed as a consequence of the attractive interaction between a heavy- and a light-hole exciton. For negative time delays when the probe pulse arrives before the pump, it is shown that the differential absorption spectra are dominated by spectral oscillations around the exciton resonance. The frequency period of these spectral oscillations is inversely proportional to the modulus of the time delay.

Also, when pumping off-resonantly below the exciton resonance, that is, in Stark effect configuration, correlations are still important. Especially for opposite circularly polarized pump and probe pulses and not too large detuning, one may find a red shift of the exciton instead of the usual blue shift. This red shift is predominantly determined by the dynamics of Coulomb correlations and more specifically is a direct signature of a Coulomb-memory effect that is *not* directly related to the presence of a bound two-exciton state. This correlation-induced red shift is observable only under certain conditions: (i) the samples need to have a large splitting between the heavy- and light-hole excitons, (ii) the detuning must be chosen properly, that is, it needs to be somewhat larger than the linewidth of the exciton and the biexciton binding energy but smaller than the splitting between the heavy- and light-hole excitons, and (iii) the pump intensity needs to be sufficiently low. With increasing detuning below the exciton, the importance of correlations generally decreases for all polarization configurations. Thus, for very large detuning the differential absorption spectra are well described within the TDHF approximation.

The results for resonant and off-resonant pumping have been extended for higher pump intensities up to the coherent $\chi^{(5)}$ limit. Also within $\chi^{(5)}$ the differential absorption spectra are strongly polarization dependent, and for resonant pumping the line shape of the pure $\chi^{(5)}$ result is approximately the negative of $\chi^{(3)}$. Thus, in the normalized spectra the amplitudes of both the bleaching and the excited state absorption are reduced with increasing pump intensity. The reduction of the excited state absorption is, however, slightly stronger than the reduction of the bleaching, such that with increasing intensity direct signatures of transitions to two excitons become less pronounced. A significant polarization dependence of the $\chi^{(5)}$ response is reported for pumping below the exciton. For opposite circularly polarized pump and probe pulses, the pure $\chi^{(5)}$ results correspond to a blue shift and thus the amplitude of the $\chi^{(3)}$ red shift is reduced with increasing pump intensity in the normalized differential absorption spectra. For cocircular polarized pump and probe pulses, the pure $\chi^{(5)}$ results corresponds to exciton broadening. Therefore, for this excitation geometry the normalized spectra change from the low intensity blue shift toward a broadening with increasing pump intensity.

If low-density incoherent occupations N are present, the six-point correlation function Z describing exciton to two-exciton transitions contributes strongly to the optical response. Compensations among the first- and higher-order Coulomb terms at the spectral positions of the exciton are obtained in the differential absorption induced by incoherent occupations. By comparing the absorption changes for excitonic or noninteracting electron–hole pairs, we find that at low temperatures the exciton bleaching induced by electron–hole pairs is stronger than for excitons. This result shows that there is a stronger nonlinear coupling of low-energy electron–hole pairs to the 1s exciton compared to the nonlinear coupling among excitons. It is further shown that at low temperatures in agreement with experiment, the exciton line shifts blue with an increasing density of incoherent occupations. This blue shift is stronger for electron–hole pairs than for exciton occupations and changes into pure bleaching at elevated temperatures.

Results on two-exciton- and especially biexciton-induced beats in TR- and TI-FWM show, in agreement with experimental observations, that no biexciton-induced beats appear in the TI-FWM of a homogeneous system within $\chi^{(3)}$. Such polarization-dependent beats are, however, present in disordered systems. For excitation with parallel-perpendicular polarized pulses, our numerical calculations show that half-period beats may appear in the TI-FWM of disordered systems for higher excitation intensities.

Most disorder-induced phenomena discussed in this chapter can be understood qualitatively by modeling the disorder as a simple correlated inhomogeneous broadening of the exciton and two-exciton transitions. To be able to properly analyze the effect of disorder-induced dephasing, however, both the many-body correlations and the energetic disorder need to be treated on a microscopic basis. We find that weak diagonal disorder may introduce a polarization-dependent decay of the TI-FWM signal. Whereas for excitation with linear-parallel polarized pulses the disorder does not lead to dephasing, some disorder-induced dephasing is obtained for linear-perpendicular excitation. These theoretical results are in qualitative agreement with experimentally observed polarization dependencies of the FWM response of disordered quantum wells, which show that the TI-FWM decays more rapidly for linear-perpendicular than for linear-parallel excitation geometry.

In conclusion, it has been shown in this chapter that Coulomb-induced many-body correlations are important for an adequate description of excitonic nonlinearities in semiconductors. They introduce significant polarization dependencies of PP and FWM signals and are relevant even for higher excitation intensities and in the presence of disorder and incoherent occupations.

VII. Outlook

Although nowadays many aspects of correlation-induced signatures in nonlinear optical spectroscopy on semiconductors are understood, this field is far from being exhausted. Especially a proper unified theoretical approach is missing that is able to describe correlations adequately for arbitrary excitation intensities. Whereas for excitation with weak intensity pulses four-particle correlations seem to be very important, they become less dominant with increasing excitation intensity. As a consequence of the dephasing and scattering induced by the many-body Coulomb interaction, the dynamical action of correlations is modified with increasing intensity. Therefore, a unified theory needs to be developed that is able to adequately describe correlations in both limits of weak and high-intensity excitation and properly interpolates between both limits. The derivation of such a description has to include the low-density equations discussed here, but also needs to incorporate excitonic and carrier populations, screening of the Coulomb interaction, Coulomb scattering, and the transition between coherent and incoherent regimes.

Another important step that would lead toward a better understanding of correlation-induced signatures is the proper inclusion of the exciton–photon coupling, that is, the simultaneous treatment of the coupled dynamics of the carrier system and the light field. The exciton–photon coupling has been shown to induce radiative decay (Stroucken *et al.*, 1996; Hübner *et al.*, 1996) of the optical transitions and leads in extended systems to polaritonic effects for which the normal mode resonances in a microcavity are one example (Jahnke et al., 1996, 1997). It is therefore obvious that the dynamics of the carrier system and the light field are strongly coupled and should be treated simultaneously and self-consistently. To be able to address the important field of excitonic luminescence in such a description, a quantized light field should be used (Kira et al., 1998, 1999; Thränhardt et al., 2000).

Also, concerning the inclusion of disorder, further theoretical investigations are highly desirable. Even within the one-dimensional model used in this chapter it would be interesting to study larger systems with increased exciton Bohr radii, to be able to investigate effects induced by spatial correlations between the disorder of the valence and the conduction bands. Furthermore, in order to directly connect to experimental observations on quantum wells, theoretical results are needed that microscopically include disorder within a two-dimensional model. Since currently it is not possible to perform such numerical calculations even on the fastest available computers, it is necessary to find ways of reducing the numerical requirements of this problem. One possibility could be to project the full single- and two-exciton dynamics onto the relevant states. It is, however, not clear whether this can be done avoiding the explicit solution of the full two-exciton problem.

Besides the points just discussed, which need further development of the theoretical description, there are some other interesting areas of research that could be addressed using the presently available description of correlations in the low-density limit. These include the detailed theoretical investigation of correlations in nondegenerate (Kim et al., 1998; Ahn et al., 1999) and partly nondegenerate (Cundiff et al., 1996; Euteneuer et al., 1999) FWM, where the incident pulses have different center frequencies, and/or different temporal envelopes. Another important area is the field of coherent control of nonlinear optical signals that is performed with sequences of phase stabilized pulses. The coherent control signals in the nonlinear PP and FWM response, as presented, for example, in Heberle et al. (1995), are surely influenced by correlations and depend on the polarization configuration of the incident pulses. Also, the influence of correlations on the nonlinear optical response in the presence of external fields still needs more work. Whereas for the case of magnetic fields applied to quantum wells, there are some theoretical results on four-particle correlations (see Kner et al., 1997, 1998; Chernyak et al., 1998b; and Yokojima et al., 1999), we are not aware of an equivalent treatment of such correlation-induced effects with externally applied electric fields. This area is highly interesting, since the electric field introduces an anisotropy of the system resulting in the possibility of coherent photoinduced currents due to Bloch oscillating carriers (Dignam et al., 1993; Meier et al., 1995, 1998, 2000b). One may expect that the correlation-induced signatures obtained for a homogeneous system are strongly modified in the presence of a symmetry-breaking electric field. For example, one would expect that the biexciton is destroyed by much weaker electric fields than the ones required to ionize excitons (von Plessen *et al.*, 1996).

Clearly, these are only few of the many interesting open problems in this active field of research. However, they already show that more detailed experimental and theoretical studies are needed.

ACKNOWLEDGMENTS

We acknowledge many stimulating discussions and collaborations with Prof. P. Thomas. Furthermore we thank our co-workers and students, Dr. A. Knorr, Dr. F. Jahnke, C. Sieh, S. Weiser, and our experimental colleagues Profs. H. M. Gibbs, G. Khitrova, W. W. Rühle, Dr. J. Kuhl, and co-workers for many fruitful and ongoing collaborations. This work is supported by the Deutsche Forschungsgemeinschaft through the Leibniz prize and the Sonderforschungsbereich 383. We wish to acknowledge the John von Neumann Institut für Computing (NIC), Forschungszentrum Jülich, Germany, for extended CPU time grants on their supercomputer systems.

REFERENCES

Ahn, Y. H., Yahng, J. S., Sohn, J. Y., Yee, K. J., Hohng, S. C., Woo, J. C., Kim, D. S., Meier, T., Koch, S. W., Lim, Y. S., and Kim, E. K. (1999). *Phys. Rev. Lett.* **82**, 3879–3882.
Albrecht, T. F., Bott, K., Meier, T., Schulze, A., Koch, M., Cundiff, S. T., Feldmann, J., Stolz, W., Thomas, P., Koch, S. W., and Göbel, E. O. (1996). *Phys. Rev. B* **54**, 4436–4439.
Allen, L., and Eberly, J. H. (1975). *Optical Resonance and Two-Level Atoms*. Wiley, New York.
Axt, V. M., and Stahl, A. (1994a). *Z. Phys. B* **93**, 195–204.
Axt, V. M., and Stahl, A. (1994b). *Z. Phys. B* **93**, 205–211.
Axt, V. M., Victor, K., and Stahl, A. (1996). *Phys. Rev. B* **53**, 7244–7258.
Axt, V. M., Victor, K., and Kuhn, T. (1998). *phys. stat. sol.* (b) **206**, 189–196.
Bányai, L., and Koch, S. W. (1986). *Z. Phys. B* **63**, 283–291.
Bányai, L., Vu, Q. T., Mieck, B., and Haug, H. (1998). *Phys. Rev. Lett.* **81**, 882–885.
Bar-Ad, S. and Bar-Joseph, I. (1992). *Phys. Rev. Lett.* **68**, 349–352.
Bartels, G., Axt, V. M., Victor, K., Stahl, A., Leisching, P., and Köhler, K. (1995). *Phys. Rev. B* **51**, 11217–11220.
Bartels, G., Cho, G. C., Dekorsy, T., Kurz, H., Stahl, A., and Köhler, K. (1997). *Phys. Rev. B* **55**, 16404–16413.
Bayer, M., Gutbrod, T., Forchel, A., Kulakovskii, V. D., Gorbunov, A., Michel, M., Steffen, R., and Wang, K. H. (1998). *Phys. Rev. B* **58**, 4740–4753.
Bennhardt, D., Thomas, P., Weller, A., Lindberg, M., and Koch, S. W. (1991). *Phys. Rev. B* **43**, 8934–8945.
Bennhardt, D., Thomas, P., Eccleston, R., Mayer, E. J., and Kuhl, J. (1993). *Phys. Rev. B* **47**, 13485–13490.

Bigot, J.-Y., Daunois, A., Oberle, J., and Merle, J.-C. (1993). *Phys. Rev. Lett.* **71**, 1820–1823.
Binder, R., Koch, S. W., Lindberg, M., Peyghambarian, N., and Schäfer, W. (1990). *Phys. Rev. Lett.* **65**, 899–902.
Binder, R., Koch, S. W., Lindberg, M., Schäfer, W., and Jahnke, F. (1991). *Phys. Rev. B* **43**, 6520–6529.
Binder, R., Scott, D., Paul, A. E., Lindberg, M., Henneberger, K., and Koch, S. W. (1992). *Phys. Rev. B* **44**, 1107–1115.
Bott, K., Heller, O., Bennhardt, D., Cundiff, S. T., Thomas, P., Mayer, E. J., Smith, G. O., Eccleston, R., Kuhl, J., and Ploog, K. (1993). *Phys. Rev. B* **48**, 17418–17426.
Bott, K., Mayer, E. J., Smith, G. O., Heuckeroth, V., Hübner, M., Kuhl, J., Meier, T., Schulze, A., Lindberg, M., Koch, S. W., and Thomas, P. (1996). *J. Opt. Soc. Am. B* **13**, 1026–1030.
Carmel O., and Bar-Joseph, I. (1993). *Phys. Rev. B* **47**, 7606–7609.
Chernyak, V., Zhang, W. M., and Mukamel, S. (1998a). *J. Chem. Phys.* **109**, 9587–9601.
Chernyak, V., Yokojima, S., Meier, T., and Mukamel, S. (1998b). *Phys. Rev. B* **58**, 4496–4516.
Chow, W. W., Wright, A. F., Girndt, A., Jahnke, F., and Koch, S. W. (1997). *Appl. Phys. Lett.* **71**, 2608–2610.
Combescot, M., and Combescot, R. (1988). *Phys. Rev. Lett.* **61**, 117–120.
Cundiff, S. T., Wang, H., and Steel, D. G. (1992). *Phys. Rev. B* **46**, 7248–7251.
Cundiff, S. T., Koch, M., Knox, W. H., Shah, J., and Stolz, W. (1996). *Phys. Rev. Lett.* **77**, 1107–1110.
Dignam, M., Sipe, J. E., and Shah, J. (1993). *Phys. Rev. B* **49**, 10502–10513.
Erland, J., and Balslev, I. (1993). *Phys. Rev. A* **48**, 1765–1768.
Euteneuer, A., Finger, E., Hofmann, M., Stolz, W., Meier, T., Thomas, P., Koch, S. W., Rühle, W. W., Hey, R., and Ploog, K. (1999). *Phys. Rev. Lett.* **83**, 2073–2076.
Fan, X., Wang, H., Hou, H. Q., and Hammons, B. E. (1998). *Phys. Rev. B* **57**, 9451–9454.
Feldmann, J., Meier, T., von Plessen, G., Koch, M., Göbel, E. O., Thomas, P., Bacher, G., Hartmann, C., Schweizer, H., Schäfer, W., and Nickel, H. (1993). *Phys. Rev. Lett.* **70**, 3027–3030.
Feuerbacher, B. F., Kuhl, J., and Ploog, K. (1991). *Phys. Rev. B* **43**, 2439–2441.
Finkelstein, G., Bar-Ad, S., Carmel, O., Bar-Joseph, I., and Levinson, Y. (1993). *Phys. Rev. B* **47**, 12964–12967.
Fluegel, B., Peyghambarian, N., Olbright, G., Lindberg, M., Koch, S. W., Joffre, M., Hulin, D., Migus, A., and Antonetti, A. (1990). *Phys. Rev. Lett.* **59**, 2588–2591.
Fröhlich, D., Nöthe, A., and Reiman, K. (1985). *Phys. Rev. Lett.* **55**, 1335–1337.
Girndt, A., Koch, S. W., and Chow, W. W. (1998). *Appl. Phys. A* **66**, 1–12.
Göbel, E. O., Leo, K., Damen, T. C., Shah, J., Schmitt-Rink, S., Schäfer, W., Müller, J. F., and Köhler, K. (1990). *Phys. Rev. Lett.* **64**, 1801–1804.
Guo, F., Chandross, M., and Mazumdar, S. (1995). *Phys. Rev. Lett.* **74**, 2086–2089.
Hartmann, M., Zimmermann, R., and Stolz, H. (1988). *phys. stat. sol. (b)* **146**, 357–369.
Haug, H., and Jauho, A.-P. (1996). *Quantum Kinetics in Transport and Optics of Semiconductors*. Springer Series in Solid-State Sciences, Vol. 123, Springer, Berlin.
Haug, H., and Koch, S. W. (1989). *Phys. Rev. A* **39**, 1887–1898.
Haug, H., and Koch, S. W. (1994). *Quantum Theory of the Optical and Electronic Properties of Semiconductors*. World Scientific, Singapore.
Haug, H., and Schmitt-Rink, S. (1984). *Progress in Quantum Electronics* **9**, 3–100.
Heberle, A. P., Baumberg, J. J., and Köhler, K. (1995). *Phys. Rev. Lett.* **75**, 2598–2601.
Henneberger, K., and Haug, H. (1988). *Phys. Rev. B* **38**, 9759–9770.
Hu, Y. Z., Binder, R., Koch, S. W., Cundiff, S. T., Wang, H., and Steel, D. G. (1994). *Phys. Rev. B* **49**, 14382–14386.
Hübner, M., Kuhl, J., Stroucken, T., Knorr, A., Koch, S. W., Hey, R., and Ploog, K. (1996). *Phys. Rev. Lett.* **76**, 4199–4202.
Hulin, D., and Joffre, M. (1990). *Phys. Rev. Lett.* **65**, 3425–3428.

Huhn, W., and Stahl, A. (1984). *phys. stat. sol* (b) **124**, 167–177.
Ivanov, A. L. and Keldysh, L. V. (1983). *Sov. Phys JETP* **57**, 234–244.
Jahnke, F., Koch, M., Meier, T., Feldmann, J., Schäfer, W., Thomas, P., Koch, S. W., Göbel, E. O., and Nickel, H. (1994). *Phys. Rev. B* **50**, 8114–8117.
Jahnke, F., Kira, M., Koch, S. W., Khitrova, G., Lindmark, E. K., Nelson Jr., T. R., Wick, D. V., Berger, J. D., Lyngnes, O., Gibbs, H. M., and Tai, K. (1996). *Phys. Rev. Lett.* **77**, 5257–5260.
Jahnke, F., Kira, M., and Koch, S. W. (1997). *Z. Physik B* **104**, 559–572.
Joffre, M., Hulin, D., Migus, A., and Combescot, M. (1989). *Phys. Rev. Lett.* **62**, 74–77.
Kim, D. S., Shah, J., Cunningham, J. E., Damen, T. C., Schäfer, W., Hartmann, M., and Schmitt-Rink, S. (1992a). *Phys. Rev. Lett.* **68**, 1006–1009.
Kim, D. S., Shah, J., Damen, T. C., Schäfer, W., Jahnke, F., Schmitt-Rink, S., and Köhler, K. (1992b). *Phys. Rev. Lett.* **69**, 2725–2728.
Kim, D. S., Sohn, J. Y., Yahng, J. S., Ahn, Y. H., Yee, K. J., Yee, D. S., Cho, Y. D., Hong, S. C., Kim, W. S., Woo, J. C., Meier, T., Koch, S. W., Woo, D. H., Kim, E. K., Kim, S. H., and Kim, C. S. (1998). *Phys. Rev. Lett.* **80**, 4803–4806.
Kira, M., Jahnke, F., and Koch, S. W. (1998). *Phys. Rev. Lett.* **81**, 3263–3266.
Kira, M., Jahnke, F., and Koch, S. W. (1999). *Phys. Rev. Lett.* **82**, 3544–3547.
Klingshirn, C., and Haug, H. (1981). *Phys. Rep.* **70**, 315–398.
Kner, P., Bar-Ad, S., Marquezini, M. V., Chemla, D. S., and Schäfer, W. (1997). *Phys. Rev. Lett.* **78**, 1319–1322.
Kner, P., Schäfer, W., Lövenich, R., and Chemla, D. S. (1998). *Phys. Rev. Lett.* **81**, 5386–5389.
Koch, M., Feldmann, J., von Plessen, G., Göbel, E. O., Thomas, P., and Köhler, K. (1992). *Phys. Rev. Lett.* **69**, 3631–3634.
Koch, M., Weber, D., Feldmann, J., Göbel, E. O., Meier, T., Schulze, A., Thomas, P., Schmitt-Rink, S., and Ploog, K. (1993). *Phys. Rev. B* **47**, 1532–1539.
Koch, S. W., Peyghambarian, N., and Gibbs, H. M. (1988a). *J. Appl. Phys.* (*Reviews*) **63**, R1–11.
Koch, S. W., Peyghambarian, N., and Lindberg, M. (1988b). *J. Phys. C* **21**, 5229–5249.
Koch, S. W., Sieh, C., Meier, T., Jahnke, F., Knorr, A., Brick, P., Hübner, M., Ell, C., Prineas, J., Khitrova, G., and Gibbs, H. M. (1999). *J. Luminescence* **83–84**, 1–6.
Langbein, W., Hvam, J. M., Umlauff, M., Kalt, H., Jobst, B., and Hommel, D. (1997). *Phys. Rev. B* **55**, 7383–7386.
Lee, Y. H., Chavez-Pirson, A., Koch, S. W., Gibbs, H. M., Park, S. H., Morhange, J., Peyghambarian, N., Bányai, L., Gossard, A. C., and Wiegmann, W. (1986). *Phys. Rev. Lett.* **57**, 2446–2449.
Leo, K., Wegener, M., Shah, J., Chemla, D. S., Göbel, E. O., Damen, T. C., Schmitt-Rink, S., and Schäfer, W. (1990). *Phys. Rev. Lett.* **65**, 1340–1343.
Leo, K., Shah, J., Damen, T. C., Schulze, A., Meier, T., Schmitt-Rink, S., Thomas, P., Göbel, E. O., Chuang, S. L., Luo, M. S. C., Schäfer, W., Köhler, K., and Ganser, P. (1992). *IEEE J. Quantum Electronics* **28**, 2498–2507.
Lindberg, M., and Koch, S. W. (1988). *Phys. Rev. B* **38**, 3342–3350.
Lindberg, M., Binder, R., and Koch, S. W. (1992). *Phys. Rev. A* **45**, 1865–1875.
Lindberg, M., Binder, R., Hu, Y. Z., and Koch, S. W. (1994a). *Phys. Rev. B* **49**, 16942–16952.
Lindberg, M., Hu, Y. Z., Binder, R., and Koch, S. W. (1994b). *Phys. Rev. B* **50**, 18060–18072.
Lovering, D. J., Phillips, R. T., Denton, G. J., and Smith, G. W. (1992). *Phys. Rev. Lett.* **68**, 1880–1883.
Mayer, E. J. (1995). Ph.D. thesis, Stuttgart, Germany.
Mayer, E. J., Smith, G. O., Heuckeroth, V., Kuhl, J., Bott, K., Schulze, A., Meier, T., Bennhardt, D., Koch, S. W., Thomas, P., Hey, R., and Ploog, K. (1994). *Phys. Rev. B* **50**, 14730–14733.
Mayer, E. J., Smith, G. O., Heuckeroth, V., Kuhl, J., Bott, K., Schulze, A., Meier, T., Koch, S. W., Thomas, P., Hey, R., and Ploog, K. (1995). *Phys. Rev. B* **51**, 10909–10914.
Meier, T., and Koch, S. W. (1999). *Phys. Rev. B* **59**, 13202–13208.

Meier, T., Bennhardt, D., Thomas, P., Hu, Y. Z., Binder, R., and Koch, S. W. (1994). In *Coherent Optical Interactions in Semiconductors* (R. T. Philips, Ed.), NATO ASI Series B: Physics Vol. 330, 349–354. Plenum Press, New York.
Meier, T., Rossi, F., Thomas, P., and Koch, S. W. (1995). *Phys. Rev. Lett.* **75**, 2558–2561.
Meier, T., Chernyak, V., and Mukamel, S. (1997a). *J. Phys. Chem. B* **101**, 7332–7342.
Meier, T., Chernyak, V., and Mukamel, S. (1997b). *J. Chem. Phys.* **107**, 8759–8774.
Meier, T., Thomas, P., and Koch, S. W. (1998). *Phys. Low-Dim. Struct.* **3/4**, 1–44.
Meier, T., Koch, S. W., Brick, P., Ell, C., Khitrova, G., and Gibbs, H. M. (2000a). *Phys. Rev. B*, in press.
Meier, T., Thomas, P., and Koch, S. W. (2000b). In *Ultrafast Phenomena in Semiconductors* (K. T. Tsen, Ed.), Springer, New York, to be published.
Mukamel, S. (1995). *Principles of Nonlinear Optical Spectroscopy.* Oxford, New York.
Mysyrowicz, A., Hulin, D., Antonetti, A., Migus, A., Masselink, W. T., and Morkoc, H. (1986). *Phys. Rev. Lett.* **56**, 2748–2751.
Noll, G., Siegner, U., Shevel, S., and Göbel, E. O. (1990). *Phys. Rev. Lett.* **64**, 792–795.
Östreich, T., Schönhammer, K., and Sham, L. J. (1995). *Phys. Rev. Lett.* **74**, 4698–4701.
Östreich, T., Schönhammer, K., and Sham, L. J. (1998). *Phys. Rev. B* **58**, 12920–12936.
Peyghambarian, N., Gibbs, H. M., Jewell, J. L., Antonetti, A., Migus, A., Hulin, D., and Mysyrowicz, A. (1984). *Phys. Rev. Lett.* **53**, 2433–2436.
Rappen, T., Peter, U. G., Wegener, M., and Schäfer, W. (1994). *Phys. Rev. B* **49**, 10774–10777.
Sauter-Fischer, S., Runge, E., and Zimmermann, R. (1998). *Phys. Rev. B* **57**, 4299–4303.
Schäfer, W., and Treusch, J. (1986). *Z. Phys. B* **63**, 407–426.
Schäfer, W., Jahnke, F., and Schmitt-Rink, S. (1993). *Phys. Rev. B* **47**, 1217–1220.
Schäfer, W., Kim, D. S., Shah, J., Damen, T. C., Cunningham, J. E., Goosen, K. W., Pfeiffer, L. N., and Köhler, K. (1996). *Phys. Rev. B* **53**, 16429–16443.
Schmitt-Rink, S., and Chemla, D. S. (1986). *Phys. Rev. Lett.* **57**, 2752–2755.
Schmitt-Rink, S., Chemla, D. S., and Miller, D. A. B. (1985). *Phys. Rev. B* **32**, 6601–6609.
Schmitt-Rink, S., Chemla, D. S., and Haug, H. (1988). *Phys. Rev. B* **37**, 941–955.
Schmitt-Rink, S., Chemla, D. S, and Miller, D. A. B. (1989). *Adv. Phys.* **38**, 89–188.
Schmitt-Rink, S., Bennhardt, D., Heuckeroth, V., Thomas, P., Haring, P., Maidorn, G., Bakker, H., Leo, K., Kim, D. S., Shah, J., and Köhler, K. (1992). *Phys. Rev. B* **46**, 10460–10463.
Schultheis, L., Kuhl, J., Honold, A., and Tu, C. W. (1986a). *Phys. Rev. Lett.* **57**, 1797–1800.
Schultheis, L., Honold, A., Kuhl, J., Köhler, K., and Tu, C. W. (1986b). *Phys. Rev. B* **34**, 9027–9030.
Sieh, C., Meier, T., Jahnke, F., Knorr, A., Koch, S. W., Brick, P., Hübner, M., Ell, C., Prineas, J., Khitrova, G., and Gibbs, H. M. (1999a). *Phys. Rev. Lett.* **82**, 3112–3115.
Sieh, C., Meier, T., Knorr, A., Jahnke, F., Thomas, P., and Koch, S. W. (1999b). *Eur. Phys. J. B* **11**, 407–421.
Sipe, J. E., and Ghahramani, E. (1993). *Phys. Rev. B* **48**, 11705–11722.
Smith, G. O., Mayer, E. J., Kuhl, J., and Ploog, K. (1994). *Sol. Stat. Comm.* **92**, 325–329.
Smith, G. O., Mayer, E. J., Heuckeroth, V., Kuhl, J., Bott, K., Meier, T., Schulze, A., Bennhardt, D., Koch, S. W., Thomas, P., Hey, R., and Ploog, K. (1995). *Sol. Stat. Comm.* **94**, 373–377.
Soos, Z. G., Ramasesha, R., Galvao, D. S., and Etemad, S. (1993). *Phys. Rev. B* **47**, 1742–1753.
Stroucken, T., Knorr, A., Thomas, P., and Koch, S. W. (1996). *Phys. Rev. B* **53**, 2026–2033.
Thränhardt, A., Kuckenberg, S., Knorr, A., Meier, T., and Koch, S. W. (2000). *Phys. Rev. B*, in press.
Victor, K., Axt, V. M., and Stahl, A. (1995). *Phys. Rev. B* **51**, 14164–14175.
Von Lehmen, A., Chemla, D. S., Zucker, J. E., and Heritage, J. P. (1986). *Opt. Lett.* **11**, 609–611.
von Plessen, G., Meier, T., Koch, M., Feldmann, J., Thomas, P., Koch, S. W., Göbel, E. O., Goosen, K. W., Kuo, J. M., and Kopf, R. F. (1996). *Phys. Rev. B* **53**, 13688–13693.
Wagner, H. P., Langbein, W., and Hvam, J. M. (1999). *Phys. Rev. B* **59**, 4584–4587.

Wang, H., Ferrio, K. B., Steel, D. G., Hu, Y. Z., Binder, R., and Koch, S. W. (1993). *Phys. Rev. Lett.* **71**, 1261–1264.

Wang, H., Ferrio, K. B., Steel, D. G., Berman, P. R., Hu, Y. Z., Binder, R., and Koch, S. W. (1994). *Phys. Rev. A* **49**, 1551–1554.

Webb, M. D., Cundiff, S. T., and Steel, D. G. (1991). *Phys. Rev. Lett.* **66**, 934–937.

Wegener, M., Chemla, D. S., Schmitt-Rink, S., and Schäfer, W. (1990). *Phys. Rev. A* **42**, 5675–5683.

Weiser, S., Meier, T., Möbius, J., Euteneuer, A., Mayer, E. J., Stolz, W., Hofmann, M., Rühle, W. W., Thomas, P., and Koch, S. W. (2000). *Phys. Rev. B*, **61**, 13088–13098.

Weiss, S., Mycek, M.-A., Bigot, J.-Y., Schmitt-Rink, S., and Chemla, D. S. (1992). *Phys. Rev. Lett.* **69**, 2685–2688.

Yokojima, S., Meier, T., Chernyak, V., and Mukamel, S. (1999). *Phys. Rev. B* **59**, 12584–12597.

Zhang, W. M., Meier, T., Chernyak, V., and Mukamel, S. (1998). *J. Chem. Phys.* **108**, 7763–7774.

Zimmermann. R. (1988). *Many-Particle Theory of Highly Excited Semiconductors*. Teubner, Leipzig.

Zimmermann, R. (1992). *phys. stat. sol. (b)* **173**, 129–137.

Zimmermann, R. (1999). Private communication.

CHAPTER 7

Electronic and Structural Response of Materials to Fast, Intense Laser Pulses

Roland E. Allen, Traian Dumitrică, and Ben Torralva

DEPARTMENT OF PHYSICS
TEXAS A&M UNIVERSITY
COLLEGE STATION, TEXAS

I.	INTRODUCTION	315
II.	TIGHT-BINDING ELECTRON–ION DYNAMICS	317
III.	$\epsilon(\omega)$ AND $\chi^{(2)}$ AS SIGNATURES OF A NONTHERMAL PHASE TRANSITION	327
IV.	DETAILED INFORMATION FROM MICROSCOPIC SIMULATIONS	336
	1. *A Simple Picture*	336
	2. *Excited-State Tight-Binding Molecular Dynamics*	339
	3. *Detailed Model*	343
	4. *Electronic Excitation and Time-Dependent Band Structure*	344
	5. *Dielectric Function as a Signature of the Change in Bonding*	346
	6. *Second-Order Susceptibility as a Signature of the Change in Symmetry*	348
	7. *Summary of Results for GaAs*	354
V.	FORMULA FOR THE SECOND-ORDER SUSCEPTIBILITY $\chi^{(2)}$	356
	1. *Tight-Binding Hamiltonian in an Electromagnetic Field*	357
	2. *Second-Order Susceptibility in a Tight-Binding Representation*	358
	3. *Formula for the Dielectric Function*	360
	4. *Calculation of $\chi^{(2)}(\omega)$ for GaAs*	360
VI.	RESPONSE OF Si TO FAST, INTENSE PULSES	363
VII.	DENSITY FUNCTIONAL SIMULATION FOR Si	370
VIII.	RESPONSE OF C_{60} TO ULTRAFAST PULSES OF LOW, HIGH, AND VERY HIGH INTENSITY	371
IX.	THE SIMPLEST SYSTEM, H_2^+	377
X.	BIOLOGICAL MOLECULES	382
XI.	CONCLUSION	384
	REFERENCES	386

I. Introduction

The interaction of matter with ultrafast and ultraintense laser pulses is a current frontier of science. New discoveries often result from the ability to explore a new regime. Here one is exploring both extremely short time scales (below 100 femtoseconds) and extremely high intensities (above 1 terawatt per square centimeter). The usual approximations of theoretical physics and chemistry break down under these conditions, and both electrons and atoms exhibit new kinds of behavior. The experimental techniques to achieve these conditions are rather new, and so is the theoretical approach described in this review (Allen, 1994; Graves, 1997; Graves and Allen, 1998a,b; Dumitrică *et al.*, 1998, 2000; Dumitrică and Allen, 2000; Torralva and Allen, 1998; Torralva *et al.*, 2000; Khosravi, 1997; Khosravi and Allen, 1998; Gao, 1998; Hamilton, 1999).

Our method is called tight-binding electron–ion dynamics (TED), because it permits simulations of the coupled dynamics of valence electrons and ion cores in a molecule or material, and because it employs a tight-binding representation for the electronic states. It is applicable to general nonadiabatic processes, including interactions with an intense radiation field. The vector potential $\mathbf{A}(\mathbf{x}, t)$ for this field is included in the electronic Hamiltonian $\mathbf{H}$ through a time-dependent Peierls substitution. The time-dependent Schrödinger equation is solved with an algorithm that conserves probability and satisfies the Pauli exclusion principle. The atomic forces are obtained from a generalized Hellmann–Feynman theorem, which may also be interpreted as a generalized Ehrenfest theorem. After the electronic states at time $t = 0$ are specified, TED is inherently an $O(N)$ method; that is, the computational expense is proportional to N, the number of atoms in the system, rather than N^3 or a higher power.

As reported in the following sections, calculations for GaAs, Si, C_{60}, and various molecules demonstrate that TED is a reliable and quantitative method. What is most significant, however, is the new physics that has emerged from these simulations in conjunction with the experimental studies that are discussed in this chapter (Shank *et al.*, 1983; Downer *et al.*, 1983; Tom *et al.*, 1988; Saeta *et al.*, 1991; Siegal *et al.*, 1994, 1995; Glezer *et al.*, 1995a,b; Mazur, 1996; Huang *et al.*, 1998; Callan *et al.*, 1999; Sokolowski-Tinten *et al.*, 1995; Govorkov *et al.*, 1992; Chin *et al.*, 1999; Siders *et al.*, 1999; Rischel *et al.*, 1997; Dexheimer *et al.*, 1993; Fleischer *et al.*, 1997; Hunsche *et al.*, 1996; Hansen and Echt, 1996; Hohmann *et al.*, 1994).

The principal results include the following:

- When **GaAs** and **Si** are subjected to ultrafast pulses with durations of about 70 fs, there is a **nonthermal phase transition** as a function of the intensity of the light. The existence of such a transition is already indicated by the experiments (Shank *et al.*, 1983; Downer *et al.*, 1983; Tom *et al.*, 1988; Saeta *et al.*, 1991; Siegal *et al.*, 1994, 1995; Glezer *et al.*, 1995a,b; Mazur, 1996; Huang *et al.*, 1998; Callan *et al.*, 2000; Sokolowski-Tinten *et al.*, 1995; Govorkov *et al.*, 1992; Chin *et al.*, 1999; Siders *et al.*, 1999; similar results for organic films by Rischel *et al.*, 1997) but is validated even more strongly by our microscopic simulations.

- When C_{60} is subjected to pulses with durations of order 10 fs, there is a pronounced change in the **nonlinear response as a function of intensity**. At low intensity, we observe the excitation of various optically active vibrational modes. At higher intensity, the breathing mode is by far the most dominant. At very high intensities, there is photofragmentation, with the evolution of dimers and other products. The results are in excellent agreement with the observations (Dexheimer *et al.*, 1993; Fleischer *et al.*, 1997; Hunsche *et al.*, 1996; Hansen and Echt, 1996; Hohmann *et al.*, 1994).
- TED can also be used to study the response of other systems, including biological molecules, and a wide variety of interesting phenomena are observed.

Because of space restrictions, we do not do justice to the experiments, in the sense that we do not discuss the sophisticated techniques that are used to create ultrashort pulses and extract time-resolved information. Instead we merely present some of the highest-quality results for the systems that are of interest in the present context.

II. Tight-Binding Electron–Ion Dynamics

Many processes in physics, chemistry (Ferraudi, 1988; Casey and Sundberg, 1993; Dai and Ho, 1995), and biology (Kohen *et al.*, 1995; Mathews and Van Holde, 1990; Campbell, 1990) involve the interaction of electromagnetic radiation with complex molecules and materials. Traditional treatments of this problem involve the Born–Oppenheimer approximation (in which the electrons are assumed to adiabatically follow the motion of the nuclei), the Franck–Condon principle (in which the nuclei are regarded as frozen during each electronic transition), and Fermi's Golden Rule (which is based on both first-order perturbation theory and the premise that the field varies harmonically on a long time scale). These assumptions may be difficult to employ for a complex system, and they are not necessarily valid for ultraintense and ultrashort laser pulses (Joshi and Corkum, 1995; Gommila, 1992; Bandrank, 1994; Zewail, 1992). In the most general case, one needs numerical simulations.

Tight-binding electron-ion dynamics (TED) is a technique for simulating the coupled dynamics of valence electrons and ion cores in a molecule or material. It is a generalization of tight-binding molecular dynamics (Menon and Allen, 1985, 1986, 1991; Sankey and Allen, 1985, 1986; Allen and Menon, 1986; Gryko and Allen, 1994; Sawtarie *et al.*, 1994; Andriotis *et al.*, 1996; Sankey and Niklewski, 1989), an earlier technique invented by our group for simulating the motion of atoms in the adiabatic approximation.

A first-principles formulation of our method is presented in Allen (1994), Gao (1998), and Eqs. (26)–(42) of the present paper. This approach has been used in a detailed study of the response of Si to fast intense laser pulses (Gao, 1998), with some results mentioned in Section VII. However, we find that a tight-binding representation is preferable for practical calculations: (1) The electronic excitations play a central role, so it is important that the excited states be at their proper energies. (These are fitted to experiment in a semiempirical tight-binding model, whereas they are typically too low in the local density approximation and too high in Hartree–Fock.) (2) Since the time step is of order 50 attoseconds, and the system may contain many atoms, the method must be computationally fast. (3) A tight-binding representation involves chemically meaningful basis states that are localized on the atoms, and that have the same symmetries as atomic orbitals. One can then immediately interpret the results using intuitive ideas based on ground-state and excited-state chemistry (Vogl et al., 1983; Harrison, 1989, 1999; Mailhiot et al., 1984).

The equations of TED can be obtained by simply postulating the model Lagrangian (Allen, 1994)

$$L = \sum_{\ell\alpha} \frac{1}{2} M_\ell \dot{X}_{\ell\alpha}^2 - U_{rep} + \sum_j \Psi_j^\dagger \cdot \left(i\hbar \frac{\partial}{\partial t} - \mathbf{H} \right) \cdot \Psi_j, \qquad (1)$$

but here we will give a more detailed treatment. The first term in (1) is the kinetic energy of the ions, with coordinates $X_{\ell\alpha}$, which are treated classically. The second is a summation over repulsive potentials that model the ion–ion repulsion, together with the negative of the electron–electron repulsion that is doubly counted in the third term (Allen, 1994; Menon and Allen, 1985, 1986, 1991; Sankey and Allen, 1985, 1986; Allen and Menon, 1986; Gryko and Allen, 1994; Sawtarie et al., 1994; Andriotis et al., 1996). This last term is the tight-binding version of the standard Lagrangian for particles treated in a time-dependent self-consistent-field approximation (de Shalit and Feshbach, 1974; Jackiw and Kerman, 1979). We can adopt the point of view that each electron is labeled by j and has its own time-dependent state vector Ψ_j. If there are N tight-binding basis functions in the system, Ψ_j is N-dimensional, and the time-dependent Hamiltonian $\mathbf{H}$ is $N \times N$.

Let us now consider the justification for this Lagrangian. A proper first-principles treatment of nonequilibrium problems (including many-body effects) would employ methods like those introduced by Martin and Schwinger (1959), Kadanoff and Baym (1962), or Keldysh (1965), with a self-energy that is even more complicated than that for equilibrium or quasiequilibrium problems (Fulde, 1991; Rohlfing and Louie, 1999). In the present context, however, it is a reasonable approximation to adopt a time-dependent self-consistent field picture, with an action

$$S = \int dt L \tag{2}$$

$$L = \frac{1}{2} \langle \Psi_e | \left(i\hbar \frac{\partial}{\partial t} - \mathcal{H}_e \right) | \Psi_e \rangle + \text{h.c.} + \frac{1}{2} \sum_{\ell \alpha} M_\ell \left(\frac{dX_{\ell\alpha}}{dt} \right)^2 - U_{ii} \tag{3}$$

$$\mathcal{H}_e = \sum_j H_e(\mathbf{x}_j), \qquad H_e(\mathbf{x}) = T + v_{ei}(\mathbf{x}) + \epsilon_{ee}(\mathbf{x}). \tag{4}$$

Here j labels a valence electron, ℓ labels an ion core, $\alpha = x, y, z$, "h.c." represents the Hermitian conjugate of the first term in (3), and T is the kinetic energy operator. Also, U_{ii} is a potential energy representing the repulsive Coulomb interaction between ion cores, so it is a function of their coordinates $X_{\ell\alpha}$. The electron–ion interaction $v_{ei}(\mathbf{x})$ also depends on the ion positions $X_{\ell\alpha}$, and it may be given by a nonlocal pseudopotential. The contribution $\epsilon_{ee}(\mathbf{x})$ is due to electron–electron interactions, and it is a functional of $|\Psi_e\rangle$ or the one-electron wave functions $\Psi_j(\mathbf{x}_j, t)$ defined below. We will see that the present approach is equivalent to more usual time-dependent Hartree-like treatments in which the Hamiltonian is written in the form

$$\mathcal{H}'_e = \sum_j H_1(\mathbf{x}_j) + \frac{1}{2} \sum_{jj'} H_2(\mathbf{x}_j, \mathbf{x}_{j'}). \tag{5}$$

In our approach, the ions are treated classically from the beginning. It is often inappropriate to mix a quantum and a classical treatment, but this is a quite valid approximation in the present context, since the ions have masses that are $\sim 10^4 - 10^5$ times larger than the electron mass. In addition, we assume that the core electrons in an atom move rigidly with the nucleus, so that an ion core can be treated essentially as an extended nucleus with charge Ze.

We now approximate $|\Psi_e\rangle$ by an antisymmetrized and normalized product of one-electron states: in the coordinate representation,

$$\Psi_e(\mathbf{x}_1, \mathbf{x}_2, \ldots) = \mathcal{A} \prod_j \Psi_j(\mathbf{x}_j). \tag{6}$$

Since

$$\langle \Psi_j | \Psi_{j'} \rangle = \int d^3x \Psi_j^\dagger(\mathbf{x}) \Psi_{j'}(\mathbf{x}) = \delta_{jj'}, \tag{7}$$

it follows that

$$\langle \Psi_e | \left(i\hbar \frac{\partial}{\partial t} - \mathcal{H}_e \right) | \Psi_e \rangle = \sum_j \int d^3 x \Psi_j^\dagger(\mathbf{x}) \left[i\hbar \frac{\partial}{\partial t} - H_e(\mathbf{x}) \right] \Psi_j(\mathbf{x}). \quad (8)$$

The wave functions Ψ_j are represented by a set of localized basis functions $\phi_{\ell a}$,

$$\Psi_j(\mathbf{x}, t) = \sum_{\ell a} \Psi_j(\ell a, t) \phi_{\ell a}(\mathbf{x}), \quad (9)$$

with

$$\int d^3 x \phi_{\ell a}^\dagger(\mathbf{x}) \phi_{\ell' a'}(\mathbf{x}) = S(\ell a, \ell' a') \quad (10)$$

$$\int d^3 x \phi_{\ell a}^\dagger(\mathbf{x}) H_e \phi_{\ell' a'}(\mathbf{x}) = H_e(\ell a, \ell' a'). \quad (11)$$

Let Ψ_j be the vector with components $\Psi_j(\ell a)$, and let $\mathbf{H}_e$ be the matrix with elements $H_e(\ell a, \ell' a')$. The basis functions $\phi_{\ell a}(\mathbf{x})$ move with the ion cores, so the $H_e(\ell a, \ell' a')$ are functions of the $X_{\ell a}$.

Substitution of (9) into (3) then gives

$$L = \frac{1}{2} \sum_j \Psi_j^\dagger \cdot \left(i\hbar S \frac{\partial}{\partial t} - \mathbf{H}_e \right) \cdot \Psi_j + \text{h.c.} + \frac{1}{2} \sum_{\ell a} M_\ell \left(\frac{dX_{\ell a}}{dt} \right)^2 - U_{ii}. \quad (12)$$

The overlap matrix $\mathbf{S}$, the ion–ion interaction U_{ii}, and $\mathbf{H}_e$ all depend on the atomic coordinates $X_{\ell a}$. When there is an electromagnetic field present, $\mathbf{H}_e$ will also have an explicit dependence on the time t.

As usual, the equations of motion are determined by extremalizing the action with respect to variations $\delta \Psi_j^\dagger$ and $\delta X_{\ell a}$ (after performing an integration by parts in (2)). The result is a one-electron Schrödinger equation,

$$i\hbar \frac{\partial \Psi_j}{\partial t} = \mathbf{S}^{-1} \cdot \mathbf{H} \cdot \Psi_j \quad (13)$$

or

$$-i\hbar \frac{\partial \Psi_j^\dagger}{\partial t} = \Psi_j^\dagger \cdot \mathbf{H} \cdot \mathbf{S}^{-1}, \quad (14)$$

together with a Newton's equation for the atoms,

$$M_\ell \frac{d^2 X_{\ell\alpha}}{dt^2} = -\frac{1}{2}\sum_j \Psi_j^\dagger \cdot \left(\frac{\partial \mathbf{H}}{\partial X_{\ell\alpha}} - i\hbar \frac{\partial \mathbf{S}}{\partial X_{\ell\alpha}}\frac{\partial}{\partial t}\right) \cdot \Psi_j + \text{h.c.} - \frac{\partial U_{rep}}{\partial X_{\ell\alpha}} \quad (15)$$

$$= -\sum_j \Psi_j^\dagger \cdot \left(\frac{\partial \mathbf{H}}{\partial X_{\ell\alpha}} - \frac{1}{2}\frac{\partial \mathbf{S}}{\partial X_{\ell\alpha}} \cdot \mathbf{S}^{-1} \cdot \mathbf{H} - \frac{1}{2}\mathbf{H} \cdot \mathbf{S}^{-1} \cdot \frac{\partial \mathbf{S}}{\partial X_{\ell\alpha}}\right)$$

$$\times \Psi_j - \frac{\partial U_{rep}}{\partial X_{\ell\alpha}}, \quad (16)$$

where

$$\mathbf{H} \cdot \Psi_j = \mathbf{H}_e \cdot \Psi_j + \sum_{j'} \Psi_{j'}^\dagger \cdot \left(\frac{\delta \mathbf{H}_e}{\delta \Psi_j^\dagger}\right) \cdot \Psi_{j'} \quad (17)$$

$$U_{rep} = U_{ii} - U_{ee} \quad (18)$$

$$U_{ee} = \sum_j \Psi_j^\dagger \cdot \sum_{j'} \Psi_{j'}^\dagger \cdot \left(\frac{\delta \mathbf{H}_e}{\delta \Psi_j^\dagger}\right) \cdot \Psi_{j'}, \quad (19)$$

with

$$\Psi_j^\dagger \cdot \sum_{j'}\left[\Psi_{j'}^\dagger \cdot \left(\frac{\delta \mathbf{H}_e}{\delta \Psi_j^\dagger}\right) \cdot \Psi_{j'}\right] = \sum_{\ell a} \Psi_j^\dagger(\ell a) \sum_{j'}\left[\Psi_{j'}^\dagger \cdot \left(\frac{\delta \mathbf{H}_e}{\delta \Psi_j^\dagger(\ell a)}\right) \cdot \Psi_{j'}\right] \quad (20)$$

since the Ψ_j and $X_{\ell\alpha}$ vary independently in $\mathbf{H}_e$. Let

$$\mathbf{H}' = \sum_{j''}\sum_{j'}\left[\Psi_{j'}^\dagger \cdot \left(\frac{\delta \mathbf{H}_e}{\delta \Psi_{j''}^\dagger}\right) \cdot \Psi_{j'}\right] \Psi_{j''}^\dagger \cdot \mathbf{S}. \quad (21)$$

Then (17) follows from

$$\mathbf{H} = \mathbf{H}_e + \mathbf{H}' \quad (22)$$

since

$$\Psi_j^\dagger \cdot \mathbf{S} \cdot \Psi_{j'} = \int d^3x \Psi_j^\dagger(\mathbf{x}, t)\Psi_{j'}(\mathbf{x}, t) = \delta_{jj'}. \quad (23)$$

If the Pauli principle is satisfied at time $t = 0$, it will continue to hold at later times:

$$i\hbar \frac{\partial}{\partial t}(\Psi_j^\dagger \cdot \mathbf{S} \cdot \Psi_{j'}) = \Psi_j^\dagger \cdot \mathbf{S} \cdot (\mathbf{S}^{-1} \cdot \mathbf{H} \cdot \Psi_{j'}) - (\Psi_j^\dagger \cdot \mathbf{H} \cdot \mathbf{S}^{-1}) \cdot \mathbf{S} \cdot \Psi_{j'} \quad (24)$$

$$= 0. \quad (25)$$

For example, if an excited state becomes 50% occupied by one electron, then it is 50% blocked to all the other electrons.

For better understanding of the preceding equations, and also to provide a basis for the first-principles calculations described in Section VII, let us temporarily revert to the coordinate representation and use forms that correspond to those of time-dependent density-functional theory in the local-density approximation (Gross et al., 1994):

$$\epsilon_{ee}(\mathbf{x}) = \frac{1}{2} v_e(\mathbf{x}) + \epsilon_{xc}(\mathbf{x}) \qquad (26)$$

$$v_e(\mathbf{x}) = \int d^3x' \frac{e^2 \rho(\mathbf{x}')}{|\mathbf{x} - \mathbf{x}'|} \qquad (27)$$

$$\rho(\mathbf{x}) = \sum_j \Psi_j^\dagger(\mathbf{x}) \Psi_j(\mathbf{x}), \qquad (28)$$

where $\epsilon_{xc}(\mathbf{x})$ is a parametrized function of the local density $\rho(\mathbf{x})$ (Gao, 1998). In this case we have

$$\frac{\delta H_e(\mathbf{x}')}{\delta \Psi_j^\dagger(\mathbf{x})} = \frac{\delta \epsilon_{ee}(\mathbf{x}')}{\delta \Psi_j^\dagger(\mathbf{x})} \qquad (29)$$

$$= \frac{1}{2} \frac{e^2}{|\mathbf{x}' - \mathbf{x}|} \Psi_j(\mathbf{x}) + \beta(\mathbf{x}) \epsilon_{xc}(\mathbf{x}') \rho(\mathbf{x})^{-1} \delta(\mathbf{x}' - \mathbf{x}) \Psi_j(\mathbf{x}) \qquad (30)$$

where

$$\beta(\mathbf{x}) \equiv [d \log \epsilon_{xc}(\rho)/d \log \rho]_{\rho = \rho(\mathbf{x})} \qquad (31)$$

so that

$$\int d^3x' \sum_{j'} \Psi_{j'}^\dagger(\mathbf{x}') \left(\frac{\delta H_e(\mathbf{x}')}{\delta \Psi_j^\dagger(\mathbf{x})} \right) \Psi_{j'}(\mathbf{x}') = \frac{1}{2} \int d^3x' \frac{e^2 \rho(\mathbf{x}')}{|\mathbf{x}' - \mathbf{x}|} \Psi_j(\mathbf{x}) + \beta(\mathbf{x}) \epsilon_{xc}(\mathbf{x}) \Psi_j(\mathbf{x}). \qquad (32)$$

(For example, the crudest approximation is to neglect correlation and take $\epsilon_{xc} \propto \rho^{1/3}$, in which case $\beta = \frac{1}{3}$.) From the preceding equations we then have

$$H = H_e + \frac{1}{2} v_e(\mathbf{x}) + \beta(\mathbf{x}) \epsilon_{xc}(\mathbf{x}) \qquad (33)$$

$$U_{rep} = U_{ii} - \int d^3x \left[\frac{1}{2} v_e(\mathbf{x}) + \beta(\mathbf{x}) \epsilon_{xc}(\mathbf{x}) \right] \rho(\mathbf{x}) \qquad (34)$$

$$= U_{ii} - U_{ee} \qquad (35)$$

where

$$U_{ii} = \sum_{\ell > \ell'} \frac{Z_\ell Z_{\ell'} e^2}{|\mathbf{X}_\ell - \mathbf{X}_{\ell'}|} \tag{36}$$

$$U_{ee} = U_{Coul} + U_{xc} \tag{37}$$

$$U_{Coul} = \frac{1}{2} \int d^3x \int d^3x' \frac{e^2 \rho(\mathbf{x})\rho(\mathbf{x'})}{|\mathbf{x'} - \mathbf{x}|} \tag{38}$$

$$U_{xc} = \int d^3x \beta(\mathbf{x}) \epsilon_{xc}(\mathbf{x}) \rho(\mathbf{x}) \tag{39}$$

with $\epsilon_{xc} < 0$.

If it is viewed as an effective potential energy, U_{rep} corresponds to repulsive forces between the ion cores. The electronic charge is more distributed than the ionic charge, so the Coulomb repulsion between ion cores is not fully screened as it would be for neutral atoms. Also, the strength of the electron–electron repulsion is reduced by exchange and correlation effects, since each electron is surrounded by an exchange-and-correlation hole.

In a density-functional treatment, (13) and (16) become

$$i\hbar \frac{\partial \Psi_j}{\partial t} = H \Psi_j \tag{40}$$

$$M_\ell \frac{d^2 X_{\ell\alpha}}{dt^2} = -\sum_j \int d^3x \Psi_j^\dagger \left(\frac{\partial H}{\partial X_{\ell\alpha}}\right) \Psi_j - \frac{\partial U_{rep}}{\partial X_{\ell\alpha}}. \tag{41}$$

These two equations, together with

$$T = \frac{1}{2m}\left(-i\hbar \nabla - \frac{q}{c}\mathbf{A}\right)^2, \tag{42}$$

where q is the charge of the electron, represent one way of viewing the first-principles, density-functional version of electron–ion dynamics, which was used in the calculations of Section VII. In an actual calculation, however, the forces on the ions are calculated from the ion–ion interaction U_{ii} and the electron–ion interaction v_{ei}. When the latter is given by a nonlocal pseudopotential, the Hellmann–Feynman forces are rather complicated, as they are in first-principles molecular dynamics (Payne et al., 1992; Briggs et al., 1996; Lewis et al., 1997; Godlevsky et al., 1999). Details are given elsewhere (Gao, 1998).

First-principles molecular dynamics (which was introduced just after tight-binding molecular dynamics) is ordinarily an accurate method, because the local density approximation for exchange and correlation is quite good for total energies. The calculations reported in Section VII, on the other hand, indicate that density-functional methods are not as suitable for the kind of problems addressed here, which involve excited states and nonadiabatic processes. The excited states are typically too low for semiconductors, and the simulation of nonadiabatic processes requires a time step of about 50 attoseconds or less. It is therefore prohibitively expensive to treat large systems in a true first-principles simulation.

For these reasons, tight-binding electron-ion dynamics (TED) appears to be the preferred method for simulations of the interaction of light with matter. On the other hand, as mentioned in Section VIII, the results of Torralva *et al.* indicate that a density-functional-*based* approach is more accurate than one in which the parameters are fitted more naively. The approach of Sankey and co-workers (Sankey and Niklewski, 1989; Lewis *et al.*, 1997) also seems to be promising, since it permits self-consistent calculations.

Equations (13) and (16) represent the nonorthogonal formulation of TED. These same equations can, however, be cast into an orthogonalized form if we write the Lagrangian (12) as

$$L = \frac{1}{2}\sum_j \bar{\mathbf{\Psi}}_j^\dagger \cdot \left(i\hbar\frac{\partial}{\partial t} - \bar{\mathbf{H}}_e\right)\cdot \bar{\mathbf{\Psi}}_j + \text{h.c.} + \frac{1}{2}\sum_{\ell\alpha} M_\ell \left(\frac{dX_{\ell\alpha}}{dt}\right)^2 - U_{ii} \quad (43)$$

where

$$\bar{\mathbf{\Psi}} \equiv \mathbf{S}^{1/2}\mathbf{\Psi}_j, \quad \bar{\mathbf{H}}_e \equiv \mathbf{S}^{-1/2}\mathbf{H}_e\mathbf{S}^{-1/2}. \quad (44)$$

Repetition of the foregoing arguments then leads to

$$i\hbar\frac{\partial \bar{\mathbf{\Psi}}_j}{\partial t} = \bar{\mathbf{H}}\cdot\bar{\mathbf{\Psi}}_j \quad (45)$$

$$M_\ell \frac{d^2 X_{\ell\alpha}}{dt^2} = -\sum_j \bar{\mathbf{\Psi}}_j^\dagger \cdot \frac{\partial \bar{\mathbf{H}}}{\partial X_{\ell\alpha}}\cdot \bar{\mathbf{\Psi}}_j - \frac{\partial \bar{U}_{rep}}{\partial X_{\ell\alpha}}. \quad (46)$$

In a semiempirical or density-functional-based tight-binding scheme, one can fit either $\bar{\mathbf{H}}$ and $\bar{U}_{rep}$ or $\mathbf{H}$ and U_{rep} to experiment and preexisting theoretical calculations. In the simulations reported here, an orthogonal tight-binding model was used for the semiconductors GaAs and Si, and in the earlier calculations for the fullerene C_{60} (Torralva and Allen, 1998). A nonorthogonal, density-functional-based model has been used in the more recent simulations for C_{60} (Torralva *et al.*).

In the following we use the notation $\mathbf{H}$ and $\mathbf{\Psi}_j$ for both the orthogonal and nonorthogonal calculations, and the context establishes which is being used. For the orthogonal case, then, the equations of motion are

$$i\hbar \partial \Psi_j/\partial t = \mathbf{H}(t)\Psi_j \qquad (47)$$

$$M\ddot{X} = -\sum_j \Psi_j^\dagger \cdot \frac{\partial \mathbf{H}}{\partial X} \cdot \Psi_j - \frac{\partial U_{rep}}{\partial X}, \qquad (48)$$

where M is the mass and X the coordinate of any ion. These are respectively the time-dependent Schrödinger equation and the Hellmann–Feynman theorem (or Ehrenfest's theorem), with the electrons treated in a tight-binding picture and the ions treated classically.

The electrons and ions are coupled in (47) and (48), because $\mathbf{H}$ is a function of the ion coordinates and the forces on the ions are influenced by the electronic states. We now need to couple the electrons to the radiation field. (One can also easily couple the ions to the electromagnetic field, but this is a minor effect if the field oscillates on a 1-fs time scale, two orders of magnitude smaller than the response time of the ions.) The most convenient way to introduce the field into the electronic Hamiltonian is to employ a time-dependent Peierls substitution (Graves, 1997; Graves and Allen, 1998a,b; Dumitrică et al., 1998, 2000; Dumitrică and Allen, 2000; Torralva and Allen, 1998; Torralva et al.; Hamilton, 1999; Peierls, 1933; Graf and Vogl, 1995). First consider the standard Hamiltonian in the coordinate representation with a time-dependent electromagnetic vector potential $\mathbf{A}$,

$$H = \left(\mathbf{p} - \frac{q}{c}\mathbf{A}\right)^2 \Big/ 2m + V(\mathbf{x}), \qquad (49)$$

where $\mathbf{p} = -i\hbar \nabla$. This is equivalent to

$$H = \exp\left(\frac{iq}{\hbar c}\int \mathbf{A} \cdot d\mathbf{x}\right) H^0 \exp\left(-\frac{iq}{\hbar c}\int \mathbf{A} \cdot d\mathbf{x}\right) \qquad (50)$$

with

$$H^0 = \mathbf{p}^2/2m + V(\mathbf{x}), \qquad (51)$$

as one can easily verify by substituting (51) into (50) and letting H operate on an arbitrary function $\Psi(\mathbf{x})$. To employ (50) in a tight-binding scheme, we recognize that the matrix elements of (50) are the same as matrix elements of (51) with the localized basis functions $\phi_a(\mathbf{x} - \mathbf{X}_\ell)$ multiplied by $\exp(-iq\int \mathbf{A} \cdot d\mathbf{x}/\hbar c)$. In this factor, it is consistent with the spirit of tight binding to take $\int \mathbf{A} \cdot d\mathbf{x} \approx \mathbf{A} \cdot \mathbf{x} \approx \mathbf{A} \cdot \mathbf{X}$, provided that $\mathbf{A}$ is slowly varying on an atomic scale (as it is for electromagnetic radiation with $\hbar\omega \sim 10\,\text{eV}$ or less). Then the matrix elements $H_{ab}(\mathbf{X} - \mathbf{X}')$ are modified by the Peierls substitution

$$H_{ab}(\mathbf{X} - \mathbf{X}') = H^0_{ab}(\mathbf{X} - \mathbf{X}') \exp\left(\frac{iq}{\hbar c} \mathbf{A} \cdot (\mathbf{X} - \mathbf{X}')\right). \quad (52)$$

This approach requires no additional parameters and in principle is valid for arbitrarily strong time-dependent fields.

Solution of the ionic equations of motion (48) is essentially the same as in tight-binding molecular dynamics (Menon and Allen, 1985, 1986, 1991; Sankey and Allen, 1985, 1986; Allen and Menon, 1986; Gryko and Allen, 1994; Sawtarie et al., 1994; Andriotis et al., 1996) and the velocity Verlet method appears to be optimal. Solution of (47) requires more care, since a naive algorithm for this first-order equation fails to conserve probability. In earlier work we followed a standard prescription and wrote the time-evolution equation in the form

$$\exp\left(\frac{i\mathbf{H}\Delta t}{2\hbar}\right) \cdot \mathbf{\Psi}_j(t + \Delta t) = \exp\left(\frac{-i\mathbf{H}\Delta t}{2\hbar}\right) \cdot \mathbf{\Psi}_j(t). \quad (53)$$

If the exponential is approximated by its first two terms, this gives the Cayley algorithm

$$\mathbf{\Psi}_j(t + \Delta t) = \left(1 + \frac{i\mathbf{H}\Delta t}{2\hbar}\right)^{-1} \cdot \left(1 - \frac{i\mathbf{H}\Delta t}{2\hbar}\right) \cdot \mathbf{\Psi}_j(t). \quad (54)$$

Then probability and orthogonality are preserved because

$$\mathbf{\Psi}^\dagger_j(t + \Delta t) \cdot \mathbf{\Psi}_{j'}(t + \Delta t) = \mathbf{\Psi}^\dagger_j(t + \Delta t) \cdot \left(1 + \frac{i\mathbf{H}\Delta t}{2\hbar}\right) \cdot \left(1 - \frac{i\mathbf{H}\Delta t}{2\hbar}\right)^{-1}$$
$$\times \left(1 + \frac{i\mathbf{H}\Delta t}{2\hbar}\right)^{-1} \cdot \left(1 - \frac{i\mathbf{H}\Delta t}{2\hbar}\right) \cdot \mathbf{\Psi}_{j'}(t) \quad (55)$$
$$= \mathbf{\Psi}^\dagger_j(t) \cdot \mathbf{\Psi}_{j'}(t). \quad (56)$$

More recently a still better method has been introduced by Torralva et al. (1999b). In this approach, the first-order term in a Dyson-like series for the time evolution operator $U(t + \Delta t, t)$ is written in unitary form,

$$U(t + \Delta t, t) = \left(1 + \frac{i}{2\hbar}\int_t^{t+\Delta t} dt' \mathbf{H}(t')\right)^{-1} \left(1 - \frac{i}{2\hbar}\int_t^{t+\Delta t} dt' \mathbf{H}(t')\right), \quad (57)$$

so that

$$\mathbf{U}^\dagger(t+\Delta t,t)\cdot\mathbf{U}(t+\Delta t,t) = \left(1+\frac{i}{2\hbar}\int_t^{t+\Delta t}dt'\mathbf{H}(t')\right)$$
$$\times\left(1-\frac{i}{2\hbar}\int_t^{t+\Delta t}dt'\mathbf{H}(t')\right)^{-1}$$
$$\times\left(1-\frac{i}{2\hbar}\int_t^{t+\Delta t}dt'\mathbf{H}(t')\right)^{-1}$$
$$\times\left(1-\frac{i}{2\hbar}\int_t^{t+\Delta t}dt'\mathbf{H}(t')\right)$$
$$=\mathbf{1}. \qquad (58)$$

After evaluating each element of $\mathbf{U}(t+\Delta t,t)$ with Simpson's rule (for example), one then obtains the electron states from

$$\Psi_j(t+\Delta t) = \mathbf{U}(t+\Delta t,t)\cdot\Psi_j(t). \qquad (59)$$

With this algorithm, unitarity (i.e., orthonormality of the one-electron states Ψ_j) is preserved to the machine accuracy of better than 10^{-12}.

III. $\epsilon(\omega)$ and $\chi^{(2)}$ as Signatures of a Nonthermal Phase Transition

Using the method just outlined, we have performed calculations for the electronic and structural response of semiconductors to ultraintense and ultrashort laser pulses (Shank et al., 1983; Downer et al., 1983; Tom et al., 1988; Saeta et al., 1991; Siegal et al., 1994, 1995; Glezer et al., 1995a,b; Mazur, 1996; Huang et al., 1998; Callan et al., 1999; Sokolowski-Tinten et al., 1995; Govorkov et al., 1992; Chin et al., 1999; Siders et al., 1999; Rischel et al., 1997).

The time dependence of the electronic states and ionic positions was calculated as described earlier, and the imaginary part of the dielectric function was obtained from the formula (Graf and Vogl, 1995)

$$\mathrm{Im}\,\epsilon(\omega) \propto \frac{1}{\omega^2}\sum_{n,m,\mathbf{k}}[f_n(\mathbf{k})-f_m(\mathbf{k})]\mathbf{p}_{nm}(\mathbf{k})\cdot\mathbf{p}_{mn}(\mathbf{k})\delta(\omega-\omega_{mn}(\mathbf{k})). \qquad (60)$$

For numerical reasons, the slightly different form given as Eq. (94) in this chapter was used in the actual calculations. All the notation is defined and explained in Section V. This result of linear response theory is valid in the probe phase of a pump-probe experiment. (In the present context, it is legitimate to define a susceptibility that is a function of frequency and that is also time-dependent, because the relevant time scales differ by two orders of magnitude.) The preceding formula also yields good agreement with more usual optical measurements (Aspnes and Studna, 1983), as illustrated later in Fig. 29 for GaAs and Si (or in Graf and Vogl, 1995, for GaAs with a slightly different tight-binding model). In Fig. 29, Im $\epsilon(\omega)$ was calculated in arbitrary units, and the height of the theoretical curve was then adjusted for a better comparison with the measurements. Without this adjustment, theory and experiment still agree to within a factor of two (Graf and Vogl, 1995). A set of 512 sample **k**-points were used in calculating the time-dependent dielectric function during the simulations.

The tight-binding model of Vogl *et al.* (1983) was employed, together with Harrison's r^{-2} scaling for the interatomic matrix elements (Harrison, 1999). We used a nonstandard repulsive potential with the form

$$u(r) = \frac{\alpha}{r^4} + \frac{\beta}{r^6} + \frac{\gamma}{r^8}. \tag{61}$$

The three parameters α, β, and γ were fitted to the experimental values of the cohesive energy, interatomic spacing, and bulk modulus — properties associated with the zeroth, first, and second derivatives of the total energy. A cubical cell containing eight atoms was used in the early simulations (Graves, 1997; Graves and Allen, 1998a,b; Dumitrică and Allen, 1999a) and sixty-four atoms in more recent work (Dumitrică and Allen). Periodic boundary conditions were imposed on the motion of the ions. The electronic states are then Bloch states corresponding to this large unit cell. Further details are given below, in Section IV.

Representative results for the time dependence of Im $\epsilon(\omega)$, during and after the application of a laser pulse, are shown for GaAs in Fig. 1 (from a 64-atom simulation with a fluence just above the threshold for structural change) and Fig. 2 (from 8-atom simulations at various fluences). Notice the difference in the behavior of $\epsilon(\omega)$ for pulses of lower and higher intensity. For the pulse of lower intensity, Im $\epsilon(\omega)$ is zero at all times for $\hbar\omega$ less than the bandgap, demonstrating that the material remains a semiconductor and there is no absorption within this range of energies. In addition, the structural features of Fig. 29 persist, showing that the original band structure remains intact. For the pulse of higher intensity, on the other hand, below-bandgap absorption is observed soon after the pulse, indicating a transition to metallic behavior. The original structural features in Im $\epsilon(\omega)$ are also washed out following the pulse.

The second-order nonlinear susceptibility $\chi^{(2)}$ was also calculated (Dumitrică and Allen, 2000), using the formulas derived in Dumitrică *et al.* (1998) and given here as (88) and (89). Results are shown in Figs. 6–8.

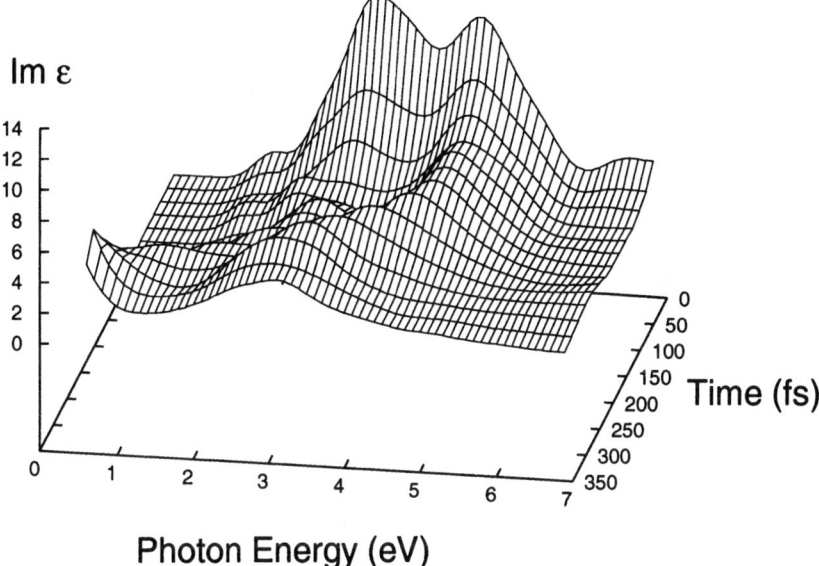

FIG. 1. Imaginary part of the time-dependent dielectric function (in arbitrary units) for an intensity just above threshold, in a 64-atom simulation for GaAs (Dumitrică and Allen). In all the simulations for GaAs and Si described here, the pump pulse had a FWHM duration of 70 fs with a photon energy $\hbar\omega$ equal to 1.95 eV. In this figure the pulse was applied between $t = 0$ and $t = 140$ fs, and the value of A_0 was 2.51 G cm. As mentioned in the text, $A_0 = 1.00$ G cm corresponds to a fluence of 0.815 kJ/m^2, and the fluence is proportional to A_0^2 for a given pulse shape and duration. The loss of the original structural features in Im $\epsilon(\omega)$ signals a loss of the original tetrahedral bonding. Im $\epsilon(\omega)$ also becomes nonzero for photon energies below the original bandgap energy of about 1.4 eV, and in fact begins to exhibit Drude-like behavior at low energy. This metallic behavior signals a collapse of the bandgap beyond about 250 fs. Notice the "hump" in Im $\epsilon(\omega)$ that persists even after the bandgap has collapsed, and that appears to indicate that there are still bonding-to-antibonding transitions, even after the long-range crystalline order has been lost. Compare all of these features with the above-threshold experimental measurements of Fig. 5.

It is, of course, extremely interesting to compare with the recent data of Callan *et al.*, as well as the earlier measurements of Huang *et al.* (1998), Sokolowski-Tinten *et al.* (1995), and the other excellent studies over the years (Saeta *et al.*, 1991; Siegal *et al.*, 1994, 1995; Glezer *et al.*, 1995a,b). As Figs. 1–10, 15, and 19–24 indicate, there is good general agreement between the simulations and the experiments.

In order to obtain some microscopic insight into the nature of the reversible and irreversible transitions represented by Figs. 1–10, one can calculate the average displacement of the atoms from their equilibrium positions as a function of time. (This is the average over all atoms at a given

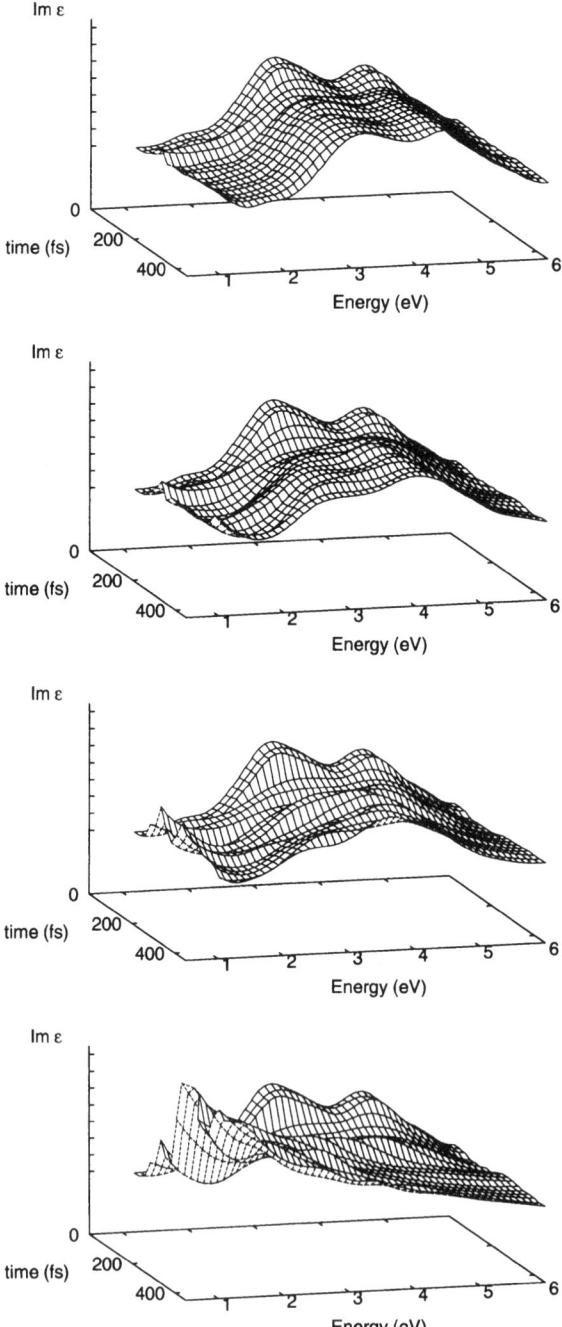

FIG. 2. Time-dependent dielectric function in eight-atom simulations for GaAs at four intensities: $A_0 = 1.73$, 2.00, 2.45, and 2.83 G cm, from top to bottom. In these runs the pulse was applied between 50 and 190 fs. After Graves and Allen (1998b).

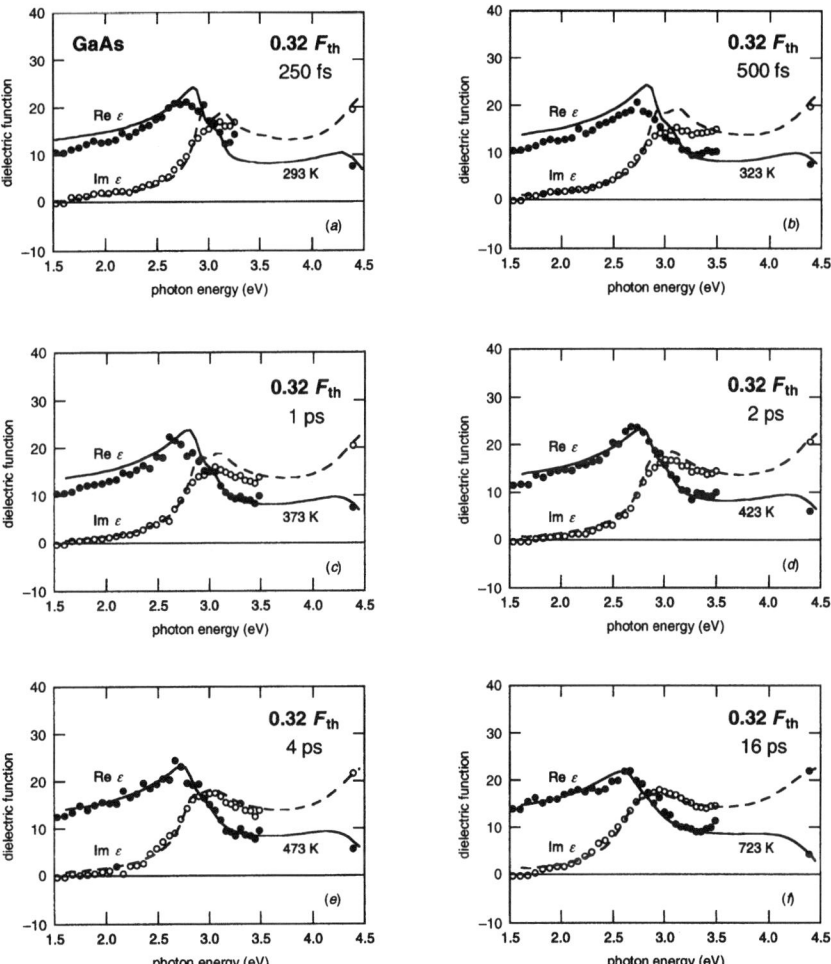

FIG. 3. Temporal evolution of the dielectric function of GaAs [solid circles, Re $\epsilon(\omega)$; open circles, Im $\epsilon(\omega)$] after excitation by a pulse of fluence 0.32 F_{th}, with $F_{th} = 1.0\,\text{kJ/m}^2$. The curves show Re $\epsilon(\omega)$ (solid line) and Im $\epsilon(\omega)$ (dashed line) for GaAs heated to various temperatures derived from fits to the data: (a) 293 K, (b) 323 K, (c) 373 K, (d) 423 K, (e) 473 K, (f) 723 K. The curves in (e) and (f) show $\epsilon(\omega)$ for amorphous GaAs at room temperature. After Callan et al. (1999). Reprinted with permission.

time t.) Some results are displayed in Figs. 11 and 12, for a range of values of A_0. There is a remarkably sharp transition from reversible to irreversible behavior for a critical intensity or fluence corresponding to $A_0 \approx 2.5\,\text{G cm}$. As discussed in Section IV, the precise value of the threshold intensity is far from accurate, but the interesting feature is the existence of such a well-defined threshold.

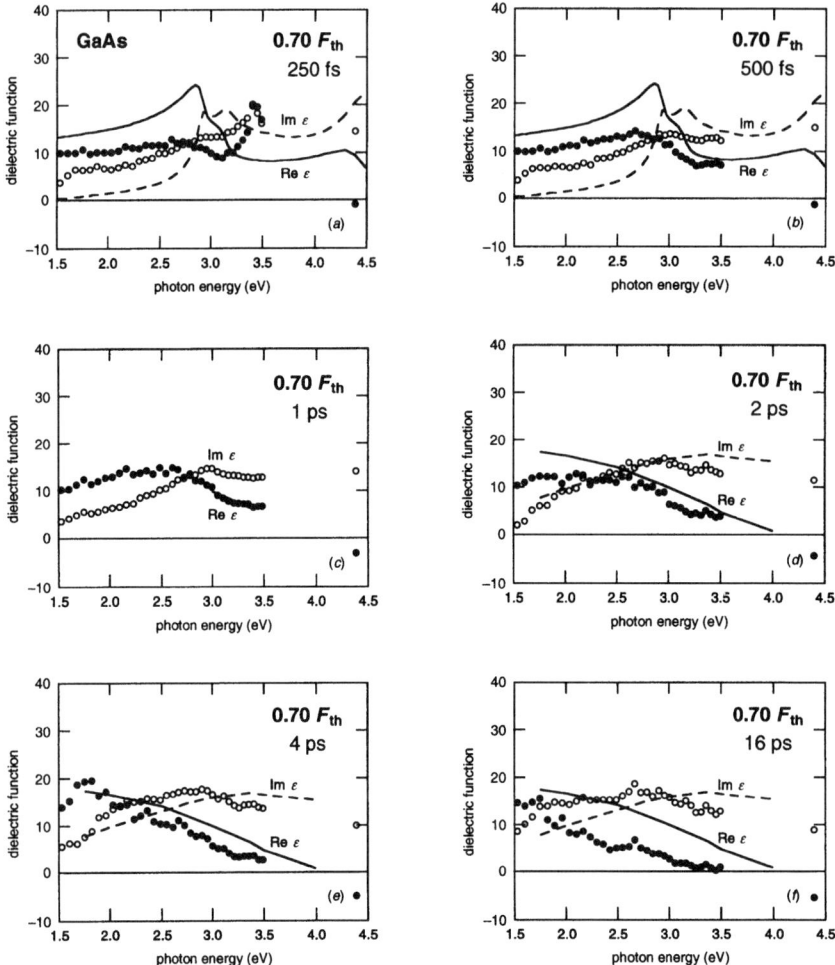

FIG. 4. Temporal evolution of the dielectric function of GaAs [solid circles, Re $\epsilon(\omega)$; open circles, Im $\epsilon(\omega)$] for a pump fluence of 0.70 F_{th}. The curves in (a) and (b) show Re $\epsilon(\omega)$ (solid line) and Im $\epsilon(\omega)$ (dashed line) for crystalline GaAs at room temperature. The curves in (e) and (f) show $\epsilon(\omega)$ for amorphous GaAs at room temperature. After Callan et al. (2000). Reprinted with permission.

The abrupt transition in the behavior of the system seen in Fig. 13 suggests that a true phase transition occurs as the intensity of the pulse is varied.

In Fig. 13, notice that the thermal oscillations from equilibrium are continuously amplified in duration and magnitude as the intensity of the pulse is increased. For the subcritical value of $A_0 = 2.50390\,\text{G cm}$, there is a large departure from equilibrium, with $R_{avg} \approx 0.12\,\text{Å}$, but the original

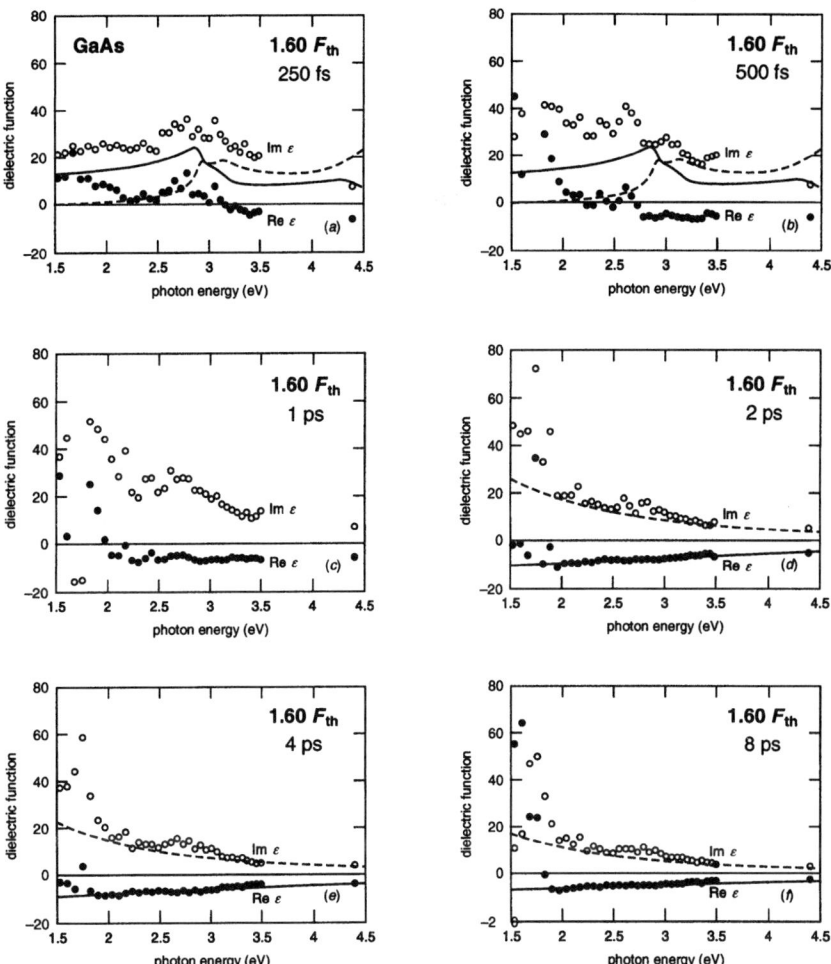

FIG. 5. Temporal evolution of the dielectric function of GaAs [solid circles, Re $\epsilon(\omega)$; open circles, Im $\epsilon(\omega)$] for a pump fluence of 1.60 F_{th}. The curves in (a) and (b) show Re $\epsilon(\omega)$ (solid line) and Im $\epsilon(\omega)$ (dashed line) for crystalline GaAs at room temperature. The curves in (d), (e), and (f) show Drude model dielectric functions with 0.18 fs relaxation time and plasma frequencies of 13.0, 12.0, and 10.5 eV, respectively. After Callan *et al.* (2000). Reprinted with permission.

structure is ultimately regained. On the other hand, a tiny increase to $A_0 = 2.50393$ G cm results in sudden destabilization of the structure, after about 375 fs. The fact that the observed destabilization occurs *after* the maximum in R_{avg} suggests that the transition is not a collective phenomenon, but rather an initially local disruption that spreads to the whole structure on a short time scale.

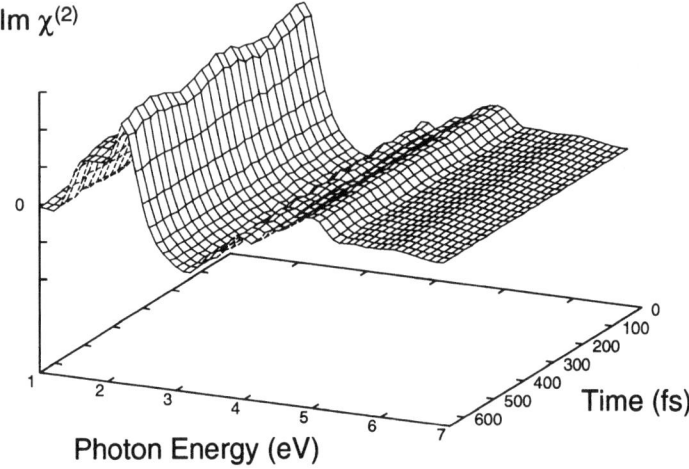

FIG. 6. Time-dependent Im $\chi^{(2)}(\omega)$ (in arbitrary units) for GaAs as a function of the photon energy $\hbar\omega$, well below threshold, calculated using Dumitrică's formula [Eq. (89)]. The amplitude of the vector potential for the pump pulse is $A_0 = 0.5$ G cm. The pulse begins at $t = 0$ and, as in all the simulations for GaAs and Si described here, has a FWHM duration of 70 fs with $\hbar\omega = 1.95$ eV. The results of this figure and of Figs. 7 and 8 were obtained in eight-atom simulations. Notice that the original structural features are retained with this subthreshold fluence. After Dumitrică and Allen (2000).

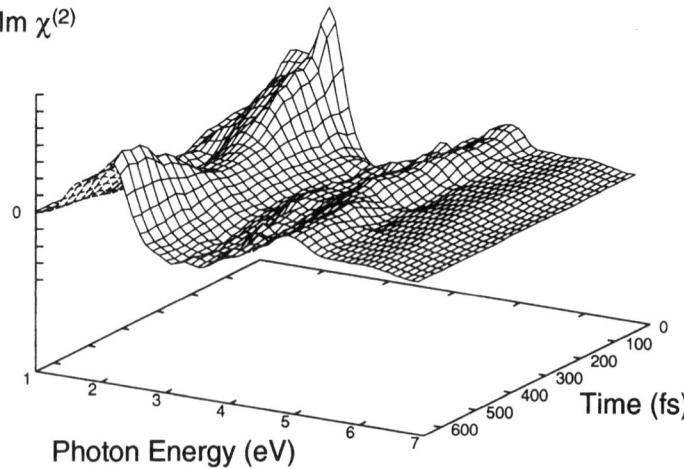

FIG. 7. Im $\chi^{(2)}(\omega)$ for GaAs just below threshold, with $A_0 = 2.50$ G cm (Dumitrică and Allen, 2000). Notice the "elastic" behavior, with recovery after a minimum. Compare with the experimental results of Fig. 9 below threshold.

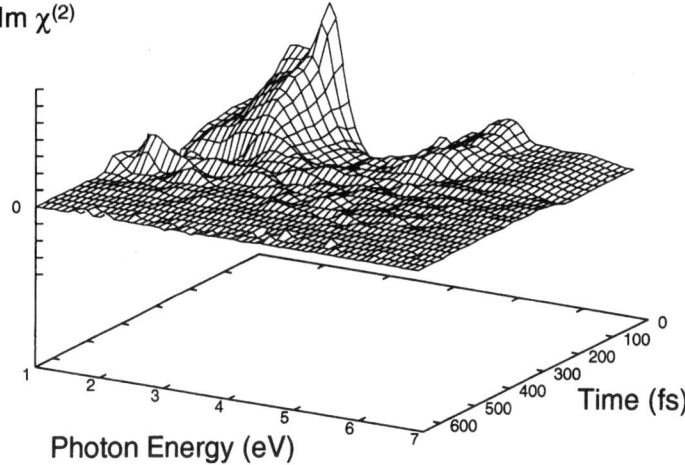

FIG. 8. Im $\chi^{(2)}(\omega)$ for GaAs just above threshold, with $A_0 = 2.51$ G cm (Dumitrică and Allen, 2000). Notice that Im $\chi^{(2)}(\omega)$ falls to zero over the entire range of photon energies, signaling a loss of the original symmetry of the GaAs lattice. Compare with the experimental results of Fig. 9 above threshold.

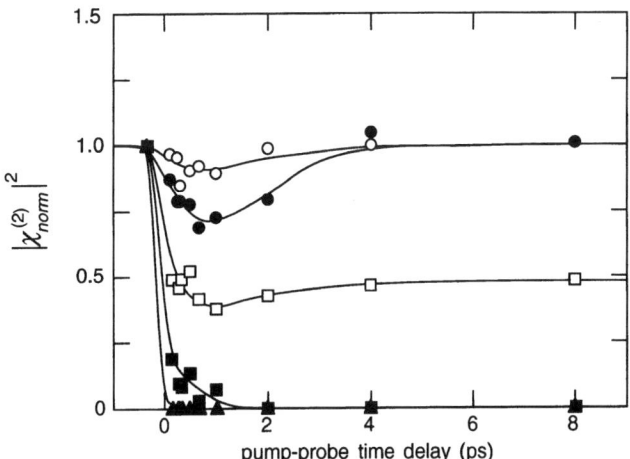

FIG. 9. Square of the second-order susceptibility for GaAs vs. pump-probe time delay for various pump fluences. The curves are drawn to guide the eye. Open circles, $0.2 \, kJ/m^2$; solid circles, $0.4 \, kJ/m^2$; open squares, $0.6 \, kJ/m^2$; solid squares, $0.8 \, kJ/m^2$; triangles, $1.5 \, kJ/m^2$. The probe photon energy was 2.2 eV. After Glezer et al. (1995b). Reprinted with permission.

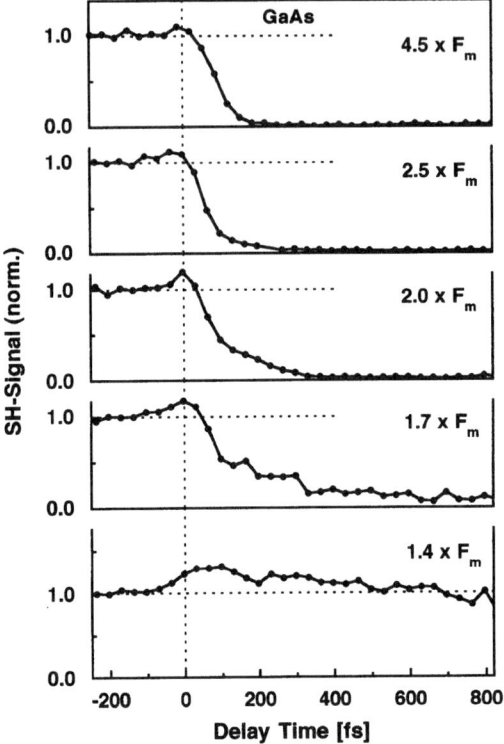

FIG. 10. s-Polarized second harmonic of GaAs as a function of delay time for various pump fluences, which are normalized to $F_m = 0.17\,\text{kJ/m}^2$. After Sokolowski-Tinten et al. (1995). Reprinted with permission.

A more detailed analysis of the lattice destabilization, bandgap collapse, and modification of the dielectric function is presented in the next section.

IV. Detailed Information from Microscopic Simulations

1. A Simple Picture

There are two distinct mechanisms through which an intense laser pulse can destabilize the structure of a molecule or material. On a relatively long time scale ($\sim 2\,\text{ps}$), the energy of excited electrons can be transferred to thermal motion of the atoms. On a shorter time scale ($\sim 100\,\text{fs}$), the promotion of electrons to antibonding states immediately leads to repulsive interatomic forces and the possibility of nonthermal disruption.

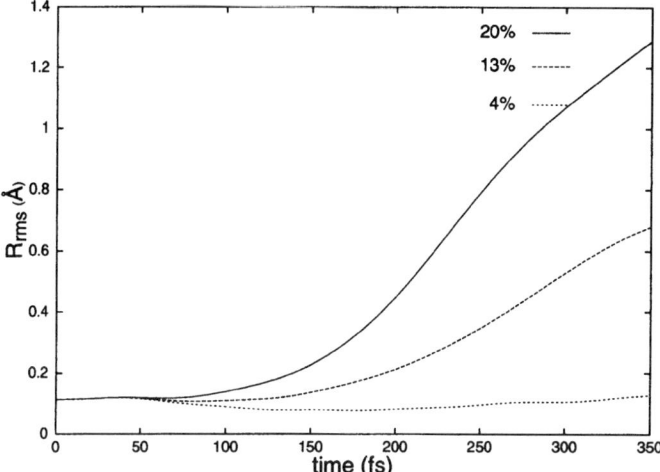

FIG. 11. Root-mean-square displacement of atoms for three different intensities, in 64-atom simulations for GaAs. The fraction of electrons promoted to the conduction bands is given for each curve, as determined from (73). After Dumitrică and Allen, to be published.

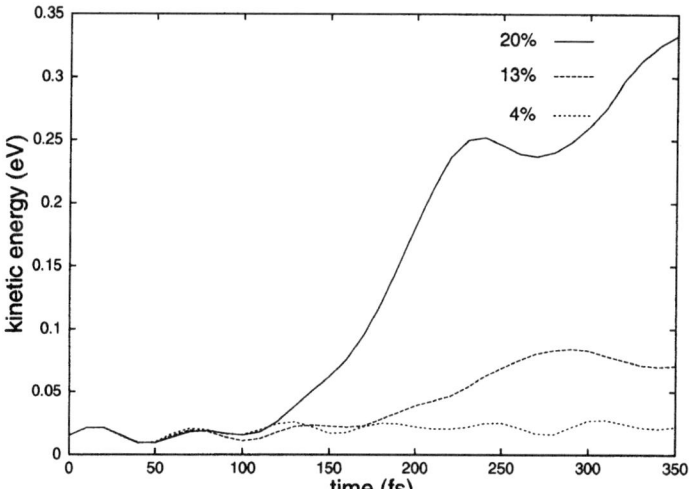

FIG. 12. Average kinetic energy of atoms in 64-atom simulations for GaAs. The fraction of the electrons promoted to excited states is given for each curve, labeled as in Fig. 11. A comparison of this figure with Fig. 11 demonstrates that the atoms are undergoing large displacements from their original positions *at the same time* that the kinetic energy of the atoms is increasing. This is one of the features in the microscopic simulations that leads us to conclude that the structural transformation is nonthermal. Equally important, however, is the short time scale on which the atoms are observed to undergo large displacements—a few hundred femtoseconds. Perhaps most importantly, the behavior in these graphs clearly arises from repulsive forces between atoms, and not from a release of the energy of excited electrons to the lattice. After Dumitrică and Allen, to be published.

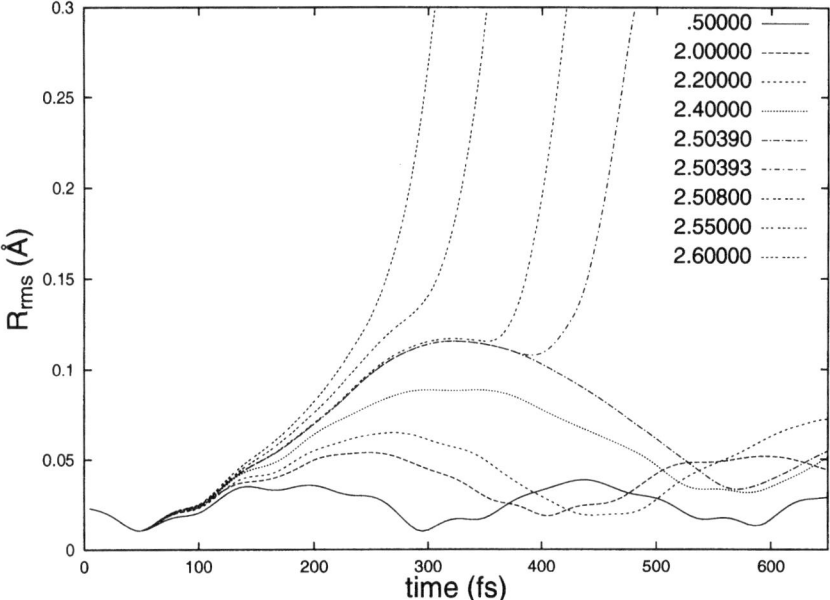

FIG. 13. Root-mean-square displacement of atoms from their equilbrium positions for various intensities of the applied laser pulse, obtained in eight-atom simulations for GaAs. The value of A_0 is given for each curve. After Dumitrică and Allen (2000).

Consider, for example, a two-atom tight-binding model with one orbital per atom. The Hamiltonian is

$$H(r) = \begin{pmatrix} \epsilon_1 & V(r) \\ V(r) & \epsilon_2 \end{pmatrix} \tag{62}$$

so the bonding and antibonding states have energies

$$\epsilon_\pm = \tfrac{1}{2}(\epsilon_1 + \epsilon_2) \pm \tfrac{1}{2}[(\epsilon_1 - \epsilon_2)^2 + 4V(r)^2]^{1/2}. \tag{63}$$

Suppose that we assume the Harrison scaling rules (Harrison, 1980)

$$V(r) = a/r^2, \quad u(r) = b/r^4 \tag{64}$$

where $u(r)$ is the repulsive atom–atom interaction. Since the total energy is $n_+\epsilon_+ + n_-\epsilon_- + u(r)$, where $n_\pm$ represents the occupancies of the states, the force on one atom is

$$F(r) = 2(n_+ - n_-)\left[1 + \left(\frac{\epsilon_1 - \epsilon_2}{2V(r)}\right)^2\right]^{-1/2} \frac{|V(r)|}{r} + 4\frac{u(r)}{r}. \tag{65}$$

In the ground state, with $n_+ - n_- = -2$, an equilibrium separation can be found. But if one electron is excited to the antibonding state, making $n_+ - n_- = 0$, the force becomes purely repulsive and the atoms will dissociate.

During the past 20 years, there has been considerable interest in the analogous problem for tetrahedral semiconductors: destabilization of the covalent bonding as electrons are excited across the bandgap (Shank et al., 1983; Downer et al., 1983; Tom et al., 1988; Saeta et al., 1991; Siegal et al., 1994, 1995; Glezer et al., 1995a,b; Mazur, 1996; Huang et al., 1998; Callan et al.; Sokolowski-Tinten et al., 1995; Govorkov et al., 1992; Chin et al., 1999; Siders et al., 1999; Van Vechten, 1984; Van Vechten et al., 1979; Bok, 1981; Combescot and Box, 1982; Stampfli and Bennemann, 1990, 1992; Kim et al., 1994; Das Sarma and Senna, 1994; Silvestrelli et al., 1996, 1997). Early experiments employed pulses with durations longer than a picosecond, but more recent work at Harvard, Berkeley, MIT, Essen, and other laboratories has used pulses with durations that are comparable to 100 fs or less. In addition, careful measurements of the linear and second-order nonlinear susceptibilities, during the probe phase of a pump-probe experiment, allows the response to be monitored on a subpicosecond time scale.

Motivated by the experiments just described, we have calculated various physical properties that show in detail how the electrons and ions in GaAs respond to fast intense laser pulses (with durations of order 100 fs and intensities of order 1–10 TW/cm^2). The population of excited electrons and band structure are calculated as functions of time, during and after application of each pulse, in addition to the atomic displacements, atomic pair-correlation function, and imaginary part of the dielectric function. We will see that threshold intensity corresponds to promotion of about 10% of the electrons to the conduction band. Above this intensity, the lattice is destabilized and the bandgap collapses to zero. As discussed earlier, this is clearly revealed in the dielectric function $\epsilon(\omega)$, which exhibits metallic behavior and loses its structural features after about 200 fs, and the second-order nonlinear susceptibility $\chi^{(2)}$, which signals a loss of the original crystal structure by decreasing to zero, but the other properties are also of considerable interest. For example, the pair correlation function of Fig. 14 demonstrates the loss of crystalline order above threshold, and the displacements of Fig. 18 also signal structural changes.

2. Excited-State Tight-Binding Molecular Dynamics

Before investigating the full response of electrons and ions to an intense laser pulse, let us first consider a much simpler problem: the dynamics of the atoms when some fraction of the electrons are artificially promoted to excited states. We use the standard sp^3s^* tight-binding Hamiltonian (Vogl et al., 1983) and a nonstandard repulsive potential with the form (61). The

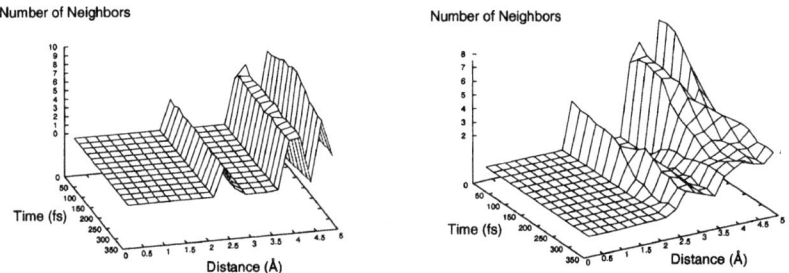

FIG. 14. Pair correlation function for an intensity below threshold (left) and above threshold (right) in 64-atom simulations for GaAs. The values of A_0 are respectively 0.5 and 2.51 G cm. After Dumitrică and Allen, to be published.

total energy (leaving aside the kinetic energy of the ions) is then

$$E = \sum_k n_k \epsilon_k + \sum_{l>l'} u(R_{ll'}) \tag{66}$$

where n_k is the occupancy of the electronic state labeled by k (which includes the spin index) and $R_{ll'}$ is the separation of ions l and l'. This is simply the generalization of the expression for E used in tight-binding molecular dynamics for the ground state (Menon and Allen, 1985, 1986, 1991; Sankey and Allen, 1985, 1986; Allen and Menon, 1986; Gryko and Allen, 1994; Sawtarie et al., 1994; Andriotis et al., 1996). As usual, the second term in (66) represents $U_{ii} - U_{ee}$, where U_{ii} is the ion–ion repulsion and U_{ee} corresponds to the electron–electron repulsion, which is too strongly represented in the first term of (66). [See (35). The Coulomb interaction is double-counted in the one-electron Hamiltonian H; the negative exchange-and-correlation energy is also overcounted, but this is a weaker effect.] For spherically symmetrical and well-separated neutral atoms, we have $U_{ii} - U_{ee} \approx 0$, so $u(r)$ should fall off rapidly with distance. In (61), we have modified the basic Harrison scaling of (64) by adding two higher-order terms. We also multiply by a cutoff function $C(r)$, which is taken to have the form of a Fermi function:

$$C(r) = [\exp((r - r_c)/r_w) + 1]^{-1}. \tag{67}$$

The cutoff distance r_c was chosen to be midway between $1.2\, r_1$ and r_2, where $r_1 = 2.35\,\text{Å}$ and $r_2 = 3.84\,\text{Å}$ are respectively the first and second neighbor distances: $r_c = (1.2\, r_1 + r_2)/2$. The cutoff width r_w was chosen to be 0.1 Å. With these choices, the cutoff function has little effect for bond-length changes up to 30% (so that the initial stages of destabilization will be reliably described), but falls to nearly zero at the second-neighbor distance (so that there are no unphysical distant interactions). The matrix

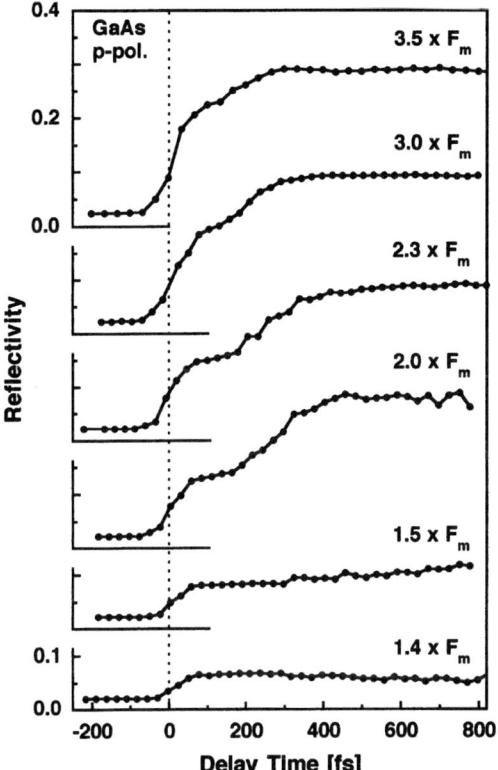

FIG. 15. p-Polarized reflectivity of GaAs as a function of delay time for various pump fluences, which are normalized to $F_m = 0.17$ kJ/m^2. Like the changes in the dielectric function of Figs. 1–5, these changes in the reflectivity are a signature of bandgap collapse and the onset of metallic behavior. After Sokolowski-Tinten *et al.* (1995). Reprinted with permission.

elements of the tight-binding Hamiltonian are taken to have the Harrison scaling (64) and the same cutoff function,

$$H^0_{\alpha\beta}(ll') = \bar{H}_{\alpha\beta}(ll')(r_1/R(ll'))^2 C(R(ll'))/C(r_1), \qquad (68)$$

where here α and β represent orbitals on atoms l and l'. $\bar{H}_{\alpha\beta}(ll')$ is the Hamiltonian obtained from the parameters of Table I, using the usual Slater–Koster rules (Harrison, 1989). The superscript "0" indicates that there is not yet an applied electromagnetic field.

The parameters α, β, and γ of (61) were determined by fitting the cohesive energy, lattice spacing, and bulk modulus to experiment. Details of the fitting procedure are given elsewhere (Graves, 1997). The resulting values for both GaAs and Si are listed in Table II.

TABLE I

Tight-Binding Parameters in the sp^3s^* Model for GaAs and Si[a]

	GaAs	Si
ϵ_{s_a}	−2.657	−4.200
ϵ_{p_a}	3.669	1.715
$\epsilon_{s_a^*}$	6.739	6.685
ϵ_{s_c}	−8.343	−4.200
ϵ_{p_c}	1.041	1.715
$\epsilon_{s_c^*}$	8.591	6.685
$\eta_{s_a s_c \sigma}$	−1.271	−1.504
$\eta_{s_a p_c \sigma}$	1.529	1.798
$\eta_{p_a s_c \sigma}$	−1.974	−1.798
$\eta_{p_a p_c \sigma}$	2.386	1.969
$\eta_{p_a p_c \pi}$	−0.6153	−0.5182
$\eta_{s_a s_c^* \sigma}$	0.0	0.0
$\eta_{s_a^* s_c \sigma}$	0.0	0.0
$\eta_{s_a^* p_c \sigma}$	−1.640	−1.687
$\eta_{p_a s_c^* \sigma}$	1.652	1.687
$\eta_{s_a^* s_c^* \sigma}$	0.0	0.0

[a] Taken from Vogl et al. (1983). The dimensionless coefficients η are defined in Harrison (1980). Here c and a, respectively, denote the cation and anion.

In the simulations of this section, either an 8-atom or a 64-atom cubic cell was used. With five orbitals per atom, the Hamiltonian matrix is then 40×40 or 320×320. Each atom interacts with all other atoms within the cell and their replicas outside the cell. The motion of the atoms is taken to satisfy periodic boundary conditions, so the electronic states are Bloch states corresponding to this large unit cell. In calculating the Hellmann–Feynman forces on the atoms, we used the special point $\mathbf{k} = (\frac{1}{4}, \frac{1}{4}, \frac{1}{4})(2\pi/a)$, together

TABLE II

Repulsive Potential Parameters for GaAs and Si[a]

	α	β	γ
GaAs	263.7	−1227.5	3653.1
Si	263.2	−1027.0	2631.8

[a] After Graves and Allen (1998b). These values are appropriate when distances are measured in angstroms and energies in electron volts.

with the other points that are related to it through symmetry transformations. For the GaAs interactions, we used the parameters and cutoff function described earlier. For Ga–Ga and As–As interactions (which are irrelevant in the initial stages of destabilization), we used the same parameters, but replaced the Fermi function cutoff (67) by a theta function cutoff $\theta(r_1 - R(ll'))$. The velocity Verlet algorithm was used (Gould and Tobochnik, 1988; Heerman and Burkitt, 1991), with a time step of 0.05 fs. Energy is then typically conserved to about one part in 10^6 for low excitation of the electrons, or one part in 10^4 at high excitations that cause more violent atomic motion. The expression (66) leads to the usual Hellmann–Feynman theorem of tight-binding molecular dynamics (Menon and Allen, 1985, 1986; Sankey and Allen, 1985, 1986; Allen and Menon, 1986),

$$M\ddot{X} = -\sum_k n_k \Psi_k^\dagger \cdot \frac{\partial \mathbf{H}^0}{\partial X} \cdot \Psi_k - \frac{\partial U}{\partial X} \tag{69}$$

where X and M are any ion coordinate and mass,

$$U = \sum_{l>l'} u(R(ll')), \tag{70}$$

and $\mathbf{H}^0$ is the Hamiltonian matrix of (68).

Figure 30 shows results for Si when the atoms are given an initial kinetic energy corresponding to 300 K, but some fraction of the electrons are artificially promoted from the top of the valence band to the bottom of the conduction band. If 25% of the electrons are promoted, the atoms are observed to move far from their equilibrium positions in the original tetrahedral structure, so the lattice has definitely been destabilized.

3. Detailed Model

Let us now turn to full simulations of the coupled dynamics of electrons and ions in material that is subjected to an intense laser pulse. The vector potential is taken to have the time dependence

$$\mathbf{A}(t) = A_0 \cos\left(\frac{\pi(t - t_0/2)}{t_0}\right) \cos(\omega t), \quad 0 \leqslant t \leqslant t_0. \tag{71}$$

This form (i) closely resembles a Gaussian (Graves, 1997), (ii) clips the pulse to zero at beginning and end, (iii) gives zero slope for $A(t)^2$ at beginning and end, and (iv) gives a full-width at half-maximum (FWHM) duration for $A(t)^2$ of exactly half the total pulse time t_0.

The usual Hellmann–Feynman theorem of (69) is no longer valid when the electronic states Ψ_j are no longer eigenstates of the Hamiltonian.

However, (69) can be replaced by the generalized Hellmann–Feynman theorem (or generalized Ehrenfest theorem) of (16), (41), or (48). (These same two equations can be obtained through the arguments of Allen, 1994, which are different from the ones given in the present paper.) The equation for the ion dynamics is coupled to the time-dependent Schrödinger equation (13), (40), or (47) for the electron dynamics (Allen, 1994). In a TED simulation, the electrons are in turn coupled to the radiation field through the time-dependent Peierls substitution (52) (Graves, 1997; Graves and Allen, 1998a,b; Dumitricǎ et al., 1998, 2000; Dumitricǎ and Allen, 2000; Torralva and Allen, 1998; Torralva et al.; Hamilton, 1999). [In a first-principles simulation (Gao, 1998) the electrons are coupled to the radiation field through (49).] The direct force of the electromagnetic field on the ions is omitted, since this force oscillates on a 1-fs time scale, two orders of magnitude shorter than the response time of the ions.

The second-order equation (48) was solved with the velocity Verlet algorithm, which preserves phase space. In the earlier work, the first-order equation (47) was solved with the Cayley algorithm (54), which conserves probability and ensures that the Pauli exclusion principle is satisfied at all times. In later work we used the still better algorithm (57) introduced by Torralva et al. (1999b).

4. Electronic Excitation and Time-Dependent Band Structure

In Figs. 11 and 12, we have already seen the atomic motion that results when laser pulses of various intensities are applied to GaAs in thermal equilibrium at 300 K (after an equilibration period of 2000 fs). In each case the FWHM pulse duration was 70 fs, with $\hbar\omega = 1.95$ eV, and a polarization in the (1.7, 1.0, 0) direction referenced to the cube edges (Graves, 1997).

As shown in Graves (1997), an amplitude $A_0 = 1.00$ G cm corresponds to a fluence of 0.815 kJ/m². The threshold for permanent structural change is about 2.00 G cm, or 3.26 kJ/m². This is about three times as large as the experimental threshold (Siegal et al., 1994, 1995; Glezer et al., 1995a,b; Mazur, 1996; Huang et al., 1998; Callan et al.; Sokolowski-Tinten et al., 1995). Since the present theory yields a dielectric function that is roughly half that observed experimentally (Graf and Vogl, 1995), one expects the nonlinear response to also be underestimated, so this level of agreement is quite satisfactory. In other words, it is not unexpected that the absolute cross section for absorption is not accurately calculated with the present model.

Although $\Psi_j(t)$ can be interpreted as the physical state for the jth electron, one can also define eigenvectors $\Phi_m(\mathbf{k})$ of the time-dependent Hamiltonian:

$$\mathbf{H} \cdot \Phi_n(\mathbf{k}) = \epsilon_n(\mathbf{k})\Phi_n(\mathbf{k}). \tag{72}$$

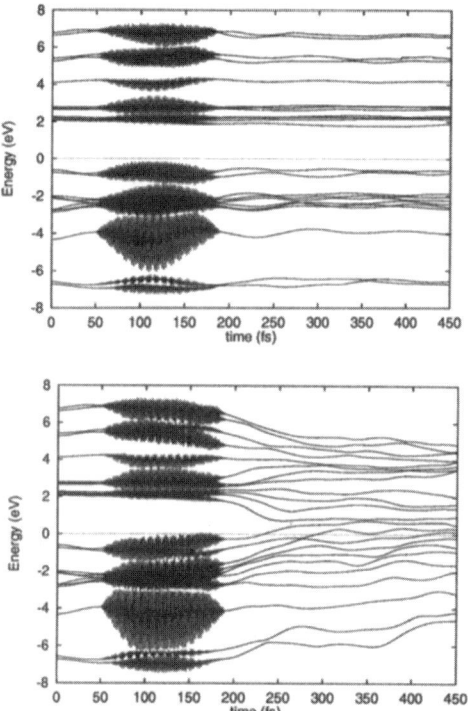

FIG. 16. Electronic energy eigenvalues for GaAs at the $(2\pi/a)(\frac{1}{4},\frac{1}{4},\frac{1}{4})$ point as a function of time, with $A_0 = 1.73$ G cm (top) and 2.83 G cm (bottom) in eight-atom simulations. The bandgap at this particular **k**-point is larger than the fundamental bandgap, but at the higher intensity it has collapsed to zero, demonstrating that the material is now metallic rather than semiconducting.

The eigenvalues $\epsilon_n(\mathbf{k})$ at the special point $\mathbf{k} = (\frac{1}{4}, \frac{1}{4}, \frac{1}{4})(2\pi/a)$ are plotted as functions of time in Fig. 16 for two different intensities. Notice that the bandgap at this point (which is larger than the fundamental band gap at $(0,0,0)$) exhibits only thermal oscillations for $A_0 = 1.00$ G cm, but has completely closed up for $A_0 = 2.83$ G cm because of the large atomic displacements associated with lattice destabilization. The rapid oscillations during application of the pulse are due to the Peierls factor in (52).

The occupancy of the kth state is given by

$$n_k = \sum_j |\mathbf{\Psi}_j^\dagger \cdot \mathbf{\Phi}_k|^2 \qquad (73)$$

where $k \leftrightarrow \mathbf{k}, n$. The total occupancy of all the conduction bands (again at

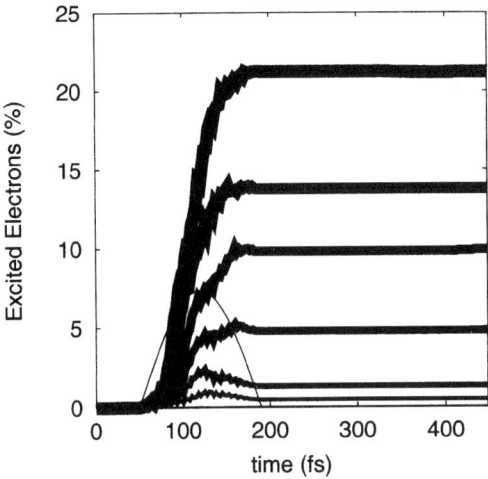

FIG. 17. The percentage of valence electrons promoted to excited states is shown as a function of time for GaAs with varying pulse intensities. The pulse is represented by a solid curve. The amplitudes A_0 are given in Fig. 18. Comparison of the two figures shows that the threshold intensity (or fluence) corresponds to about 10% occupation of the excited conduction-band states. After Graves and Allen, (1998b). In the later work of Dumitrică and Allen (2000), shown in Fig. 11, the threshold intensity was more precisely pinned down, and was also found to be at about 2.5 G cm rather than 2 G cm. The threshold occupancy of excited states was also found to be higher. Recent work with 64-atom simulations indicates that the threshold occupancy is approximately 17% for both GaAs and Si (Dumitrică and Allen, to be published).

the special point) is plotted as a function of time in Fig. 17, where it is expressed as a percentage of the total number of valence electrons. Since our model does not include carrier interactions, n_k remains constant after the pulse is turned off.

In Fig. 14, the pair correlation function was plotted as a function of time for two intensities (in 64-atom simulations), above and below the threshold for lattice destabilization. The loss of order confirms that the higher intensity leads to permanent structural change.

5. Dielectric Function as a Signature of the Change in Bonding

As discussed in the preceding section, the most direct comparison with experiment is provided by the imaginary part of the dielectric function, which can be calculated from the formula (60), where $\omega_{mn}(\mathbf{k}) = [\epsilon_m(\mathbf{k}) - \epsilon_n(\mathbf{k})]/\hbar$, $f_m(\mathbf{k})$ is the same as the occupancy of (73) (with $k \leftrightarrow \mathbf{k}, n$), and $\mathbf{p}_{nm}(\mathbf{k})$ is given in (90) (using the notation of Graf and Vogl, 1995, to

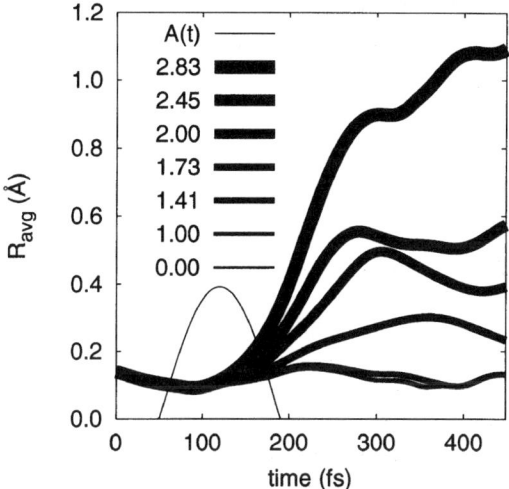

FIG. 18. Average distance moved by an atom, during and following a laser pulse. The amplitude A_0 of the vector potential is given in gauss centimeter. After Graves and Allen (1998b).

avoid confusion). In the summation of (60), the following **k**-points were included:

$$\mathbf{k}_{n_1,n_2,n_3} = \tfrac{1}{16}(n_1, n_2, n_3) \qquad (74)$$

with

$$n_1, n_2, n_3 = \pm 1, \pm 3, \pm 5, \pm 7. \qquad (75)$$

Also, the δ-function was approximated by a Gaussian,

$$\delta(\omega) \approx \frac{1}{\sqrt{\pi}} \frac{e^{-(\hbar\omega/\delta\epsilon)^2}}{\delta\epsilon/\hbar}, \qquad (76)$$

with $\delta\epsilon = 0.3$ eV.

The panels of Fig. 2 show the imaginary part of the dielectric function for $0.5 \text{ eV} \leqslant \hbar\omega \leqslant 6.0 \text{ eV}$ and for different intensities, ranging from $A_0 = 1.73$ to $A_0 = 2.83$ G cm. (The corresponding fluences range up to 6.5 kJ/m^2.) At low intensities there is no absorption for $\hbar\omega$ less than the bandgap of 1.5 eV (i.e., Im $\epsilon(\omega)$ is zero in this range) and the structural features in Im $\epsilon(\omega)$ persist at all times. At high intensities, one can observe metallic behavior (with subbandgap absorption) and the structural features are washed out. These conclusions are consistent with the measurements (Shank *et al.*, 1983;

Downer *et al.*, 1983; Tom *et al.*, 1988; Saeta *et al.*, 1991; Siegal *et al.*, 1994, 1995; Glezer *et al.*, 1995a,b; Mazur, 1996; Huang *et al.*, 1998; Callan *et al.*; Sokolowski-Tinten *et al.*, 1995).

6. Second-Order Susceptibility as a Signature of the Change in Symmetry

Using TED, we have also calculated the evolution of the nonlinear susceptibility $\chi^{(2)}(\omega)$ in GaAs during the first few hundred femtoseconds following an ultrafast and ultraintense laser pulse. Above a threshold fluence, our simulations show that $\chi^{(2)}(\omega)$ drops to zero, in agreement with the experimental measurements. The results indicate a rapid nonthermal transition from the original tetrahedral structure to a disordered structure, and support the conclusion that structural changes following ultrashort pulses are a direct consequence of bond destabilization.

In recent experiments (Saeta *et al.*, 1991; Glezer *et al.*, 1995b; Sokolowski-Tinten *et al.*, 1995), the time evolution of the second-order nonlinear susceptibility $\chi^{(2)}$ has been measured in GaAs, using a 2.2-eV probe pulse after excitation with an intense pump pulse with a photon energy of about 1.9 eV. As discussed earlier, the observations indicate that the response of a semiconductor to an ultrafast laser pulse, with a duration of 100 fs or less, is fundamentally different from its response to a pulse with a duration of 1 ps or more. Whereas the longer pulses appear to produce ordinary heating of the sample by phonon emission, there is convincing evidence that ultrafast pulses induce a structural transition by directly destabilizing the atomic bonds.

When an ultrashort laser pulse is applied, valence electrons are promoted to the conduction bands on a time scale that is short compared to that for atomic motion (~ 10–100 fs versus ~ 100–1000 fs). Electronic relaxation subsequently occurs through a combination of carrier scattering, phonon emission, Auger recombination, radiative recombination, and carrier diffusion. Our model does not include any of these relaxation processes, so the occupancy of the states n_k remains constant after the end of the laser pulse. However, this should not be a significant limitation in the present work, which focuses on the initial electronic and structural response rather than the subsequent behavior at longer times.

Our TED simulations discussed in the preceding sections show that there is a structural change in GaAs above a threshold fluence that corresponds to promotion of $\sim 15\%$ of the valence electrons to conduction-band states, on a time scale of 100–200 fs. When 15% of the electrons are promoted from bonding to antibonding states, the effect is roughly the same as removing 30% of the bonds. There are consequently strong repulsive interactions in

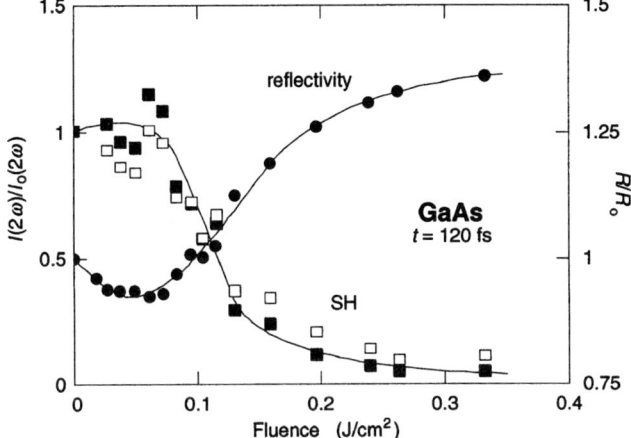

FIG. 19. Fluence dependence of the relative reflected second-harmonic intensity (squares) and reflectivity (circles) of a (110) GaAs surface. The data were taken at 120 fs delay with 100 fs pulses of 620 nm wavelength at an incident angle of 45°. The open squares are the second-harmonic data divided by $[(1 - R)/(1 - R_0)]^2$ to correct for changes in reflectivity. After Saeta et al. (1991). Reprinted with permission.

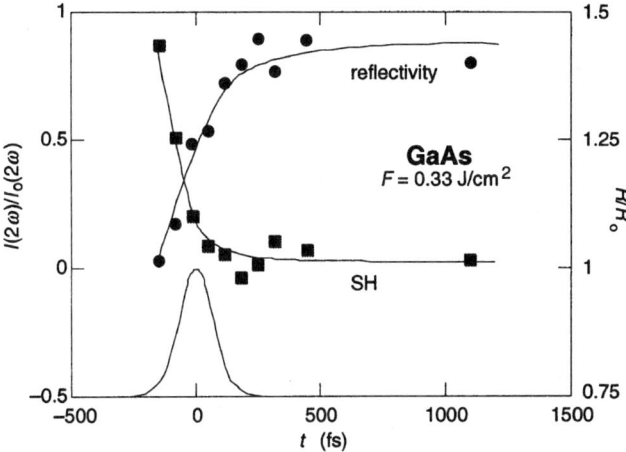

FIG. 20. Time dependence of the relative second-harmonic intensity (squares) and reflectivity (circles) at a fluence of 0.33 J/cm². The curves are fits by exponentials yielding 1/e times of 90 fs for the second-harmonic decay and 170 fs for the reflectivity rise. The peak at $t = 0$ shows the duration of the laser pulse. After Saeta et al. (1991). Reprinted with permission.

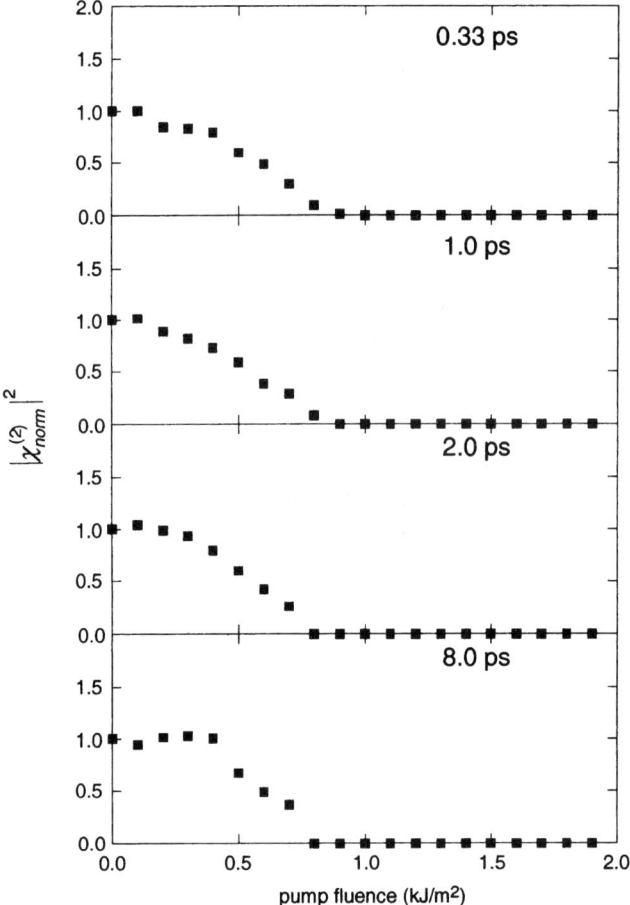

FIG. 21. Square of the second-order susceptibility of GaAs versus pump fluence for four different pump-probe time delays. For fluences that are above the threshold, and for a probe photon energy of 2.2 eV, notice that $\chi^{(2)}$ falls to zero after only 330 fs. Compare with the simulations of Figs. 6–8. After Glezer et al. (1995b). Reprinted with permission.

the initial atomic geometry, which produce massive disruption of this geometry on a time scale of a few hundred femtoseconds.

Because of its sensitivity to the crystal symmetry, the second-order susceptibility $\chi^{(2)}(\omega)$ can provide direct information about structural changes in noncentrosymmetric materials such as GaAs, for which $\chi^{(2)}$ is nonvanishing in the usual bulk geometry. One of us (Dumitrică) has developed a formalism for calculating $\chi^{(2)}(\omega)$ with a tight-binding Hamiltonian (Dumitrică et al., 1998), which will be described in more detail in the next section. Employing an sp^3s^* orbital basis, together with the analytical expression for $\chi^{(2)}$ derived in Dumitrică et al. (1998) and discussed in Section V, we obtain results that are in agreement with the best available first-principles calculations and experimental measurements for the unperturbed GaAs crystal (Dumitrică et al., 1998).

In the case of the present time-dependent simulations, the imaginary part of the second-order susceptibility tensor was calculated after each 25 fs during the first 650 fs of a run. Just as for the usual first-order susceptibility (or dielectric function $\epsilon(\omega)$), a Kramers–Krönig transform relates the real and imaginary parts of $\chi^{(2)}$. It follows that a decrease in the imaginary part of $\chi^{(2)}(\omega)$ over the whole frequency range also means a decrease in the real part.

Figures 6–8 show the evolution of the imaginary part of $\chi^{(2)}(\omega)$ when a 70 fs pump laser pulse is applied to GaAs, at three distinct fluences. For relatively low fluences, we find that $\chi^{(2)}$ exhibits no significant changes. This can be clearly seen in the results of Fig. 6, for a fluence corresponding to an amplitude $A_0 = 0.5$ G cm. One can relate A_0 to the fluence by using the conversion of Graves and Allen (1998b): For a pulse with $A_0 = 1.0$ G cm, and a full-width-at-half-maximum (FWHM) of 70 fs, the fluence is 0.815 kJ/m^2. The intensity, or fluence for fixed pulse duration, then varies

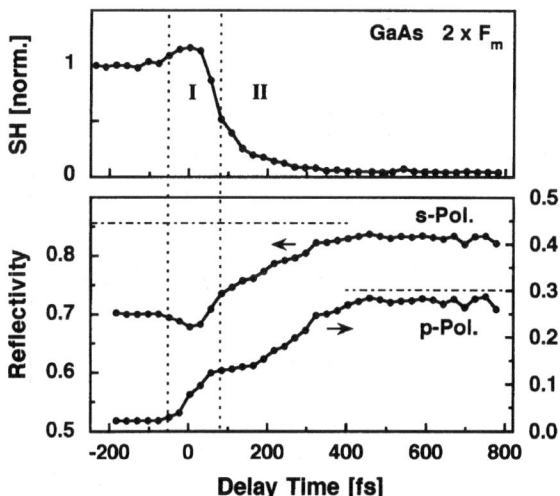

FIG. 22. Measured s- and p-polarized reflectivity and reflected second harmonic of GaAs ($\lambda = 625$ nm, $\Theta_{in} = 70°$) as a function of delay time. The pump fluence is 0.34 kJ/m^2 ($\approx 2F_m$). The different stages of signal evolution are marked by the dashed lines. The dash-dotted lines indicate the reflectivities of liquid GaAs. Compare the time scale with that in the simulations of Figs. 1 and 2 (for the dielectric function, which is related to the reflectivity) and Figs. 6–8 (for the second-order nonlinear susceptibility, which is related to the second harmonic (SH)). After Sokolowski-Tinten et al. (1995). Reprinted with permission.

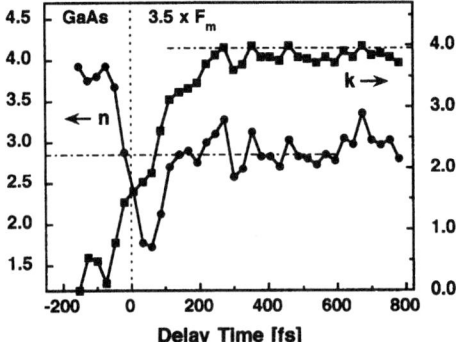

FIG. 23. Optical constants of GaAs as a function of time, derived from the reflectivity data for $F = 3.5F_m$. Dash-dotted lines, optical constants of liquid GaAs. For unperturbed crystalline GaAs, $n = 3.87$ and $k = 0.2$. Again, compare the time scales in the experiments and simulations. After Sokolowski-Tinten et al. (1995). Reprinted with permission.

as A_0^2. It should be reiterated that the threshold value of A_0 obtained in the present simulations is substantially higher than that obtained in the experiments, because the present method does not accurately predict absolute cross-sections for absorption.

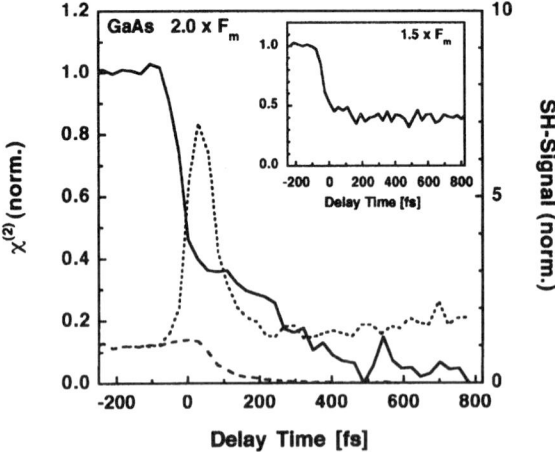

FIG. 24. Measured reflected second harmonic of the probe pulse (dashed curve), calculated second harmonic (dotted curve), and the measured nonlinear susceptibility $\chi^{(2)}$ of GaAs as a function of time for $F = 2F_m$. The inset shows the measured time dependence of $\chi^{(2)}$ for a lower pump fluence, $F = 1.5\ F_m$. Compare with the behavior of $\chi^{(2)}$ in the simulations of Figs. 6–8. After Sokolowski-Tinten et al. (1995). Reprinted with permission.

As the threshold fluence is approached, the material exhibits "elastic" behavior: A strong decrease in the imaginary part of $\chi^{(2)}$ is followed by a rapid recovery, as can be seen in Fig. 7. The time at which $\chi^{(2)}$ reaches a minimum (near 400 fs) approximately corresponds to maximum departure from the original tetrahedral bonding. (This can be seen in Fig. 12.) The value of A_0 in Fig. 7 is 2.50 G cm, corresponding to a fluence that is only very slightly below threshold.

Even slightly above threshold, on the other hand, there is an irreversible structural change. In Fig. 8, with $A_0 = 2.51$ G cm, it is apparent that $\chi^{(2)}$ is nearly zero over the whole frequency range after about 400 fs. This behavior indicates a change of symmetry in the material, and its persistence implies that a permanent structural change has occurred.

It is interesting that all three regimes represented by Figs. 6–8—no significant change in $\chi^{(2)}$ at low fluence, large change followed by recovery at subthreshold fluence, and decrease to zero at large fluence—have been observed in experiments (Saeta et al., 1991; Glezer et al., 1995b; Sokolowski-Tinten et al., 1995). The time scale for a near-threshold recovery is longer in the experiments, presumably because the simulations involve a closed system and the experiments an open system. That is, the atoms in the simulations are in a very restricted region (a cell containing 8 or 64 atoms), and they all experience the laser pulse. The sample of material in an experiment, on the other hand, is much larger, and only those atoms in a relatively small central region experience the pulse. It is therefore natural that the time scale for an "elastic" recovery in the experiments should be larger.

The calculations discussed in this subsection are complementary to those of the preceding subsection, where it was emphasized that the first-order susceptibility $\epsilon(\omega)$ evolves from semiconducting to metallic behavior. The results for the nonlinear susceptibility provide independent evidence for a structural transformation, from an ordered structure with a nonzero value of $\chi^{(2)}$ to a disordered structure with $\chi^{(2)} = 0$.

Let us now summarize this subsection: Using an analytical expression for the second-order nonlinear susceptibility (Dumitrică et al., 1998), we have simulated the interaction of ultrashort and ultraintense laser pulses with GaAs and calculated the evolution of $\chi^{(2)}(\omega)$ on a femtosecond time scale. Above a threshold fluence, $\chi^{(2)}$ drops to zero on a time scale of about 200 fs. This is the same threshold fluence, and same time scale, associated with the appearance of metallic behavior in the dielectric function $\epsilon(\omega)$. The time evolution of $\chi^{(2)}$ indicates that the original symmetry of the GaAs crystal is lost, and the time evolution of ϵ indicates that there is a loss of the original tetrahedral bonding. Both of these complementary results are consistent with the experimental measurements of $\chi^{(2)}(\omega)$ and $\epsilon(\omega)$, and they independently support the conclusion that ultrashort laser pulses induce a nonthermal phase transition in this material.

7. SUMMARY OF RESULTS FOR GaAs

To summarize this whole section, we have performed simulations that show in detail how the electrons and ions in GaAs respond to fast intense laser pulses. As discussed earlier, we employed TED and included the radiation field through a time-dependent Peierls substitution. The population of excited electrons, atomic displacements, atomic pair-correlation function, band structure, imaginary part of the dielectric function, and imaginary part of the second-order nonlinear susceptibility were all calculated as functions of time, during and after application of each pulse. Above the threshold intensity, the lattice is destabilized and the bandgap collapses to zero. This is clearly revealed in the dielectric function $\epsilon(\omega)$, which exhibits metallic behavior and loses its structural features. It is also clearly revealed by the second-order susceptibility $\chi^{(2)}$, which goes to zero and thus demonstrates that the symmetry of the system has changed.

We close this section by mentioning a very new frontier: ultrafast X-ray diffraction experiments, in which the atomic structure can be probed on a subpicosecond time scale. Results from two pioneering groups are shown in Figs. 25–27.

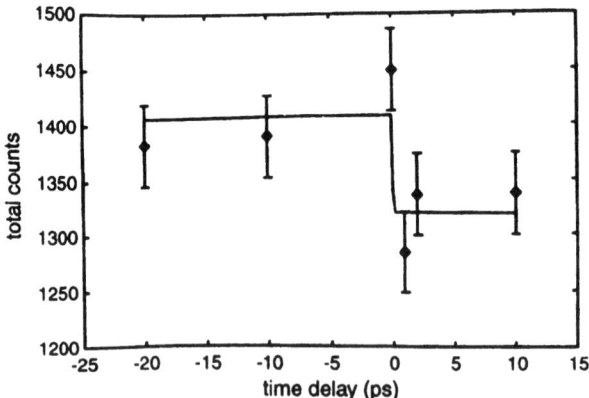

FIG. 25. Direct evidence for nonthermal ultrafast disordering in InSb, with 100 fs, 800 nm laser pump pulses followed by 300 fs, 0.4 Å X-ray probe pulses. These ultrafast diffraction experiments can detect both changes in the lattice parameter and the onset of disorder, on a time scale of only a few hundred femtoseconds. This figure shows normalized integrated X-ray diffracted photons as a function of time delay. After Chin et al. (1999). Reprinted with permission.

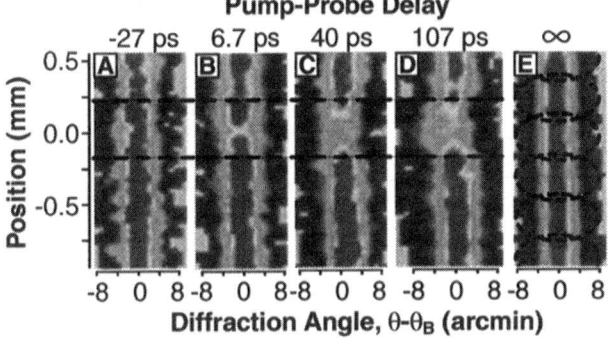

FIG. 26. Evidence for an ultrafast and nonthermal solid-to-liquid phase transition in Ge, followed by recrystallization. These X-ray diffraction probe measurements employed 1.54 Å photons, following 100 fs, 800 nm pump pulses. The figure shows images obtained from a photon-counting X-ray area detector for five pump-probe time delays. The horizontal axis corresponds to the diffraction angle, shown relative to the Bragg angle for Ge. The vertical axis corresponds to the position on the semiconductor wafer. The optical pump photoexcites only a portion [indicated by dotted lines in (**A**) through (**D**)] of the entire X-ray-probed area. The image at infinite time delay (**E**), including six single-shot damage regions (indicated by dotted circles), was taken nonrastered and with the optical pump blocked during the exposure. Reprinted with permission from Siders *et al.* (1999). Copyright 1999 American Association for the Advancement of Science.

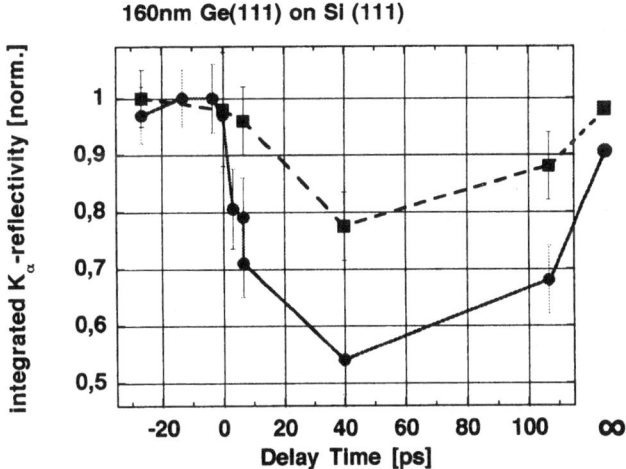

FIG. 27. Evidence for ultrafast melting in the region subjected to the laser pulses of Fig. 26. In contrast, nearby regions undergo a slower disordering that is consistent with ordinary thermal melting. The figure shows time-resolved K_α X-ray reflectivity from a 160-nm Ge(111) film, integrated over the central pumped region (solid line) and over a region vertically displaced ~0.2 mm from the center (dashed lines). Reprinted with permission from Siders *et al.* (1999). Copyright 1999 American Association for the Advancement of Science.

V. Formula for the Second-Order Susceptibility $\chi^{(2)}$

In this section we explain the origin of the expressions that were used to calculate the dielectric function ϵ (or $\chi^{(1)}$) and the second-order susceptibility $\chi^{(2)}$. We use the notation of Dumitrică et al. (1998), in which an expression for $\chi^{(2)}$ was first obtained in a tight-binding representation. This is the same as the notation in Graf and Vogl (1995), where the corresponding expression for ϵ was derived.

Nonlinear optical phenomena in semiconductors are of considerable interest for both applications and understanding of the fundamental physics (see, for example, Fraser et al., 1999). For this reason there have been several previous theoretical studies of the second-order nonlinear susceptibility $\chi^{(2)}(\omega)$ (Moss et al., 1987; Hughes and Sipe, 1996; Ghahramani et al., 1991; Huang and Ching, 1992). An additional motivation involves the primary focus of this chapter: experiments in which semiconductors are subjected to intense subpicosecond laser pulses (Shank et al., 1983; Downer et al., 1983; Tom et al., 1988; Saeta et al., 1991; Siegal et al., 1994, 1995; Glezer et al., 1995a,b; Mazur, 1996; Huang et al., 1998; Callan et al., 2000; Sokolowski-Tinten et al., 1995). As discussed in the previous section, measurements of second harmonic generation (SHG) provide information about the dynamics of the structural changes that take place in the material between pump and probe pulses. It is clearly very useful to have a formalism that permits calculations of nonlinear effects from a tight-binding Hamiltonian, since tight-binding methods provide a versatile approach to many problems involving real materials (Harrison, 1980, 1999).

Our calculation is based on the formalism introduced by Graf and Vogl (1995), who first recognized that a time-dependent Peierls substitution can be used to couple electrons to an electromagnetic field without the need of any additional parameters. In their approach, each element of the unperturbed tight-binding Hamiltonian is multiplied by a phase factor containing the vector potential associated with an arbitrarily intense and time-dependent electromagnetic field. They employed this idea in obtaining an analytical expression for the linear dielectric function in a tight-binding representation, and performing a calculation for GaAs that yielded satisfactory agreement with the experimental measurements. In Subsection 1, we summarize the essential features of their formalism. In Subsection 2, we extend it to obtain an analytical expression for the second-order nonlinear susceptibility $\chi^{(2)}(\omega)$. This expression is then employed in Subsection 4, where the imaginary part of $\chi^{(2)}(\omega)$ is determined for GaAs. (The real part can be seen in Dumitrică et al., 1998.) These results are in good agreement with previous calculations and with the available experimental data.

1. TIGHT-BINDING HAMILTONIAN IN AN ELECTROMAGNETIC FIELD

Let us begin with a Bloch sum over the localized Löwdin orbitals $|\alpha, L\rangle$:

$$|\alpha, \mathbf{k}\rangle = \frac{1}{\sqrt{N}} \sum_L e^{i\mathbf{k} \cdot \mathbf{R}_{\alpha L}} |\alpha, L\rangle. \tag{77}$$

Here L labels the unit cell and α labels a specific atomic orbital on a specific site. There are N unit cells labeled by lattice vectors $\mathbf{R}_{\alpha L}$. The matrix elements of the tight-binding Hamiltonian are

$$\langle \alpha', \mathbf{k}|H|\alpha, \mathbf{k}\rangle = \sum_L e^{i\mathbf{k} \cdot (\mathbf{R}_{\alpha'L} - \mathbf{R}_{\alpha})} t_{\alpha',\alpha}(\mathbf{R}_{\alpha',L} - \mathbf{R}_{\alpha}) + \epsilon_\alpha \delta_{\alpha',\alpha} \tag{78}$$

where $\mathbf{R}_\alpha = \mathbf{R}_{\alpha,0}$. Here $t_{\alpha',\alpha}$ and ϵ_α are the usual off-site and on-site matrix elements. Each eigenstate $|n, \mathbf{k}\rangle$ is a superposition of Bloch sums (77) with appropriate coefficients $C_\alpha(n\mathbf{k})$,

$$|n\mathbf{k}\rangle = \sum_\alpha C_\alpha(n\mathbf{k})|\alpha\mathbf{k}\rangle, \tag{79}$$

where n is the band index.

When $\mathbf{k} \cdot \mathbf{p}$ theory is adapted to the tight-binding form, an effective momentum operator $\mathbf{p}$ and a "kinetic energy" operator $\mathbf{T}$ can be defined (Graf and Vogl, 1995). In matrix form these operators are

$$\mathbf{p}_{n,n'}(\mathbf{k}) = \frac{m_o}{\hbar} \mathbf{C}^\dagger(n\mathbf{k}) \nabla_\mathbf{k} \mathbf{H}(\mathbf{k}) \mathbf{C}(n'\mathbf{k}) \tag{80}$$

$$\mathbf{T}_{n,n'}(\mathbf{k}) = \frac{m_o}{\hbar^2} \mathbf{C}^\dagger(n\mathbf{k}) \nabla_\mathbf{k} \nabla_\mathbf{k} \mathbf{H}(\mathbf{k}) \mathbf{C}(n'\mathbf{k}). \tag{81}$$

Here $\mathbf{H}(\mathbf{k})$ is the Hamiltonian matrix whose elements are defined in (78), and $\mathbf{C}_\alpha(n\mathbf{k})$ is the vector whose components are defined in (79).

Interaction with an electromagnetic field requires an appropriate modification of the Hamiltonian (78). The most efficient approach in a tight-binding picture is to use the Peierls substitution (Peierls, 1933), which has long been a useful tool for time-independent fields and has been generalized to the time-dependent case (Graf and Vogl, 1995). The familiar minimal coupling substitution $\mathbf{p} \to \mathbf{p} - e\mathbf{A}/c$, where e is the charge of the electron and $\mathbf{A}$ is the vector potential, is equivalent to the replacement

$$t_{\alpha',\alpha}(\mathbf{R}' - \mathbf{R}) = t^0_{\alpha',\alpha}(\mathbf{R}' - \mathbf{R}) \exp\left(-\frac{ie}{2\hbar c}(\mathbf{R}' - \mathbf{R}) \cdot \mathbf{A}(t)\right). \tag{82}$$

(To avoid confusion, we use the notation of Dumitrică et al. (1998) and Graf and Vogl (1995) throughout the present section.) The tight-binding expression for the current density operator **J** can be written in terms of the matrices defined in the preceding (Graf and Vogl, 1995):

$$\mathbf{J}_{n',n} = \frac{e}{m_o}\mathbf{p}_{n',n} + \frac{e^2}{m_o c}\mathbf{T}_{n',n}(\mathbf{k})\cdot\mathbf{A}(t). \tag{83}$$

2. Second-Order Susceptibility in a Tight-Binding Representation

In an intense field, the macroscopic current density contains contributions to all orders in the electric field. In particular, the second-order contribution will be related to the electric fields $E_\beta(\omega_1)$ and $E_\gamma(\omega_2)$ via the second-order conductivity tensor $\sigma^{(2)}_{\alpha\beta\gamma}(\omega_1, \omega_2)$:

$$J^{(2)}_\alpha(\omega_1, \omega_2) = \sigma^{(2)}_{\alpha\beta\gamma}(\omega_1, \omega_2)E_\beta(\omega_1)E_\gamma(\omega_2). \tag{84}$$

Here α, β, and γ represent Cartesian coordinates. The second-order susceptibility tensor is related to the conductivity tensor by

$$\chi^{(2)}_{\alpha\beta\gamma}(\omega_1, \omega_2) = \frac{i}{2\omega}\sigma^{(2)}_{\alpha\beta\gamma}(\omega_1, \omega_2). \tag{85}$$

For simplicity, we will limit the calculation to second-harmonic generation when the two frequencies ω_1, ω_2 are equal.

Standard response theory extended to second order in the interaction (see, for example, Callaway, 1974) involves the thermodynamic average of the current density operator:

$$\langle J_\alpha(t)\rangle = \langle J_\alpha(t)\rangle_0 + \frac{i}{c\hbar}\int_{-\infty}^t dt_1 \langle [\tilde{J}_\alpha(t), \tilde{J}_\beta(t_1)]\rangle_0 A_\beta(t_1)$$
$$+ \left(\frac{i}{c\hbar}\right)^2 \int_{-\infty}^t dt_1 \int_{-\infty}^{t_1} dt_2 \langle [[\tilde{J}_\alpha(t), \tilde{J}_\beta(t_1)], \tilde{J}_\gamma(t_2)]\rangle_0 A_\beta(t_1)A_\gamma(t_2). \tag{86}$$

Tildes are used to denote operators in the interaction picture, and $\langle\rangle_0$ indicates an average in the unperturbed state.

If the completeness relation satisfied by the eigenvectors $|n,\mathbf{k}\rangle$ is inserted on the right-hand side of (86), one obtains a product of current density matrices. We note that the current operators on the right-hand side of (86) result from the interaction Hamiltonian (Callaway, 1974)

$$H' = -\frac{1}{c} J_\alpha A_\alpha. \quad (87)$$

This expression is correct to only first order in the vector potential **A**; however, the term that is neglected (involving $\mathbf{A}^2$) does not give rise to electronic transitions in the long-wavelength approximation, since it can be eliminated through a unitary transformation (Schlicher et al., 1984; see Eqs. (2.17)–(2.20)). Only the last term in (86) gives a contribution that is second-order in the electric field, since the subscript "0" indicates that the current operator (86) is evaluated in the unperturbed system, with $\mathbf{A} = 0$. The second-order susceptibility tensor must be symmetric (Shen, 1968) in the last two Cartesian coordinates β and γ, so we permute the times t_1 and t_2 in (86). The resulting expression is

$$\chi^{(2)}_{\alpha\beta\gamma} = \frac{i}{2\hbar^2 \Omega}\left(\frac{e}{m_o}\right)^3 \sum_{m,n,l,\mathbf{k}} \frac{p^\alpha_{nm}[p^\beta_{ml} p^\gamma_{ln}]}{\omega^3(2\omega - \omega_{mn})}\left(\frac{f_{ln}}{\omega - \omega_{ln}} + \frac{f_{ml}}{\omega - \omega_{ml}}\right) \quad (88)$$

where Ω is the crystal volume and $[p^\beta_{nl} p^\gamma_{ln}]$ indicates a symmetrized form. Expression (88) is a general one, in the sense that it is not simplified by any symmetry of the material. As in the case of the linear dielectric tensor (Graf and Vogl, 1995), the tight-binding expression for the second-order susceptibility is similar to the classical one, but the matrix elements $\mathbf{p}_{nm}$ are given by (80). A general result of $\mathbf{k} \cdot \mathbf{p}$ theory is that the average of $\nabla_\mathbf{k} H(\mathbf{k})$ for any Bloch state (with $H(\mathbf{k})$ the $\mathbf{k} \cdot \mathbf{p}$ Hamiltonian) gives the average of the momentum operator $-i\hbar \nabla$. The preceding result is therefore not unexpected.

Invoking time-reversal symmetry, and adding an infinitesimal imaginary part to the frequency, $\omega + i\eta$ with $\eta \to 0$, one can separate the real and imaginary parts of $\chi^{(2)}(\omega)$; in particular,

$$\text{Im}[\chi^{(2)}_{\alpha\beta\gamma}] = \frac{i\pi}{2\hbar^2 \Omega}\left(\frac{e}{m_o \omega}\right)^3 \sum_{m,n,l,\mathbf{k}} p^\alpha_{nm}[p^\beta_{ml} p^\gamma_{ln}]$$

$$\times \left(f_{ln} \frac{\delta(2\omega - \omega_{mn}) - \delta(\omega - \omega_{ln})}{\omega - \omega_{ml}} + f_{ml} \frac{\delta(2\omega - \omega_{mn}) - \delta(\omega - \omega_{ml})}{\omega - \omega_{ln}} \right). \quad (89)$$

The apparent divergence at $\omega = 0$ can be cured in the same way as it will be in (94): One replaces ω^{-3} by the appropriate ω_{ij}^{-3} in each term of (89).

3. FORMULA FOR THE DIELECTRIC FUNCTION

Here we briefly review the corresponding ideas for the linear susceptibility ϵ. First, the matrix elements of the momentum operator $\mathbf{p}$ of (80) are given by

$$\mathbf{p}_{n,n'}(\mathbf{k}) = \frac{m_o}{\hbar} \sum_{\alpha',\alpha} C_{\alpha'}^* \sum_L i(\mathbf{R}_{\alpha',L} - \mathbf{R}_\alpha) \cdot e^{i\mathbf{k}\cdot(\mathbf{R}_{\alpha',L} - \mathbf{R}_\alpha)} t_{\alpha',\alpha} C_\alpha. \quad (90)$$

The macroscopic current density in linear response theory is connected to the conductivity through the relation

$$J_\alpha^{(1)}(\omega) = \sigma_{\alpha\beta}^{(1)}(\omega) E_\beta(\omega) \quad (91)$$

and $\sigma^{(1)}$ is related to the linear dielectric susceptibility via

$$\chi_{\alpha\beta}^{(1)}(\omega) = i\sigma_{\alpha\beta}(\omega)/\omega. \quad (92)$$

Using the first two terms of the thermodynamic average (86) of the current density operator $\langle J(t) \rangle$ up to first order in the field, one obtains

$$\operatorname{Im} \chi_{\alpha\beta}^{(1)}(\omega) = \frac{e^2 \pi}{\Omega \hbar m_o^2 \omega^2} \sum_{n,m,k} f_{nm}(k) p_{nm}^\alpha p_{nm}^\beta \delta(\omega - \omega_{mn}) \quad (93)$$

or in a nondivergent form that was utilized in our calculations

$$\operatorname{Im} \chi_{\alpha\beta}^{(1)}(\omega) = \frac{e^2 \pi}{\Omega \hbar m^2} \sum_{n,m,k} \frac{1}{\omega_{nm}^2} f_{nm}(k) \cdot p_{nm}^\alpha p_{nm}^\beta \delta(\omega - \omega_{mn}). \quad (94)$$

4. CALCULATION OF $\chi^{(2)}(\omega)$ FOR GaAs

A susceptibility function is calculated from the eigenvalues and eigenvectors at many points in the Brillouin zone. In addition, both the ground state and relevant excited states are important. For this reason we follow Vogl et al. (1983) in extending the minimal sp^3 orbital basis with an additional orbital s^*, which models the manifold of d-states and other higher-lying excited states that are omitted in a minimal basis.

Because of its noncentrosymmetric structure, it is appropriate to apply our formalism to GaAs. The analytic formula for $\chi^{(2)}(\omega)$ involves virtual electron processes (with n = valence-band state and m, l = conduction-band states) and virtual hole processes (with n, m = valence-band states and

l = conduction-band state). Aspnes (1972) showed that the latter type of contribution can be neglected. We therefore include only the virtual electron transitions. We also choose to evaluate the imaginary part (89), and then use the Kramers–Kronig relation to obtain the real part:

$$\text{Re}[\chi^{(2)}(\omega)] = \frac{2}{\pi} \mathbf{P} \int_0^\infty \frac{\omega'}{\omega'^2 - \omega^2} \text{Im}[\chi^{(2)}(\omega')] d\omega'. \tag{95}$$

The numerical integration employs an adaptation of the method used to calculate the linear dielectric function. Details of this method are the same as in Section IV (Graves, 1997; Graves and Allen, 1998a,b) and will not be repeated here. The results for the imaginary part of $\chi^{(2)}_{xyz}$ (the only independent component) are presented in Fig. 28. Our results compare well with those calculated using other methods (Moss et al., 1987; Hughes and Sipe, 1996; Ghahramani et al., 1991; Huang and Ching, 1992). The major features in the structure of our calculated curves clearly resemble those obtained in a first-principles calculation (Huang and Ching, 1992). There is also good agreement with experiment (Parsons and Chang, 1971; Bethune et al., 1975). In the data for $|\chi^{(2)}(\omega)|$, the first peak at 1.5 eV agrees with the present results (Dumitrică et al., 1998), and the deep minimum and the second peak at 2.3 eV appear to be only slightly shifted.

The results presented in Fig. 28 are in arbitrary units. As in the case of the linear dielectric function (Graf and Vogl, 1995), shown in Fig. 29 for both GaAs and Si, the overall scale is too low by about a factor of two.

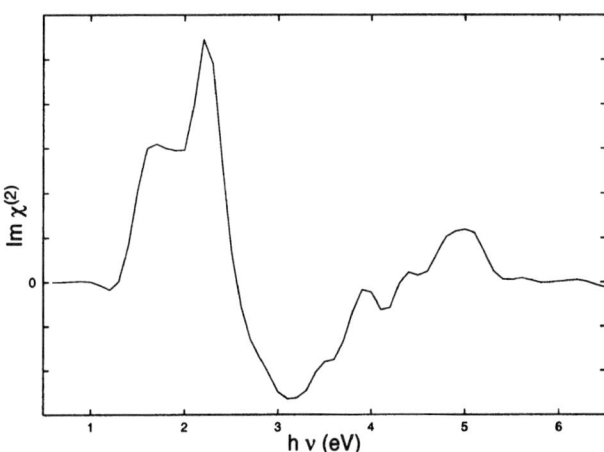

FIG. 28. Imaginary part of $\chi^{(2)}(\omega)$, calculated with the formula of Dumitrică et al. (1998) or (89).

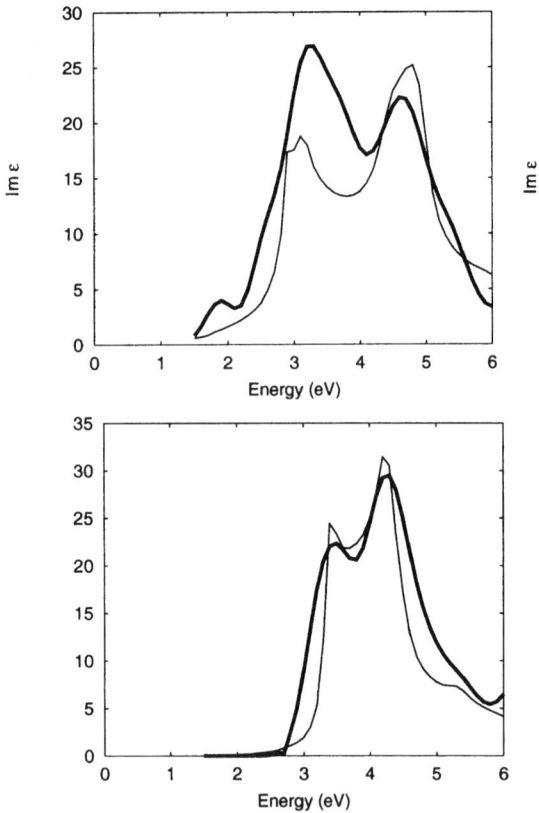

FIG. 29. Im $\epsilon(\omega)$ calculated using the Peierls substitution (heavy curves), compared with the measurements of Aspnes and Studna (1983) (light curves). Upper panel: GaAs. Lower panel: Si. After Graves and Allen (1998b).

However, the energy dependence of $\chi^{(2)}$ is correct, as noted earlier. Another check is provided by a set of sum rules (Scandolo and Bassani, 1995). Those that weight the higher frequencies heavily (by some power of ω) are not well satisfied by the present model, which is only valid for excitations with energies up to a few electron volts. On the other hand, the sum rule

$$\int_0^\infty \mathrm{Re}[\chi^{(2)}(\omega)]d\omega = 0 \qquad (96)$$

is a valid test. Our numerical results give

$$\int_0^\infty \mathrm{Re}[\chi^{(2)}(\omega)]d\omega = 0.35, \qquad (97)$$

which is quite satisfactory for a function that varies over a range of about -20 to $+10$ (Dumitrică et al., 1998).

To summarize this section, we have obtained an expression for the second-order nonlinear susceptibility $\chi^{(2)}(\omega)$ that can be employed with a tight-binding Hamiltonian. This expression has been tested for GaAs, and the results agree with previous calculations and experiment.

VI. Response of Si to Fast, Intense Pulses

In the foregoing, we discussed simulations of the response of GaAs (the most important compound semiconductor) to ultraintense and ultrashort laser pulses. Here we describe precisely the same kind of simulations for Si (the most important elemental semiconductor).

The results are shown in Figs. 31–35. For the four lowest intensities the original lattice is preserved, but there is a threshold at about $A_0 = 2.0$ G cm for lattice destabilization with large atomic excursions from the initial positions. (As in Graves and Allen, 1998b, these initial positions correspond to thermal equilibrium at 300 K, after an equilibration run of 2000 fs.) The corresponding levels of excitation are shown in Fig. 32. Comparison with Fig. 17 shows that a laser pulse of given intensity excites a considerably larger fraction of the valence electrons in Si than in GaAs. The reason for this appears to be the difference in the bandgaps: 1.1 eV for Si versus 1.4–1.5 eV for GaAs (over the temperature range 0–300 K). The gap involving the high-density p-derived states is still larger in the case of GaAs. As a result, the joint density of states for $\hbar\omega \approx 2$ eV is considerably larger for Si.

Figure 30 shows the results when some fraction of the electrons are artificially promoted from the top of the valence band to the bottom of the conduction band. If 25% of the electrons are promoted, the atoms are observed to move far from their equilibrium positions in the original tetrahedral structure, and the lattice has certainly been destabilized.

Figure 31, on the other hand, shows the atomic motion in a true TED simulation. For each intensity the FWHM pulse duration was 70 fs, with $\hbar\omega = 1.95$ eV, and a polarization in the (1.7, 1.0, 0) direction referenced to the cube edges (Graves, 1997).

Recall that an amplitude of $A_0 = 1.00$ G cm corresponds to a fluence of 0.815 kJ/m^2. The threshold for the structural transformation is about 2.00 G cm, or 3.26 kJ/m^2. This is several times large as the experimental threshold for Si (Sokolowski-Tinten et al., 1995). As mentioned earlier for GaAs, since the present theory yields a dielectric function that is roughly half that observed experimentally (Graf and Vogl, 1995), one expects the nonlinear response to also be underestimated, so this level of agreement is quite satisfactory.

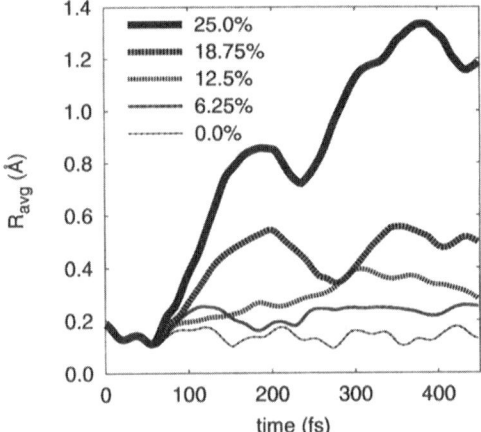

FIG. 30. Excited-state molecular dynamics simulations for Si. The fraction of valence electrons promoted to the conduction bands is indicated for each run. This is the kind of simulation that was done prior to the invention of tight-binding electron–ion dynamics: The interaction with the radiation field was not treated, and electrons were merely placed in excited states.

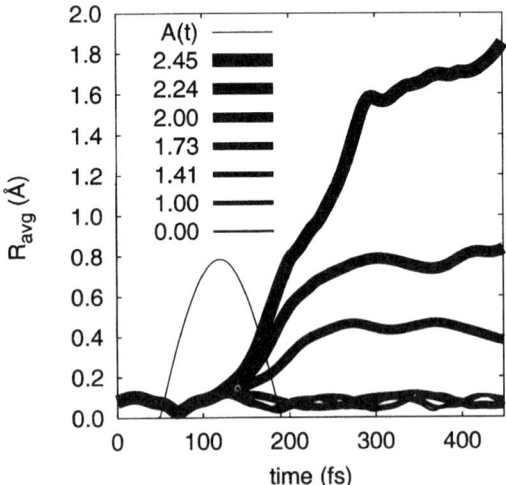

FIG. 31. Average distance moved by an atom, during and following a laser pulse. The amplitude A_0 of the vector potential is given in gauss centimeters.

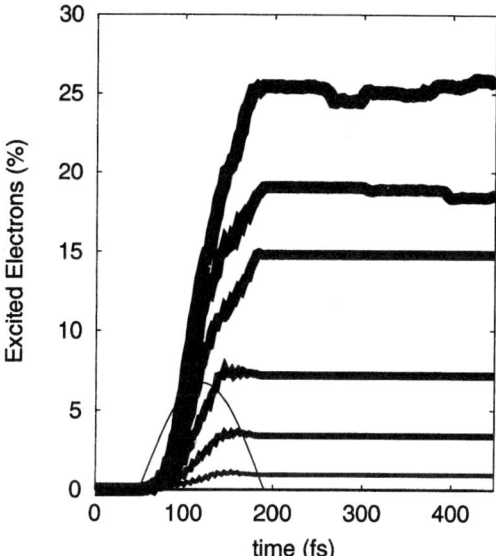

FIG. 32. The percentage of valence electrons promoted to excited states in Si is shown as a function of time for varying pulse intensities. The pulse is represented by a solid curve. The amplitudes A_0 are given in Fig. 31.

The eigenvalues $\epsilon_n(\mathbf{k})$ of (72) at the special point $\mathbf{k} = (\frac{1}{4}, \frac{1}{4}, \frac{1}{4})(2\pi/a)$ are plotted as functions of time in Fig. 34, for two different intensities. Notice that the bandgap at this point (which is larger than the fundamental bandgap at $(0, 0, 0)$) exhibits only thermal oscillations for $A_0 = 1.00$ G cm, but has completely closed up for $A_0 = 2.45$ G cm because of the large atomic displacements associated with lattice destabilization.

The occupancy for the kth state is given by (73). The total occupancy of all the conduction bands (again at the special point) is plotted as a function of time in Fig. 32, where it is expressed as a percentage of the total number of valence electrons. Comparison with Fig. 31 shows that the threshold for permanent structural change corresponds to excitation of about 15% of the valence electrons.

In Fig. 33, the pair correlation function is plotted as a function of time. The structural order remains intact for $A_0 = 1.00$ G cm, but is lost after about 2.00 fs for $A_0 = 2.00$ G cm, confirming that this higher intensity leads to permanent structural change.

For Si, the most direct comparison with experiment is again provided by the imaginary part of the dielectric function, calculated exactly as for GaAs. The panels of Fig. 35 show Im ϵ for 0.5 eV $\leqslant \hbar \omega \leqslant 6.0$ eV and for six different intensities, ranging from $A_0 = 1.73$ to 2.45 G cm. (The corresponding fluences range up to 6.5 kJ/m^2.) At low intensities there is no absorption

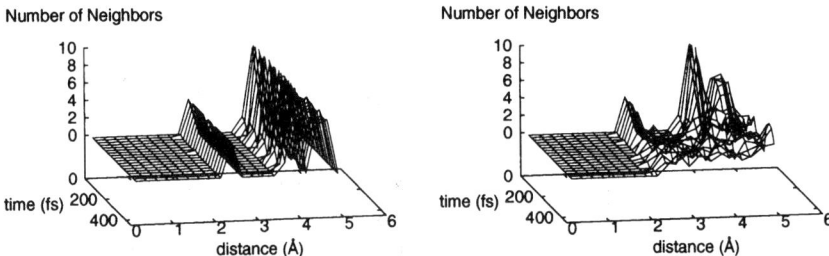

FIG. 33. Time evolution of the pair correlation function for a field strength of $A_0 = 1.00$ G cm (left) and 2.45 G cm (right).

for $\hbar\omega$ less than the bandgap of 1.5 eV (i.e., Im $\epsilon(\omega)$ is zero in this range) and the structural features in Im $\epsilon(\omega)$ persist at all time. At high intensities, one can observe metallic behavior (with subbandgap absorption) and the structural features are washed out.

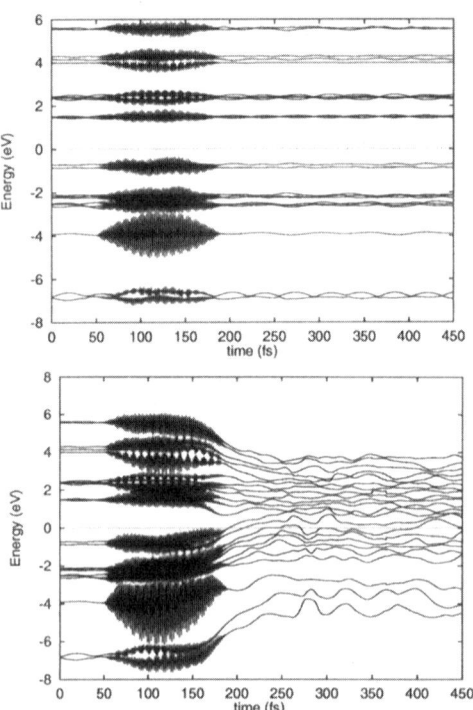

FIG. 34. Electronic energy eigenvalues for Si at the $(2\pi/a)(\frac{1}{4}, \frac{1}{4}, \frac{1}{4})$ point as a function of time, with $A_0 = 1.00$ G cm (top) and 2.45 G cm (bottom).

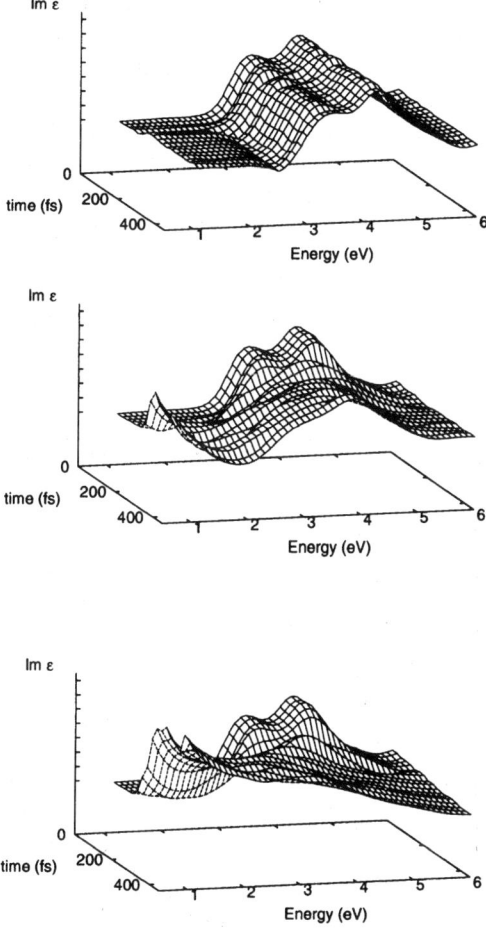

FIG. 35. Time evolution of the dielectric function of Si for four intensities: $A_0 = 1.73$, 2.00, 2.24, and 2.45 G cm, from top to bottom. Im $\epsilon(\omega)$ is shown for a time interval of 450 fs, with the pulse applied between 50 and 190 fs. These and all the other results of this section were obtained in the 8-atom simulations of Graves and Allen (1998b).

The Essen group has performed detailed studies for Si (Sokolowski-Tinten et al., 1995), and their measurements are represented by Figs. 36 and 37. Earlier studies were reported by Shank et al. (1983), and their results are displayed in Figs. 38 and 39.

The results of the simulations are again in good general agreement with the experiments. In addition, however, the simulations nicely complement the experiments, by providing a detailed microscopic understanding of the behavior of both electrons and ions. As can be seen in Fig. 32, the density of excited electrons increases with the intensity (or fluence) of the pulse; for $A_0 = 2.0$ G cm, about 15% of the valence electrons are excited into the conduction bands. Figure 31 shows that this represents the approximate threshold for atoms to perform large excursions from their initial positions,

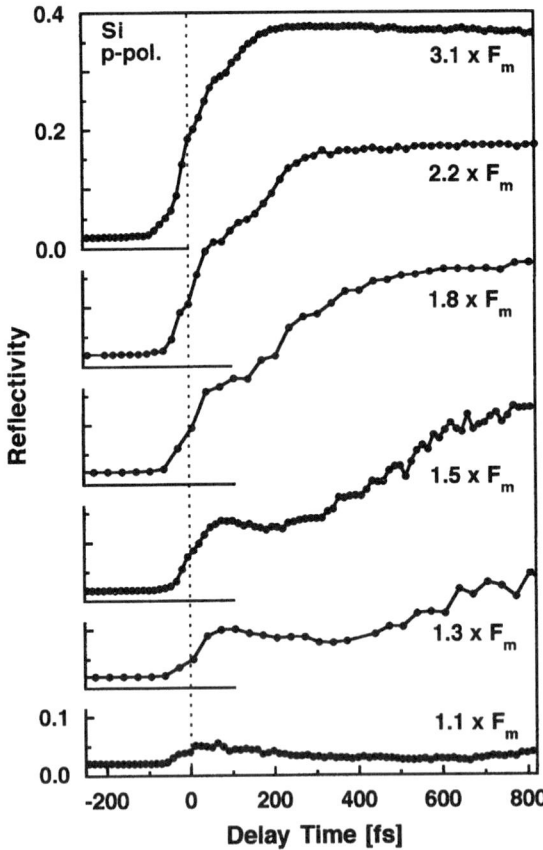

FIG. 36. p Polarized reflectivity of silicon as a function of delay time for various pump fluences, which are normalized to $F_m = 0.17$ kJ/m². After Sokolowski-Tinten et al. (1995). Reprinted with permission.

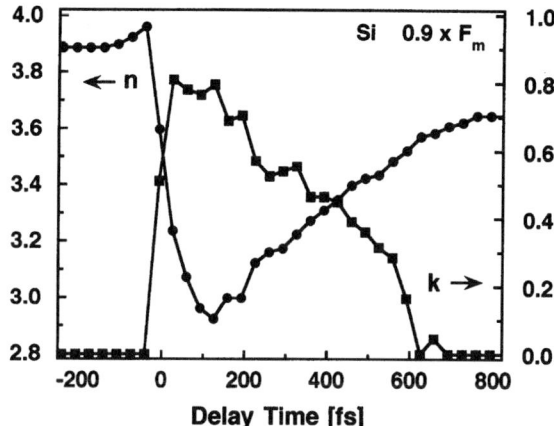

FIG. 37. Optical constants of silicon as a function of time, derived from the reflectivity data for $F = 0.9 F_m$. For unperturbed crystalline silicon, $n = 3.885$ and $k = 0.017$. Once again, compare the time scales in the experiments and simulations. After Sokolowski-Tinten *et al.* (1995). Reprinted with permission.

corresponding to permanent structural change. The fact that there is a structural transformation is confirmed by calculation of the pair correlation function, shown in Fig. 33.

Accompanying the structural transformation at higher intensities, there is a collapse of the bandgap, which can be clearly seen in Fig. 34 at the single

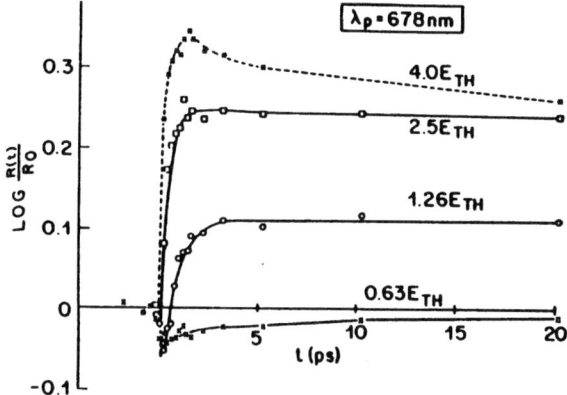

FIG. 38. Early measurements of the reflectivity of Si, indicating a transition from semiconducting to metallic behavior for this material after a 90 fs, 620 nm (or 2.0 eV) laser pulse. After Shank *et al.* (1983). Reprinted with permission.

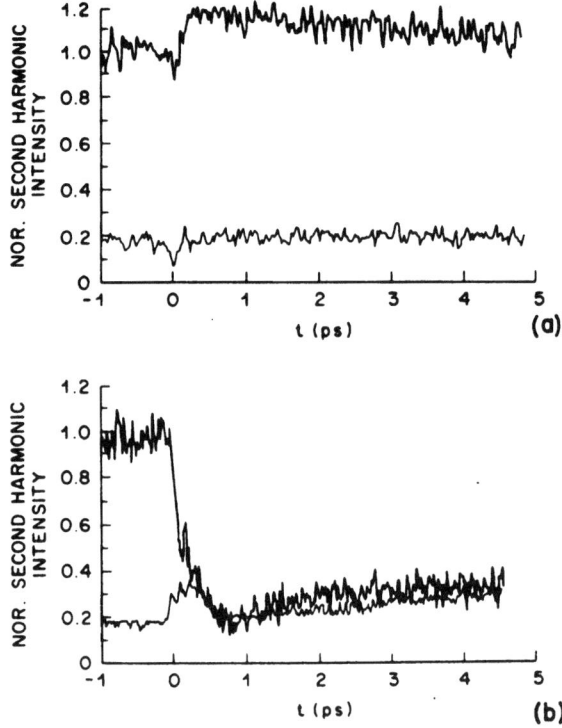

FIG. 39. Early measurements of the second-harmonic intensity of a ⟨111⟩ Si surface, indicating a change of symmetry after a 90 fs, 2.0 eV pulse. After Shank *et al.* (1983). Reprinted with permission.

k-point $(\frac{1}{4}, \frac{1}{4}, \frac{1}{4})(2\pi/a)$. A better measure of the onset of metallic behavior is provided by the dielectric function $\epsilon(\omega)$. As can be seen in Fig. 35, Im $\epsilon(\omega)$ becomes nonzero within the bandgap (i.e., for $\hbar\omega < 1.5$ eV) when A_0 exceeds the threshold value of about 2.0 G cm.

The simulations predict that the threshold fluence is about the same for Si and GaAs, in agreement with the experiments (Sokolowski-Tinten *et al.*, 1995). This is not a trivial conclusion, because Si has stronger bonding, and the lattice destabilization in Si appears to require that a larger number of electrons be excited.

VII. Density Functional Simulation for Si

Our group has also performed first-principles self-consistent simulations of the response of Si to ultrashort and ultraintense laser pulses (Gao, 1998). In addition to the usual complications of a static calculation, with a nonlocal pseudopotential, self-consistent solution of the Kohn–Sham equation, etc., we included (1) the motion of the ions as determined by Hellmann–Feynman forces, (2) the dynamics of the electrons as determined by an effective time-dependent Schrödinger equation, and (3) the coupling of the electrons to the radiation field through the time-dependent vector potential. There are numerous subtleties that are discussed in Gao (1998), along with other details of the rather involved calculations. Because of computational limitations, however, the simulations employed only two atoms per unit cell.

The principal conclusion of this work is that a laser pulse with a duration of 70 fs and an intensity of a few terawatts causes a substantial number of electrons to be excited from the bonding valence bands of silicon to the antibonding conduction bands, and this in turn causes repulsive Hellmann–Feynman forces that lead to large displacements of the atoms from their initial equilibrium positions, on a time scale of a few hundred femtoseconds. For an intensity of 4.0×10^{12} W/cm^2, we found that about 20% of the valence electrons were excited across the bandgap, and that the resulting destabilization of the bonding led to large ion displacements. As a consequence, the bandgap collapsed to zero, yielding a metal-like rather than semiconducting system, about 200–300 fs after the beginning of the pulse. The band structure at 200 fs is shown in Fig. 40, together with the fraction of electrons excited to the conduction bands as a function of time.

On the other hand, a pulse with an intensity of 4.0×10^{11} W/cm^2 excited only 2–3% of the electrons across the bandgap, produced much smaller ion displacements, and had no significant effect on the bandgap. These and many other results are shown in Gao (1998).

All these results of the simulations are in qualitative agreement with the experiments. However, there are clearly two problems with a density-functional calculation: First, the bandgap is unphysically small (roughly half the correct value in the case of Si). Second, the enormous complexity of the calculations (together with a required time step of 50 attoseconds or less) places severe restrictions on the size of the system that can be treated.

For these two reasons, we conclude that tight-binding electron–ion dynamics (TED) is the preferred method for treating the coupled dynamics of electrons and ions in a material.

VIII. Response of C_{60} to Ultrafast Pulses of Low, High, and Very High Intensity

During the past few years, vibrational excitation (Dexheimer *et al.*, 1993; Fleischer *et al.*, 1997) and photofragmentation (Hunsche *et al.*, 1996; Hansen and Echt, 1996; Hohmann *et al.*, 1994) have been observed for C_{60}

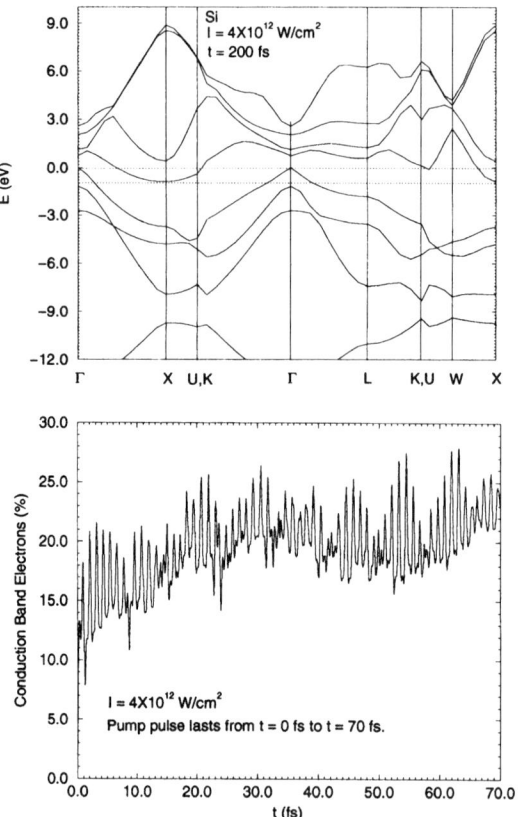

FIG. 40. Perturbation of band structure (top) and excitation of electrons (bottom) in density-functional calculation for Si. After Gao (1998). Reprinted with permission.

molecules subjected to laser pulses with durations as short as 12 fs. Motivated by these experiments, and by potential applications to the photochemistry of fullerenes, we have performed simulations at both moderate and high fluences. We have also studied collisions of atoms with C_{60} molecules following such ultrashort and ultraintense pulses, to consider the possibility of laser-promoted encapsulation. Here, however, we focus primarily on vibrational excitation.

Details of the calculations are given elsewhere (Torralva and Allen, 1998; Torralva et al., 1999a,b). Here we mention only that the technique is TED, that the tight-binding model of Xu et al. (1992) was used in the calculations of Figs. 41 and 42, and that a density-functional-based approach (Porezag et al., 1995) was employed in the calculations of Figs. 43–45.

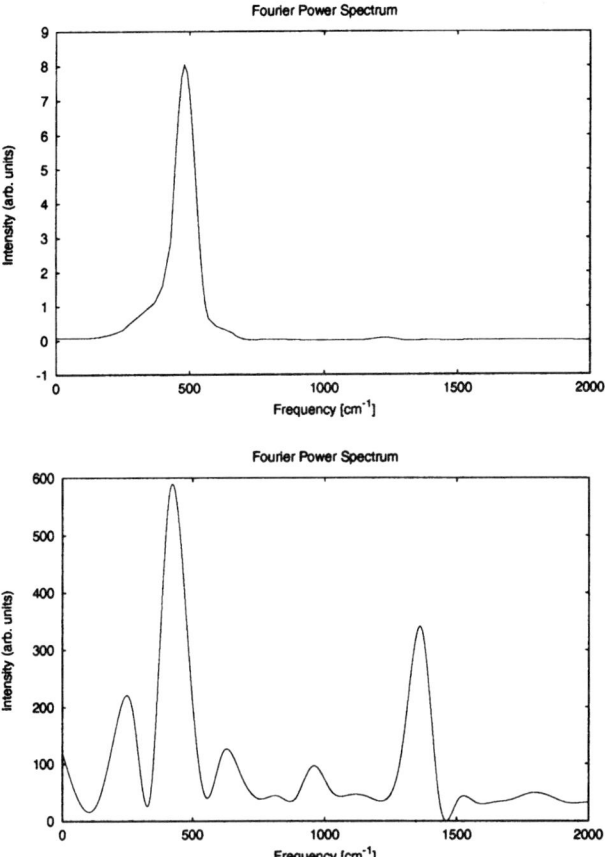

FIG. 41. Fourier power spectrum for a C_{60} molecule responding to a 12 fs FWHM, 2.0 eV laser pulse with a fluence of 0.07 kJ/m² at $T = 0$ (top) and 1.10 kJ/m² at $T = 300$ K (bottom). The results of this figure and Fig. 42 were obtained with the tight-binding model of Xu et al. (1992) and were reported in Torralva and Allen (1998).

Results for several different fluences at two different temperatures are shown in Figs. 41 and 42. The first simulation, in the top panel of Fig. 41, is for a fluence of 0.07 kJ/m². The temperature T was chosen to be 0 K, so that the effect of this lower fluence laser pulse would not be intermingled with the effects of thermal vibrations. One can see that there is a sharp peak near 500 cm^{-1}, a frequency that is in very good agreement with the experimental value for the breathing mode of C_{60}. Furthermore, a computer animation of the motion in real space and time shows that the response of the molecule to the laser pulse is indeed a symmetric breathing motion.

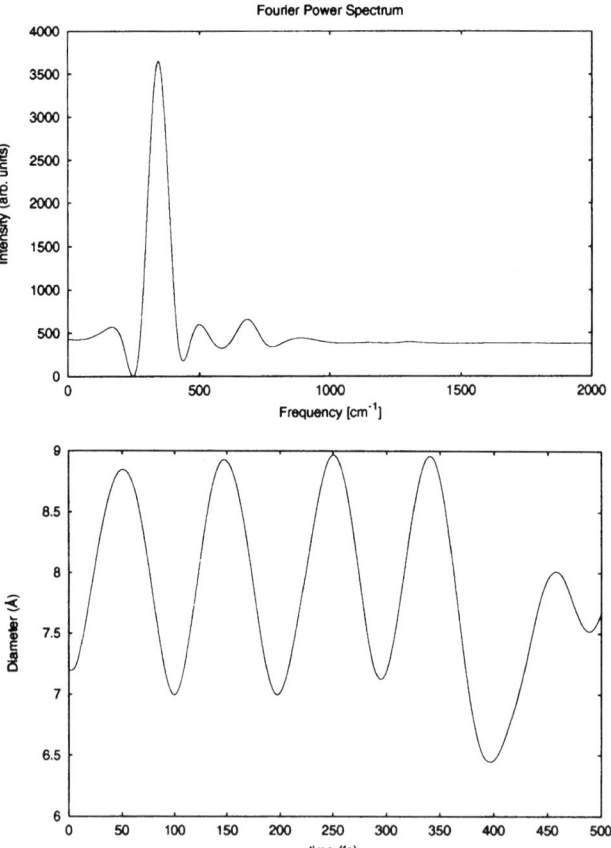

FIG. 42. Spectrum for C_{60} responding to a 12 fs, 2.0 eV pulse with a fluence of 4.40 kJ/m² and $T = 300$ K (top). Diameter of C_{60} molecule versus time following this same pulse (bottom). After Torralva and Allen (1998).

Analysis of the results shows that the electrons undergo a symmetric promotion from bonding to antibonding molecular states. It is the consequent bond softening that then leads to vibrational excitation.

The bottom panel of Fig. 41 shows the response at 300 K to a more intense pulse. Although the spectrum is complicated by thermal vibrations, the most prominent peak is again associated with the breathing mode. A comparison with the power spectrum before the pulse (not shown here) demonstrates that the dominant new strength comes from this mode.

The top panel of Fig. 42 shows the response to a still more intense pulse. Despite the thermal vibrations, it is clear that it is again the breathing mode that is primarily excited. Notice, however, that the frequency is shifted

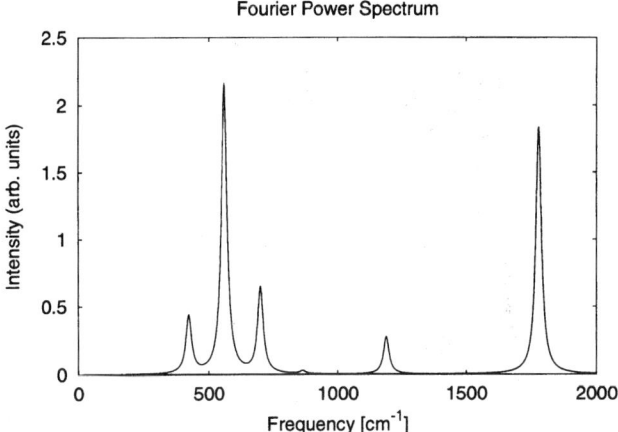

FIG. 43. Spectrum for C_{60} responding to a laser pulse. The FWHM duration of this pulse and that of the following figure was 5 fs, the photon energy was 2.0 eV, and $T = 0$ initially. In the present figure, the fluence was 0.005 kJ/m^2. The results of this figure and of Figs. 44 and 45 were obtained using a density-functional-based approach (Torralva et al., to be published). Notice that various optically active vibrational modes are excited in the nearly linear regime coresponding to this low fluence.

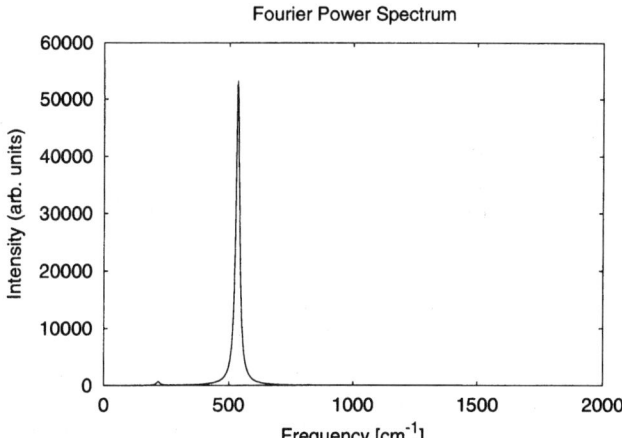

FIG. 44. Spectrum for C_{60} following a 5 fs, 2.0 eV pulse with a fluence of 0.25 kJ/m^2. Notice that the breathing mode is completely dominant, to the point that the other modes cannot even be seen in the figure, although a detailed examination of the output shows that they are excited with nonzero amplitude. Compare the results of this simulation with the experimental results of Figs. 46 and 47. After Torralva et al., to be published.

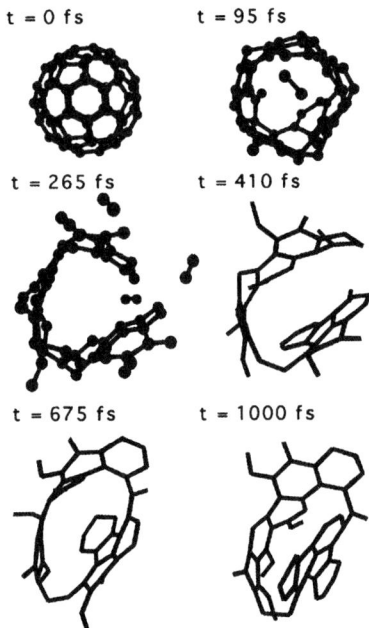

FIG. 45. Fragmentation of C_{60} following a 30 fs FWHM pulse with a fluence of 1.8 kJ/m². Notice that the carbon atoms tend to detach in the form of dimers, a result that is consistent with experiment (Hursche *et al.*, 1996; Hansen and Echt, 1996; Hohmann *et al.*, 1994). This result, together with the lower value of the fluence required for fragmentation, indicate that the density-functional-based approach used in this simulation is more quantitative than a simple orthogonal tight-binding model. After Torralva *et al.*, to be published.

because of anharmonic effects, which result from the large amplitude of the vibrations. It is also clear that the response of the electrons is quite nonlinear. That is, the electronic excitations involve multiphoton processes.

The bottom panel of Fig. 42 shows the remarkably large amplitude vibrations whose Fourier spectrum is given in the top panel. The unusual behavior beyond 400 fs is a precursor of fragmentation.

The results of Figs. 43–45 were obtained using the density-functional-based appraoch of Porezag *et al.* (1995), and the information needed to reproduce the calculations can be found in this paper. As discussed elsewhere (Torralva *et al.*) the rather precise agreement with experiment indicates that this is a promising approach to quantitative simulations for various chemical systems. For example, the vibrational modes are well predicted, and one obtains the quite nontrivial result that the carbon atoms during photofragmentation tend to detach as dimers. Furthermore, this same model was used in the simulation for butadiene shown in Section X, and another very nontrivial result was obtained: The molecule straightens

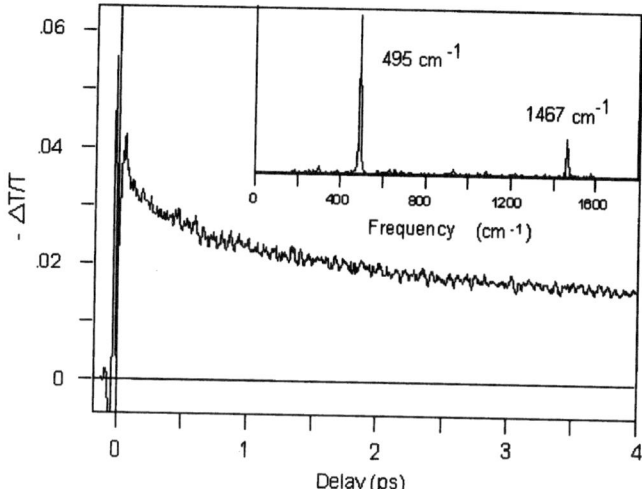

FIG. 46. Excitation of the breathing mode in C_{60}, and the pentagonal pinch mode at higher frequency, following an ultrafast laser pulse. The figure shows the light-induced negative differential transmittance $(-\Delta T/T)$ of a C_{60} thin film detected at 580 nm as a function of time delay between 12 fs pump and probe pulses centered on 620 nm. The Fourier power spectrum of the oscillatory part of the response is shown in the inset. After Dexheimer et al. (1993). Reprinted with permission.

out when it absorbs light, in a conformational change that is reminiscent of that exhibited by retinal molecules in the primary photochemical process responsible for vision.

There have been two experimental studies of the response of C_{60} molecules to ultrafast laser pulses, one using transmittance measurements for C_{60} films (Dexheimer et al., 1993) and the other using reflectance measurements for K_3C_{60} (Fleischer et al., 1997). The results are shown in Figs. 46 and 47. Since the principal modes observed are intramolecular modes of the molecule itself, it is appropriate to compare these experiments with the simulations for an isolated molecule. As can be seen from a comparison of the measurements with either the lower-fluence tight-binding simulation of Fig. 41 or the density-functional-based simulation of Fig. 44, there is truly excellent agreement between the simulations and the measurements.

IX. The Simplest System, H_2^+

We have employed a method that is equivalent to TED to treat the coupled dynamics of electrons and nuclei in a simple molecule that is subjected to electromagnetic radiation (Khosravi, 1997; Khosravi and Allen, 1998). The method leads to quantitatively accurate simulations for H_2^+.

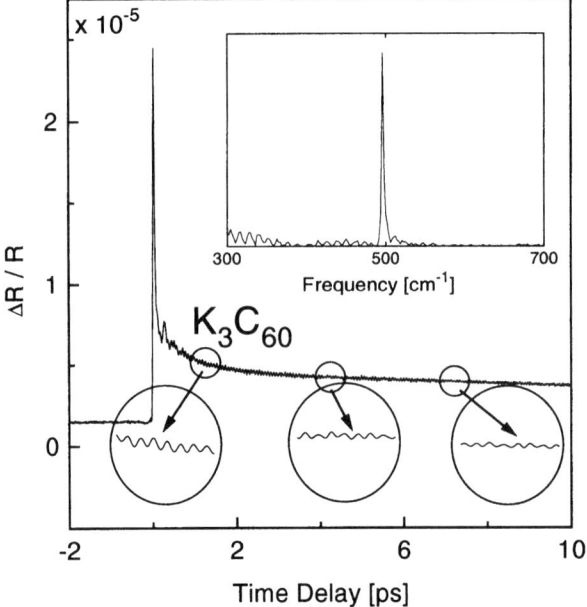

FIG. 47. Excitation of the breathing mode in C_{60} following an ultrafast 800 nm laser pulse. The figure demonstrates coherent phonon oscillations in K_3C_{60} at 300 K. The pump-probe data were taken in reflectivity with a single wavelength pump-probe setup having a time resolution of about 20 fs. The three circular insets show the minute oscillations superimposed on the decay. The larger inset shows the Fourier transform power spectrum with a sharp peak at 492.5 ± 0.25 cm^{-1} obtained for K_3C_{60}. After Fleischer et al. (1997). Reprinted with permission.

H_2^+ is the simplest molecule in nature, but it exhibits remarkably rich behavior when subjected to an ultrashort laser pulse. It is, in fact, an excellent test system for understanding both the standard and the more exotic processes that transpire when a molecule interacts with an intense and time-dependent radiation field. In our detailed simulations, we have observed many interesting phenomena, including the following (Khosravi, 1997; Khosravi and Allen, 1998).

- Photodissociation. The promotion of an electron from a bonding state to an antibonding state can produce a repulsive interaction that leads to dissociation.
- Bond softening. A very intense laser field deforms each adiabatic potential curve in the vicinity of a multiphoton crossing, causing the lower curve to be flattened or "softened." This effect can produce a dramatic lowering of the barrier to dissociation.

- Above-threshold dissociation. The nonlinear effects associated with intense radiation can permit dissociation via various paths: for example, absorption of a single photon, absorption of three photons, or absorption of three photons followed by emission of one photon. The products will thus emerge with a set of different kinetic energies.
- Ion-population trapping. The radiation field can create an adiabatic potential well that is deep enough to trap the ions.
- Electron-population trapping. An intense laser field can hold an electron in an excited state.
- Sudden electronic transitions. These are observed at multiphoton avoided crossings. Even though two dressed-state levels do not cross, there is an exchange of character when they come near each other. As the electron adiabatically follows one curve, it is observed to suddenly undergo either an upward or a downward transition, corresponding to the absorption or emission of one or more photons.
- Rabi flopping. In an intense field, an electron can rapidly oscillate back and forth between two accessible states.
- Harmonic generation. Photons are radiated at frequencies that are multiples of the fundamental laser frequency.

Because of space limitations, we do not show all of the above effects here, and further details are given elsewhere (Khosravi, 1997). The main idea, as before, is to solve the time-dependent Schrödinger equation (for the electrons in the molecule), together with Newton's equation of motion (for the atomic nuclei). Suppose for simplicity that there is only one electron (as there is in the case considered here). One then has the following equations for the coupled dynamics of this electron and the nuclei:

$$M \frac{d^2}{dt^2} X = -\sum_n |c_n(t)|^2 \frac{\partial \epsilon_n}{\partial X} - \sum_{mn} c_m^*(t) c_n(t) \exp(i\epsilon_{mn} t) \frac{\partial}{\partial X} \langle m|V_{int}(t)|n \rangle$$

$$- \frac{\partial V_{ii}}{\partial X} - \frac{\partial V_{ext}(t)}{\partial X} \quad (98)$$

$$i \frac{d}{dt} c_m(t) = \sum_m \langle m|V_{int}(t)|n \rangle \exp(i\epsilon_{mn} t) c_n(t). \quad (99)$$

Here, $\epsilon_{mn} = \epsilon_m - \epsilon_n$ and

$$H_{el}|n\rangle = \epsilon_n |n\rangle, \quad |\Psi\rangle = \sum_n a_n(t)|n\rangle, \quad a_n(t) = c_n(t) \exp(-i\epsilon_n t). \quad (100)$$

Also, X is any nuclear coordinate. The effective classical Hamiltonian for the molecule has the form

$$H = \sum_j \langle \Psi(t)|[H_{el} + V_{int}(t)]|\Psi(t)\rangle + V_{ii} + \sum_\ell P_\ell^2/2M_\ell + V_{ext}(t) \quad (101)$$

where H_{el} is the one-electron Hamiltonian with no field, V_{int} is the electron–field interaction, V_{ii} is the ion–ion repulsion, V_{ext} is the nuclei–field interaction, and the nuclei are labeled by ℓ.

For the present problem of an H_2^+ molecule, detailed considerations (Khosravi, 1997) give the following results for the energies of the lowest electronic states, and for their interaction with a field of intensity I and frequency ω:

$$\epsilon_1(R) = 0.1025\{\exp[-1.44(R-2)] - 2.00\exp[-0.72(R-2)]\} \quad (102)$$

$$\epsilon_2(R) = 0.1025\{\exp[-1.44(R-2)] - 2.22\exp[-0.72(R-2)]\}, \quad (103)$$

$$\langle 1|V_{int}(t)|2\rangle = \gamma \cos(\omega t) \quad (104)$$

$$\gamma = 5.3416 \times 10^{-9}\sqrt{I(\text{W/cm}^2)}\mu \quad (105)$$

$$\mu = \langle 1|z|2\rangle = -8.93 + 10\exp[0.0396(R-2)]. \quad (106)$$

All these quantities are given in atomic units, with 0.529 Å as the unit of length and 1 hartree as the unit of energy. Under the usual experimental conditions, only these two states (the bonding and antibonding states derived from the 1s atomic orbitals) are important, and the only important nuclear degree of freedom is the internuclear separation R.

Figures 48–50 show the results for a simulation in which there was one-photon absorption leading to dissociation. The initial internuclear separation was 1.3 Å, the initial total energy was 1.95 eV, the photon energy is 2.33 eV, the peak intensity was 1.0×10^{13} W/cm², and the pulse duration was 100 fs. At about 20 fs, the electron becomes partially excited, as can be seen in Fig. 48. There are then rather erratic oscillations in the amplitude $a_1(t)$, or probability $|a_1(t)|^2 = 1 - |a_2(t)|^2$, until it becomes dominantly in the excited state $|2\rangle$ at about 80 fs. Since this is an antibonding state, the molecule then dissociates, after executing three vibrations in Fig. 49. The difference between final and initial energies is close to the photon energy of 2.33 eV, as one can see in Fig. 50.

Notice that there is a richness of detail that is missing in conventional treatments based on approximations such as Fermi's Golden Rule. It is also interesting that the usual result of one-photon absorption emerges automatically, even though the radiation field is treated semiclassically and without the usual approximations. In other simulations, we have observed multiphoton dissociation (Khosravi, 1997).

7 MATERIAL RESPONSE TO FAST, INTENSE LASER PULSES   381

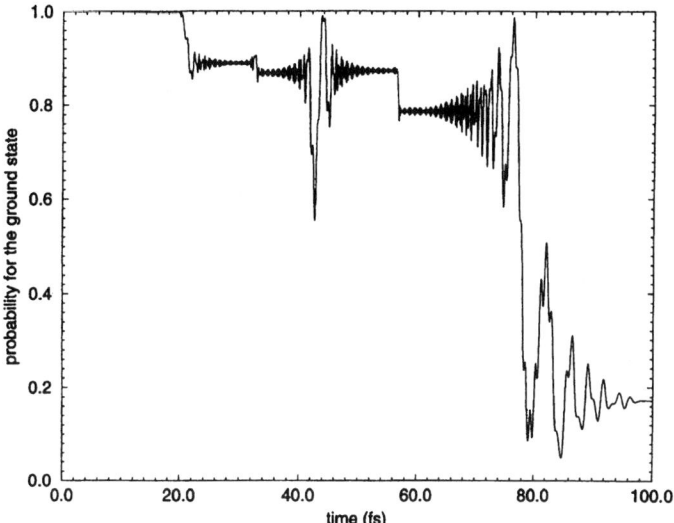

FIG. 48. Square of the amplitude for the ground state as a function of time, when an H_2^+ molecule is subjected to an ultrafast laser pulse that is specified in the text.

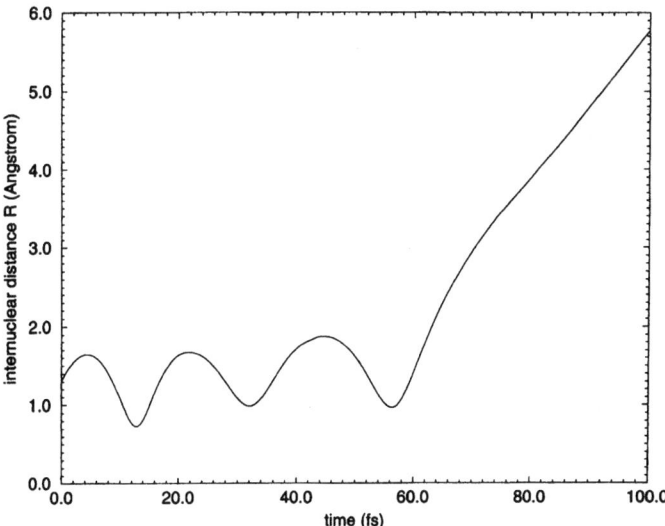

FIG. 49. Internuclear separation as a function of time, demonstrating one-photon absorption, in the simulation of Fig. 48.

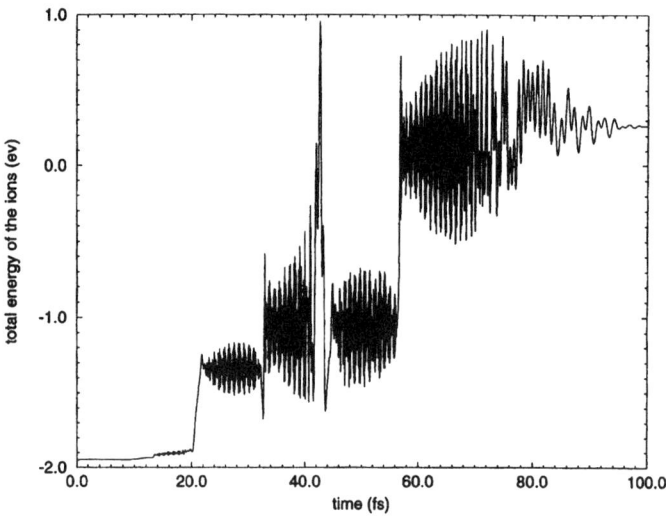

FIG. 50. Total energy as a function of time for the simulation of Figs. 48 and 49.

Very unusual behavior has been seen in many simulations. For example, Fig. 51 provides a particularly dramatic example of ion-population trapping: The nuclei exhibit an extremely large excursion from their equilibrium distance, but do not dissociate. Instead, they are trapped by the effective potential energy barrier associated with the adiabatic curves for the dressed states. In this simulation, the molecule was assumed to be prepared in an initial state with an internuclear distance of 2.3 Å and an energy of -0.1 eV. The intensity and photon energy were again 1.0×10^{13} W/cm^2 and 2.33 eV, and the pulse duration was 80 fs.

In short, using an approach similar to tight-binding electron–ion dynamics, we have observed many interesting phenomena for the H_2^+ molecule interacting with ultrafast and ultraintense laser pulses. Two of these are shown here: one-photon dissociation and ion-population trapping.

X. Biological Molecules

It is of considerable interest that one can also use TED to treat the response of molecules to light, including molecules chemisorbed on surfaces and, in a very different context, biological molecules.

For example, we have studied the response of a single chlorophyll molecule (Hamilton, 1999), and a few results are shown in Fig. 52. Many more results, and the numerical techniques, are provided in Hamilton (1999).

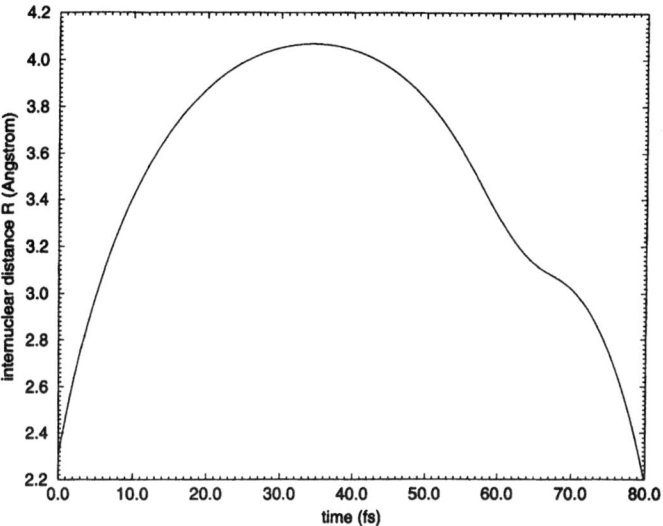

FIG. 51. Internuclear separation as a function of time, demonstrating ion-population trapping in an intense field.

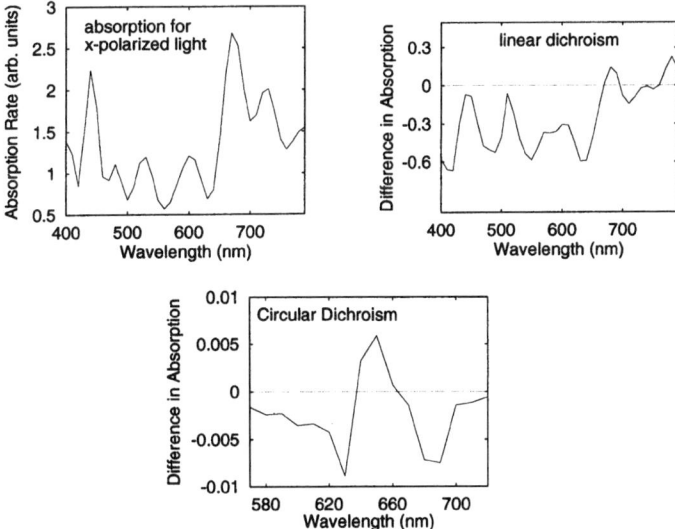

FIG. 52. Simulations of the absorption of light by chlorophyll using tight-binding electron–ion dynamics. The upper left panel shows the absorption as a function of wavelength for light polarized along the x axis (as defined in Hamilton, 1999). There is a principal peak in the red end of the spectrum. The upper right panel shows the difference in absorption for light linearly polarized in the x and y directions. The bottom panel shows the difference in absorption for light that is left- and right-circularly polarized.

Over a range of wavelengths, the absorption was calculated for unpolarized, linearly polarized, and circularly polarized light. The findings are consistent with previous experiments, although detailed comparisons are not possible because the experiments involve chlorophyll molecules in more complicated environments. For unpolarized light, the absorption peaks in the red part of the visible spectrum (as it does in many organisms employing photosynthesis, causing them to be green by reflected light). There is a secondary shoulder in the blue. For linearly polarized light, the absorption depends on wavelength and the direction of polarization. This can be understood as arising from the joint density of states for transitions at each photon energy, together with matrix-element effects (both of which are included in the present formulation). For circular polarization, the dichroism as a function of wavelength is slightly more subtle, but again can be understood in terms of matrix elements for the states involved in a transition at a given photon energy. We also found that an "effective helicity" of the molecule is useful in understanding the circular dichroism (Hamilton, 1999).

One of the great strengths of TED is that the structural response is treated at the same time as the electronic response. We are currently undertaking studies of the conformational change of retinal, in the primary photochemical process that leads to vision, and of the damage to melanin from ultraviolet radiation. Although no results are yet available for these systems, in Fig. 53 we show the conformational change of a butadiene (C_4H_6) molecule that was obtained in a simulation employing a density-functional-based approach (Torralva et al.).

TED is a promising approach for biological molecules and complex materials because it is inherently an $O(N)$ method. That is, after the electronic states at time $t = 0$ are specified, the computational expense is proportional to N, the number of atoms in the system, rather than N^p with $p \geqslant 3$.

XI. Conclusion

This chapter has reviewed some theoretical and experimental studies of the response of matter to light, with an emphasis on semiconductors and fast intense laser pulses.

Our simulations employ the new technique of tight-binding electron–ion dynamics (TED). Comparison with experiment shows good agreement in all important respects. In the case of the semiconductors GaAs and Si, there is a nonthermal phase transition when the intensity is varied at fixed pulse duration, as demonstrated by both experimentally accessible optical properties — $\epsilon(\omega)$ and $\chi^{(2)}$ — and by the many other properties that can be studied in microscopic simulations.

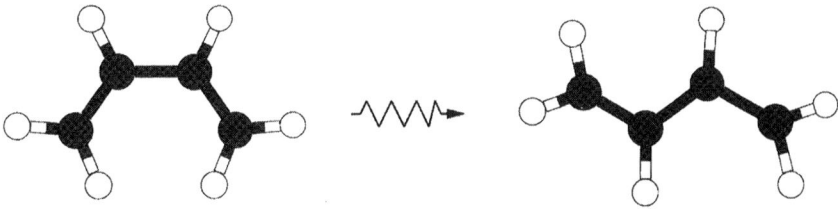

FIG. 53. Simulation of the response of a butadiene (C_4H_6) molecule to irradiation with light, using tight-binding electron–ion dynamics. There is a structural transformation of the molecule when light is absorbed and electrons are consequently excited. The photon energy was 3.8 eV, the fluence 0.16 kJ/m^2, and the pulse duration 10 fs FWHM. After Torralva et al., to be published.

These "hidden" properties include the pair correlation function, which is related to the femtosecond-scale X-ray diffraction measurements of Figs. 25–27; the band structure, which exhibits bandgap collapse above a threshold fluence; the atomic displacements, which become very large on a time scale that is too short for an ordinary thermal transition; the kinetic energy of the atoms, which increases in a manner that suggests strong repulsive interactions; and the population of excited conduction-band states, which at threshold is found to be larger for Si than for GaAs, presumably because Si has tighter bonding.

The comparison between simulations and experiment even appears to show some detailed agreement. The dielectric function $\epsilon(\omega)$ exhibits a characteristic "hump" above the transition, like that seen in Callan et al. The second-order nonlinear susceptibility $\chi^{(2)}$ shows an "elastic" response just below the transition, like that in Glezer et al. (1995b). The time scales and other features in the experimental data of the following references are in good agreement with the results of the simulations: Shank et al. (1983); Downer et al. (1983); Tom et al. (1988); Saeta et al. (1991); Siegal et al. (1994, 1995); Glezer et al. (1995a,b); Mazur (1996); Huang et al. (1998); Callan et al.; Sokolowski-Tinten et al. (1995); Govorkov et al. (1992); Chin et al. (1999); Siders et al. (1999).

The response of the fullerene C_{60} has also been studied as a function of the intensity of the laser pulse. At low intensity, various optically active vibrational modes are excited. At higher intensity, the breathing mode is by far the most dominant, in agreement with the measurements of Dexheimer et al. (1993) and Fleischer et al. (1997). At very high intensities, there is photofragmentation, with the evolution of dimers and other products.

Finally, we presented results for various interesting phenomena involving the response of H_2^+ to fast intense laser pulses, for the absorption and dichroism of chlorophyll, and for a conformational change in butadiene that results from the absorption of light.

ACKNOWLEDGMENTS

This work was supported by the Robert A. Welch Foundation. We have greatly benefitted from discussions, visits, and e-mail exchanges with Eric Mazur, Paul Callan, Klaus Sokolowski-Tinten, Charles Shank, Susan Dexheimer, Erich Ippen, Mildred Dresselhaus, Siegfried Fleischer, and Henry van Driel.

REFERENCES

R. E. Allen, *Phys. Rev. B.* **50**, 18 629 (1994).
R. E. Allen and M. Menon, *Phys. Rev. B* **33**, 5611 (1986).
A. N. Andriotis, N. N. Lathiotakis, and M. Menon, *Europhys. Lett.* **36**, 37 (1996).
D. E. Aspnes, *Phys. Rev. B* **6**, 4648 (1972).
D. E. Aspnes and A. A. Studna, *Phys. Rev. B* **27**, 985 (1983).
Molecules in Laser Fields, edited by A. D. Bandrank (Marcel Dekker, New York, 1994).
D. Bethune, A. J. Shmidt, and Y. R. Shen, *Phys. Rev. B* **11**, 3867 (1975).
J. Bok, *Phys. Lett.* **84A**, 448 (1981).
E. L. Briggs, D. J. Sullivan, and J. Bernholc, *Phys. Rev. B* **52**, R5471 (1995) and **54**, 14362 (1996).
J. P. Callan, A. M.-T. Kim, L. Huang, and E. Mazur, *Chem. Phys.* **251**, 167–172 (2000).
J. Callaway, *Quantum Theory of the Solid State* (Academic Press, New York, 1974).
N. A. Campbell, *Biology*, 2nd edition (Benjamin/Cummings, Menlo Park, 1990).
F. A. Casey and R. J. Sundberg, *Advanced Organic Chemistry* (Plenum, New York, 1993).
A. H. Chin, R. W. Schoenlein, T. E. Glover, P. Balling, W. P. Leemans, and C. V. Shank, *Phys. Rev. Lett.* **83**, 336 (1999).
M. Combescot and J. Bok, *Phys. Rev. Lett.* **48**, 1413 (1982).
H.-L. Dai and W. Ho, *Laser Spectroscopy and Photochemistry on Metal Surfaces* (World Scientific, Singapore, 1995), Vols. 1 and 2.
S. Das Sarma and J. R. Senna, *Phys. Rev. B* **49**, 2443 (1994).
A. de Shalit and H. Feshbach, *Theoretical Nuclear Physics* (Wiley, New York, 1974), Vol. 1, p. 530.
S. L. Dexheimer, D. M. Mittleman, R. W. Schoenlein, W. Vareka, X.-D. Xiang, A. Zettl, and C. V. Shank, in *Ultrafast Phenomena VIII*, edited by J. L. Martin, A. Migus, G. A. Mourou, A. H. Zewail (Springer-Verlag, Berlin, 1993).
M. C. Downer, R. L. Fork, and C. V. Shank, *J. Opt. Soc. Am. B* **2**, 595 (1983).
T. Dumitrică and R. E. Allen, *Solid State Commun.* **113**, 653 (2000).
T. Dumitrică and R. E. Allen, to be published.
T. Dumitrică, J. S. Graves, and R. E. Allen, *Phys. Rev. B* **58**, 15340 (1998).
T. Dumitrică, B. Torralva, and R. E. Allen, to be published in the proceedings of the MRS Symposium on Computational Approaches to Predicting the Optical Properties of Materials (Fall, 1999), edited by J. Chelikowsky, S. Louie, G. Martinez, and E. Shirley (Materials Research Society, Warrendale, Pennsylvania, 2000).
G. J. Ferraudi, *Elements of Inorganic Photochemistry* (Wiley, New York, 1988).
S. B. Fleischer, B. Pevzner, D. J. Dougherty, H. J. Zeiger, G. Dresselhaus, M. S. Dresselhaus, E. P. Ippen, and A. F. Hebard, *Appl. Phys. Lett.* **71**, 2734 (1997).
J. M. Fraser, A. I. Shkrebtii, J. E. Sipe, and H. M. van Driel, *Phys. Rev. Lett.* **83**, 4192 (1999).
P. Fulde, *Electron Correlations in Molecules and Solids* (Springer-Verlag, Berlin, 1991).
Q. Gao, Ph.D. thesis, Texas A&M University, 1998.

E. Ghahramani, D. J. Moss, and J. E. Sipe, *Phys. Rev. B* **43**, 8990 (1991).
E. N. Glezer, Y. Siegal, L. Huang, and E. Mazur, *Phys. Rev. B* **51**, 6959 (1995a).
E. N. Glezer, Y. Siegal, L. Huang, and E. Mazur, *Phys. Rev. B* **51**, 9589 (1995b).
V. V. Godlevsky, M. Jain, J. J. Derby, and J. R. Chelikowsky, *Phys. Rev. B* **60**, 8640 (1999) and references therein.
Atoms in Intense Laser Fields, edited by M. Gommila (Academic Press, Boston, 1992).
H. Gould and J. Tobochnik, *An Introduction to Computer Simulation Methods* (Addison-Wesley, New York, 1988).
S. V. Govorkov, Th. Schröder, I. L. Shumay, and P. Heist, *Phys. Rev. B* **46**, 6864 (1992) and references therein.
M. Graf and P. Vogl, *Phys. Rev. B* **51**, 4940 (1995).
J. S. Graves, Ph.D. thesis, Texas A&M University, 1997.
J. S. Graves and R. E. Allen, in *Tight-Binding Approach to Computational Materials Science*, edited by P. E. A. Turchi, A. Gonis, and L. Colombo (Materials Research Society, Warrendale, Pennsylvania, 1998a).
J. S. Graves and R. E. Allen, *Phys. Rev. B* **58**, 13 627 (1998b) and to be published.
E. K. U. Gross, C. A. Ullrich, and U. J. Gossmann, in *Density Functional Theory*, edited by E. K. U. Gross and R. M. Dreizler (Plenum, New York, 1994).
J. Gryko and R. E. Allen, in *High Performance Computing and Its Application in the Physical Sciences*, edited by D. A. Brown et al. (World Scientific, Singapore, 1994).
R. Hamilton, Ph.D. thesis, Texas A&M University, 1999.
K. Hansen and O. Echt, *Phys. Rev. Lett.* **78**, 2337 (1996).
W. A. Harrison, *Electronic Structure and the Properties of Solids* (Freeman, San Francisco, 1980; also published by Dover Publications. New York, 1989).
W. A. Harrison, *Elementary Electronic Structure* (World Scientific, Singapore, 1999).
D. W. Heerman and A. N. Burkitt, *Parallel Algorithms in Computational Science* (Springer-Verlag, New York, 1991).
H. Hohmann, C. Callegari, S. Furrer, D. Grosenick, E. E. B. Cambell, and I. V. Hertel, *Phys. Rev. Lett.* **73**, 1919 (1994).
M. Z. Huang and W. Y. Ching, *Phys. Rev. B* **45**, 8738 (1982).
L. Huang, J. P. Callan, E. N. Glezer, and E. Mazur, *Phys. Rev. Lett.* **80**, 185 (1998).
J. P. Hughes and J. E. Sipe, *Phys. Rev. B* **53**, 10751 (1996).
S. Hunsche, T. Starczewski, A. l'Huillier, A. Persson, C.-G. Wahlström, B. van Linden van den Heuvell, and S. Svanberg, *Phys. Rev. Lett.* **77**, 1966 (1996).
R. Jackiw and A. Kerman, *Phys. Lett.* **71A**, 158 (1979).
C. J. Joshi and P. B. Corkum, *Physics Today* **48**, 36 (1995).
L. P. Kadanoff and G. Baym, *Quantum Statistical Mechanics* (Addison-Wesley, Reading, Massachusetts, 1962).
L. V. Keldysh, *Soviet Physics JETP* **20**, 1018 (1965).
Don H. Kim, H. Ehrenreich, and E. Runge, *Solid State Commun.* **89**, 119 (1994).
S. Khosravi, Ph.D. thesis, Texas A&M University, 1997.
S. Khosravi and R. E. Allen, in *Tight-Binding Approach to Computational Materials Science*, edited by P. E. A. Turchi, A. Gonis, and L. Colombo (Materials Research Society, Warrendale, Pennsylvania, 1998).
E. Kohen, R. Santus, and J. G. Hirschberg, *Photobiology* (Academic Press, Boston, 1995).
J. P. Lewis, P. Ordejón, and O. F. Sankey, *Phys. Rev. B* **55**, 680 (1997).
C. Mailhiot, C. B. Duke, and D. J. Chadi, *Phys. Rev. Lett.* **53**, 2114 (1984).
P. C. Martin and J. Schwinger, *Phys. Rev.* **115**, 1342 (1959).
C. K. Mathews and K. E. Van Holde, *Biochemistry*, 2nd edition (Benjamin/Cummings, Menlo Park, 1990).
E. Mazur, in *Spectroscopy and Dynamics of Collective Excitations in Solids*, edited by B. D. Bartolo (Plenum, NATO ASI series, New York, 1996).

M. Menon and R. E. Allen, *Bull. Am. Phys. Soc.* **30**, 362 (1985).
M. Menon and R. E. Allen, *Phys. Rev. B* **33**, 7099 (1986).
M. Menon and R. E. Allen, *Phys. Rev. B* **44**, 11293 (1991).
D. J. Moss, J. E. Sipe, and H. M. van Driel, *Phys. Rev. B* **36**, 9708 (1987).
F. G. Parsons and R. K. Chang, *Opt. Commun.* **3**, 173 (1971).
M. C. Payne, M. P. Teter, D. C. Allan, T. A. Arias, and J. D. Joannopoulos, *Rev. Mod. Phys.* **64**, 1045 (1992) and references therein.
R. Peierls, *Z. Phys.* **80**, 763 (1933).
D. Porezag, Th. Frauenheim, Th. Höhler, G. Seifert, and R. Kaschner, *Phys. Rev. B* **51**, 12947 (1995).
C. Rischel, A. Rousse, I. Uschmann, P.-A. Albouy, J.-P. Geindre, P. Audebert, J.-C. Gauthier, E. Förster, J. L. Martin, and A. Antonetti, *Nature* **390**, 490 (1997).
M. Rohlfing and S. G. Louie, *Phys. Rev. Lett.* **82**, 1959 (1999) and references therein.
P. Saeta, J. K. Wang, Y. Siegal, N. Bloembergen, and E. Mazur, *Phys. Rev. Lett.* **67**, 1023 (1991).
O. F. Sankey and R. E. Allen, *Bull. Am. Phys. Soc.* **30**, 362 (1985).
O. F. Sankey and R. E. Allen, *Phys. Rev. B* **33**, 7164 (1986).
O. F. Sankey and D. J. Niklewski, *Phys. Rev. B* **40**, 3979 (1989). This approach permits first-principles calculations with some of the favorable features of a tight-binding picture.
M. Sawtarie, M. Menon, and K. R. Subbaswammy, *Phys. Rev. B* **49**, 7739 (1994).
S. Scandolo and F. Bassani, *Phys. Rev. B* **51**, 6925 (1995).
R. R. Schlicher, W. Becker, J. Bergou, and M. O. Scully in *Quantum Electrodynamics and Quantum Optics*, edited by A. O. Barut (Plenum, New York, 1984). See Eqs. (2.17)–(2.20).
C. V. Shank, R. Yen, and C. Hirlimann, *Phys. Rev. Lett.* **50**, 454 (1983) and **51**, 900 (1983).
Y. R. Shen, *Phys. Rev. B* **167**, 818 (1968).
C. W. Siders, A. Cavalleri, K. Sokolowski-Tinten, Cs. Tóth, T. Guo, M. Kammler, M. Horn von Hoegen, K. R. Wilson, D. von der Linde, and C. P. J. Barty, *Science* **286**, 1340 (1999).
Y. Siegal, E. N. Glezer, and E. Mazur, *Phys. Rev. B* **49**, 16403 (1994).
Y. Siegal, E. N. Glezer, L. Huang, and E. Mazur, *Ann. Rev. Mat. Sci.* **25**, 223 (1995).
P. Silvestrelli, A. Alavi, M. Parrinello, and D. Frenkel, *Phys. Rev. Lett.* **77**, 3149 (1996).
P. L. Silvestrelli, A. Alavi, M. Parrinello, and D. Frenkel, *Phys. Rev.* **56**, 3806 (1997).
K. Sokolowski-Tinten, J. Bialkowski, and D. von der Linde, *Phys. Rev. B* **51**, 14186 (1995).
P. Stampfli and K. H. Bennemann, *Phys. Rev. B* **42**, 7163 (1990).
P. Stampfli and K. H. Bennemann, *Phys. Rev. B* **46**, 10686 (1992).
H. W. K. Tom, G. D. Aumiller, and C. H. Brito-Cruz, *Phys. Rev. Lett.* **60**, 1438 (1988).
B. Torralva and R. E. Allen, in the Proceedings of the 24th International Conference on the Physics of Semiconductors (Jerusalem, August 1998).
B. Torralva, R. Hamilton, and R. E. Allen, to be published.
B. Torralva *et al.*, to be published.
J. A. Van Vechten, R. Tsu, F. W. Saris, *Phys. Lett.* **74a**, 422 (1979).
J. A. Van Vechten, in *Semiconductors Probed by Ultrafast Laser Spectroscopy*, edited by R. R. Alfano (Academic Press, London, 1984).
P. Vogl, H. P. Hjalmarson, and J. D. Dow, *J. Phys. Chem. Solids* **44**, 365 (1983).
C. H. Xu, C. Z. Wang, C. T. Chan, and K. M. Ho, *J. Phys. Condens. Matter* **4**, 6047 (1992).
The Chemical Bond, Structure and Dynamics, edited by A. Zewail (Academic Press, Boston, 1992).

CHAPTER 8

Coherent THz Emission in Semiconductors

E. Gornik

INSTITUT FÜR FESTKÖRPERELEKTRONIK
TECHNISCHE UNIVERSITÄT WIEN
WIEN, AUSTRIA

R. Kersting

DEPARTMENT OF PHYSICS, APPLIED PHYSICS, AND ASTRONOMY
RENSSELAER POLYTECHNIC INSTITUTE
TROY, NEW YORK

I. INTRODUCTION	390
II. PRINCIPLES OF PULSED THz EMISSION AND DETECTION	392
1. *Dipole Emission*	392
2. *Temporal and Spatial Coherence*	394
3. *Detection of THz Pulses*	395
III. MACROSCOPIC CURRENTS AND POLARIZATIONS	400
1. *Drude–Lorentz Model of Carrier Transport*	400
2. *The Instantaneous Polarization*	403
3. *Phases of Transport and Polarization*	404
4. *Emission from Field-Induced Transport*	405
5. *Bandwidth Increase by Ultrafast Currents*	409
6. *Emission from an Instantaneous Polarization*	410
IV. COHERENT CHARGE OSCILLATIONS	410
1. *Plasma Oscillations of Photoexcited Carriers*	411
2. *Plasma Oscillations of Extrinsic Carriers*	414
3. *Plasma Oscillations in 2D Layers*	417
4. *Coherent Phonons and Plasmon–Phonon Coupling*	418
V. THz EMISSION FROM QUANTUM STRUCTURES	420
1. *Quantum Beats*	420
2. *Bloch Oscillations*	422
3. *Intersubband Transitions*	426
VI. APPLICATIONS IN SEMICONDUCTOR SPECTROSCOPY	427
1. *Optical Pump–THz Probe Spectroscopy*	428
2. *THz Time-Domain Spectroscopy*	429
VII. CONCLUSIONS	433
REFERENCES	434

I. Introduction

Today, nearly forty years after their first demonstration, semiconductor lasers have become the most used coherent light source. Both the variety of semiconductor materials and semiconductor device technologies have made it possible to achieve coherent light emission over nearly the entire optical spectrum. Extreme wavelengths have been reached, for instance, close to the ultraviolet by gallium nitride lasers (Nakamura et al., 1996), in the infrared by quantum cascade lasers (Faist et al., 2000) and by lead salt lasers (Tacke, 1999), and in the far-infrared by p-germanium lasers (Gornik and Andronov, 1991). In the past decade the accessible optical spectrum has been extended through the whole THz range by the generation of coherent THz light pulses in various materials such as semiconductors, dielectrics, and even organic compounds. Most of the applied generation techniques have in common the fact that the THz emission follows an ultrafast laser excitation of carriers in matter. Within a few years the spectrum of far-infrared pulses has been extended from a few hundred gigahertz to several tens of THz. Thus, today a new optical band is accessible with frequencies much higher than those that can be reached by conventional electronic techniques such as Gunn oscillators (Gunn, 1963; Eisele and Haddad, 1995).

The rapid development in the THz field has been stimulated and triggered by the evolution of short-pulse laser sources, which enabled a better understanding of the ultrafast carrier dynamics in semiconductors. Milestones in this evolution have been the development of ultrafast dye lasers, such as the colliding pulse mode-locked laser (Fork et al., 1981) and the mode-locked Ti:Sapphire laser (Spence et al., 1991). Having pulse durations of less than 100 fs, these laser sources give the opportunity to investigate ultrafast carrier dynamics in semiconductors (Shah, 1996). The study of transient processes such as the dynamics of hot carrier distributions or the ultrafast lattice–carrier interaction turned out to be of essential importance for the development of ultrafast devices. The understanding of the dynamics of these processes laid down the basis for the generation of pulsed THz emission.

The historical roots of the generation of pulsed far-infrared radiation go back to 1971 when Yang et al. (1971) reported the observation of FIR radiation by focusing a high-intensity picosecond laser beam into lithium niobate. In other transparent electro-optic media such as lithium tantalate, bursts of far-infrared radiation have been generated by the inverse electro-optic effect, leading to a THz Cherenkov cone (Auston et al., 1984b). Although the spectral range spanned by THz Cherenkov radiation extends as far as 5 THz, the extraction of the pulses into free air turned out to be rather limited by the high index of refraction of the materials used.

In semiconductors, the first optoelectronic switching was achieved in silicon devices giving picosecond pulses traveling on transmission lines

(Auston, 1975). Coupling the pulses to antenna structures enabled the generation of freely propagating microwave pulses (Mourou *et al.* 1981). A bandwidth of 150 GHz was achieved by integrating a photoconductor into an antenna structure (Pastol *et al.*, 1988). The bandwidth was significantly increased by applying microstructuring techniques for the fabrication of the antennas (Defonzo *et al.*, 1987). Hertzian dipole antennas (Auston *et al.*, 1984a) enabled emission with a broad angular distribution and led to a maximum bandwidth of about 5 THz (Pedersen *et al.*, 1993). The collimation of the emission was achieved by refractive silicon lenses (van Exter and Grischkowsky, 1990c) or by sapphire half-spheres (Fattinger and Grischkowsky, 1989; Jepsen and Keiding, 1995) that were attached to the antennas. These techniques lead to highly directional THz beams, thus opening the way for first applications in spectroscopy.

Because of the low photon energy, the intensities in the THz region are usually quite low as compared with laser sources operating at wavelengths in the visible spectrum. Although the intensities are usually below 1 μW, a multitude of promising applications have been proposed (Nuss and Orenstein, 1998). Most of the applications make use of strong absorptive transitions in the far-infrared that enable sensitive THz transmission spectroscopy (Nuss, 1996). Such applications are, for instance, chemical analysis, process control, and wafer characterization. THz ranging studies on miniaturized models are expected to give similar results as on full-size objects (Cheville and Grischkowsky, 1995). One of the most promising applications might be tomography by using THz pulses for imaging (Hu and Nuss, 1995). THz pulses are particularly attractive for the tomography of biological tissues because the very low photon energy avoids ionization processes.

The main purpose of this chapter is to give a state-of-the art overview on the most common techniques of THz generation in semiconductors. From the variety of ultrafast processes that have been exploited for the generation of pulsed THz radiation, we focus on those that have a high potential for further improvement and are based on semiconductors. These are, in particular, the ultrafast currents and plasma oscillations that can be designed by band structure and device engineering. The variety of processes and techniques is large, and they can also be applied to other materials.

The contribution focuses on those processes in semiconductors that can be initiated by ultrafast laser excitation and lead to the emission of THz pulses, in particular the ultrafast acceleration of photoexcited carriers within a field region of the semiconductor and resonant phenomena as plasma or phonon oscillations. The chapter begins with an analysis of the emission properties of oscillating dipoles in matter that includes the spatial and temporal coherence of the emitted field. We present and compare the most often used THz detection techniques that take advantage of the coherence. In Section III we discuss the carrier dynamics that follow an impulsive laser excitation in terms of a Drude–Lorentz model. The discussion focuses on the characteristic THz pulse shapes that result from the ultrafast acceleration of charges in a field and the instantaneous polarization. Section IV is devoted to coherent charge oscillations that emit THz radiation, particularly plasma oscillations and coherent oscillations of phonons. The generation of THz pulses in semiconductor quantum structures is discussed in Section V. Examples of THz spectroscopy on semiconductors are presented in Section VI.

II. Principles of Pulsed THz Emission and Detection

Semiconductors are an ideal material system for generating THz pulses by initiating ultrafast currents. Two main properties have to be fulfilled: a high mobility of the charge carriers together with a low intrinsic carrier concentration in the field region. These two features enable a high current contrast between a steady-state photoconductive device and the photocurrent initiated by ultrafast laser excitation.

1. Dipole Emission

As the efficiency of the THz generation process is very small, a dipole approximation is justified to describe the properties of the radiation field. In general, the emission of a THz field results from the buildup of a transient polarization that is related to the vector of the dipole moment d and its temporal derivatives. The emitted field is (Born and Wolf, 1980; Jackson, 1998)

$$E_{em} = \frac{1}{4\pi\varepsilon\varepsilon_0}\left[-\frac{d''}{r} + \frac{(d''\cdot r)\cdot r}{r^3} - \frac{d'}{r^2} + \frac{3(d'\cdot r)\cdot r}{r^4} - \frac{d}{r^3} + \frac{3(d\cdot r)\cdot r}{r^5}\right]. \quad (1)$$

The wavelengths of the THz radiation in air are of the order of 100 μm and thus are considerably longer than the wavelengths of visible light. As the distance between the THz source and the detector is usually longer than the wavelength or the diameter of the coherently emitting source, the far-field approximation can be used for the description of the emitted field:

$$E_{em\perp} = -\frac{1}{4\pi\varepsilon\varepsilon_0}\frac{d''}{r}. \quad (2)$$

Within the framework of classical electrodynamics, a single oscillating electron emits its energy through the radiation field with a damping rate of

(Meystre and Sargent III, 1998)

$$\gamma_{rad} = \frac{e^2\omega^2}{12\pi\varepsilon_0 c^3 m} = \frac{\omega^2 r_0}{3c} \tag{3}$$

where r_0 is the classical electron radius of $e^2/4\pi\varepsilon_0 mc^2 = 2.8 \times 10^{-15}$ m. A similar but somewhat simpler result is obtained from a quantum mechanical calculation. For frequencies of the THz band the radiative lifetimes are of the order of microseconds to milliseconds, whereas, for example, dipoles emitting in the visible spectrum have radiative lifetimes of the order of nanoseconds. In general the dipole will lose its energy via nonradiative channels with a rate γ_{nr}, resulting in an efficiency for the radiative emission of

$$\eta_{rad} = \frac{\gamma_{rad}}{\gamma_{rad} + \gamma_{nr}}. \tag{4}$$

In condensed matter, nonradiative scattering events are much faster than the emission rate at THz frequencies. Figure 1 shows the spectral dependence of the single dipole emission rate and efficiencies for different nonradiative lifetimes. In the visible spectrum, at frequencies of around

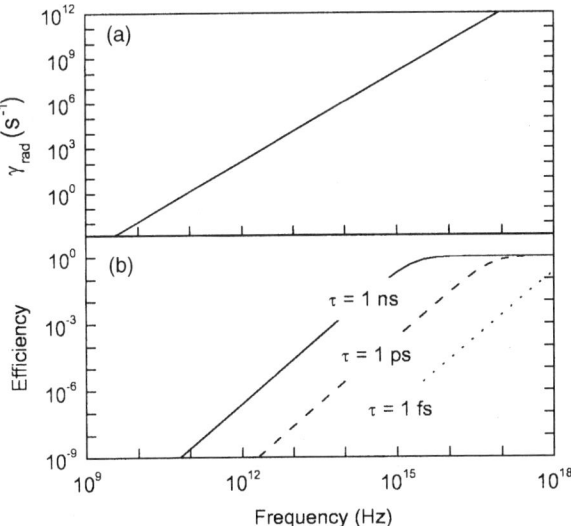

FIG. 1. (a) Radiative lifetime for a single electron excitation as a function of the transition frequency. (b) Corresponding radiative efficiencies, including nonradiative recombination processes of 1 ns, 1 ps, and 1 fs, respectively.

6×10^{14} Hz, the efficiencies are close to 1. The efficiencies drop by orders of magnitude going to the far-infrared. At THz frequencies the efficiencies are as low as 10^{-6} even if nonradiative lifetimes as long as 1 ns are assumed. Figure 1 illustrates how difficult it is to achieve THz radiation of significant intensity with isolated oscillating dipoles.

2. Temporal and Spatial Coherence

One well-known and efficient way out of this dilemma is used in the radio frequency regime. Here, macroscopic current oscillations are used that result from the collective oscillation of a high density of electrons. The resulting macroscopic dipoles scale with the number of electrons N contributing to the collective oscillation $d = Ne\Delta x$. They cause, according to Eq. (2), a linear increase of the emitted field with N and thus an intensity proportional to the square of N. The analogous effect of coherently oscillating dipoles is described as superradiant emission in optics (Dicke, 1954). Emission rates are increased in this case by orders of magnitude as compared to isolated dipoles. Ultrafast laser excitation is a new method that enables the generation of macroscopic currents in semiconductors by enforcing a certain phase onto the motion of the individual charge carriers in a frequency region that has so far not been penetrated by radio-frequency techniques. The generated coherent dynamics of the carriers leads to a macroscopic current and as a consequence to enhanced emission of THz radiation.

The beam pattern of the THz radiation is related to the spatial extent of the coherent currents within the device. In a single dipole antenna the current is limited by the dipole length, which is on the order of the wavelength. Emission profiles of antennas have a broad angular distribution and are well known from electrodynamics. In contrast, coherent current oscillations can be generated over large surface regions in semiconductors. Diameters of several wavelengths can be easily achieved by using, for instance, a laser excitation across several millimeters of a semiconductor surface as shown in Fig. 2. The coherence between different points within the excitation area results from the time delay of the exciting pulse. The phase is given by the wavefront of the exciting laser pulse and depends only on the angle of incidence ϕ. The THz emission occurs, according to Snell's law, in the direction of $-\phi$. In addition, the coherence of the current leads to a diffraction limited emission (Born and Wolf, 1980) within a solid angle $\Omega = 4\pi \sin^2 \phi/2 \approx \lambda^2/F$, limited by the uncertainty principle. Small beam divergences of a few degrees can be achieved by initiating coherent currents across a surface area that is considerably larger than the wavelength. Planar large aperture emitters with electrode spacings up to 10 mm have been reported for the generation of highly directional THz pulses (Hu et al., 1989; Darrow et al., 1990). Furthermore, the use of large-aperture antennas

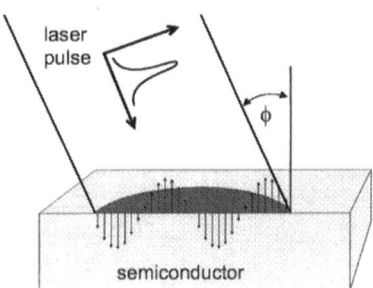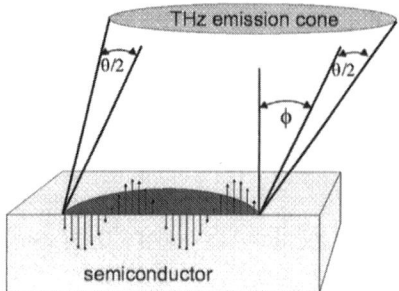

FIG. 2. Generation of coherent current surges on a semiconductor surface with a diameter larger than the wavelength of the THz emission. The solid angle of the THz cone is $\Omega = 4\pi\theta$.

enables the generation of high-energy pulses of far-infrared radiation when excited by laser pulses extracted from amplifier systems. Because of the enormous dipole moments the THz pulses have energies close to 1 μJ (You et al., 1993).

An efficient outcoupling of THz radiation from the semiconductor into the half-sphere above requires an impedance matching between semiconductor and air. The standard method is using microstructured metallic gratings and antennas that are fabricated on top of the semiconductor surface. Although the sources of the radiation are always the ultrafast currents and polarizations in the semiconductor, the antenna structures influence the THz emission significantly because of their intrinsic bandwidth, dispersion, and directionality. These modifications due to the coupling to an antenna can be described by linear electrodynamics. The scale invariance of these properties relative to the wavelength enables the direct application of concepts that have been elaborated for longer wavelengths. Therefore, we limit the discussion in this contribution to the generation of ultrafast charge dynamics that arise in the semiconductors and leave out those modifications induced by the antenna structures. A more detailed discussion can be found in Nuss and Orenstein (1998) and references therein.

3. Detection of THz Pulses

Historically, the access to the far-infrared spectrum has been hindered by the lack of suitable emission sources and detectors. The sources most often used were incoherent light sources such as globars emitting thermal radiation. The weak spectral energy density of blackbody radiation required sophisticated detection techniques for the far-infrared spectrum that enabled at least spectroscopic applications. Considerably higher sensitivities can be achieved if the THz radiation is coherent and pulsed. On one hand, the short duration of the pulses enables a time-gated detection, which leads to an increased discrimination of the thermal background. On the other hand, the coherence of the radiation enables the use of such highly sensitive detection methods as free-space electro-optic sampling or the detection of the THz field using antenna techniques. Most of the THz detection methods measure the electromagnetic field. In contrast to intensity measurements in the visible part of the spectrum, THz techniques provide the amplitude and phase information of the electromagnetic radiation. However, the detection of THz radiation depends strongly on the characteristic properties of the technique used, such as bandwidth or dispersion of the detector.

a. Antenna Detectors

The first antenna detection techniques for pulsed THz radiation were applied by Auston and Mourou (Mourou *et al.*, 1981; Auston *et al.*, 1984a). Although a variety of antenna shapes have been used in the past decade for detection, most recent work has been performed using dipole antennas. However, the detection principles are independent of the geometrical shape of the antenna used. Figure 3 shows a scheme illustrating the detection of a THz pulse using a photoconductive dipole antenna. The structure can be fabricated on substrates that show a fast response to illumination by femtosecond laser pulses, such as radiation-damaged silicon on sapphire (Smith *et al.*, 1988). Other suitable photoconductors are ion-bombarded InP

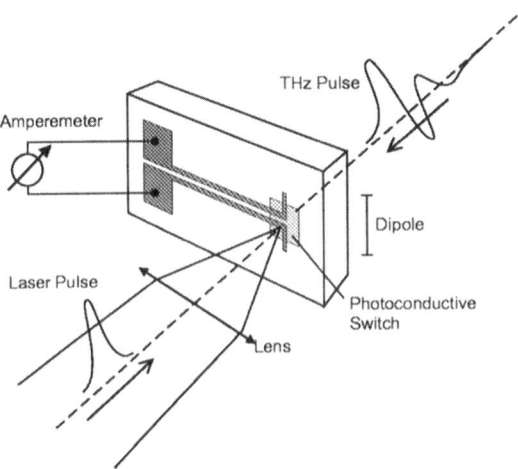

FIG. 3. Photoconducting dipole antenna, after Smith *et al.* (1988), Nuss and Orenstein (1998). Copyright © 1988 IEEE. Reprinted with permission.

(Karin et al., 1986) or low-temperature-grown GaAs (Chwalek et al., 1991). The THz pulse is detected by shining a femtosecond laser pulse on the photoconductor between the arms of the antenna. As long as the photoconductor is not illuminated, it reveals a high resistance that drops over orders of magnitude when photocarriers are injected by a femtosecond laser pulse. For a short time interval the gap becomes conductive and the THz field focused onto the antenna induces a charge separation between the arms and thus a measurable current. For the case that the time interval of laser-induced conductance is small compared to the oscillation time of the incident field, the THz pulse can be time resolved. The shape of the THz field can be mapped by repetitive photoconductive sampling at different time delays between the laser pulse and the THz pulse.

To increase the sensitivity of antenna systems, many different antenna shapes have been used. Logarithmic periodic antennas, for instance, reveal a sensitivity that is up to an order of magnitude higher than those of comparable dipole antennas (Dykaar et al., 1991b; Chwalek et al., 1991). However, in logarithmic periodic antennas the propagation time to reach the feed point of the antenna depends on the wavelength (Dykaar et al., 1991a). Thus, detecting broad band THz pulses with antennas may influence the shape of the detected pulse. Other distortions of the detected pulse may arise from the spatiotemporal properties of a focused THz pulse (Hunsche et al., 1999). Today, antenna techniques enable the detection of THz radiation up to frequencies of about 5 THz. However, most of the antennas have a maximum response between 1 and 2 THz.

b. Electro-optic Detection

The experimental difficulties in directly detecting THz fields have led to the idea of modulating an optical beam with the THz pulse. This electro-optic modulation requires a switching process in a nonlinear optical material. The switching can be achieved by the Pockels effect (Saleh and Teich, 1991), where a quasi DC field, the THz field, modulates the properties of an optical wave. In the presence of the DC field, the polarization of the optical beam is rotated in materials exhibiting nonlinear and birefringent properties. Since the rotation is proportional to the magnitude and polarization of the DC field, this technique enables the direct measurement of the THz beam by detecting the polarization of the optical pulse. For the case that the switched optical pulse is much shorter than the THz pulse, the amplitude and phase of the THz pulse can be mapped.

Picosecond electro-optical sampling was demonstrated in lithium niobate by resolving an ultrafast electrical pulse generated on a GaAs photodetector (Valdmanis et al., 1982). Subpicosecond time resolution was achieved in lithium tantalate, since this material makes it possible to achieve a better

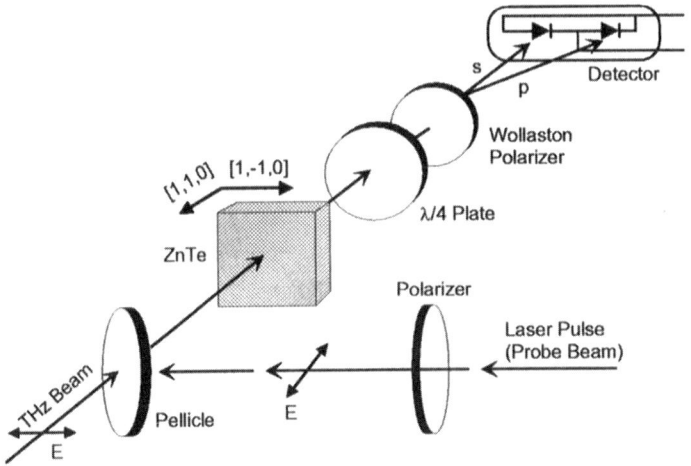

FIG. 4. Free space electro-optic sampling, after Lu et al. (1997). Reprinted with permission of American Institute of Physics.

velocity matching of electrical and optical pulses (Valdmanis et al., 1983). The achieved time resolutions enabled THz time-domain spectroscopy. Although the applications were limited to the measurement of local fields, the technique gave valuable insights into excitation dynamics in matter, such as the generation of Cherenkov radiation (Auston and Nuss, 1988) and the ultrafast carrier dynamics in semiconductors (Nuss et al., 1987).

One of the most important steps for the application of electro-optic switching was the development of free-space electro-optic sampling (Wu and Zhang, 1995; Jepsen et al., 1996b). Figure 4 shows a typical setup. The THz beam and the sampling laser beam are collinearly superimposed at a pellicle beam splitter. Because of the Pockels effect, the polarization of the laser pulse is rotated in the nonlinear crystal. The wollaston prism separates both polarization components spatially to enable the detection of both polarizations on individual photodiodes. Although the first techniques used lithium tantalate as the electro-optic crystal, further studies focused on zinc telluride (Wu and Zhang, 1996) or gallium phosphide (Wu and Zhang, 1997).

The time resolution that can be achieved by electro-optic sampling is only limited by the group velocity dispersion between the THz pulse and the sampling optical pulse within the crystal (Bakker et al., 1998). For high THz intensities this problem may be overcome by using thin crystals. This increases the bandwidth of the method, but significantly decreases the signal intensity because of the shortened interaction length of the THz pulse and the optical beam. Bandwidths as high as 37 THz (Wu and Zhang, 1997) and 70 THz (Leitenstorfer et al., 1999a) have been reported. Typical signal-to-noise ratios are about 10.000:1. Although higher signal-to-noise ratios can be achieved with antenna techniques, electro-optic sampling is attractive, in particular at frequencies above 3 THz (Cai et al., 1998). Novel potential techniques of electro-optic sampling are two-dimensional imaging (Wu et al., 1996) and chirped pulse detection (Sun et al., 1998; Jiang and Zhang, 1998).

c. *Interferometric Detection*

A more fundamental technique for the characterization of THz radiation is the interferometric measurement of the THz autocorrelation. This technique gives only an amplitude spectrum of the emitted THz pulses. The phase information is not accessible since only the intensity of the interferogram is recorded. However, THz interferometry has the outstanding advantage that detectors such as 4 K bolometers, which have a frequency-independent response function, can be used. This enables a reliable characterization of the emission spectrum up to frequencies as high as 10 THz. The standard setup is a Michelson interferometer using a wire grid beam splitter. Although pulse durations as short as 160 fs have been measured (Greene et al., 1991), incoherent emission may also contribute to the detected correlation signal. An alternative setup is shown in Fig. 5. The interfering THz beams are generated by a sequence of two laser pulses. Thus, the correlation signal is free from contributions of incoherent thermal light coming from the semiconductor (Kersting et al., 1998). Assuming negligible interaction between the two laser excitations in the semiconductor, the measured signal intensity is

$$I_{det}(\tau) \sim \int_{-\infty}^{+\infty} [E_{rad,1}(t-\tau) + E_{rad,2}(t)]^2 dt \tag{5}$$

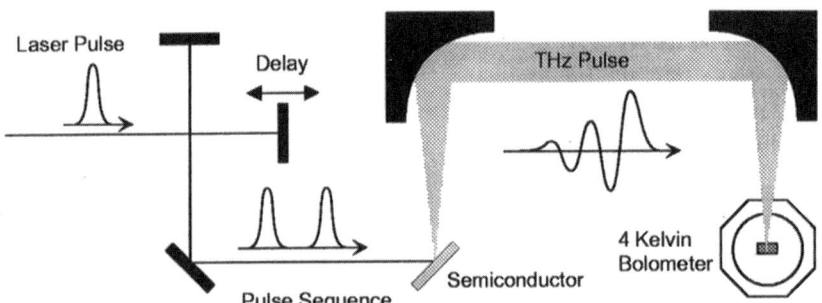

FIG. 5. Setup of a THz autocorrelation technique for the characterization of the THz emission from an optically excited semiconductor. From Kersting et al., 1998.

where $E_{rad,i}$ are the THz fields generated by each of the exciting laser pulses. Although the sensitivity of bolometers is smaller than that achievable with antenna detection and electro-optic sampling, the advantage of the technique is the constant response function up to frequencies of about 10 THz. This ensures measurements free from absorption and dispersion effects, which are very difficult to achieve with the techniques described above.

III. Macroscopic Currents and Polarizations

According to Maxwell's equations, an electromagnetic field can be emitted by changes either of currents or of transient polarizations in matter. In semiconductors, both transport currents and the generation of a transient polarization can be used for the generation of coherent THz pulses. The following sections address the origin of both field sources and the characteristic emission properties.

1. Drude–Lorentz Model of Carrier Transport

In a first approximation, the motion of the photocarriers in an electric field may be described by a Drude–Lorentz model of carrier transport that takes into account the ultrafast generation of charge carriers in the field region of a semiconductor. The model addresses the time scale where the photoexcitations have lost their coherence with the laser field and are accelerated by the field present in the semiconductor. Applying the model one has to consider the buildup of the density $n(t)$ of free photocarriers given by the temporal shape of the laser pulse $g(t)$ (Jepsen et al., 1996a). Since the free carriers may be trapped and recombine, a loss rate has to be considered:

$$\frac{dn(t)}{dt} = g(t) - \frac{n(t)}{\tau_{rec}}. \qquad (6)$$

Responding to the local field E_{loc}, the average velocity of the charge carriers is

$$\frac{dv(t)}{dt} = -\frac{v(t)}{\tau_{sc}} + \frac{qE_{loc}(t)}{m^*} \qquad (7)$$

where τ_{sc} is the scattering time and q the charge of the carrier. The resulting photocurrent $j(t) = en(t)v(t)$ induces a polarization that develops in accordance with

$$\frac{dP(t)}{dt} = \frac{-P(t)}{\tau_p} + j(t) \tag{8}$$

where τ_p is a phenomenological decay time of the polarization. A decay may arise when the carriers have been transported out of the field region which is equivalent to the recharging of a semiconductor device. Equations (7) and (8) are coupled because the transient polarization screens the external field and reduces the local field, accelerating the carriers $E_{loc}(t) = E_{ext} - P(t)/\varepsilon$.

For a solution of the equations just shown, the dynamics of photo-generated electrons and holes have to be taken into account. A simple but instructive solution will show the dynamics in a qualitative way. More detailed calculations require the consideration of the dispersion of the conduction and valence bands as well as energy-dependent scattering rates (Nag, 1980). Detailed discussions of transport and scattering properties following ultrafast changes of fields or carrier densities can be found in Wysin *et al.* (1988), Constant (1985), and references therein.

The qualitative analysis displayed in Fig. 6 shows transients of the photocurrent, the polarization, and the emitted field for three different situations that may arise in semiconductors after optical excitation. We start the discussion with the case where high scattering rates of the carriers are present. After their photogeneration the carriers are accelerated by the field. Because of fast scattering the carrier motion soon changes from an accelerated to a drift motion characterized by a constant velocity. During their

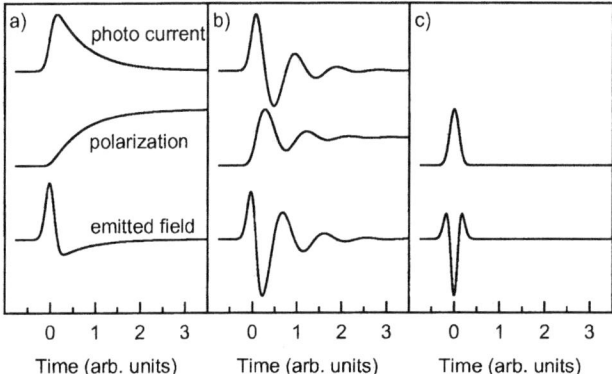

FIG. 6. Transients of the photocurrent, the polarization, and the emitted field for (a) high scattering rates of the photocarriers and (b) for low scattering rates; (c) shows only the polarization due to a virtual excitation and the emitted field.

motion within the field region, the carriers build up a polarization that screens the external field. The screening reduces the current, which finally drops to zero when the field is completely screened. The polarization remains until the carriers have been transported out of the region where the bias has been applied. Because of the rise and decay of the photocurrent the carriers emit a far field, as given by Eq. (2). In case of strong scattering the emitted field would consist only of one oscillation cycle. Whereas the first half of the slope is given by the onset of charge transport in the semiconductor, the second half results from the decay of the photocurrent. The first part strongly depends on the pulse width of the exciting laser and the time behavior of the charge acceleration by the field. Small effective carrier masses cause a rapidly increasing photocurrent. Therefore, in most semiconductors mainly the electrons contribute to the THz emission because of their smaller effective masses compared to holes. An efficient emission can therefore be expected for those semiconductors where the effective electron mass is small, such as InAs and InP. Beside the intraband scattering rate τ_{sc}, other mechanisms may limit the drift velocities. In the case of direct semiconductors such as GaAs, the maximum drift velocity is limited by intervalley scattering. The second half of the THz field slope results from the decay of the photocurrent. At high carrier densities the screening of the field reduces the drift velocities of the charges and thus the photocurrent. Another mechanism is the fast trapping of free photocarriers, leading to a reduction of the photocurrent. A detailed discussion of the different phases of carrier transport is given in Subsection 3.

A different picture arises when the charge carriers have small momentum relaxation rates as illustrated in Fig. 6b. In this case the carriers can achieve a significant macroscopic momentum up to the time when the field is screened. In the following time the continuing motion of the carriers is accompanied by a buildup of restoring Coulomb forces, which initiate the plasma oscillation of the charge carriers. Considering only one species of oscillating charges, they oscillate with a frequency given by the fundamental plasma frequency ω_p and the damping time τ_{sc}:

$$\omega = \sqrt{\omega_p^2 - \frac{1}{4\tau_{sc}^2}}, \qquad \omega_p = \sqrt{\frac{ne^2}{\varepsilon\varepsilon_0 n^*}}. \qquad (9)$$

Similarly, as already discussed for the strongly damped case, the leading slope of the emitted field results from the onset of the photocurrent and is mainly influenced by the duration of the exciting laser pulse, the mobility of the charge carriers, and the maximum drift velocity. After the plasma oscillations have started, the emitted field oscillates with the same frequency and the decay of the amplitude is given by the macroscopic momentum relaxation rate.

2. THE INSTANTANEOUS POLARIZATION

Independently, Yamanishi (1987) and Chemla *et al.* (1987) proposed the generation of ultrashort electrical pulses with a subbandgap photoexcitation in biased semiconductor heterostructures. In heterostructures the motion of charge carriers is quantized because of the nanometer dimensions of the two-dimensional semiconductor layers (Bastard, 1988). The wave functions can be made asymmetric when the quantum wells (QWs) are biased by an external electric field. Figure 7a shows a scheme of the band structure and an excitation process between the nearest states of conduction and valence band. Electrons and holes are pushed against opposite walls of the QW. This leads to the so-called quantum confined Stark effect (Miller *et al.*, 1984) and to a red shift in the absorption spectra.

An excitation below the lowest transition would not result in photogenerated electrons and holes after laser excitation. However, the quantum well responds with the buildup of a transient polarization to the driving laser field. The process of photoexcitation can be understood in the framework of perturbation theory using the Heisenberg picture. In these terms the electrons are driven by the laser field into a virtual state $\psi_{virt} = a_1(t)\psi_1 + a_2(t)\psi_2$, which is a linear combination of the wave functions ψ_1 and ψ_2. The wave functions ψ_1 and ψ_2 are the asymmetric wave functions of electrons and holes within the QW states. It is evident that the excited virtual state is accompanied by the buildup of a polarization. The polarization disappears instantaneously with the decay of the laser field

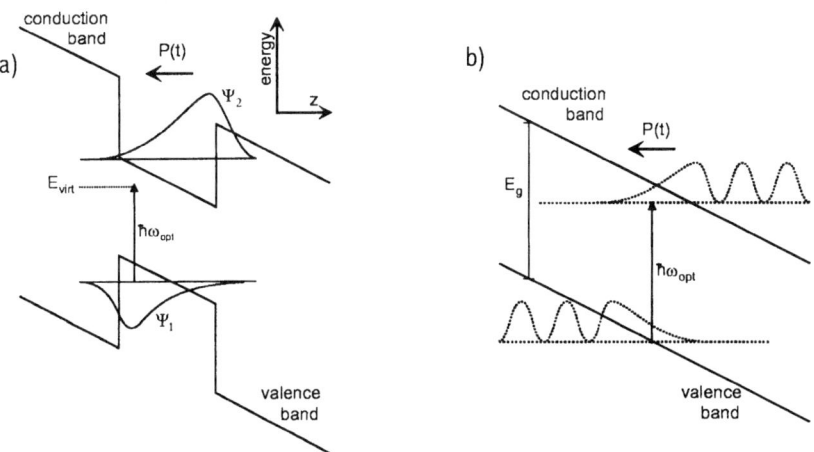

FIG. 7. (a) Subbandgap photoexcitation into a virtual state below the interband transition of a semiconductor heterostructure. (b) Subbandgap photoexcitation in a biased bulk semiconductor. The photoexcitation can be understood in terms of the inverse Franz–Keldysh effect.

since no real excitations are generated. The second time derivative of the transient polarization gives the emitted field, as shown in Fig. 6. If the laser pulses are of the order of 100 fs, the result is a broadband THz signal. The symmetrical shape of the emitted field is remarkable. The exciting laser produces a bipolar waveform with a negative center peak and positive leading and trailing lobes. Another important property is that the emitted THz power is proportional to the square of the intensity of the exciting laser pulse.

The same concept of generating a THz polarization by virtual excitation was extended to bulk semiconductors (Yablonovitch et al., 1989). The virtual excitation is found to generate a third-order nonlinear polarization that is the source of the emitted THz radiation,

$$P^{(3)}(\omega_0) = \chi^{(3)}(-\omega_0, \omega_0, 0, 0) \frac{|E(0)|^2}{2} E(\omega_0), \qquad (10)$$

where $\chi^{(3)}$ is the nonlinear susceptibility due to the Franz–Keldysh effect, ω_0 is the frequency of the laser pulse, and $E(0)$ is the DC field, which may be a surface depletion field or an applied bias. The instantaneous polarization also arises in the case of a photoexcitation above the bandgap of a biased semiconductor. However, in contrast to excitation below the bandgap, the polarization persists after the laser pulse has decayed. A detailed discussion of displacement and transport currents can be found in Kuznetsov and Stanton (1993).

3. Phases of Transport and Polarization

During the presence of the laser pulse, the carriers can be described by a state that is coherent to the exciting laser wave. This phase ends with the decay of the laser pulse, leading to a persistent carrier density in case of sufficient photon energy of the laser pulse. In the following, the carriers are accelerated by the applied field, which leads to a ballistic motion. Until the carriers undergo scattering their displacement increases quadratically with time. With the onset of sufficient scattering, the carriers move with constant velocity. Later the mass of the carriers may increase because of scattering into side valleys, which decreases the response of the carriers to a further acceleration by the electric field.

The different phases of charge transport can be identified by the THz emission signal, since each change of the carrier velocity induces a characteristic change of the emitted THz signal. Experiments performed by Hu and co-workers give a complete picture of the dynamics that lead to the emission of THz radiation (Hu et al., 1995). An experimental time resolution of 10 fs allows the separation of the distinct phases even on time scales before the charges undergo their first scattering events. Investigations on different semiconductors show how the carrier transport depends on band structure and the effective mass of the charge carriers. In a bulk GaAs Schottky diode a negative slope of the emitted THz signal was observed during photoexcitation as shown in Fig. 6. This signal results from the instantaneous electron–hole polarization as described before and decays with the laser pulse. In the following, only charge transport of photogenerated carriers contributes to the THz generation. After excitation the charge carriers perform a nearly ballistic motion. Within the first 70 fs the quadratic slope of the measured signal indicates the continuous acceleration. The observed THz signal mainly results from the transport of electrons because of their smaller effective mass. Later, the slope becomes linear while scattering sets in. The estimated values for the drift velocity show that at this time electrons reach a velocity overshoot. Finally, on a time scale of about 120 fs, the saturation drift velocity is achieved. The carriers have been transferred into the side valleys.

Besides the different phases of charge transport, the experiments show the influence of the band structure on the emitted THz signal. In quantum structures, as in the investigated multiple quantum well (MQW), the drift of photocarriers is limited by the band discontinuities of the quantum wells. Thus, the signal shows mainly a contribution from the instantaneous polarization. As discussed above, the inherent property of these excitations is their coherence with the driving laser field. They cannot be observed in indirect semiconductors such as silicon (see Fig. 8). In indirect semiconductors the photoexcitation is mediated by scattering processes such as phonon scattering that destroy the coherence with the laser field and therefore prevent the generation of an instantaneous polarization.

4. EMISSION FROM FIELD-INDUCED TRANSPORT

The generation of THz emission due to an ultrafast charge drift has been the focus of several experimental studies, for example, Zhang *et al.* (1990). In these experiments the photocarriers have been injected either into the depletion field of a semiconductor surface or in the intrinsic region of a *p-i-n* structure by photoexcitation with an ultrafast laser pulse. Figure 9 shows the waveform of a THz pulse emitted from an InP wafer that was excited with 70 fs pulses at around 620 nm. The observed THz pulse is attributed to the acceleration of photocarriers within the depletion field close to the surface of the InP wafer. At the excitation wavelength the absorption length of the laser light is on the order of a few hundred nanometers. Thus, a significant amount of photocarriers are generated in the surface depletion region. The Fermi level is pinned at the surface because of surface states,

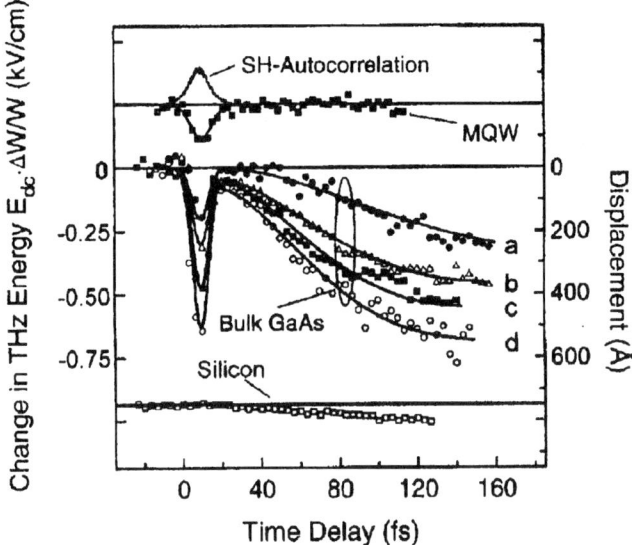

FIG. 8. Measured THz energy as a function of time delay relative to the prepulse for three samples: bulk GaAs, GaAs-MQW, and silicon. The bias field for curves a, b, c, and d are 15.8, 17.8, 20, and 23 kV/cm, respectively. The bias field in the intrinsic region of the *p-i-n* diode is roughly 50 kV/cm. The built-in field of the MQW sample is 20 kV/cm. The solid curves are numerical fits to the experimental data. For reference the autocorrelation trace of the optical pulse is also plotted. After Hu *et al.* (1995).

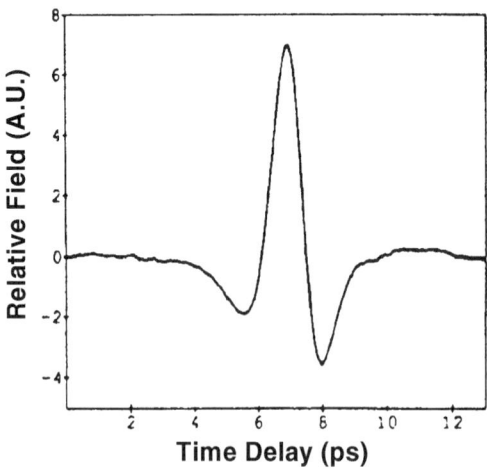

FIG. 9. Waveform of the outward electromagnetic radiation from a semi-insulating InP with $\Theta = 45°$. After Zhang *et al.* (1990). Reprinted with permission of American Institute of Physics.

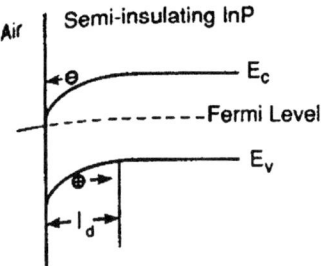

FIG. 10. Band diagram of semi-insulating Fe:InP. Surface states near the conduction band edge causes Fermi level pinning at the interface. Photocarriers are swept across the depletion width l_d by the built-in field. After Zhang and Auston (1992). Reprinted with permission of American Institute of Physics.

which gives a band structure as illustrated in Fig. 10. Close to the surface the band is bent and depleted of mobile carriers. In a first approximation the shape of the potential slope depends on the extrinsic carrier density and the built-in potential V_{bi}. In case of a structure doped at a concentration of N_d one obtains for the width of the depletion layer (Sze, 1985)

$$l_d = \sqrt{\frac{2\varepsilon_s}{qN_d} V_{bi}}. \tag{11}$$

The drop of the electric field within the depletion region is linear:

$$|E(z)| = \frac{qN_D}{\varepsilon_s}(l_d - z). \tag{12}$$

Even in the case of moderately doped semiconductors, the surface fields can be as high as 100 kV/cm. Photocarriers that are generated within the depletion zone are therefore accelerated strongly by the surface field and emit a THz light pulse during their acceleration.

As shown in Fig. 11, two THz beams are generated. One beam propagates in the direction of the reflected optical pulse and is called the pseudoreflected THz beam. The other propagates in the direction in which the exciting laser pulse would have propagated, if it were not absorbed by the semiconductor. Both propagation directions result directly from the coherence of the excited dipoles in the semiconductor surface, which is given by the wave front of the incident laser pulse. Thus, the directions of THz emission are independent of material properties (for example, the indices of refraction). The polarization of the THz beams depends on the orientation of the electric field relative to the surface. As the surface depletion field is normal to the surface; the photoinduced currents are also directed perpendicular to the

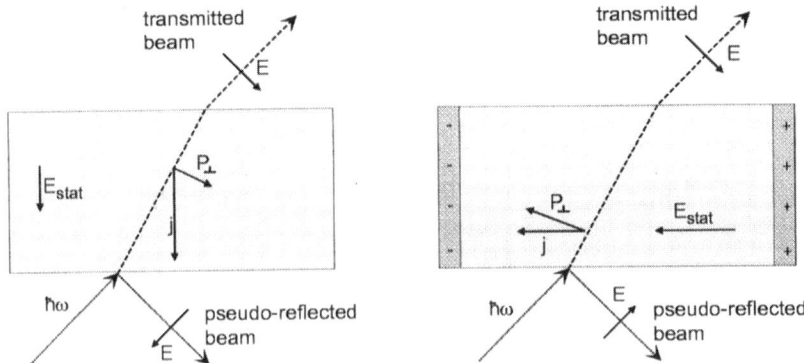

FIG. 11. Polarization of the transmitted and pseudoreflected THz beams generated in semiconductors with fields perpendicular (left) and parallel to the surface (right). After Li et al. (1995). Reprinted with permission of American Institute of Physics.

surface. As shown in Li et al. (1995), this geometry leads to antiparallel polarizations of transmitted and pseudoreflected THz beam (see Fig. 11). In case of an external field that is applied parallel to the semiconductor surface, for instance in large-aperture emitters, both THz beams have parallel polarization.

The intensities of the pseudoreflected and transmitted THz beams show a strong dependence on the angle of incidence (Fig. 12). At normal incidence the induced photocurrent is parallel to the propagation direction of the exciting light and thus the field amplitude is zero. A maximum THz emission is achieved close to the Brewster angle of InP at about 74°. According to

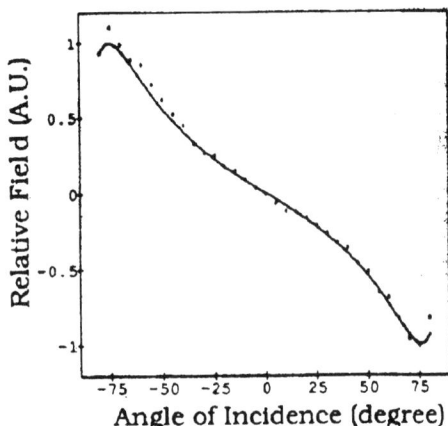

FIG. 12. Transmitted amplitude of radiated field E_1 as a function of the angle of incidence. After Zhang et al. (1990). Reprinted with permission of American Institute of Physics.

the Fresnel equations the amplitude of the THz emission field strongly depends on the relative index of refraction and the angle of emission relative to the semiconductor surface. As shown by Zhang and co-workers (1990), the amplitudes of the pseudoreflected field E_1, the inward radiated field E_2, and the transmitted radiated field E_t can be expressed as

$$E_1 = \eta J_s \frac{\sin \Theta_1}{\cos \Theta_1 + n \cos \Theta_2},$$

$$E_2 = -\frac{1}{n} E_1, \qquad (13)$$

$$E_t = t(\Theta_2) E_2,$$

where η is the characteristic impedance of the semiconductor. The inward and outward angles of the radiated fields are Θ_1 and Θ_2, respectively. The relative index of refraction n is given by the air/semiconductor interface and causes an angle-dependent transmission $t(\Theta_2)$. J_s is the photocurrent normal to the surface of the semiconductor and depends on the angle of incidence of the optical beam (Θ_1) and its reflection $R(\Theta_1)$ from the semiconductor surface:

$$J_s \sim [1 - R(\Theta_1)] \cos \Theta_1. \qquad (14)$$

5. Bandwidth Increase by Ultrafast Currents

The bandwidth of the generated THz pulses can be increased by tailoring the rise and decay times of the photocurrents. Ultrafast switch-on times of the photocurrent can be achieved using laser pulses with durations of about 100 fs, which are today the standard of most commercial laser systems. The switch-off time of the photocurrent depends on the carrier lifetime within the structure used and may be of the order of tens to hundreds of picoseconds in most commonly used semiconductors. An alternative is offered by materials such as radiation-damaged silicon on sapphire (SOS) (Doany et al., 1987) or low-temperature-grown GaAs (LT-GaAs) (Gupta et al., 1992). These materials are similar in that free carriers are efficiently captured into defect states. Carrier lifetimes of 200 fs have been reported for LT-GaAs (Prabhu et al., 1997) and 250 fs for SOS (Ketchen et al., 1986). In particular LT-GaAs exhibits promising properties for THz applications. Despite an excess of arsenic of about 1%, the material reveals a relatively high mobility of 1000 cm^2/Vs (Gupta et al., 1992) and a high resistivity that limits the dark current in biased emitters (Gupta et al., 1991).

6. EMISSION FROM AN INSTANTANEOUS POLARIZATION

Obviously the highest THz bandwidth should be achieved if the lifetime of photocarriers and the resulting current is as short as the laser pulse itself. Thus, high THz bandwidths can be expected when the exciting laser beam creates only a virtual population during the presence of the laser pulse. Experiments performed by Hu and co-workers on semi-insulating GaAs showed a THz signal that was attributed to virtual excitations. The signal observed at a detuning of $\delta\mathscr{E} = -20$ meV is attributed to the field emitted by the nonlinear polarization due to virtual excitation. Figure 13 shows how the THz signal changes when the laser excitation is tuned below the bandgap of GaAs. Most important is the change of sign of the main peak of the emitted signal (Hu *et al.*, 1994). This experimental observation agrees with the expected behavior as illustrated in Figs. 6a and 6c. Above the bandgap, the photoexcitation leads to a population of states that can contribute to a photocurrent and the buildup of the polarization. In case of a subbandgap excitation, the generated polarization decays with the laser pulse. According to Eq. (2), a field with a symmetric transient and a negative main peak is emitted.

IV. Coherent Charge Oscillations

A charge plasma responds in a collective motion when an external field drives it out of equilibrium. This collective response is related to the ability of the charge carriers to follow the time behavior of the external field. As the plasma frequency sets an upper limit for ultrafast electronic applications, it is of inherent importance for the usability of any solid-state material. However, plasma oscillations are rarely dominating semiconductor properties, since plasma frequencies are low because of the much smaller carrier concentrations as compared to metals. In general, impurity or phonon scattering is responsible for the dephasing, which is in most cases faster than the plasma oscillation period. Consequently, the frequency response due to the collective plasma oscillation is mostly structureless and does not exhibit a sharp resonance. Plasma oscillations are well suited for the generation of coherent THz radiation for two main reasons: First, the emission efficiency is increased, as already discussed in Subsection 1 of Section II. Since the coherent motion of the charge carriers is an inherent property of a plasma excitation, the emission intensity is proportional to the square of the number of coherently oscillating carriers. Thus, an increase of the emission intensity should be expected when increasing the number of oscillating charges within a coherence volume. Second, the collective motion of the electrons leads to a strongly enhanced frequency response. Compared to a pure structureless

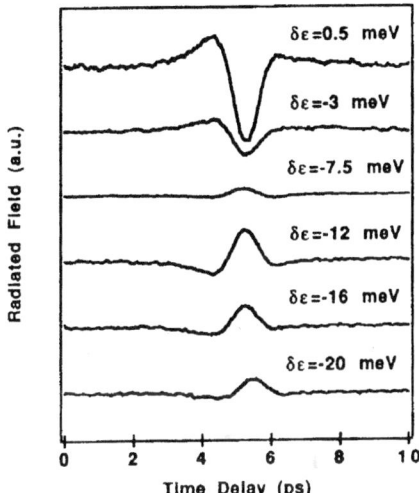

FIG. 13. Emitted THz fields from an undoped semi-insulating GaAs sample. The sample temperature is kept below 130 K. After Hu *et al.* (1991).

Drude-like response, a plasma oscillation persists over several oscillation periods, leading to prolonged coherence lengths of the emitted THz radiation. The benefits of the temporal and spatial coherence of the emission are obvious: Plasma oscillations are expected to emit more efficiently because of their increased emission time. In case of an emission from a semiconductor surface, a higher spatial coherence and thus a higher directionality are the result.

With the use of ultrafast laser pulses, plasma oscillations can be initiated in semiconductors by impulsive excitation. Because of this short-time excitation, these oscillations exhibit coherent properties along the illuminated surface. In the following the term *coherent plasmon* will address a macroscopic excitation across the surface of a semiconductor that is large compared to the wavelength of the plasma oscillation. Both the spatial coherence of the plasma and the resonance at the plasma frequency lead to attractive new THz emission properties.

1. PLASMA OSCILLATIONS OF PHOTOEXCITED CARRIERS

A first experimental indication of coherent plasma oscillations in semiconductors was observed by Sha *et al.* (1995). The experiments were performed on GaAs *p-i-n* structures where a bias was applied between the *n*-doped cap layer and the *p*-doped substrate of the device. Photocarriers were generated in the device by near-bandgap excitation with 100 fs laser pulses. As the

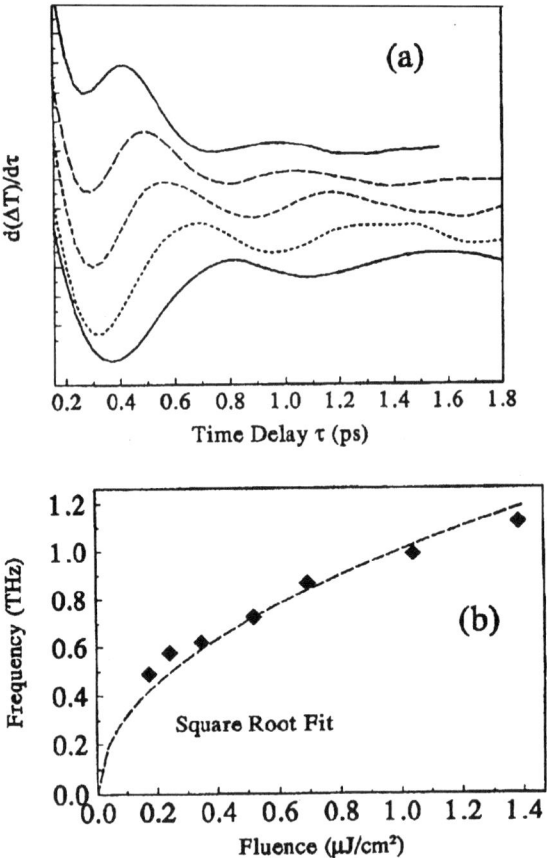

FIG. 14. (a) The time derivative of the change in electroabsorption vs. pump-probe time delay for fluences of 1.0, 0.7, 0.5, 0.35, and 0.25 μJ cm^2 (top to bottom, respectively), and (b) the frequency of the oscillating field (as extracted from the electroabsorption measurements) as a function of incident optical fluence. The dashed curve is a least square fit by a square-root function (Sha et al., 1995).

n-doped GaAs top layer had a thickness of a few nanometers, most of the photocarriers were generated in the field region of the intrinsic zone. Photogenerated electrons and holes are accelerated by the field in opposite directions. The buildup of the polarization leads to a restoring force that causes the plasma to oscillate. The transients of the electric field within the device were mapped by time-resolved electroabsorption measurements. Oscillations of the electric field arise as shown in Fig. 14. They are attributed to the oscillating polarization of the photogenerated electron–hole plasma. The observed oscillation frequencies followed a square root dependence on the excitation density as given by Eq. (9). Similar oscillations were observed by Fischler et al. (1996) after photoexcitation of GaAs Schottky diodes.

The square root dependence of the THz emission frequency was confirmed by measurements of the THz emission from photoexcited plasma oscillations (Kersting et al., 1997b). Time-resolved autocorrelation experiments as described in Subsection 3c of Section II enabled the measurements of the coherence length of the emitted THz pulses. The observed emission spectra were broad, indicating that the coherence of the plasma is lost on the time scale of a few oscillation cycles because of scattering. Electron phonon scattering is found to be the major scattering mechanism leading to the coherence loss of the plasma (Fischler et al., 1996). The acceleration of the electrons in the electric field leads to a considerable occupation of energy levels above the LO-phonon energy and thus to strong phonon scattering. The hot carrier distribution depends on the initially excited distribution of carriers within the band, that is, the excess energy of the photogenerated carriers (Wysin et al., 1988). Measurements using a photoexcitation close to the bandgap reveal shorter damping times than excitations high in the band (Fischler et al., 1996). The amplitude reached by the charge carrier oscillation during their initial displacement is limited by the built-in field of the device and the dominant scattering. The upper limit set by the built-in field E_B is

$$x_{max} = 2\varepsilon E_B / e n_{exc} \qquad (15)$$

where E_B is the built-in field in the device and n_{exc} is the density of photogenerated carriers. This can lead to oscillation amplitudes as high as 200 nm for fields of 3.5 kV/cm (Sha et al., 1995). Assuming a ballistic motion until the carriers reach an energy where phonon emission sets in gives an estimate of the amplitude:

$$x_{sc} = \frac{\Delta E_{ph}}{e|E|}. \qquad (16)$$

For a phonon energy ΔE_{ph} of 36 meV and a field of 16 kV/cm a displacement of about 20 nm is obtained for GaAs. This agrees with the value experimentally observed by Hu et al. (1995) for the maximum displacement before scattering sets in. The difference between x_{max} and x_{sc} determines the gain of energy in the electric field, which leads to a heating of the carrier distributions. This heating limits the dipole moment that can be achieved because of the increased scattering rates that occur in hot carrier distributions. Since the THz emission power is proportional to the square of the dipole moment and thus to the oscillation amplitude, the heating of the electron distribution leads to a significant loss in the coherent THz emission intensity (Kersting et al., 1997a).

2. PLASMA OSCILLATIONS OF EXTRINSIC CARRIERS

The reduced emission efficiency of the hot plasma has stimulated efforts to keep the electron temperature low and prevent the carriers from inelastic scattering. An alternative approach has been found in n-doped semiconductors. Here coherent oscillations of cold extrinsic electrons can be initiated by the ultrafast dynamics of the surface field that follows a femtosecond laser excitation (Kersting et al., 1997b).

The plasma oscillations are mapped in time-resolved autocorrelation experiments on bulk n-doped GaAs. As shown in Fig. 15, an overdamped oscillation of the carriers is observed at low doping concentrations. However, with increasing doping concentration the plasma frequency increases, which leads to several oscillations before the collective motion of the carriers is damped out by scattering. The observed center frequency of the THz emission is given directly by the doping concentration, indicating that the source of the THz emission is the collective oscillation of the extrinsic electrons. Thus, n-doped GaAs can be used as a material for the generation of THz radiation in which the frequency of emission can be set by the doping. Additionally, emission intensities as high as 100 nW can be gener-

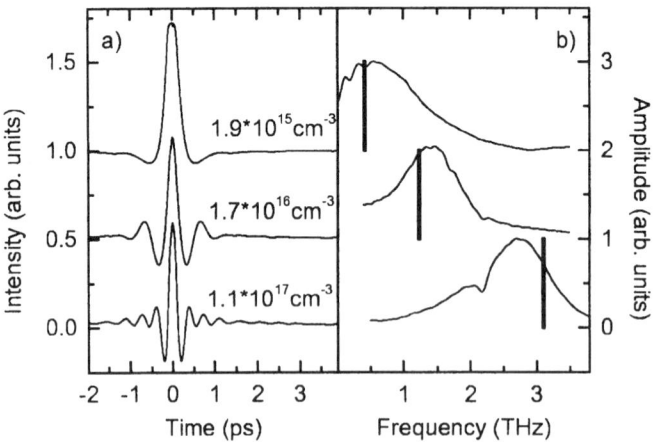

FIG. 15. (a) Correlation data recorded on n-doped GaAs structures. The excitation density is always smaller than the doping concentration. (b) Corresponding Fourier spectra. The bars indicate the calculated plasma frequencies. From Kersting et al., 1998.

ated. Since the extrinsic electrons are cold compared to photoexcited carriers, they undergo less scattering, leading to an increased emission intensity.

a. Excitation of Cold Plasma Oscillations

Initially the extrinsic bulk electrons are confined between the surface depletion field and the undoped GaAs substrate. Their plasma oscillation is started by the ultrafast field change after photoexcitation (see Fig. 16). A similar mechanism has been reported previously for the excitation of coherent phonon oscillations in GaAs (Cho et al., 1990). Calculations of the photocarrier dynamics in the depletion field (Dekorsy et al., 1993) have shown that besides the ultrafast drift transport, the diffusion of photocarriers also contributes significantly to the field dynamics. Dember fields (Dember, 1931) arise from differences in the diffusion of electrons and holes and lead to ultrafast field changes in the region of the depletion zone. Detailed descriptions of models applied for the analysis of the field dynamics can be found in Dekorsy et al. (1993) and Rosenwaks et al. (1994). Figure 16 shows the dynamics of both the photocarrier generation and the depletion field at the surface of an n-doped GaAs epilayer (Kersting et al., 1998). The density of photoexcited carriers is 10^{16} cm^{-3} and the doping concentration 1.7×10^{16} cm^{-3}. At this doping concentration the depletion zone has a

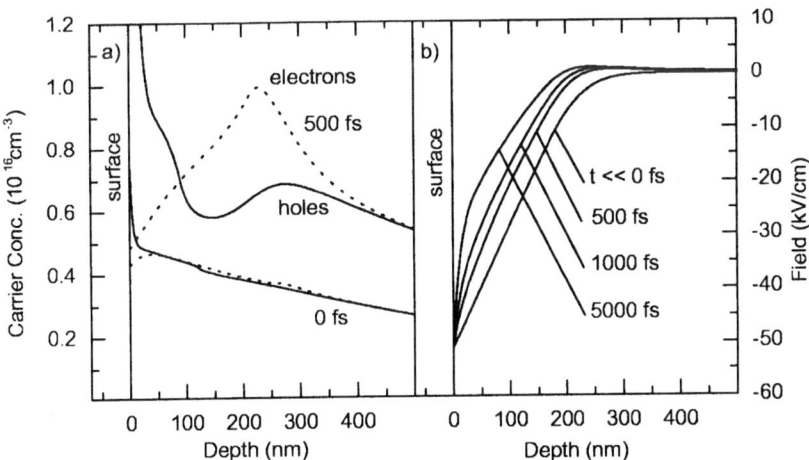

FIG. 16. (a) Calculated densities of photogenerated electrons (dashed lines) and holes (solid lines) in a structure n-doped at 1.7×10^{16} cm^{-3} during excitation (0 fs) and 500 fs after excitation. The excitation density is 10^{16} cm^{-3}. (b) Change of the surface depletion field after photoexcitation. From Kersting et al., 1998.

width of about 250 nm. Even 500 fs after excitation by an ultrafast laser pulse, the photogenerated electrons and holes have been separated by the surface depletion field. The accompanying screening leads to significant field changes within the former depletion zone. Positive fields arise due to the Dember effect, where the depletion zone ends and the density of extrinsic electrons begins. The positive field change accelerates the extrinsic electrons toward the surface and initiates their plasma oscillation (Kersting et al., 1998).

b. Damping

As shown in Subsection 1 of Section II, the efficiency of THz emission depends strongly on the ratio between the radiative and the nonradiative lifetime of the plasma excitation. Thus, an understanding of the damping mechanism is of vital importance for the use of coherent charge oscillations for the generation of THz radiation. Since plasma oscillations are collective excitations, their nonradiative lifetime is the relaxation time of the macroscopic momentum. Elastic scattering processes such as electron–electron and hole–hole scattering do not contribute to the relaxation, since the macroscopic momentum is conserved. In unipolar plasmas, the electron–hole scattering can also be neglected, and the damping times observed in THz experiments can therefore be directly related to scattering times connected to the DC mobility $\mu = e\langle\tau\rangle/m^*$.

Extensive studies have shown that in n-doped III–V compounds the room-temperature mobility is mainly limited by optical phonon scattering (Stillman et al., 1970). At high doping concentrations, impurity scattering contributes significantly to the dephasing. The analysis of the temperature-dependent damping time of plasma oscillations leads to a similar result. Experiments on n-doped GaAs showed that for doping concentrations of up to 10^{16} cm^{-3}, the coherence time of the plasmon is mainly limited by optical phonon scattering, which is in agreement with results from mobility measurements (Kersting et al., 1998). At higher doping densities, shorter damping rates are observed, indicating the dominance of impurity scattering. To find the regions of reduced damping is important, since the emitted THz pulse energy W depends on the damping time according to

$$W \propto (\tau E)^2. \qquad (17)$$

Hu and co-workers found the predicted increase of emission energy in several semiconductors (Hu et al., 1990). Phonon scattering was reduced by decreasing the lattice temperature to 80 K. In n-doped GaAs an increase of the emission intensity by a factor of 3.4 was measured and attributed to the increased mobility.

3. Plasma Oscillations in 2D Layers

In two-dimensional structures the THz emission from charge oscillations is quite different from those in 3D bulk semiconductors. The main difference arises from the quantization of the carrier motion perpendicular to the layer, such as that present in surface inversion layers (Stern and Howard, 1967). The reduced number of degrees of freedom leads to longitudinal plasma waves having a wave vector that has only components parallel to the 2D layer (Stern, 1967; Ritchie, 1973). The dispersion of these longitudinal waves depends strongly on the geometrical properties of the structure. For strictly two-dimensional plasmas, the dispersion for small wave vectors is given by $\omega_{pl} \propto k_{\|}^{1/2}$ (Stern, 1967). For layered carrier systems the dispersion is linear (Allen et al., 1977; Olego et al., 1982). However, in general the slope of the plasmon dispersion is smaller than the slope of the light line. This prevents an intersection between the two dispersions and excludes a direct coupling between the 2D plasmon and freely propagating light. An efficient coupling between the plasmon and the light line can be achieved with metallic gratings of micrometer periodicity deposited on the semiconductor surface.

Confined two-dimensional electron gases have found considerable interest as tunable far-infrared emitters because the plasma frequency can be controlled by tuning of the carrier density. Intense studies have been performed on silicon metal-oxide-semiconductor field effect transistors (MOSFETs) (Tsui et al., 1980; Höpfel et al., 1982b) and AlGaAs/GaAs heterostructures (Höpfel et al., 1982a; Hirakawa et al., 1995) by measurements of continuous wave emission in the THz frequency band. Coherent THz emission from a 2D plasmon was observed after creation of a nonequilibrium plasma distribution in a selectively doped AlGaAs/GaAs single-interface heterojunction (Hirakawa et al., 1996; Vosseburger et al., 1996). The short-time excitation process was induced by shining 100 fs laser pulses through the metallic grating and creating photocarriers in the GaAs layer. The time-resolved THz signals show a strong beating (Fig. 17), indicating an emission originating from different transitions. The observed emission frequencies were found to agree with the first- and second-order plasma frequencies, indicating that the observed coherent emission is due to the radiative decay of 2D plasmons. Temperature-resolved measurements have shown that the coherence loss of the 2D plasmon is similar to that of bulk plasmons. At temperatures above 100 K a fast dephasing is observed. Below 100 K the plasmon decay is considerably slower, but still faster than expected from the cw mobility, indicating that the presence of optical excitations induce additional relaxation channels.

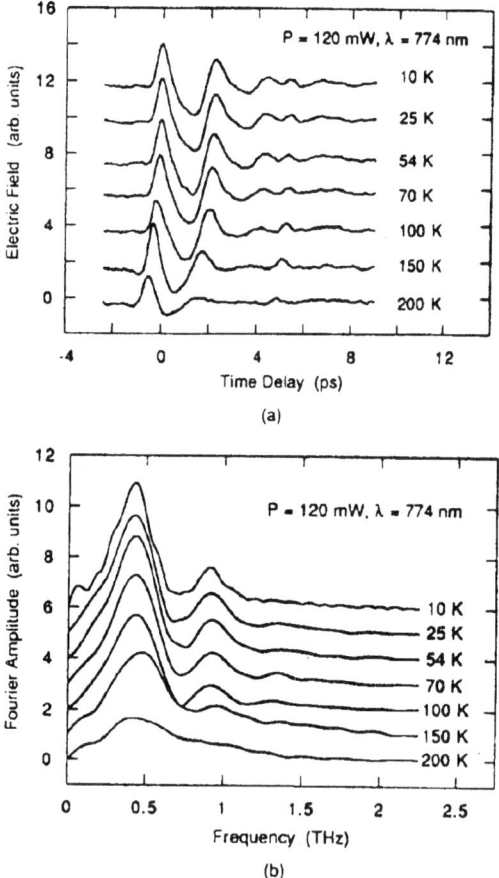

FIG. 17. Temperature dependence of the waveforms (a) and the Fourier spectra (b) of the coherent THz radiation from 2D plasmons excited by optical pulses with a wavelength of 774 nm. Each curve is shifted for clarity. From Hirakawa *et al.* (1996). Reprinted with permission from Elsevier Science.

4. Coherent Phonons and Plasmon–Phonon Coupling

Coherent phonon oscillations can be initiated when the field change resulting from a laser excitation is faster than the oscillation period of the phonon. In GaAs coherent phonons were found to be initiated by the screening of the surface depletion field (Cho *et al.*, 1990). In tellurium coherent phonon oscillations arise because of an ultrafast buildup of a Dember field (Dekorsy *et al.*, 1995). Other mechanisms are summarized by Kütt *et al.* (1992). In case of infrared-active phonon modes, the collectively oscillating phonons lead to the emission of a coherent THz pulse. A recent

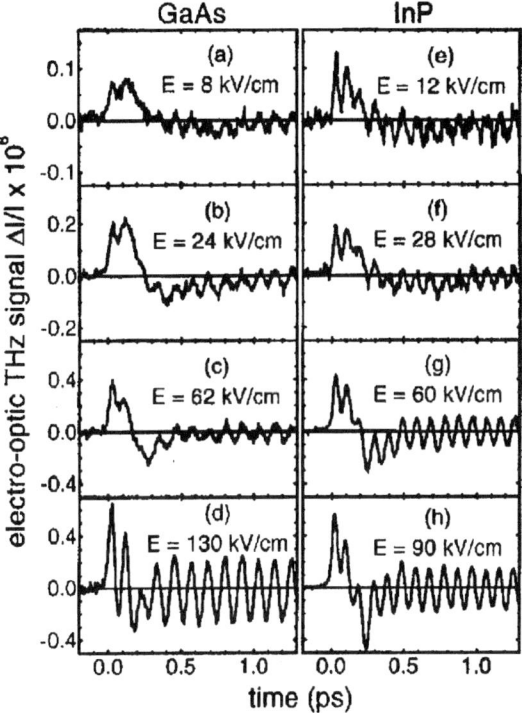

FIG. 18. THz transients from GaAs (a–d) and InP (e–h) *p-i-n* diodes for different bias fields E and at 300 K. Note the different scales of the abscissa. From Leitenstorfer *et al.* (1999b).

work on GaAs and InP (Leitenstorfer *et al.*, 1999b) showed that the field of the THz pulse is given by the second derivative of the carrier displacement $D(t)$ reduced by the lattice polarization $P(t)$:

$$E_{THz} \propto \frac{\partial^2}{\partial t^2}[D(t) - P(t)]. \tag{18}$$

As a result, THz signals consisting of a contribution from the LO phonon (8.8 THz in GaAs and 10.3 THz in InP) are observed (Fig. 18). While the contributions due to the charge dynamics decay within a picosecond, the phonon oscillations reveal a significantly larger decay of the order of 5 ps and thus lead to sharp structures in the emission spectra.

At high carrier densities where the plasma frequency comes close to the optical phonon frequencies, the plasmon–phonon coupling has to be considered. The coupled plasmon–phonon modes can be described by their coupled equations of motion (Kuznetsov and Stanton, 1995),

$$\ddot{P} + \gamma_{pl}\dot{P} + \omega_{pl}^2 P = \frac{e^2 N}{\varepsilon_\infty \mu}[E_{ext} - 4\pi\gamma_{12}Q] \quad (19)$$

$$\ddot{Q} + \gamma_{ph}\dot{Q} + \omega_{LO}^2 Q = \frac{\gamma_{12}}{\varepsilon_\infty}[E_{ext} - 4\pi P], \quad (20)$$

where P is the electronic polarization, Q is the normalized lattice displacement, $\omega_{pl,ph}$ are the plasmon and phonon frequencies, E_{ext} is the driving field, and γ_{12} is the electric-field phonon coupling. Coherent plasmon phonon modes have been observed in bulk n-doped GaAs (Cho et al., 1996) and GaAs/AlGaAs heterostructures containing a 2D electron plasma (Baumberg and Williams, 1996).

V. THz Emission from Quantum Structures

1. Quantum Beats

In coupled quantum wells, charge oscillations arise as an inherent property of the quantized carrier states within the wells. Charge oscillations can be realized in coupled quantum wells as shown in Fig. 19. One important requirement is that the separating barrier be thin enough to enable a strong coupling between the wells by tunneling. At the resonance of the single-well states $|1\rangle$ and $|2\rangle$, new mixed eigenstates arise, giving a symmetric and antisymmetric wave function

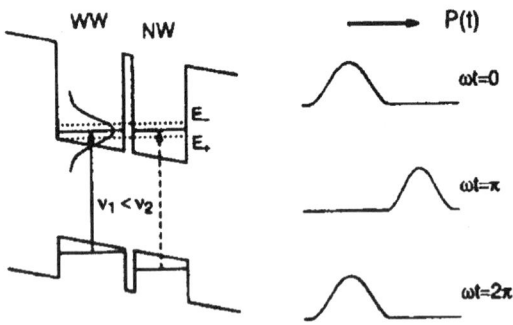

FIG. 19. Schematic band diagram of the asymmetric coupled-quantum-well structure. Interband transitions in the wide well (WW) occur at lower photon energies than in the narrow well (NW), so that a short laser pulse with a width $\Delta v > (E_- - E_+)/h$ can excite an electronic wave packet that is mostly localized in the WW. A time-varying polarization $p(t)$ accompanies the tunneling of the wave packet back and forth between both wells that follows the excitation in the WW. After Roskos et al. (1992).

$$|\pm\rangle = \frac{1}{\sqrt{2}}(|1\rangle \pm |2\rangle). \quad (21)$$

The corresponding energy eigenvalues E_+ and E_- are shown in Fig. 19. After an electron has been put at time $t = 0$ in either one of the quantum wells, corresponding, for example, to an initial state $|2\rangle$, this state evolves according to (Luryi, 1991)

$$|t\rangle = e^{-iE_0 t/\hbar}[|2\rangle \cos(\omega t/2) - i|1\rangle \sin(\omega t/2)], \quad (22)$$

where the oscillation frequency ω corresponds to the splitting $\Delta E = E_- - E_+$. In the absence of scattering the electron will perform oscillations between the wells, leading to an oscillating dipole, which results in the emission of radiation.

Coherent charge oscillations of a wave packet were first observed using degenerate four-wave mixing in a GaAs/AlGaAs structure containing asymmetric quantum wells (Leo *et al.*, 1991). The asymmetry enables the selective excitation of the wide well by subpicosecond laser pulses because of its lower interband transition energy as compared to the narrow well. The beating frequency can be controlled by applying an electric field across the structures. The emission of coherent THz radiation generated on a similar structure was achieved by Roskos *et al.* (1992). To prevent a fast dephasing, the structures were held at cryogenic temperatures. Figure 20 shows THz

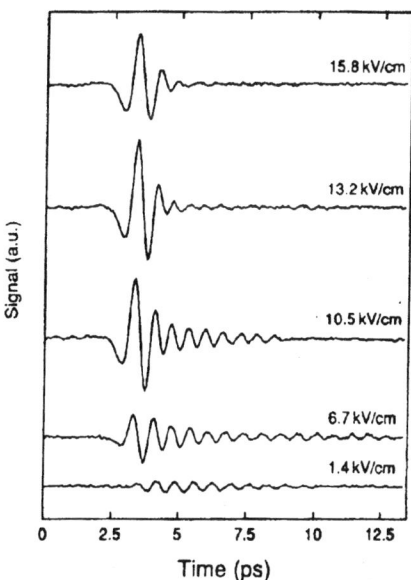

FIG. 20. Measured coherent electromagnetic transients emitted from the coupled quantum well for different bias fields for a photon energy of 1.53 eV. After Roskos *et al.* (1992).

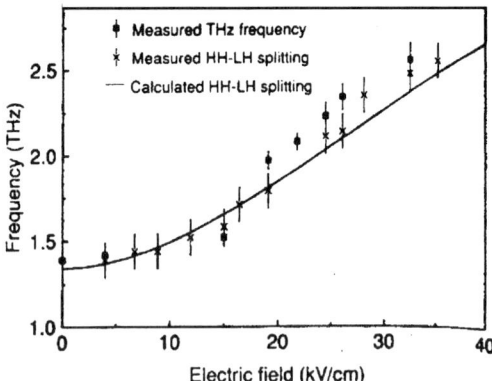

FIG. 21. Measured THz frequencies, lh–hh splittings taken from photocurrent spectra, and calculated lh–hh splittings as a function of electric field in the MQW. The calculation includes excitonic effects. After Planken et al. (1992).

transients recorded at different electric fields corresponding to different resonance conditions. A direct correlation of the emission frequency on the applied field was not observed. This observation was attributed to excitonic effects that arise because the photogenerated holes remain localized while the electrons perform the oscillation (Luo et al., 1994).

A tunable THz emitter was demonstrated by Planken et al. (1992) using a biased single quantum well. In this case the charge oscillations are the result of the time evolution of the superposition of light-hole and heavy-hole exciton states. As shown in Fig. 21, the emission frequency could be tuned between 1.4 and 2.6 THz. A similar concept for the generation of a coherent superposition of states is used in asymmetric step-quantum wells. Here, the photoexcitation can be performed from a heavy-hole state into the first two quantum states of the electrons. Frequencies as high as 30 THz have been reported (Bonvalet et al., 1996). At these high frequencies a coherent THz emission can be observed even at room temperature, since several oscillations are performed before scattering destroys the coherence.

2. BLOCH OSCILLATIONS

Bloch oscillations, one of the fundamental concepts of solid-state physics, describe the motion of electrons in a periodic potential. Although the theoretical framework was set up by F. Bloch (Bloch, 1928) and C. Zener (Zener, 1934), it was more than 60 years later that these oscillations were observed experimentally. Bloch oscillations arise when electrons are accelerated by a field in a periodic potential. The periodicity of the potential leads

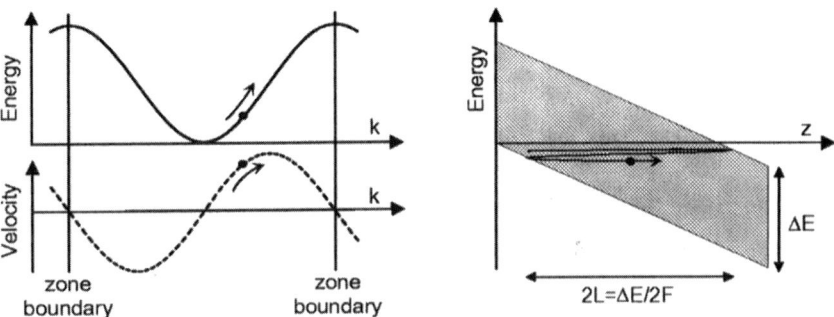

FIG. 22. Semiclassical picture of Bloch oscillations: (Left) Dispersion relation and real space group velocity. (Right) Associated real space oscillation.

to a dispersion of the band structure as illustrated in Fig. 22. Within the semiclassical picture of Esaki and Tsu (1970), the accelerated electron is reflected every time it reaches the boundary of the Brillouin zone. In the absence of scattering, the electron will then perform a periodic motion with a Bloch frequency given by

$$v_{Bloch} = \frac{a}{2\pi}\frac{dk}{dt} = \frac{eFd}{h} \tag{23}$$

where a is the lattice constant of the periodic potential and F the electric field. This motion is associated with an oscillation in real space. For the amplitude of the oscillation one can find a value derived in this semiclassical model,

$$L = \frac{\Delta E}{2eF}, \tag{24}$$

where ΔE is the width of the band as illustrated in Fig. 22. An alternative description of Bloch oscillations can be found using the Wannier–Stark picture, which is described in detail in Haring Bolivar et al. (2000). Describing the electrons in terms of wave functions of Wannier–Stark states leads to delocalized wave functions of the electrons in the case of small fields (Wannier, 1960, 1962). For higher fields a split-up into a series of discrete levels, the Wannier–Stark ladder, arises:

$$E_n = E_0 + neaF. \tag{25}$$

The difference frequency of these levels can then be related to the Bloch frequency of Eq. (23).

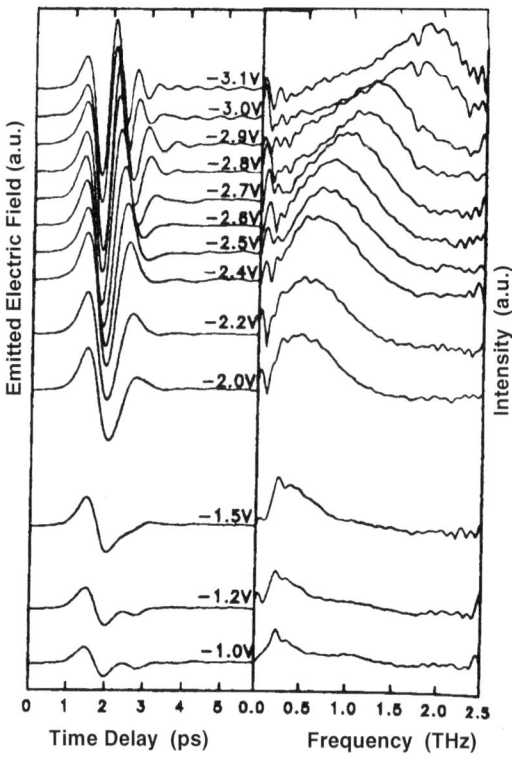

FIG. 23. (Left side) Measured coherent electromagnetic transients from the superlattice for different reverse-bias voltages. (Right side) Fourier transforms of the time-domain data. The spectra are corrected for the frequency dependence of the detection system. (After Waschke et al., 1993).

The short lattice periodicity of common solids would require enormous fields to achieve a Bloch oscillation before scattering interrupts the motion of the charge carrier. Such fields are far beyond electrical breakdown between the bandgap. However, semiconductor superlattices enable a delocalization of electrons over a large number of lattice periods and thus the formation of a miniband (Esaki and Tsu, 1970). The much longer periodicity enables a complete Bloch oscillation at reasonable fields. Bloch oscillations were first observed in time-resolved experiments in GaAs/AlGaAs superlattices (Feldmann et al., 1992; Leo et al., 1992). Waschke et al. demonstrated that electrons generated by a femtosecond laser pulse within a superlattice emit THz radiation during their Bloch oscillation (Waschke et al., 1993). As shown in Fig. 23, the oscillation frequency ranges from about 0.5 to 2 THz.

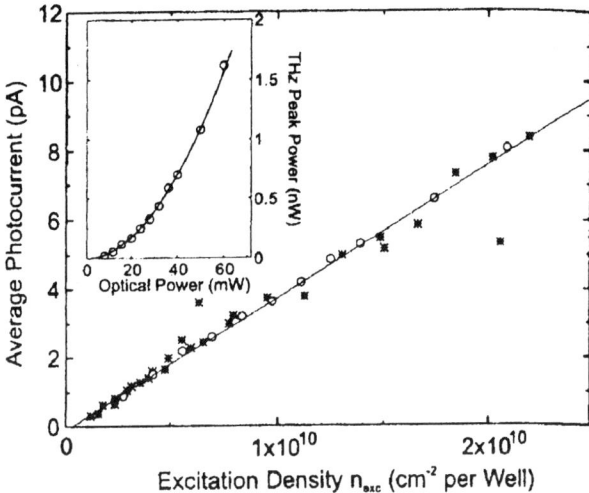

FIG. 24. Amplitude of the THz radiation as a function of the density of photoexcited carriers. Data for two excitation photon energies are shown: 1.542 eV (stars) and 1.557 eV (circles). The inset shows the calculated THz peak power as a function of the optical power for the excitation energy of 1.557 eV. The 1.542 eV data scatter more strongly because of the lower signal-to-noise ratio in the particular measurements. The solid lines are guides to the eye. After Martini et al. (1996).

The clearly observed dependence of the emission frequency on the applied superlattice bias illustrates the potential of Bloch oscillators as frequency tunable sources of coherent THz emission. However, applications are hampered by the low conversion efficiencies, which are typically below 10^{-7} (Haring Bolivar et al., 2000).

Novel concepts use the superradiant properties of coherently oscillating electrons as they are present in a Bloch oscillator (Victor et al., 1994). As discussed in Subsection 2 of Section II, the emission intensity scales with the square of the number of optically excited electrons. This leads to a drastic increase of the THz emission power for times shorter than the coherence time of the electron motion. Investigations on the excitation dependence (Fig. 24) of the THz emission power have clearly shown the expected quadratic dependence on the excitation density of photogenerated electrons (Martini, et al., 1996). Although the observed properties are promising, a further enhancement is hindered by the loss of coherence due to increased scattering at higher carrier densities mainly due to an increase in carrier–carrier scattering. At highest densities the emission power is limited to about 1.75 nW (Martini et al., 1996).

3. Intersubband Transitions

Modulation n-doped quantum structures possess low scattering rates (Dingle et al., 1978), which makes them attractive for efficient THz generation. Here, the transition of electrons between neighboring intersubband states can be used for the emission of far-infrared radiation. For a discussion of intersubband transitions, the single-particle picture of an electron, which is bound within a QW, is not sufficient. Collective phenomena arise because the radiation field within the quantum well is screened by the two-dimensional electron gas (Ando et al., 1982). The interaction with radiation not only causes intersubband transitions, it also modulates the charge density within the QW and impresses a plasmonic character onto the intersubband transition. The so-called depolarization shift leads to increased absorption and emission frequencies that are, in approximation (Allen et al., 1976),

$$\omega = \sqrt{\omega_0^2 + \omega_p^2}, \tag{26}$$

where ω_0 is the intersubband frequency of the single particle intersubband transition. ω_p can be understood in terms of a three-dimensional plasma frequency. Other many-body effects, such as exchange correlation and exciton interaction, are summarized in Helm (2000). In general, they lead to much smaller modifications of the intersubband transition compared to the depolarization shift.

Far-infrared luminescence resulting from incoherent intersubband transitions has been observed after optical pumping (Lavon et al., 1995) or electrical excitation (e.g., Gornik and Tsui, 1978; Helm et al., 1988; Maranowski et al., 1996). The coherent response of intersubband transitions was investigated by Heyman et al. (1998) using as excitation a broadband THz pulse. By absorption of the THz pulse electrons are excited from the lower into the upper intersubband state as shown in the inset of Fig. 25. Since the transition is accompanied by an oscillating polarization, the electrons emit an electric field during their coherent motion that is characterized by their intersubband frequency and coherence time. As shown in Fig. 25, the electrons perform several oscillations before their collective motion is damped out. These prolonged decay times result from the reduced scattering within the modulation-doped QW, which is held at 6 K. In consequence, the electrons emit a THz signal with a sharp amplitude spectrum.

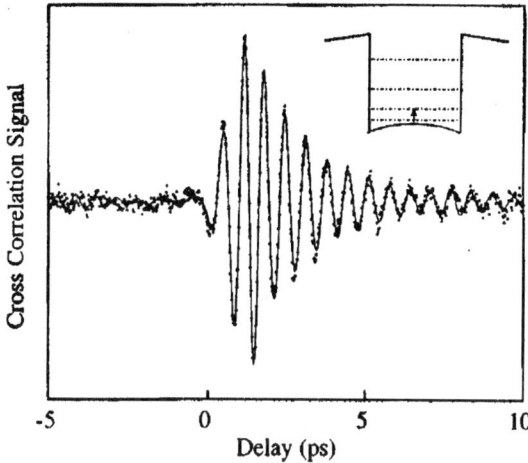

FIG. 25. Measured cross-correlation signal (dots) obtained by modulating the charge density in the 510 Å quantum well sample, and recording the change in transmission with a lock-in amplifier. The THz pulse excites electrons into a coherent superposition of states in the lowest subbands (see sketch in the inset). The solid line is a simulation calculated from a single oscillator model of the quantum well response. After Heyman *et al.* (1998). Reprinted with permission of American Institute of Physics.

VI. Applications in Semiconductor Spectroscopy

Triggered by the demand for ultrafast semiconductor devices, time-resolved spectroscopy has become an important tool for the investigation of charge transport in semiconductors. A very interesting spectral range is the far-infrared. Here, the frequencies of electromagnetic waves become comparable to the scattering rates of carriers, which makes THz spectroscopy an attractive method for investigations of ultrafast charge transport in semiconductors (van Exter and Grischkowsky, 1990a). Although the intensities of pulsed THz radiation are much higher than those that are achievable with thermal light, applications of THz pulses are still limited to linear spectroscopy. In this case the coherent THz radiation is used as a probe of the linear dielectric function, which results from the charge carriers within the semiconductor.

According to the Drude model, the dielectric constant and complex conductivity are given by

$$\varepsilon(\omega) = \varepsilon_\infty + \frac{i\sigma}{\omega \varepsilon_0} = \frac{\omega_p^2}{\omega(\omega + i\gamma)} \qquad (27)$$

$$\sigma(\omega) = \sigma_{DC} \frac{i\gamma}{\omega + i\gamma} = \frac{i\varepsilon_0 \omega_p^2}{\omega + i\gamma} \qquad (28)$$

where $\gamma = 1/\tau_{sc}$ and ε_∞ results from the contribution of the dielectric. The angular frequency of the plasma ω_p is defined according to Eq. (9) and depends on the density of mobile carriers n. The DC conductivity σ_{DC} is given by $\sigma_{DC} = e\mu n$, where μ is the DC mobility and n the concentration of mobile carriers.

1. OPTICAL PUMP–THz PROBE SPECTROSCOPY

Linear THz spectroscopy (e.g., transmission of reflection spectroscopy) provides the complex index of refraction $\tilde{n}(\omega) = \sqrt{\varepsilon(\omega)}$ and thus an estimate of the mean scattering rate and the complex conductivity. Transient changes of scattering rates can be measured by optical pump–THz probe spectroscopy where photocarriers are impulsively generated by the absorption of a femtosecond laser pulse. One example is the change of the electron mobility after photoexcitation in GaAs (Nuss et al., 1987). In these experiments, the photoexcitations were generated in GaAs by absorption of 60 fs laser pulses having a center wavelength of 625 nm. The transport properties of the photogenerated electrons were probed by the reflection of a THz pulse. This technique is primarily sensitive to electrons due to their smaller effective

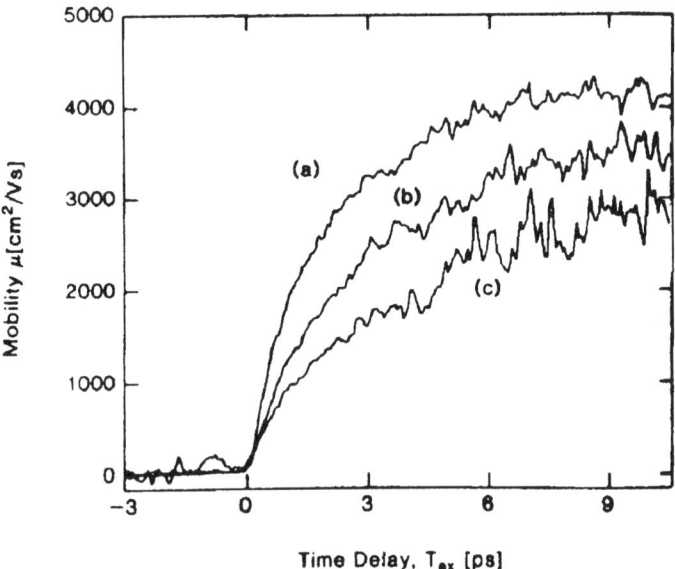

FIG. 26. Change of the electron mobility extracted from THz reflection measurements in GaAs. The traces a–c correspond to injected carrier densities of 5×10^{17}, 5×10^{18}, and $5 \times 10^{19}\,\text{cm}^{-3}$, respectively. After Nuss et al. (1987).

masses compared to holes. Figure 26 shows the changes of the electron mobility after photoexcitation. At the high photon energies used in this experiment, most photogenerated electrons are scattered within a few hundred femtoseconds into the satellite valleys of GaAs where the electrons have a low mobility and show a small response to the probing THz pulse. With a time constant of about 2 ps the electrons return into the Γ-valley, which becomes visible due to the increase in mobility. Measurements performed at higher photoexcitation densities show a reduced mobility that is attributed to increased electron–hole scattering. The increased rise time may result from a nonequilibrium LO phonon distribution, which prolongs the energy relaxation of the electrons.

Another application of optical pump–THz probe spectroscopy is the measurement of carrier lifetimes in semiconductors. The dynamics can be measured because every decrease of the carrier density is accompanied by a decrease of the plasma frequency, which changes the dielectric function. The short durations of the THz pulses have made THz transmission spectroscopy a useful tool to deduce ultrafast free carrier lifetimes. THz spectroscopy on LT-GaAs have been used to determine carrier life times as short as 200 fs and gave insight into the relation between the capture time and the growth and annealing conditions of LT-GaAs (Prabhu et al., 1997).

2. THz Time-Domain Spectroscopy

THz time-domain spectroscopy (THz-TDS) is a new technique to measure the response of electrons when driven by a THz field. Although the technique is analogous to conventional far-infrared Fourier spectroscopy, a much higher signal-to-noise ratio can be achieved, since the pulsed THz radiation enables a gated detection. Furthermore, the spatial coherence of the THz pulses makes transmission experiments easy, because of the highly directional beams and small spot sizes that can be realized. A standard setup using photoconducting antennas as emitters and detectors is shown in Fig. 27. THz pulses generated in the transceiver are collimated and transmitted through the sample. The transmitted THz signal is time-resolved by delaying the reference laser pulse, which excites the antenna with respect to the THz pulse. THz-TDS has been applied to characterize the far-infrared properties of a variety of semiconductors. The Drude dielectric function has been probed by THz-TDS in bulk doped semiconductors such as silicon (van Exter and Grischkowsky, 1990b) and GaAs (Katzenellenbogen and Grischkowsky, 1992).

Figure 28 shows results achieved on n-doped silicon. The transmission signal reveals a main structure at a delay of about 12 ps that is due to the THz pulse transmitted directly through the wafer. The second pulse at about 20 ps results from a Fabry–Perot type reflection within the wafer leading to

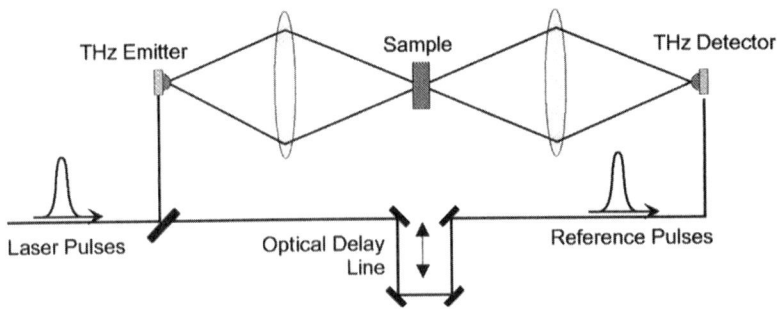

FIG. 27. Schematic diagram of a THz-TDS setup using photoconductive THz transmitters and receivers.

a delayed signal. The wafer's complex index of refraction $\tilde{n}$ can be deduced by normalizing the transmission to the signal acquired without sample. The frequency-dependent transmission is given by

$$t(\omega) = \frac{4\tilde{n}}{(\tilde{n}+1)^2} \exp\left[ik_0 \int_0^d \tilde{n}(z)dz\right] \quad (29)$$

where k_0 is the wave vector of the radiation in vacuum, and the first factor describes the losses due to reflection at the front and back side of the wafer (Hunsche and Nuss, 1998). Fabry–Perot effects are not considered in Eq. (29). Applying a Drude model enables one to determine the dependence of the complex conductivity σ on frequency and thus parameters such as carrier scattering rate and plasma frequency (van Exter and Grischkowsky, 1990a).

In semiconductor heterostructures THz-TDS can be applied to measure the dielectric response of electrons performing transitions between quantized states. Measurements of electron gases in magnetic fields have successfully shown the contribution of cyclotron oscillations to the dielectric response of an electron gas (Some and Nurmikko, 1994). Of particular interest are measurements of Heyman et al. (1998), which were performed on semiconductor devices that contain a modulation-doped GaAs/AlGaAs heterostructure. The devices enable the electronic control of the electron density within the quantum wells and thus the absorption because of intersubband transitions. THz-modulation spectroscopy shows directly the response of the electrons while performing an intersubband transition, and enables one to deduce the dielectric function of the isolated electron gas.

Continuing work on modulation-doped parabolic quantum wells addresses the quantum dynamics of two-level intersubband transitions (Kersting et al., 2000). For the first time, THz time-domain spectroscopy gives the opportunity to study optical excitations in two-level systems with a time

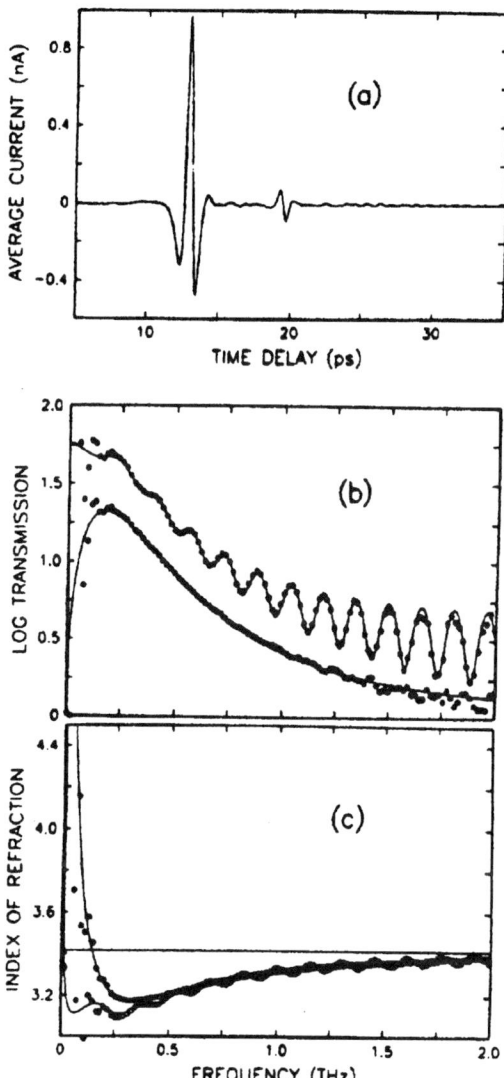

FIG. 28. (a) THz pulse transmitted through a 283-μm thick wafer of 1.15 Ωcm, n-type silicon at room temperature; (b) logarithm of the amplitude transmission (circles) and the amplitude transmission numerically corrected for reflections (dots); (c) effective index of refraction, derived by division of the measured phase shift of the transmitted terahertz pulses by the length of the sample (circles) the index of refraction corrected for reflections (dots). After van Exter and Grischkowsky (1990a).

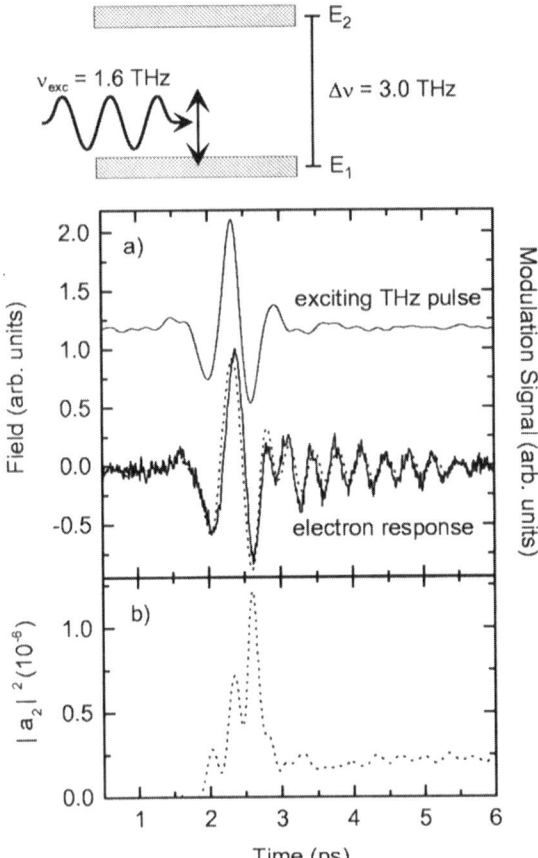

FIG. 29. The scheme shows a Jablonski diagram of the excitation process. The intersubband transition at 3.0 THz is driven by an off-resonant 1.6 THz pulse. (a) Driving THz pulse and response of the electrons. (b) Population probability as deduced from the model calculation shown by dotted lines in (a) and (b). From Kersting et al., 2000. Reprinted with permission of Optical Society of America.

resolution shorter than the oscillation time of the exciting light pulse. The response of the electrons was measured by applying modulation spectroscopy as described in Heyman et al. (1998). The modulation signal is proportional to the motion of the electrons. Figure 29 shows measurements performed using a nonresonant excitation where the center frequency of the exciting THz pulse is much lower than the frequency of the intersubband transition. Although the frequency of the exciting THz pulse does not fit the eigenfrequency of the intersubband transition, the electrons follow this oscillation during the presence of the exciting pulse. As soon as the driving pulse has decayed, the electrons relax into a faster oscillation characterized by their intersubband frequency.

The dynamics of the two-level system can be treated using the density matrix formalism

$$i\hbar \frac{d}{dt}\rho = [H_0, \rho] + [H_H, \rho] + [H_{int}, \rho] + i\hbar \frac{\partial \rho}{\partial t}\bigg|_{relax} \quad (30)$$

where H_0 is the reduced unperturbed Hamiltonian and H_{int} describes the interaction between the radiation field and the quantum system. H_H addresses the Hartree field responsible for the collective dynamics of the electrons and the depolarization shift. Whereas the diagonal matrix elements of ρ give the population of the involved states, the off-diagonal elements give the polarization. In case of low excitation densities, the population dynamics can be deduced from the polarization data as shown in Fig. 29b. During the presence of the exciting THz pulse, an increased population of the upper level can be observed. This population results from the virtual state into which the electrons are excited by the THz pulse. In the framework of the Heisenberg picture this state is given by a linear combination of the lower and the upper state wave functions $\psi(t) = a_1(t)\psi_1 + a_2(t)\psi_2$. Since the excitation of this state decays with the exciting THz pulse, the population probability of the upper state $|a_2|^2$ decays on the same time scale. After the excitation process, the probability shows a value that results from the spectral overlap between intersubband transitions and the spectrum of the exciting THz pulse. Although the emission intensities of intersubband transitions are still below 1 nW, detailed study of the electron dynamics gives insight into the polarization and population dynamics. The knowledge of these properties is valuable and important for the design of new coherent far-infrared emitters following the concepts of cascade lasers (Faist *et al.*, 1994; Capasso *et al.*, 1997).

VII. Conclusions

We present an introductory overview on the ultrafast carrier dynamics in semiconductors that provides the basis for the emission of pulsed THz radiation. Three cases of charge transport dynamics that lead to the emission of coherent radiation in bulk semiconductors are discussed in a qualitative manner: the acceleration of photogenerated carriers by an electric field, plasma oscillations, and the generation of an instantaneous polarization by displacement currents. Each generation process induces specific properties of the generated THz radiation, such as the frequency spectrum or the coherence length. The techniques based on bulk semiconductors have in common that a significant power of THz radiation can be generated at room temperature that is high enough to allow first applications. Efficient THz emission from low-dimensional quantum structures seems to be limited to cryogenic temperatures. However, these systems provide the attractive feature of frequency tunability as demonstrated for coherent charge oscillations in quantum wells.

The main limitation on further development of THz sources is the presently low intensity. Therefore the main challenge will be to increase the emission efficiency. One of the most attractive ways is to increase the macroscopic dephasing time by band-structure engineering in quantum structures. In this context modulation-doped structures may lead to increased emission intensities. Another possibility is to use the collective manner of the charge oscillations, which could lead to an emission similar to superradiant processes.

Semiconductor THz emitters have opened up a new route for high-frequency material and device characterization, despite the low beam intensities that are accessible today. For the first time, the quantum dynamics of electrons can be studied in heterostructures. Other studies reveal the dielectric properties of doped semiconductors or the isolated electron gas in heterostructures. Besides their academic interest, these studies are of importance for ultrafast semiconductor devices and may pave the way toward novel laser sources operating in the far-infrared.

Acknowledgments

We are grateful for enlightening discussions with many researchers in the field, especially K. Unterrainer of TU Vienna, J. N. Heyman of Macalester College, A. Leitenstorfer of TU Munich, G. Strasser of TU Vienna, G. C. Cho of Rensselaer, and X.-C. Zhang of Rensselaer. This work was partially supported by the TMR INTERACT program of the European Community, the Austrian Society for Microelectronics, and the "Fonds zur Förderung der wissenschaftlichen Forschung" Austria, Wittgenstein Award.

References

Allen, S. Jr., Tsui, D., and Vinter, B. (1976). On the absorption of infrared radiation by electrons in semiconductor inversion layers. *Solid State Commun.*, 20:425.

Allen, S. Jr., Tsui, D., and Logan, R. (1977). Observation of the two-dimensional plasmon silicon inversion layers. *Phys. Rev. Lett.*, 38:980.

Ando, T., Fowler, A., and Stern, F. (1982). Electronic properties of two-dimensional systems. *Rev. Mod. Phys.*, 54:437.

Auston, D. (1975). Picosecond optoelectronic switching and gating in silicon. *Appl. Phys. Lett.*, 26:101.

Auston, D., and Nuss, M. (1988). Electrooptic-generation and detection of femtosecond electrical transients. *IEEE Journ. Quant. Electr.*, 24:184.

Auston, D., Cheung, K., and Smith, P. (1984a). Picosecond photoconducting Hertzian dipoles. *Appl. Phys. Lett.*, 45:284.

Auston, D., Cheung, K., Valdmanis, J., and Kleinman, D. (1984b). Cherenkov radiation from femtosecond optical pulses in electro-optic media. *Phys. Rev. Lett.*, 53:1555.

Bakker, H., Cho, G., Kurz, H., Wu, Q., and Zhang, X. (1998). Distortion of THz pulses in electro-optic sampling. *J. Opt. Soc. Am. B*, 15:1795.

Bastard, G. (1988). *Wave Mechanics Applied to Semiconductor Heterostructures*. Les Editions de Physique.

Baumberg, J., and Williams, D. (1996). Coherent phonon plasmon interactions in GaAs:AlGaAs heterostructures. *Phys. Rev. B*, 53:16140.

Bloch, F. (1928). Über die Quantenmechanik der Elektronen in Kristallgittern. *Z. für Physik*, 52:555.

Bonvalet, A., Nagle, J., Berger, V., Migus, A., Martin, J.-L., and Joffre, M. (1996). Femtosecond infrared emission resulting from coherent charge oscillations in quantum wells. *Phys. Rev. Lett.*, 76:4392.

Born, M., and Wolf, E. (1980). *Principles of Optics*. Cambridge University Press.

Cai, Y., Brener, I., Lopata, J., Wynn, J., Pfeiffer, L., Stark, J., Wu, Q., Zhang, X., and Federici, J. (1998). Coherent terahertz radiation detection: Direct comparison between free-space electro-optic sampling and antenna detection. *Appl. Phys. Lett.*, 73:444.

Capasso, F., Faist, J., Sirtori, C., and Cho, A. (1997). Infrared (4–11 μm) quantum cascade lasers. *Solid State Commun.*, 102:231.

Chemla, D., Miller, D., and Schmitt-Rink, S. (1987). Generation of ultrashort electrical pulses through screening by virtual populations in biased quantum wells. *Phys. Rev. Lett.*, 59:1018.

Cheville, R., and Grischkowsky, D. (1995). Time domain terahertz impulse ranging studies. *Appl. Phys. Lett.*, 67:1960.

Cho, G., Kütt, W., and Kurz, H. (1990). Subpicosecond time-resolved coherent phonon oscillations in GaAs. *Phys. Rev. Lett.*, 65:764.

Cho, G., Dekorsy, T., Bakker, H., Hövel, R., and Kurz, H. (1996). Generation and relaxation of coherent majority plasmons. *Phys. Rev. Lett.*, 77:4062.

Chwalek, J., Withaker, J., and Mourou, G. (1991). Low temperature epitaxially grown GaAs as a high-speed photoconductor for terahertz spectroscopy. In Sollner, T. and Shah, J., editors, *Picosecond Electronics and Optoelectronics*, p. 15. Optical Society of America.

Constant, E. (1985). Non-steady-state carrier-transport in semiconductors in perspective with submicrometer devices. In Reggiani, L., editor, *Hot Electron Transfer in Semiconductors*, p. 227. Springer Verlag, Berlin.

Darrow, J., Hu, B., Zhang, X.-C., and Auston, D. (1990). Subpicosecond electromagnetic pulses from large-aperture photoconducting antennas. *Opt. Lett.*, 15:323.

DeFonzo, A., Jarwala, M., and Lutz, C. (1987). Transient response of planar integrated optoelectronic antennas. *Appl. Phys. Lett.*, 50:1155.

Dekorsky, T., Pfeifer, T., Kütt, W., and Kurz, H. (1993). Subpicosecond carrier transport in GaAs surface-space-charge fields. *Phys. Rev. B*, 47.

Dekorsy, T., Auer, H., Waschke, C., Bakker, H., Roskos, H., Kurz, H., Wagner, V., and Grosse, P. (1995). Emission of submillimeter electromagnetic waves by coherent phonons. *Phys. Rev. Lett.*, 74.

Dember, H. (1931). Über eine photoelektromotorische Kraft in Kupferoxydul-Kristallen. *Physik. Zeitschr.*, 32:554.

Dicke, R. (1954). Coherence in spontaneous radiation processes. *Phys. Rev.*, 93:99.

Dingle, R., Störmer, H., Gossard, A., and Wiegmann, W. (1978). Electron mobilities in modulation-doped semiconductor heterojunction superlattices. *Appl. Phys. Lett.*, 33:665.

Doany, F., Grischkowsky, D., and Chi, C.-C. (1987). Carrier lifetime versus ion-implantation dose in silicon on sapphire. *Appl. Phys. Lett.*, 50:460.

Dykaar, D., Greene, B., Federici, J., Levi, A., Pfeiffer, L., and Kopf, R. (1991a). Log-periodic antennas for pulsed terahertz radiation. *Appl. Phys. Lett.*, 59:262.

Dykaar, D., Greene, B., Federici, J., Levi, A., Pfeiffer, L., and Kopf, R. (1991b). Broadband periodic antennas for detecting picosecond terahertz radiation. In Sollner, T. and Shah, J., editors, *Picosecond Electronics and Optoelectronics*, p. 20. Optical Society of America.

Eisele, H., and Haddad, G. (1995). High-performance InP Gunn devices for fundamental-mode operation in D-Band (110-170 GHz). *IEEE Microw. Guid. Wave Lett.*, 5:385.

Esaki, L., and Tsu, R. (1970). Superlattices and negative differential conductivity in semiconductors. *IBM J. Res. Develop.*, 14:61.

Faist, J., Capasso, F., Sirtori, C., Sivco, D., and Cho, A. (2000). Quantum cascade lasers. In Willardson, R., and Weber, E., editors, *Semiconductors and Semimetals, Intersubband Transitions in Quantum Wells: Physics and Device Applications II*, Vol. 66, p. 1. Academic Press.

Faist, J., Capasso, F., Sivco, D., Sirtori, C., Hutchinson, A., and Cho, A. (1994). Quantum cascade laser. *Science*, 264:553.

Fattinger, C., and Grischkowsky, D. (1989). Terahertz beams. *Appl. Phys. Lett.*, 54:490.

Feldmann, J., Leo, K., Shah, J., Miller, D., Cunningham, J., Meier, T., von Plessen, G., Schulze, A., Thomas, P., and Schmitt-Rink, S. (1992). Optical investigations of Bloch oscillations in a semiconductor superlattice. *Phys. Rev. B*, 46:7252.

Fischler, W., Buchberger, P., Höpfel, R., and Zandler, G. (1996). Ultrafast reflectivity changes in photoexcited GaAs Schottky diodes. *Appl. Phys. Lett.*, 68:2778.

Fork, R. L., Greene, B. I., and Shank, C. V. (1981). Generation of optical pulses shorter than 0.1 psec by colliding pulse mode locking. *Appl. Phys. Lett.*, 38:671.

Gornik, E., and Andronov, A. (eds.) (1991). Special issue, Far-infrared semiconductor lasers. *Opt. and Quant. Electr.*, 23:1–234.

Gornik, E., and Tsui, D. (1978). Far infrared emission from hot electrons in Si-inversion layers. *Solid State Electr*, 21:139.

Greene, B., Federici, J., Dykaar, D., Jones, R., and Bucksbaum, P. (1991). Interferometric characterization of 160 fs far-infrared light pulses. *Appl. Phys. Lett.*, 59:893.

Gunn, J. B. (1963). Microwave oscillation of current in III–V semiconductors. *Solid State Commun.*, 1:88.

Gupta, S., Frankel, M., Valdmanis, J., Withaker, J., Mourou, G., Smith, F., and Calawa, A. (1991). Subpicosecond carrier lifetime in GaAs grown by molecular beam epitaxy at low temperature. *Appl. Phys. Lett.*, 59:3276.

Gupta, S., Withaker, J., and Mourou, G. (1992). Ultrafast carrier dynamics in III–V semiconductors grown by molecular beam epitaxy at very low substrate temperatures. *IEEE J. Quant. Electr.*, 28:2464.

Haring Bolivar, P., Dekorsy, T., and Kurz, H. (2000). Optically excited Bloch oscillations— Fundamentals and application perspectives. In Willardson, R., and Weber, E., editors, *Semiconductors and Semimetals, Intersubband Transitions in Quantum Wells: Physics and Device Applications II*, Vol. 66, p. 187. Academic Press.

Helm, M. (2000). The basic physics of intersubband transitions. In Willardson, R., and Weber, E., editors, *Semiconductors and Semimetals, Intersubband Transitions in Quantum Wells: Physics and Device Applications I*, Vol. 62, p. 1. Academic Press.

Helm, M., Colas, E., England, P., DeRosa, F., and Allen., S., Jr. (1988). Observation of grating-induced intersubband emission from GaAs/AlGaAs superlattices. *Appl. Phys. Lett.*, 53:1714.

Heyman, J., Kersting, R., and Unterrainer, K. (1998). Time-domain measurement of intersubband oscillations in a quantum well. *Appl. Phys. Lett.*, 72:644.

Hirakawa, K., Yamanaka, K., Grayson, M., and Tsui, D. (1995). Far-infrared emission

spectroscopy of hot two-dimensional plasmons in AlGaAs/GaAs heterojunctions. *Appl. Phys. Lett.*, 67:2326.
Hirakawa, K., Wilke, I., Yamanaka, K., Roskos, H., Vosseburger, M., Wolter, F., Waschke, C., Kurz, H., Grayson, M., and Tsui, D. (1996). Coherent submillimeter-wave emission from non-equilibrium two-dimensional free carrier plasmas in AlGaAs/GaAs heterojunctions. *Surf. Sci.*, 361/362:368.
Höpfel, R., Lindemann, G., Gornik, E., Stangl, G., Gossard, A., and Wiegmann, W. (1982a). Cyclotron and plasmon emission from two-dimensional electrons in GaAs. *Surf. Sci.*, 113:118.
Höpfel, R., Vass, E., and Gornik, E. (1982b). Thermal excitation of two-dimensional plasma oscillations. *Phys. Rev. Lett.*, 49:1667.
Hu, B., and Nuss, M. (1995). Imaging with THz waves. *Opt. Lett.*, 20:1716.
Hu, B., Darrow, J., Zhang, X.-C., Auston, D., and Smith, P. (1989). Optically steerable photoconducting antennas. *Appl. Phys. Lett.*, 56:886.
Hu, B., Zhang, X.-C., and Auston, D. (1990). Temperature dependence of femtosecond electromagnetic radiation from semiconductor surfaces. *Appl. Phys. Lett.*, 57:2629.
Hu, B., Zhang, X.-C., and Auston, D. (1991). Terahertz radiation induced by subband-gap femtosecond optical excitation of GaAs. *Phys. Rev. Lett.*, 67:2709.
Hu, B., Weling, A., Auston, D., Kusnetzow, A., and Stanton, C. (1994). dc-electric-field dependence of THz radiation induced by femtosecond excitation of bulk GaAs. *Phys. Rev. B*, 49:2234.
Hu, B. B., de Souza, E. A., Knox, W. H., Cunningham, J. E., Nuss, M. C., Kuznetsov, A. V., and Chuang, S. L. (1995). Identifying the distinct phases of carrier transport in semiconductors with 10 fs resolution. *Phys. Rev. Lett.*, 74:1689.
Hunsche, S., and Nuss, M. (1998). Terahertz "T-ray tomograph". In Afsar, M., editor, *Proceedings of SPIE, Millimeter and Submillimeter Waves IV*, Vol. 3465, p. 426.
Hunsche, S., Feng, S., Winful, H., Leitenstorfer, A., Nuss, M., and Ippen, E. (1999). Spatiotemporal focusing of light pulses. *J. Opt. Soc. Am. A*, 16:2025.
Jackson, J. (1998). *Classical Electrodynamics*. John Wiley & Sons.
Jepsen, P. U., and Keiding, S. (1995). Radiation patterns from lens-coupled terahertz antennas. *Opt. Lett.*, 20:807.
Jepsen, P. U., Jacobsen, R., and Keiding, S. (1996a). Generation, propagation and detection of terahertz radiation from biased semiconductor dipole antennas. In Svelto, O., DiSilvestri, S., and Denardo, G. (eds.). *Ultrafast Processes in Spectroscopy*, p. 637. Plenum Press.
Jepsen, P. U., Winnewisser, C., Schall, M., Schyja, V., Keiding, S., and Helm, H. (1996b). Detection of THz pulses by phase retardation in lithium tantalate. *Phys. Rev. E*, 53:R3052.
Jiang, Z., and Zhang, X. (1998). Electro-optic measurement of THz field pulses with a chirped optical beam. *Appl. Phys. Lett.*, 72:1945.
Karin, J., Downey, P., and Martin, R. (1986). Radiation from picosecond photoconductors in microstrip transmission lines. *IEEE J. Quantum. Electr.*, 22:677.
Katzenellenbogen, N., and Grischkowsky, D. (1992). Electrical characterization to 4 THz of n- and p-type GaAs using THz time-domain spectroscopy. *Appl. Phys. Lett.*, 61:840.
Kersting, R., Unterrainer, K., Strasser, G., and Gornik, E. (1997a). Coherent few-cycle THz emission from plasmons in bulk GaAs. *phys. stat. sol. (b)*, 204:67.
Kersting, R., Unterrainer, K., Strasser, G., Kauffmann, H., and Gornik, E. (1997b). Few-cycle THz emission from cold plasma oscillations. *Phys. Rev. Lett.*, 79:3038.
Kersting, R., Heyman, J., Strasser, G., and Unterrainer, K. (1998). Coherent plasmons in n-doped GaAs. *Phys. Rev. B*, 58:4553.
Kersting, R., Bratschitsch, R., Strasser, G., Unterrainer, K., and Heyman, J. (2000). Sampling a THz dipole transition with sub-cycle time-resolution. *Opt. Lett.*, 25:272.
Ketchen, M., Grischkowsky, D., Chen, T., Chi, C.-C., I, N. D., Halas, N., Halbout, J.-M., Kash, J., and Li, G. (1986). Generation of subpicosecond electrical pulses on coplanar transmission lines. *Appl. Phys. Lett.*, 48:751.

Kütt, W., Albrecht, W., and Kurz, H. (1992). Generation of coherent phonons in condensed media. *IEEE Journ. Quant. Electr.*, 28:2434.

Kuznetsov, A., and Stanton, C. (1993). Ultrafast optical generation of carriers in a dc electric field: Transient localization and photocurrent. *Phys. Rev. B*, 48:10828.

Kuznetsov, A., and Stanton, C. (1995). Coherent phonon oscillations in GaAs. *Phys. Rev. B*, 51:7555.

Lavon, Y., Sa'ar, A., Moussa, Z., Julien, F., and Planel, R. (1995). Observation of optically pumped midinfrared intersubband luminescence in a coupled quantum wells structure. *Appl. Phys. Lett.*, 67:1984.

Leitenstorfer, A., Husche, S., Shah, J., Nuss, M., and Knox, W. (1999a). Detectors and sources for ultrabroadband electro-optic sampling: Experiment and theory. *Appl. Phys. Lett.*, 74:1516.

Leitenstorfer, A., Husche, S., Shah, J., Nuss, M., and Knox, W. (1999b). Femtosecond charge transport in polar semiconductors. *Phys. Rev. Lett.*, 82:5140.

Leo, K., Shah, J., Göbel, E., Damen, T., Schmitt-Rink, S., Schäfer, W., and Köhler, K. (1991). Coherent oscillations of a wave packet in a semiconductor double-quantum-well structure. *Phys. Rev. Lett.*, 66:201.

Leo, K., Haring Bolivar, P., Brüggemann, F., Schwedler, R., and Köhler, K. (1992). Observation of Bloch oscillations in a semiconductor superlattice. *Solid State Commun.*, 84:943.

Li, M., Sun, G., Wagoner, G., Alexander, M., and Zhang, X.-C. (1995). Measurement and analysis of terahertz radiation from bulk semiconductors. *Appl. Phys. Lett.*, 67:25.

Lu, Z., Campbell, P., and Zhang, X.-C. (1997). Free-space electro-optic sampling with a high-repetition-rate regenerative amplified laser. *Appl. Phys. Lett.*, 71:593.

Luo, M., Chuang, S., Planken, P., Brener, I., Roskos, H., and Nuss, M. (1994). Generation of terahertz electromagnetic pulses from quantum-well structures. *IEEE J. Quantum Electr.*, 30:1478.

Luryi, S. (1991). Polarization oscillations in coupled quantum wells — A scheme for the generation of submillimeter electromagnetic waves. *IEEE J. Quantum Electr.*, 27:54.

Maranowski, K., Gossard, A., Unterrainer, K., and Gornik, E. (1996). Far-infrared emission from parabolically graded quantum wells. *Appl. Phys. Lett.*, 69:3522.

Martini, R., Klose, G., Roskos, H., Kurz, H., Grahn, H., and Hey, R. (1996). Superradiant emission from Bloch oscillations in semiconductor superlattices. *Phys. Rev. B*, 54:R14325.

Meystre, P., and Sargent, M. III (1998). *Elements of Quantum Optics*. Springer.

Miller, D., Chemla, D., Damen, T., Gossard, A., Wiegmann, W., Wood, T., and Burrus, C. (1984). Band-edge electroabsorption in quantum well structures: The quantum confined Stark effect. *Phys. Rev. Lett.*, 53:2173.

Mourou, G., Stancampiano, C., Antonetti, A., and Orszag, A. (1981). Picosecond microwave pulses generated with a subpicosecond laser-driven semiconductor switch. *Appl. Phys. Lett.*, 39:295.

Nag, B. (1980). *Electron Transport in Compound Semiconductors*. Springer.

Nakamura, S., Senoh, M., Nagahama, S., Iwasa, N., Yamada, T., Matsushita, T., Kiyoku, H., and Sugimoto, Y. (1996). InGaN-based multi-quantum-well-structure laser diodes. *Jpn. J. Appl. Phys.*, 35:L74.

Nuss, M. (1996). Chemistry is right for T-ray imaging. *IEEE Circu. Devices*, 12:25.

Nuss, M., and Orenstein, J. (1998). Terahertz time-domain spectroscopy. In Grüner, G., editor, *Topics in Applied Physics, Millimeter and Submillimeter Wave Spectroscopy of Solids*, Vol. 74, p. 8. Springer-Verlag, Berlin.

Nuss, M., Auston, D., and Capasso, F. (1987). Direct subpicosecond measurement of carrier mobility of photoexcited electrons in gallium arsenide. *Phys. Rev. Lett.*, 58:2355.

Olego, D., Pinczuk, A., Gossard, A., and Wiegmann, W. (1982). Plasma dispersion in a layered electron gas: A determination in GaAs-(AlGa)As heterostructures. *Phys. Rev. B*, 26:7867.

Pastol, Y., Arjavalingham, G., Halbout, J.-M., and Kopsay, G. (1988). Characterization of an optoelectronically pulsed broadband microwave antenna. *Electr. Lett.*, 24:1318.

Pedersen, J., Keiding, S., Sorensen, C., Lindelof, P., Rühle, W., and Zhou, X. (1993). 5-THz band width from a GaAs on silicon photoconductive receiver. *J. Appl. Phys.*, 74:7022.

Planken, P., Nuss, M., Brener, I., Goossen, K., Luo, M., Chuang, S., and Pfeiffer, L. (1992). Terahertz emission in single quantum wells after coherent optical excitation of light hole and heavy hole excitons. *Phys. Rev. Lett.*, 69:3800.

Prabhu, S., Ralph, S., Melloch, M., and Harmon, E. (1997). Carrier dynamics of low-temperature-grown GaAs observed via THz spectroscopy. *Appl. Phys. Lett.*, 70:2419.

Ritchie, R. (1973). Surface plasmons in solids. *Surf. Sci.*, 34:1.

Rosenwaks, Y., Thacker, B., Ahrenkiel, R., Nozik, A., and Yavneh, I. (1994). Photogenerated carrier dynamics under the influence of electric fields in III–V semiconductors. *Phys. Rev. B*, 50:1746.

Roskos, H., Nuss, M., Shah, J., Leo, K., Miller, D., Fox, A., Schmitt-Rink, S., and Köhler, K. (1992). Coherent submillimeter-wave emission from charge oscillations in a double-well potential. *Phys. Rev. Lett.*, 68:2216.

Saleh, B. E. A., and Teich, M. C. (1991). *Fundamentals of Phototonics*. Wiley.

Sha, W., Smirl, A., and Tseng, W. (1995). Coherent plasma oscillations in bulk semiconductors. *Phys. Rev. Lett.*, 74:4273.

Shah, J. (1996). *Ultrafast Spectroscopy of Semiconductors and Semiconductor Heterostructures*. Springer.

Smith, P., Auston, D., and Nuss, M. (1988). Subpicosecond photoconducting dipole antennas. *IEEE J. Quantum Electr.*, 24:255.

Some, D., and Nurmikko, A. (1994). Coherent transient cyclotron emission from photoexcited GaAs. *Phys. Rev. B*, 50:5783.

Spence, D., Kean, P., and Sibbett, W. (1991). 60-fsec pulse generation from a self-mode-locked Ti:sapphire laser. *Opt. Lett.*, 16:42.

Stern, F. (1967). Polarizability of a two-dimensional electron gas. *Phys. Rev. Lett.*, 18:546.

Stern, F., and Howard, W. (1967). Properties of semiconductor inversion layers in the electric quantum limit. *Phys. Rev.*, 163:816.

Stillman, G., Wolfe, C., and Dimmrock, J. (1970). Hall coefficient factor for polar mode scattering in *n*-type GaAs. *J. Phys. Chem. Solids*, 31:1199.

Sun, F., Jiang, Z., and Zhang, X. (1998). Analysis of terahertz pulse measurement with chirped probe beam. *Appl. Phys. Lett.*, 73:2233.

Sze, S. (1985). *Semiconductor Devices*. John Wiley & Sons.

Tacke, M. (1999). In Helm, M., editor, *Long Wavelength Infrared Emitters Based on Quantum Wells and Superlattices*. Gordon and Breach.

Tsui, D., Gornik, E., and Logan, R. (1980). Far infrared emission from plasma oscillations of Si inversion layers. *Solid State Commun.*, 35:875.

Valdmanis, J., Mourou, G., and Gabel, C. (1982). Picosecond electro-optic sampling system. *Appl. Phys. Lett.*, 41:211.

Valdmanis, J., Mourou, G., and Gabel, C. (1983). Subpicosecond electrical sampling. *IEEE J. Quantum Electr.*, 19:664.

van Exter, M., and Grischkowsky, D. (1990a). Carrier dynamics of electrons and holes in moderately doped silicon. *Phys. Rev. B*, 41:12140.

van Exter, M., and Grischkowsky, D. (1990b). Optical and electronic properties of doped silicon from 0.1 to 2 THz. *Appl. Phys. Lett.*, 56:1694.

van Exter, M., and Grischkowsky, D. R. (1990c). Characterization of an optolelectronic terahertz beam system. *IEEE Trans. Microwave Theory and Techn.*, 38:1684.

Victor, K., Roskos, H., and Waschke, C. (1994). Efficiency of submillimeter-wave generation and amplification by coherent wave-packet oscillations in semiconductors. *J. Opt. Soc. Am. B*, 11:2470.

Vosseburger, M., Roskos, H., Wolter, F., Waschke, C., Kurz, H., Hirakawa, K., Wilke, I., and Yamanaka, K. (1996). Radiative decay of optically excited coherent plasmons in a two-dimensional electron gas. *J. Opt. Soc. Am. B*, 13:1045.

Wannier, G. (1960). Wave functions and the effective Hamiltonian for Bloch electrons in an electric field. *Phys. Rev.*, 117:432.

Wannier, G. (1962). Dynamics of band electrons in electric and magnetic fields. *Rev. Mod. Phys.*, 34:645.

Waschke, C., Roskos, H., Schwedler, R., Leo, K., Kurz, H., and Köhler, K. (1993). Coherent submillimeter-wave emission from Bloch oscillations in a semiconductor superlattice. *Phys. Rev. Lett.*, 70:3319.

Wu, Q., and Zhang, X.-C. (1995). Free-space electro-optic sampling of THz beams. *Appl. Phys. Lett.*, 67:3523.

Wu, Q., and Zhang, X.-C. (1996). Ultrafast electro-optic field sensors. *Appl. Phys. Lett.*, 68:1604.

Wu, Q., and Zhang, X.-C. (1997). 7 Terahertz broadband GaP electro-optic sensor. *Appl. Phys. Lett.*, 70:1784.

Wu, Q., Hewitt, T., and Zhang, X.-C. (1996). Two-dimensional electro-optic imaging of THz beams. *Appl. Phys. Lett.*, 68:2924.

Wysin, G., Smith, D., and Redondo, A. (1988). Picosecond response of photoexcited GaAs in a uniform electric field by Monte Carlo dynamics. *Phys. Rev. B*, 38(17):12514.

Yablonovitch, E., Heritage, J., Aspnes, D., and Yafet, Y. (1989). Virtual photoconductivity. *Phys. Rev. Lett*, 63:976.

Yamanishi, M. (1987). Field-induced optical nonlinearity due to virtual transitions in semiconductor quantum-well structures. *Phys. Rev. Lett.*, 59:1014.

Yang, K., Richards, P., and Shen, Y. (1971). Generation of far-infrared radiation by picosecond light pulses in LiNbO3. *Appl. Phys. Lett.*, 19:320.

You, D., Jones, R., Bucksbaum, P., and Dykaar, D. (1993). Generation of high power sub-single-cycle 500-fs electromagnetic pulses. *Opt. Lett.*, 18:290.

Zener, C. (1934). A theory of electrical breakdown of solid dielectrics. *Proc. Roy. Soc. London A* 145:523.

Zhang, X.-C., and Auston, D. (1992). Optoelectronic measurement of semiconductor surfaces and interfaces with femtosecond optics. *J. Appl. Phys.*, 71:326.

Zhang, X.-C., Hu, B., Darrow, J., and Auston, D. (1990). Generation of femtosecond electromagnetic pulses from semiconductor surfaces. *Appl. Phys. Lett.*, 56:1011.

Index

A

Above-threshold dissociation, 379
Absorption, virtual, 10
Absorption changes for low intensities
   off-resonant, 266–274
   resonant, 254–266
Absorption changes induced by incoherent occupations, 280–289
AlGaAs/GaAs, 417
Anharmonic decay of longitudinal optical phonons in wurtzite GaN, 139–147
Antenna detection techniques, 396–397
Anti-Stokes Raman scattering, 123–124
Atomic force microscopy, 46
Auger recombination, 172–174, 193–196

B

Ballistic electron emission microscopy, 46
Band filling, 193
Bandgap renormalization, 165, 233
Band index, 158
Band structure, 158
Biexcitonic beats, 290–297
Bloch equations, 2, 158
   Hartree–Fock SBE, 236
   for pulse excitation with quantum kinetic scattering, 209–215
   semiconductor (SBE), 234, 236–237
Bloch oscillations, 422–426
Boltzmann equation, semiclassical, 2, 217, 219
Bond softening, 378
Born–Oppenheimer approximation, 7, 317

C

C_{60} response to laser pulses, 371–377
Carrier–carrier scattering, 169, 170
   suppression of, 21
Carrier confinement, 175
Carrier excitation and relaxation
   carrier–carrier scattering, 169, 170
   carrier–lattice thermalization, 172
   carrier–phonon scattering, 169, 170–172
   diffusion, 174–175
   excitation, mechanisms of, 167–169
   at low fluences, 193–196
   at medium and high fluences, 196–197
   recombination, 172–174
   redistribution, thermalization, and cooling, 169–172
   structural effects, 175–177
   summary, 177–179
   thermal versus nonthermal debate, 176–177
Carrier–phonon scattering, 169, 170–172
Cathodoluminescence spectroscopy, 40
Cayley algorithm, 326
CdTe
   intermediate coupling regime, 30–33
   longitudinal optical phonons and, 6–10
Charge-density fluctuations (CDFs), 118, 120, 121
Collection-mode geometry, 48, 49
Compton scattering, 110
Correlation terms, 164
Coulomb carrier–carrier scattering, 206–207
Coulomb correlations and optical nonlinearities

Coulomb correlations and optical nonlinearities (*Continued*)
 absorption changes induced by incoherent occupations, 280–289
 applications to FWM, 290–304
 applications to pump-probe spectroscopy, 254–279
 background information, 231–241
 biexcitonic beats, 290–297
 coherent limit, 242–248
 disorder-induced dephasing, 297–304
 higher intensities up to coherent limit, 274–279
 nonlinear optical signals, 248–250
 off-resonant absorption changes for low intensities, 266–274
 one-dimensional model system, 250–254
 outlook, 307–309
 resonant absorption changes for low intensities, 254–266
 summary, 304–307
 theoretical approach and model, 242–254
Coulomb effects, 19, 26, 120
Coulomb potential, 164, 170, 207
  Fourier transformation of retarded, 214–215
  two-time-dependent screened, 212
Crystalline solids, dielectric function of
 interband contributions, 158–160
 intraband contributions, 160–161
Crystal momentum, 158

D

Dark-field geometry, 48
de Broglie wavelength, 41
Defect and surface recombination, 172–174
Deformation potential, 171
Density matrix formalism, 3
Dielectric function
 bonding changes and, 346–348
 of crystalline solids, 158–161
 derivation, 154
 electronic configuration effects on, 163–165
 formula, 360
 of GaAs, 182–191
 lattice structure effects on, 162–163
 linear optical properties and, 154–157
 microscopic theory of, 157–158
 reasons for measuring, 165–166
Dielectric function, measuring time-resolved, 179
 extracting transient, 181–182
 transient reflectivity spectra, measuring, 180–181
 white-light probe generation, 180
Dielectric function of GaAs, time-resolved, 182
 high fluence regime, 189–191
 low fluence regime, 183–184
 medium fluence regime, 185–189
Differential transmission spectroscopy (DTS), 206, 219–222
Diffusion
 carrier, 174–175
 thermal, 175, 176
Dirac delta function, 157
Disorder-induced dephasing, 297–304
Drude–Lorentz model of carrier transport, 400–402
Drude model, 161, 162–163, 192, 196, 427
Dyson equation, 214, 217

E

Ehrenfest theorem, 316, 325, 344
Electric dipole approximation, 157
Electronic configuration effects on dielectric function, 163–165
Electronic transitions, sudden, 379
Electron–phonon interactions, in wurtzite GaN, 128–132
Electron-population trapping, 379
Electrons and atoms, dynamics of
 carrier diffusion, 174–175
 carrier excitation, 167–169
 carrier–lattice thermalization, 172
 carrier recombination, 172–174
 carrier redistribution, thermalization, and cooling, 169–172
 summary, 177–179
Electro-optic detection, 397–399
Energy-density fluctuations (EDFs), 118, 120, 121
Equilibrium many-body theory, 215
Evanescent electromagnetic waves, 49

F

Femtosecond DTS (differential transmission spectroscopy), 206, 219–222
Femtosecond FWM (four-wave mixing), 206, 217–219, 223–227
Femtosecond lasers, 152
Femtosecond microscopy, 176
Femtosecond near-field pump-and-probe spectroscopy, 89–98
Fermi–Dirac distribution functions, 134, 169, 172
Fermi Golden Rule, 2, 7, 34, 54, 124, 205, 317, 380
Fourier transform function, 116–117
Four-wave mixing (FWM), 206, 217–219, 223–227
   biexcitonic beats, 290–297
   Coulomb correlations and, 290–304
   disorder-induced dephasing, 297–304
   time-integrated versus time-resolved, 234
Franck–Condon principle, 317
Free carrier absorption, 169
Frenkel excitons, 239
Frequency-dependent dielectric function, 117
Fresnel factors, 156
Fresnel reflectivity formulas, 179
Fröhlich interactions, 124, 130
   polar-optical, 4, 7

G

GaAs (gallium arsenide)
   See also Ultrafast dynamics and phase changes in GaAs
   longitudinal optical phonons and, 3, 6–8
   response to ultraintense and ultrashort laser pulses, 356–363
   theoretical simulations, 29–30
   time- and energy-resolved transmission charges, 25–29
   ultrafast electron-phonon relaxation in, 25–30
Generalized Elliott formula, 233
Generalized Kadanoff–Baym approximation (GKBA), 206, 213
GG approximation, 207

Green function, 54, 206, 232, 233
Group velocity dispersion (GVD), 62–63

H

Hamiltonian, tight-binding, 357–358
Harmonic generation, 379
Harrison scaling, 340, 341
Hartree–Fock approximation, time-dependent (TDHF), 234, 235–236
Hartree–Fock–Coulomb renormalizations, 213
Hartree–Fock exchange effects, 219–220
Hartree–Fock SBE, 236–237
Hartree term, 165
Heisenberg representation, 117
Hellmann–Feynman theorem, 316, 323, 325, 342–344, 371

I

Illumination/collection mode geometry, 48
Illumination-mode geometry, 48, 49
Interferometric detection, 399–400
Intersubband structures, 426
Intervalley scattering, 170, 171
Intravalley scattering, 170, 171–172
Ion-population trapping, 379

K

Keldysh nonequilibrium Green's functions, 2–3, 206, 212
Kohn–Sham equation, 371
Kramers–Kronig relation, 361
Kronecker deltas, 119

L

Lagrangian model, 318
Lattice dynamics
   at low fluences, 197–198
   at medium and high fluences, 198–200
Lattice structure effects on dielectric function, 162–163

Light-hole contributions via interband selection rules, 18–21
Linear optical properties, dielectric function and, 154–157
Longitudinal optical (LO) phonons
 CdTe and, 6–10
 GaAs and, 3, 6–8
Longitudinal optical (LO) phonon scattering
 femtosecond DTS, 206, 219–222
 femtosecond FWM, 206, 217–219, 223–227
Low-temperature near-field microscopy, 59–61, 65–66

M

Mahan excition, 222
Many-body effects, 215, 232
Markovian kinetics, 205, 217
Markov limit, 269
Maxwell–Boltzmann distribution, 24, 111
Maxwell's equations, 54, 400
Mean scattering time, 161
Molecular beam epitaxy (MBE), 41
MOSFETs, 417
Multiphoton absorption (MPA), 169
Multiple-multipole-method (MMP), 53

N

Near-field optics, 49–53
Near-field probes, 56–58
Near-field scanning optical microscopy (NSOM), 41
 background information, 46–49
 collection-mode geometry, 48, 49
 conclusions, 98
 contrast mechanisms, 48–49
 dark-field geometry, 48
 description of, on semiconductor nanostructures, 53–56
 diffraction limit, 49
 illumination/collection mode geometry, 48
 illumination-mode geometry, 48, 49
 low-temperature, 59–61, 65–66
 near-field optics, 49–53
 near-field probes, 56–58
 probe-to-sample distance control, 58–59
 of quantum wires, 66–78
 schematic of, 47–48
 stationary, of semiconductor nanostructures, 65–78
 temporally and spatially resolved, 61–65
 time-resolved near-field microscopy, 79–98
Newton equation for atoms, 321, 379
Noncontact shear-force technique, 59
Nonequilibrium electron distributions and electron-longitudinal optical phonon scattering rates in, 132–139
Nonequilibrium many-body theory, 215
Nonthermal phase transition, 316

O

Occupation factor, 163
Optical nonlinearities. *See* Coulomb correlations and optical nonlinearities
Optical pump–THz probe spectroscopy, 428–429

P

Pauli blocking, 193, 232–233, 256–257
Pauli exclusion principle, 2, 19, 26, 164, 316
Pauli matrices, 115, 119
Peierls substitution, time-dependent, 316, 325–326, 344
Perturbation theory, 144–145
Phase-space filling, 232–233
Photon absorption
 multi- (MPA), 169
 single (SPA), 167–169
Phonon modes, in wurtzite GaN, 128
Photodissociation, 378
Photoluminescence (PL) studies, 65–66
 near-field spectroscopic studies using, 65–78
Plasma frequency, 161
Plasma oscillations
 damping, 416
 of extrinsic carriers, 414–416
 of photoexcited carriers, 411–414
 in 2D layers, 417

Plasma theory, 233
Pockels effect, 397, 398
Polar scattering, 171
Probe pulse, 154
Probe-to-sample distance control, 58–59
Pump-and-probe spectroscopy
  combined with NSOM, 62, 63–64
  Coulomb correlations and optical nonlinearities applications to, 254–279
  femtosecond near-field, 89–98
Pump-probe spectroscopy at shot-noise limit
  electronic amplification scheme, 16–17
  optical setup, 14–16
Pump-probe technique, 153, 154
Pump pulse, 154

Q

Quantum beats, 420–422
Quantum dots (QDs), 42, 43
  near-field spectroscopic studies of, 65–66
  nonglobal approach, 54–55
Quantum kinetic memory effects
  carrier–carrier scattering suppressed, 21
  CdTe example, 30–33
  choice of interaction process and materials, 4–10
  conclusions, 33–35
  GaAs example, 25–30
  historical research, 3–4
  hole distributions and hot phonon effects, 21–25
  pump-probe spectroscopy at shot-noise limit, 14–17
  reduction of light-hole contributions via interband selection rules, 18–21
  two-color femtosecond Ti:sapphire laser oscillator, 13–14
  ultrafast generation of nonquilibrium charge carriers, 11–13
Quantum kinetics, femtosecond spectroscopy and
  background information, 205–208
  Bloch equations, 209–215
  femtosecond DTS, 206, 219–222
  femtosecond FWM, 206, 217–219, 223–227
  low-excitation, 215–220

Quantum structures, THz emission from, 420–427
Quantum wells (QWs), 42, 43
Quantum wires (QWR), 42
  cleaved edge overgrowth, 44
  near-field spectroscopic studies of, 66–78
  quasi-one-dimensional structures, 43–46
  time-resolved near-field microscopy, 79–98
  T-shaped, 44
  V-grooved, 44–45

R

Rabi flopping, 379
Radiative recombination, 172–174
Raman spectroscopy
  classical theory, 122–124
  full quantum-mechanical approach, 112–121
  quantum-mechanical approach, 124–125
  scattering by lattice vibrations, 122–125
  scattering from carriers, 110–121
Random-phase approximation (RPA), 111, 119, 207, 212–213
Rayleigh scattering, 123
Recombination, carrier, 172–174

S

SBA, 247–248
Scanning tunneling microscopy, 46
Schrödinger equation, time-dependent, 316, 320, 325, 344, 371, 379
Screening effects, 173–174, 233
Second-order susceptibility, 348–353
  formula, 356–363
Self-phase modulation, 63
Semiconductor Bloch equations (SBE), 234, 236–237
Semiconductor nanostructures
  background information, 39–41
  description of near-field optics of, 53–56
  general aspects, 41–43
  low-temperature near-field microscopy, 59–61, 65–66
  optical studies, role of, 40
  probe-to-sample distance control, 58–59

Semiconductor nanostructures (*Continued*)
  quantum wires, structure of, 43–46
  stationary near-field spectroscopy of, 65–78
  temporally and spatially resolved near-field spectroscopy, 61–65
  time-resolved near-field microscopy, 79–98
Semiconductor-to-metal transition, 190–191
Shear-force technique, noncontact, 59
Si
  density functional simulation for, 370–371
  response to fat, intense laser pulses, 363–370
Single photon absorption (SPA), 167–169
Spin-density fluctuations (SDFs), 118, 120–121
Spin selectivity, 19–20
Stark effect, 213, 234, 403
Stationary near-field spectroscopy of semiconductor nanostructures, 65–78
Stokes Raman scattering, 123–124

T

Temporally and spatially resolved near-field spectroscopy, 61–65
Thermal diffusion, 175, 176
THz emission
  antenna detection techniques, 396–397
  applications in spectroscopy, 427–433
  background information, 390–392
  bandwidth increase due to ultrafast currents, 409
  Bloch oscillations, 422–426
  coherent phonons, 418–419
  conclusions, 433–434
  detection of pulses, 395–400
  dipole emission, 392–394
  Drude–Lorentz model of carrier transport, 400–402
  electrooptic detection, 397–399
  field-induced transport emission, 405–409
  instantaneous polarization, 403–404, 410
  interferometric detection, 399–400
  intersubband structures, 426
  phases of transport and polarization, 404–405
  plasma oscillations in 2D layers, 417
  plasma oscillations of extrinsic carriers, 414–416
  plasma oscillations of photoexcited carriers, 411–414
  plasmon–phonon coupling, 419–420
  principles of pulsed, 392–400
  quantum beats, 420–422
  from quantum structures, 420–427
  temporal and spatial coherence, 394–395
THz time-domain spectroscopy, 429–433
Tight-binding electron-ion dynamics (TED) and responses to laser pulses
  applications, 317–327
  background information, 315–317
  biological molecules and, 382–384
  bonding changes and dielectric function, 346–348
  calculations for electronic and structural responses to laser pulses, 327–336
  conclusions, 384–385
  density functional simulation for Si, 370–371
  detailed model, 343–344
  electronic excitation and time-dependent band structure, 344–346
  excited-state, 339–343
  response of C_{60}, 371–377
  response of GaAs, 356–363
  response of Si, 363–370
  second-order susceptibility, 348–353
  second-order susceptibility formula, 356–363
  simple example, 336–339
  simple molecule example, 377–382
  summary of, 354
Time-dependent Hartree–Fock approximation (TDHF), 234, 235–236
Time-resolved near-field spectroscopy
  femtosecond pump-and-probe spectroscopy, 89–98
  lateral carrier transport, 79–89
Time-resolved optical spectroscopy, 61–62
Tomonaga–Luttinger model, 8

Tuning fork setup, 59, 60
Two-color femtosecond Ti:sapphire laser oscillator, 13–14

U

Ultrafast dynamics and phase changes in GaAs
 background information, 152–153
 carrier diffusion, 174–175
 carrier dynamics, 192–197
 carrier excitation, 167–169
 carrier–lattice thermalization, 172
 carrier recombination, 172–174
 carrier redistribution, thermalization, and cooling, 169–172
 conclusions, 200–201
 dielectric function, derivation, 154
 dielectric function, electronic configuration effects on, 163–165
 dielectric function, lattice structure effects on, 162–163
 dielectric function, linear optical properties and, 154–157
 dielectric function, measuring time-resolved, 179–182
 dielectric function, microscopic theory of, 157–158
 dielectric function, reasons for measuring, 165–166
 dielectric function of crystalline solids, 158–161
 dielectric function of GaAs, 182–191
 lattice dynamics, 197–200
 pump-probe technique, 153, 154
 structureal effects, 175–177
Ultrafast electron–phonon interactions
 carrier–carrier scattering suppressed, 21
 CdTe example, 30–33
 choice of interaction process and materials, 4–10
 conclusions, 33–35
 GaAs example, 25–30
 generation of nonequilibrium charge carriers, 11–13
 historical research, 2–4
 hole distributions and hot phonon effects, 21–25
 pump-probe spectroscopy at shot-noise limit, 14–17
 reduction of light-hole contributions via interband selection rules, 18–21
 two-color femtosecond Ti:sapphire laser oscillator, 13–14
 in wide bandgap wurtzite GaN, 109–148

V

Verlet algorithm, 343, 344

W

Wannier excitons, 239
Wannier–Stark states, 423
Wigner–Weisskopf approximation, 213
Wurtzite GaN, wide bandgap
 anharmonic decay of longitudinal optical phonons in, 139–147
 conclusions and future experiments, 147–148
 electron–phonon interactions in, 128–132
 examples and setup, 125–128
 experimental results, 128–147
 nonequilibrium electron distributions and electron-longitudinal optical phonon scattering rates in, 132–139
 phonon modes in, 128
 Raman spectroscopy, use of, 110–125
 research on, 109–110

X

X-ray diffraction experiments, 198

Contents of Volumes in This Series

Volume 1 Physics of III–V Compounds

C. *Hilsum*, Some Key Features of III–V Compounds
F. *Bassani*, Methods of Band Calculations Applicable to III–V Compounds
E. O. *Kane*, The k-p Method
V. L. *Bonch-Bruevich*, Effect of Heavy Doping on the Semiconductor Band Structure
D. *Long*, Energy Band Structures of Mixed Crystals of III–V Compounds
L. M. *Roth and P. N. Argyres*, Magnetic Quantum Effects
S. M. *Puri and T. H. Geballe*, Thermomagnetic Effects in the Quantum Region
W. M. *Becker*, Band Characteristics near Principal Minima from Magnetoresistance
E. H. *Putley*, Freeze-Out Effects, Hot Electron Effects, and Submillimeter Photoconductivity in InSb
H. *Weiss*, Magnetoresistance
B. *Ancker-Johnson*, Plasma in Semiconductors and Semimetals

Volume 2 Physics of III–V Compounds

M. G. *Holland*, Thermal Conductivity
S. I. *Novkova*, Thermal Expansion
U. *Piesbergen*, Heat Capacity and Debye Temperatures
G. *Giesecke*, Lattice Constants
J. R. *Drabble*, Elastic Properties
A. U. *Mac Rae and G. W. Gobeli*, Low Energy Electron Diffraction Studies
R. *Lee Mieher*, Nuclear Magnetic Resonance
B. *Goldstein*, Electron Paramagnetic Resonance
T. S. *Moss*, Photoconduction in III–V Compounds
E. *Antoncik and J. Tauc*, Quantum Efficiency of the Internal Photoelectric Effect in InSb
G. W. *Gobeli and I. G. Allen*, Photoelectric Threshold and Work Function
P. S. *Pershan*, Nonlinear Optics in III–V Compounds
M. *Gershenzon*, Radiative Recombination in the III–V Compounds
F. *Stern*, Stimulated Emission in Semiconductors

Volume 3 Optical of Properties III–V Compounds

M. Hass, Lattice Reflection
W. G. Spitzer, Multiphonon Lattice Absorption
D. L. Stierwalt and R. F. Potter, Emittance Studies
H. R. Philipp and H. Ehrenveich, Ultraviolet Optical Properties
M. Cardona, Optical Absorption above the Fundamental Edge
E. J. Johnson, Absorption near the Fundamental Edge
J. O. Dimmock, Introduction to the Theory of Exciton States in Semiconductors
B. Lax and J. G. Mavroides, Interband Magnetooptical Effects
H. Y. Fan, Effects of Free Carries on Optical Properties
E. D. Palik and G. B. Wright, Free-Carrier Magnetooptical Effects
R. H. Bube, Photoelectronic Analysis
B. O. Seraphin and H. E. Bennett, Optical Constants

Volume 4 Physics of III–V Compounds

N. A. Goryunova, A. S. Borschevskii, and D. N. Tretiakov, Hardness
N. N. Sirota, Heats of Formation and Temperatures and Heats of Fusion of Compounds $A^{III}B^V$
D. L. Kendall, Diffusion
A. G. Chynoweth, Charge Multiplication Phenomena
R. W. Keyes, The Effects of Hydrostatic Pressure on the Properties of III–V Semiconductors
L. W. Aukerman, Radiation Effects
N. A. Goryunova, F. P. Kesamanly, and D. N. Nasledov, Phenomena in Solid Solutions
R. T. Bate, Electrical Properties of Nonuniform Crystals

Volume 5 Infrared Detectors

H. Levinstein, Characterization of Infrared Detectors
P. W. Kruse, Indium Antimonide Photoconductive and Photoelectromagnetic Detectors
M. B. Prince, Narrowband Self-Filtering Detectors
I. Melngalis and T. C. Harman, Single-Crystal Lead-Tin Chalcogenides
D. Long and J. L. Schmidt, Mercury-Cadmium Telluride and Closely Related Alloys
E. H. Putley, The Pyroelectric Detector
N. B. Stevens, Radiation Thermopiles
R. J. Keyes and T. M. Quist, Low Level Coherent and Incoherent Detection in the Infrared
M. C. Teich, Coherent Detection in the Infrared
F. R. Arams, E. W. Sard, B. J. Peyton, and F. P. Pace, Infrared Heterodyne Detection with Gigahertz IF Response
H. S. Sommers, Jr., Macrowave-Based Photoconductive Detector
R. Sehr and R. Zuleeg, Imaging and Display

Volume 6 Injection Phenomena

M. A. Lampert and R. B. Schilling, Current Injection in Solids: The Regional Approximation Method
R. Williams, Injection by Internal Photoemission
A. M. Barnett, Current Filament Formation

R. Baron and J. W. Mayer, Double Injection in Semiconductors
W. Ruppel, The Photoconductor-Metal Contact

Volume 7   Application and Devices
Part A

J. A. Copeland and S. Knight, Applications Utilizing Bulk Negative Resistance
F. A. Padovani, The Voltage-Current Characteristics of Metal-Semiconductor Contacts
P. L. Hower, W. W. Hooper, B. R. Cairns, R. D. Fairman, and D. A. Tremere, The GaAs Field-Effect Transistor
M. H. White, MOS Transistors
G. R. Antell, Gallium Arsenide Transistors
T. L. Tansley, Heterojunction Properties

Part B

T. Misawa, IMPATT Diodes
H. C. Okean, Tunnel Diodes
R. B. Campbell and Hung-Chi Chang, Silicon Junction Carbide Devices
R. E. Enstrom, H. Kressel, and L. Krassner, High-Temperature Power Rectifiers of $GaAs_{1-x}P_x$

Volume 8   Transport and Optical Phenomena

R. J. Stirn, Band Structure and Galvanomagnetic Effects in III–V Compounds with Indirect Band Gaps
R. W. Ure, Jr., Thermoelectric Effects in III–V Compounds
H. Piller, Faraday Rotation
H. Barry Bebb and E. W. Williams, Photoluminescence I: Theory
E. W. Williams and H. Barry Bebb, Photoluminescence II: Gallium Arsenide

Volume 9   Modulation Techniques

B. O. Seraphin, Electroreflectance
R. L. Aggarwal, Modulated Interband Magnetooptics
D. F. Blossey and Paul Handler, Electroabsorption
B. Batz, Thermal and Wavelength Modulation Spectroscopy
I. Balslev, Piezoptical Effects
D. E. Aspnes and N. Bottka, Electric-Field Effects on the Dielectric Function of Semiconductors and Insulators

Volume 10   Transport Phenomena

R. L. Rhode, Low-Field Electron Transport
J. D. Wiley, Mobility of Holes in III–V Compounds
C. M. Wolfe and G. E. Stillman, Apparent Mobility Enhancement in Inhomogeneous Crystals
R. L. Petersen, The Magnetophonon Effect

Volume 11   Solar Cells

H. J. Hovel, Introduction; Carrier Collection, Spectral Response, and Photocurrent; Solar Cell Electrical Characteristics; Efficiency; Thickness; Other Solar Cell Devices; Radiation Effects; Temperature and Intensity; Solar Cell Technology

Volume 12   Infrared Detectors (II)

W. L. Eiseman, J. D. Merriam, and R. F. Potter, Operational Characteristics of Infrared Photodetectors
P. R. Bratt, Impurity Germanium and Silicon Infrared Detectors
E. H. Putley, InSb Submillimeter Photoconductive Detectors
G. E. Stillman, C. M. Wolfe, and J. O. Dimmock, Far-Infrared Photoconductivity in High Purity GaAs
G. E. Stillman and C. M. Wolfe, Avalanche Photodiodes
P. L. Richards, The Josephson Junction as a Detector of Microwave and Far-Infrared Radiation
E. H. Putley, The Pyroelectric Detector—An Update

Volume 13   Cadmium Telluride

K. Zanio, Materials Preparations; Physics; Defects; Applications

Volume 14   Lasers, Junctions, Transport

N. Holonyak, Jr. and M. H. Lee, Photopumped III–V Semiconductor Lasers
H. Kressel and J. K. Butler, Heterojunction Laser Diodes
A Van der Ziel, Space-Charge-Limited Solid-State Diodes
P. J. Price, Monte Carlo Calculation of Electron Transport in Solids

Volume 15   Contacts, Junctions, Emitters

B. L. Sharma, Ohmic Contacts to III–V Compounds Semiconductors
A. Nussbaum, The Theory of Semiconducting Junctions
J. S. Escher, NEA Semiconductor Photoemitters

Volume 16   Defects, (HgCd)Se, (HgCd)Te

H. Kressel, The Effect of Crystal Defects on Optoelectronic Devices
C. R. Whitsett, J. G. Broerman, and C. J. Summers, Crystal Growth and Properties of $Hg_{1-x}Cd_xSe$ alloys
M. H. Weiler, Magnetooptical Properties of $Hg_{1-x}Cd_xTe$ Alloys
P. W. Kruse and J. G. Ready, Nonlinear Optical Effects in $Hg_{1-x}Cd_xTe$

Volume 17   CW Processing of Silicon and Other Semiconductors

J. F. Gibbons, Beam Processing of Silicon
A. Lietoila, R. B. Gold, J. F. Gibbons, and L. A. Christel, Temperature Distributions and Solid Phase Reaction Rates Produced by Scanning CW Beams

A. Leitoila and J. F. Gibbons, Applications of CW Beam Processing to Ion Implanted Crystalline Silicon
N. M. Johnson, Electronic Defects in CW Transient Thermal Processed Silicon
K. F. Lee, T. J. Stultz, and J. F. Gibbons, Beam Recrystallized Polycrystalline Silicon: Properties, Applications, and Techniques
T. Shibata, A. Wakita, T. W. Sigmon, and J. F. Gibbons, Metal-Silicon Reactions and Silicide
Y. I. Nissim and J. F. Gibbons, CW Beam Processing of Gallium Arsenide

Volume 18 Mercury Cadmium Telluride

P. W. Kruse, The Emergence of $(Hg_{1-x}Cd_x)Te$ as a Modern Infrared Sensitive Material
H. E. Hirsch, S. C. Liang, and A. G. White, Preparation of High-Purity Cadmium, Mercury, and Tellurium
W. F. H. Micklethwaite, The Crystal Growth of Cadmium Mercury Telluride
P. E. Petersen, Auger Recombination in Mercury Cadmium Telluride
R. M. Broudy and V. J. Mazurczyck, (HgCd)Te Photoconductive Detectors
M. B. Reine, A. K. Soad, and T. J. Tredwell, Photovoltaic Infrared Detectors
M. A. Kinch, Metal-Insulator-Semiconductor Infrared Detectors

Volume 19 Deep Levels, GaAs, Alloys, Photochemistry

G. F. Neumark and K. Kosai, Deep Levels in Wide Band-Gap III–V Semiconductors
D. C. Look, The Electrical and Photoelectronic Properties of Semi-Insulating GaAs
R. F. Brebrick, Ching-Hua Su, and Pok-Kai Liao, Associated Solution Model for Ga-In-Sb and Hg-Cd-Te
Y. Ya. Gurevich and Y. V. Pleskon, Photoelectrochemistry of Semiconductors

Volume 20 Semi-Insulating GaAs

R. N. Thomas, H. M. Hobgood, G. W. Eldridge, D. L. Barrett, T. T. Braggins, L. B. Ta, and S. K. Wang, High-Purity LEC Growth and Direct Implantation of GaAs for Monolithic Microwave Circuits
C. A. Stolte, Ion Implantation and Materials for GaAs Integrated Circuits
C. G. Kirkpatrick, R. T. Chen, D. E. Holmes, P. M. Asbeck, K. R. Elliott, R. D. Fairman, and J. R. Oliver, LEC GaAs for Integrated Circuit Applications
J. S. Blakemore and S. Rahimi, Models for Mid-Gap Centers in Gallium Arsenide

Volume 21 Hydrogenated Amorphous Silicon
Part A

J. I. Pankove, Introduction
M. Hirose, Glow Discharge; Chemical Vapor Deposition
Y. Uchida, di Glow Discharge
T. D. Moustakas, Sputtering
I. Yamada, Ionized-Cluster Beam Deposition
B. A. Scott, Homogeneous Chemical Vapor Deposition

F. J. Kampas, Chemical Reactions in Plasma Deposition
P. A. Longeway, Plasma Kinetics
H. A. Weakliem, Diagnostics of Silane Glow Discharges Using Probes and Mass Spectroscopy
L. Gluttman, Relation between the Atomic and the Electronic Structures
A. Chenevas-Paule, Experiment Determination of Structure
S. Minomura, Pressure Effects on the Local Atomic Structure
D. Adler, Defects and Density of Localized States

Part B

J. I. Pankove, Introduction
G. D. Cody, The Optical Absorption Edge of a-Si:H
N. M. Amer and W. B. Jackson, Optical Properties of Defect States in a-Si:H
P. J. Zanzucchi, The Vibrational Spectra of a-Si:H
Y. Hamakawa, Electroreflectance and Electroabsorption
J. S. Lannin, Raman Scattering of Amorphous Si, Ge, and Their Alloys
R. A. Street, Luminescence in a-Si:H
R. S. Crandall, Photoconductivity
J. Tauc, Time-Resolved Spectroscopy of Electronic Relaxation Processes
P. E. Vanier, IR-Induced Quenching and Enhancement of Photoconductivity and Photo luminescence
H. Schade, Irradiation-Induced Metastable Effects
L. Ley, Photoelectron Emission Studies

Part C

J. I. Pankove, Introduction
J. D. Cohen, Density of States from Junction Measurements in Hydrogenated Amorphous Silicon
P. C. Taylor, Magnetic Resonance Measurements in a-Si:H
K. Morigaki, Optically Detected Magnetic Resonance
J. Dresner, Carrier Mobility in a-Si:H
T. Tiedje, Information about band-Tail States from Time-of-Flight Experiments
A. R. Moore, Diffusion Length in Undoped a-Si:H
W. Beyer and J. Overhof, Doping Effects in a-Si:H
H. Fritzche, Electronic Properties of Surfaces in a-Si:H
C. R. Wronski, The Staebler-Wronski Effect
R. J. Nemanich, Schottky Barriers on a-Si:H
B. Abeles and T. Tiedje, Amorphous Semiconductor Superlattices

Part D

J. I. Pankove, Introduction
D. E. Carlson, Solar Cells
G. A. Swartz, Closed-Form Solution of I–V Characteristic for a a-Si:H Solar Cells
I. Shimizu, Electrophotography
S. Ishioka, Image Pickup Tubes

P. G. LeComber and W. E. Spear, The Development of the a-Si:H Field-Effect Transistor and Its Possible Applications
D. G. Ast, a-Si:H FET-Addressed LCD Panel
S. Kaneko, Solid-State Image Sensor
M. Matsumura, Charge-Coupled Devices
M. A. Bosch, Optical Recording
A. D'Amico and G. Fortunato, Ambient Sensors
H. Kukimoto, Amorphous Light-Emitting Devices
R. J. Phelan, Jr., Fast Detectors and Modulators
J. I. Pankove, Hybrid Structures
P. G. LeComber, A. E. Owen, W. E. Spear, J. Hajto, and W. K. Choi, Electronic Switching in Amorphous Silicon Junction Devices

Volume 22 Lightwave Communications Technology
Part A

K. Nakajima, The Liquid-Phase Epitaxial Growth of InGaAsP
W. T. Tsang, Molecular Beam Epitaxy for III–V Compound Semiconductors
G. B. Stringfellow, Organometallic Vapor-Phase Epitaxial Growth of III–V Semiconductors
G. Beuchet, Halide and Chloride Transport Vapor-Phase Deposition of InGaAsP and GaAs
M. Razeghi, Low-Pressure Metallo-Organic Chemical Vapor Deposition of $Ga_xIn_{1-x}AsP_{1-y}$ Alloys
P. M. Petroff, Defects in III–V Compound Semiconductors

Part B

J. P. van der Ziel, Mode Locking of Semiconductor Lasers
K. Y. Lau and A. Yariv, High-Frequency Current Modulation of Semiconductor Injection Lasers
C. H. Henry, Special Properties of Semiconductor Lasers
Y. Suematsu, K. Kishino, S. Arai, and F. Koyama, Dynamic Single-Mode Semiconductor Lasers with a Distributed Reflector
W. T. Tsang, The Cleaved-Coupled-Cavity (C^3) Laser

Part C

R. J. Nelson and N. K. Dutta, Review of InGaAsP InP Laser Structures and Comparison of Their Performance
N. Chinone and M. Nakamura, Mode-Stabilized Semiconductor Lasers for 0.7–0.8- and 1.1–1.6-μm Regions
Y. Horikoshi, Semiconductor Lasers with Wavelengths Exceeding 2 μm
B. A. Dean and M. Dixon, The Functional Reliability of Semiconductor Lasers as Optical Transmitters
R. H. Saul, T. P. Lee, and C. A. Burus, Light-Emitting Device Design
C. L. Zipfel, Light-Emitting Diode-Reliability
T. P. Lee and T. Li, LED-Based Multimode Lightwave Systems
K. Ogawa, Semiconductor Noise-Mode Partition Noise

Part D

F. *Capasso*, The Physics of Avalanche Photodiodes
T. P. *Pearsall and M. A. Pollack*, Compound Semiconductor Photodiodes
T. *Kaneda*, Silicon and Germanium Avalanche Photodiodes
S. R. *Forrest*, Sensitivity of Avalanche Photodetector Receivers for High-Bit-Rate Long-Wavelength Optical Communication Systems
J. C. *Campbell*, Phototransistors for Lightwave Communications

Part E

S. *Wang*, Principles and Characteristics of Integrable Active and Passive Optical Devices
S. *Margalit and A. Yariv*, Integrated Electronic and Photonic Devices
T. *Mukai, Y. Yamamoto, and T. Kimura*, Optical Amplification by Semiconductor Lasers

Volume 23 Pulsed Laser Processing of Semiconductors

R. F. *Wood, C. W. White, and R. T. Young*, Laser Processing of Semiconductors: An Overview
C. W. *White*, Segregation, Solute Trapping, and Supersaturated Alloys
G. E. *Jellison, Jr.*, Optical and Electrical Properties of Pulsed Laser-Annealed Silicon
R. F. *Wood and G. E. Jellison, Jr.*, Melting Model of Pulsed Laser Processing
R. F. *Wood and F. W. Young, Jr.*, Nonequilibrium Solidification Following Pulsed Laser Melting
D. H. *Lowndes and G. E. Jellison, Jr.*, Time-Resolved Measurement During Pulsed Laser Irradiation of Silicon
D. M. *Zebner*, Surface Studies of Pulsed Laser Irradiated Semiconductors
D. H. *Lowndes*, Pulsed Beam Processing of Gallium Arsenide
R. B. *James*, Pulsed CO_2 Laser Annealing of Semiconductors
R. T. *Young and R. F. Wood*, Applications of Pulsed Laser Processing

Volume 24 Applications of Multiquantum Wells, Selective Doping, and Superlattices

C. *Weisbuch*, Fundamental Properties of III–V Semiconductor Two-Dimensional Quantized Structures: The Basis for Optical and Electronic Device Applications
H. *Morkoc and H. Unlu*, Factors Affecting the Performance of (Al, Ga)As/GaAs and (Al, Ga)As/InGaAs Modulation-Doped Field-Effect Transistors: Microwave and Digital Applications
N. T. *Linh*, Two-Dimensional Electron Gas FETs: Microwave Applications
M. *Abe et al.*, Ultra-High-Speed HEMT Integrated Circuits
D. S. *Chemla, D. A. B. Miller, and P. W. Smith*, Nonlinear Optical Properties of Multiple Quantum Well Structures for Optical Signal Processing
F. *Capasso*, Graded-Gap and Superlattice Devices by Band-Gap Engineering
W. T. *Tsang*, Quantum Confinement Heterostructure Semiconductor Lasers
G. C. *Osbourn et al.*, Principles and Applications of Semiconductor Strained-Layer Superlattices

Volume 25 Diluted Magnetic Semiconductors

W. Giriat and J. K. Furdyna, Crystal Structure, Composition, and Materials Preparation of Diluted Magnetic Semiconductors

W. M. Becker, Band Structure and Optical Properties of Wide-Gap $A_{1-x}^{II} Mn_x B_{IV}$ Alloys at Zero Magnetic Field

S. Oseroff and P. H. Keesom, Magnetic Properties: Macroscopic Studies

T. Giebultowicz and T. M. Holden, Neutron Scattering Studies of the Magnetic Structure and Dynamics of Diluted Magnetic Semiconductors

J. Kossut, Band Structure and Quantum Transport Phenomena in Narrow-Gap Diluted Magnetic Semiconductors

C. Riquaux, Magnetooptical Properties of Large-Gap Diluted Magnetic Semiconductors

J. A. Gaj, Magnetooptical Properties of Large-Gap Diluted Magnetic Semiconductors

J. Mycielski, Shallow Acceptors in Diluted Magnetic Semiconductors: Splitting, Boil-off, Giant Negative Magnetoresistance

A. K. Ramadas and R. Rodriquez, Raman Scattering in Diluted Magnetic Semiconductors

P. A. Wolff, Theory of Bound Magnetic Polarons in Semimagnetic Semiconductors

Volume 26 III–V Compound Semiconductors and Semiconductor Properties of Superionic Materials

Z. Yuanxi, III–V Compounds

H. V. Winston, A. T. Hunter, H. Kimura, and R. E. Lee, InAs-Alloyed GaAs Substrates for Direct Implantation

P. K. Bhattacharya and S. Dhar, Deep Levels in III–V Compound Semiconductors Grown by MBE

Y. Ya. Gurevich and A. K. Ivanov-Shits, Semiconductor Properties of Supersonic Materials

Volume 27 High Conducting Quasi-One-Dimensional Organic Crystals

E. M. Conwell, Introduction to Highly Conducting Quasi-One-Dimensional Organic Crystals

I. A. Howard, A Reference Guide to the Conducting Quasi-One-Dimensional Organic Molecular Crystals

J. P. Pouquet, Structural Instabilities

E. M. Conwell, Transport Properties

C. S. Jacobsen, Optical Properties

J. C. Scott, Magnetic Properties

L. Zuppiroli, Irradiation Effects: Perfect Crystals and Real Crystals

Volume 28 Measurement of High-Speed Signals in Solid State Devices

J. Frey and D. Ioannou, Materials and Devices for High-Speed and Optoelectronic Applications

H. Schumacher and E. Strid, Electronic Wafer Probing Techniques

D. H. Auston, Picosecond Photoconductivity: High-Speed Measurements of Devices and Materials

J. A. Valdmanis, Electro-Optic Measurement Techniques for Picosecond Materials, Devices, and Integrated Circuits.

J. M. Wiesenfeld and R. K. Jain, Direct Optical Probing of Integrated Circuits and High-Speed Devices

G. Plows, Electron-Beam Probing

A. M. Weiner and R. B. Marcus, Photoemissive Probing

Volume 29   Very High Speed Integrated Circuits: Gallium Arsenide LSI

M. Kuzuhara and T. Nazaki, Active Layer Formation by Ion Implantation
H. Hasimoto, Focused Ion Beam Implantation Technology
T. Nozaki and A. Higashisaka, Device Fabrication Process Technology
M. Ino and T. Takada, GaAs LSI Circuit Design
M. Hirayama, M. Ohmori, and K. Yamasaki, GaAs LSI Fabrication and Performance

Volume 30   Very High Speed Integrated Circuits: Heterostructure

H. Watanabe, T. Mizutani, and A. Usui, Fundamentals of Epitaxial Growth and Atomic Layer Epitaxy
S. Hiyamizu, Characteristics of Two-Dimensional Electron Gas in III–V Compound Heterostructures Grown by MBE
T. Nakanisi, Metalorganic Vapor Phase Epitaxy for High-Quality Active Layers
T. Nimura, High Electron Mobility Transistor and LSI Applications
T. Sugeta and T. Ishibashi, Hetero-Bipolar Transistor and LSI Application
H. Matsueda, T. Tanaka, and M. Nakamura, Optoelectronic Integrated Circuits

Volume 31   Indium Phosphide: Crystal Growth and Characterization

J. P. Farges, Growth of Discoloration-free InP
M. J. McCollum and G. E. Stillman, High Purity InP Grown by Hydride Vapor Phase Epitaxy
T. Inada and T. Fukuda, Direct Synthesis and Growth of Indium Phosphide by the Liquid Phosphorous Encapsulated Czochralski Method
O. Oda, K. Katagiri, K. Shinohara, S. Katsura, Y. Takahashi, K. Kainosho, K. Kohiro, and R. Hirano, InP Crystal Growth, Substrate Preparation and Evaluation
K. Tada, M. Tatsumi, M. Morioka, T. Araki, and T. Kawase, InP Substrates: Production and Quality Control
M. Razeghi, LP-MOCVD Growth, Characterization, and Application of InP Material
T. A. Kennedy and P. J. Lin-Chung, Stoichiometric Defects in InP

Volme 32   Strained-Layer Superlattices: Physics

T. P. Pearsall, Strained-Layer Superlattices
F. H. Pollack, Effects of Homogeneous Strain on the Electronic and Vibrational Levels in Semiconductors
J. Y. Marzin, J. M. Gerárd, P. Voisin, and J. A. Brum, Optical Studies of Strained III–V Heterolayers
R. People and S. A. Jackson, Structurally Induced States from Strain and Confinement
M. Jaros, Microscopic Phenomena in Ordered Superlattices

Volume 33   Strained-Layer Superlattices: Materials Science and Technology

R. Hull and J. C. Bean, Principles and Concepts of Strained-Layer Epitaxy
W. J. Schaff, P. J. Tasker, M. C. Foisy, and L. F. Eastman, Device Applications of Strained-Layer Epitaxy

S. T. Picraux, B. L. Doyle, and J. Y. Tsao, Structure and Characterization of Strained-Layer Superlattices
E. Kasper and F. Schaffer, Group IV Compounds
D. L. Martin, Molecular Beam Epitaxy of IV–VI Compounds Heterojunction
R. L. Gunshor, L. A. Kolodziejski, A. V. Nurmikko, and N. Otsuka, Molecular Beam Epitaxy of II–VI Semiconductor Microstructures

Volume 34   Hydrogen in Semiconductors

J. I. Pankove and N. M. Johnson, Introduction to Hydrogen in Semiconductors
C. H. Seager, Hydrogenation Methods
J. I. Pankove, Hydrogenation of Defects in Crystalline Silicon
J. W. Corbett, P. Deák, U. V. Desnica, and S. J. Pearton, Hydrogen Passivation of Damage Centers in Semiconductors
S. J. Pearton, Neutralization of Deep Levels in Silicon
J. I. Pankove, Neutralization of Shallow Acceptors in Silicon
N. M. Johnson, Neutralization of Donor Dopants and Formation of Hydrogen-Induced Defects in n-Type Silicon
M. Stavola and S. J. Pearton, Vibrational Spectroscopy of Hydrogen-Related Defects in Silicon
A. D. Marwick, Hydrogen in Semiconductors: Ion Beam Techniques
C. Herring and N. M. Johnson, Hydrogen Migration and Solubility in Silicon
E. E. Haller, Hydrogen-Related Phenomena in Crystalline Germanium
J. Kakalios, Hydrogen Diffusion in Amorphous Silicon
J. Chevalier, B. Clerjaud, and B. Pajot, Neutralization of Defects and Dopants in III–V Semiconductors
G. G. DeLeo and W. B. Fowler, Computational Studies of Hydrogen-Containing Complexes in Semiconductors
R. F. Kiefl and T. L. Estle, Muonium in Semiconductors
C. G. Van de Walle, Theory of Isolated Interstitial Hydrogen and Muonium in Crystalline Semiconductors

Volume 35   Nanostructured Systems

M. Reed, Introduction
H. van Houten, C. W. J. Beenakker, and B. J. van Wees, Quantum Point Contacts
G. Timp, When Does a Wire Become an Electron Waveguide?
M. Büttiker, The Quantum Hall Effects in Open Conductors
W. Hansen, J. P. Kotthaus, and U. Merkt, Electrons in Laterally Periodic Nanostructures

Volume 36   The Spectroscopy of Semiconductors

D. Heiman, Spectroscopy of Semiconductors at Low Temperatures and High Magnetic Fields
A. V. Nurmikko, Transient Spectroscopy by Ultrashort Laser Pulse Techniques
A. K. Ramdas and S. Rodriguez, Piezospectroscopy of Semiconductors
O. J. Glembocki and B. V. Shanabrook, Photoreflectance Spectroscopy of Microstructures
D. G. Seiler, C. L. Littler, and M. H. Wiler, One- and Two-Photon Magneto-Optical Spectroscopy of InSb and $Hg_{1-x}Cd_xTe$

Volume 37 The Mechanical Properties of Semiconductors

A.-B. Chen, A. Sher and W. T. Yost, Elastic Constants and Related Properties of Semiconductor Compounds and Their Alloys
D. R. Clarke, Fracture of Silicon and Other Semiconductors
H. Siethoff, The Plasticity of Elemental and Compound Semiconductors
S. Guruswamy, K. T. Faber and J. P. Hirth, Mechanical Behavior of Compound Semiconductors
S. Mahajan, Deformation Behavior of Compound Semiconductors
J. P. Hirth, Injection of Dislocations into Strained Multilayer Structures
D. Kendall, C. B. Fleddermann, and K. J. Malloy, Critical Technologies for the Micromachining of Silicon
I. Matsuba and K. Mokuya, Processing and Semiconductor Thermoelastic Behavior

Volume 38 Imperfections in III/V Materials

U. Scherz and M. Scheffler, Density-Functional Theory of sp-Bonded Defects in III/V Semiconductors
M. Kaminska and E. R. Weber, El2 Defect in GaAs
D. C. Look, Defects Relevant for Compensation in Semi-Insulating GaAs
R. C. Newman, Local Vibrational Mode Spectroscopy of Defects in III/V Compounds
A. M. Hennel, Transition Metals in III/V Compounds
K. J. Malloy and K. Khachaturyan, DX and Related Defects in Semiconductors
V. Swaminathan and A. S. Jordan, Dislocations in III/V Compounds
K. W. Nauka, Deep Level Defects in the Epitaxial III/V Materials

Volume 39 Minority Carriers in III–V Semiconductors: Physics and Applications

N. K. Dutta, Radiative Transitions in GaAs and Other III–V Compounds
R. K. Ahrenkiel, Minority-Carrier Lifetime in III–V Semiconductors
T. Furuta, High Field Minority Electron Transport in p-GaAs
M. S. Lundstrom, Minority-Carrier Transport in III–V Semiconductors
R. A. Abram, Effects of Heavy Doping and High Excitation on the Band Structure of GaAs
D. Yevick and W. Bardyszewski, An Introduction to Non-Equilibrium Many-Body Analyses of Optical Processes in III–V Semiconductors

Volume 40 Epitaxial Microstructures

E. F. Schubert, Delta-Doping of Semiconductors: Electronic, Optical, and Structural Properties of Materials and Devices
A. Gossard, M. Sundaram, and P. Hopkins, Wide Graded Potential Wells
P. Petroff, Direct Growth of Nanometer-Size Quantum Wire Superlattices
E. Kapon, Lateral Patterning of Quantum Well Heterostructures by Growth of Nonplanar Substrates
H. Temkin, D. Gershoni, and M. Panish, Optical Properties of $Ga_{1-x}In_xAs$/InP Quantum Wells

Volume 41 High Speed Heterostructure Devices

F. Capasso, F. Beltram, S. Sen, A. Pahlevi, and A. Y. Cho, Quantum Electron Devices: Physics and Applications
P. Solomon, D. J. Frank, S. L. Wright, and F. Canora, GaAs-Gate Semiconductor–Insulator–Semiconductor FET
M. H. Hashemi and U. K. Mishra, Unipolar InP-Based Transistors
R. Kiehl, Complementary Heterostructure FET Integrated Circuits
T. Ishibashi, GaAs-Based and InP-Based Heterostructure Bipolar Transistors
H. C. Liu and T. C. L. G. Sollner, High-Frequency-Tunneling Devices
H. Ohnishi, T. More, M. Takatsu, K. Imamura, and N. Yokoyama, Resonant-Tunneling Hot-Electron Transistors and Circuits

Volume 42 Oxygen in Silicon

F. Shimura, Introduction to Oxygen in Silicon
W. Lin, The Incorporation of Oxygen into Silicon Crystals
T. J. Schaffner and D. K. Schroder, Characterization Techniques for Oxygen in Silicon
W. M. Bullis, Oxygen Concentration Measurement
S. M. Hu, Intrinsic Point Defects in Silicon
B. Pajot, Some Atomic Configurations of Oxygen
J. Michel and L. C. Kimerling, Electical Properties of Oxygen in Silicon
R. C. Newman and R. Jones, Diffusion of Oxygen in Silicon
T. Y. Tan and W. J. Taylor, Mechanisms of Oxygen Precipitation: Some Quantitative Aspects
M. Schrems, Simulation of Oxygen Precipitation
K. Simino and I. Yonenaga, Oxygen Effect on Mechanical Properties
W. Bergholz, Grown-in and Process-Induced Effects
F. Shimura, Intrinsic/Internal Gettering
H. Tsuya, Oxygen Effect on Electronic Device Performance

Volume 43 Semiconductors for Room Temperature Nuclear Detector Applications

R. B. James and T. E. Schlesinger, Introduction and Overview
L. S. Darken and C. E. Cox, High-Purity Germanium Detectors
A. Burger, D. Nason, L. Van den Berg, and M. Schieber, Growth of Mercuric Iodide
X. J. Bao, T. E. Schlesinger, and R. B. James, Electrical Properties of Mercuric Iodide
X. J. Bao, R. B. James, and T. E. Schlesinger, Optical Properties of Red Mercuric Iodide
M. Hage-Ali and P. Siffert, Growth Methods of CdTe Nuclear Detector Materials
M. Hage-Ali and P Siffert, Characterization of CdTe Nuclear Detector Materials
M. Hage-Ali and P. Siffert, CdTe Nuclear Detectors and Applications
R. B. James, T. E. Schlesinger, J. Lund, and M. Schieber, $Cd_{1-x}Zn_xTe$ Spectrometers for Gamma and X-Ray Applications
D. S. McGregor, J. E. Kammeraad, Gallium Arsenide Radiation Detectors and Spectrometers
J. C. Lund, F. Olschner, and A. Burger, Lead Iodide
M. R. Squillante, and K. S. Shah, Other Materials: Status and Prospects
V. M. Gerrish, Characterization and Quantification of Detector Performance
J. S. Iwanczyk and B. E. Patt, Electronics for X-ray and Gamma Ray Spectrometers
M. Schieber, R. B. James, and T. E. Schlesinger, Summary and Remaining Issues for Room Temperature Radiation Spectrometers

Volume 44   II–IV Blue/Green Light Emitters: Device Physics and Epitaxial Growth

J. Han and R. L. Gunshor, MBE Growth and Electrical Properties of Wide Bandgap ZnSe-based II–VI Semiconductors

S. Fujita and S. Fujita, Growth and Characterization of ZnSe-based II–VI Semiconductors by MOVPE

E. Ho and L. A. Kolodziejski, Gaseous Source UHV Epitaxy Technologies for Wide Bandgap II–VI Semiconductors

C. G. Van de Walle, Doping of Wide-Band-Gap II–VI Compounds — Theory

R. Cingolani, Optical Properties of Excitons in ZnSe-Based Quantum Well Heterostructures

A. Ishibashi and A. V. Nurmikko, II–VI Diode Lasers: A Current View of Device Performance and Issues

S. Guha and J. Petruzello, Defects and Degradation in Wide-Gap II–VI-based Structures and Light Emitting Devices

Volume 45   Effect of Disorder and Defects in Ion-Implanted Semiconductors: Electrical and Physiochemical Characterization

H. Ryssel, Ion Implantation into Semiconductors: Historical Perspectives

You-Nian Wang and Teng-Cai Ma, Electronic Stopping Power for Energetic Ions in Solids

S. T. Nakagawa, Solid Effect on the Electronic Stopping of Crystalline Target and Application to Range Estimation

G. Müller, S. Kalbitzer and G. N. Greaves, Ion Beams in Amorphous Semiconductor Research

J. Boussey-Said, Sheet and Spreading Resistance Analysis of Ion Implanted and Annealed Semiconductors

M. L. Polignano and G. Queirolo, Studies of the Stripping Hall Effect in Ion-Implanted Silicon

J. Stoemenos, Transmission Electron Microscopy Analyses

R. Nipoti and M. Servidori, Rutherford Backscattering Studies of Ion Implanted Semiconductors

P. Zaumseil, X-ray Diffraction Techniques

Volume 46   Effect of Disorder and Defects in Ion-Implanted Semiconductors: Optical and Photothermal Characterization

M. Fried, T. Lohner and J. Gyulai, Ellipsometric Analysis

A. Seas and C. Christofides, Transmission and Reflection Spectroscopy on Ion Implanted Semiconductors

A. Othonos and C. Christofides, Photoluminescence and Raman Scattering of Ion Implanted Semiconductors. Influence of Annealing

C. Christofides, Photomodulated Thermoreflectance Investigation of Implanted Wafers. Annealing Kinetics of Defects

U. Zammit, Photothermal Deflection Spectroscopy Characterization of Ion-Implanted and Annealed Silicon Films

A. Mandelis, A. Budiman and M. Vargas, Photothermal Deep-Level Transient Spectroscopy of Impurities and Defects in Semiconductors

R. Kalish and S. Charbonneau, Ion Implantation into Quantum-Well Structures

A. M. Myasnikov and N. N. Gerasimenko, Ion Implantation and Thermal Annealing of III-V Compound Semiconducting Systems: Some Problems of III-V Narrow Gap Semiconductors

Volume 47 Uncooled Infrared Imaging Arrays and Systems

R. G. Buser and M. P. Tompsett, Historical Overview
P. W. Kruse, Principles of Uncooled Infrared Focal Plane Arrays
R. A. Wood, Monolithic Silicon Microbolometer Arrays
C. M. Hanson, Hybrid Pyroelectric-Ferroelectric Bolometer Arrays
D. L. Polla and J. R. Choi, Monolithic Pyroelectric Bolometer Arrays
N. Teranishi, Thermoelectric Uncooled Infrared Focal Plane Arrays
M. F. Tompsett, Pyroelectric Vidicon
T. W. Kenny, Tunneling Infrared Sensors
J. R. Vig, R. L. Filler and Y. Kim, Application of Quartz Microresonators to Uncooled Infrared Imaging Arrays
P. W. Kruse, Application of Uncooled Monolithic Thermoelectric Linear Arrays to Imaging Radiometers

Volume 48 High Brightness Light Emitting Diodes

G. B. Stringfellow, Materials Issues in High-Brightness Light-Emitting Diodes
M. G. Craford, Overview of Device issues in High-Brightness Light-Emitting Diodes
F. M. Steranka, AlGaAs Red Light Emitting Diodes
C. H. Chen, S. A. Stockman, M. J. Peanasky, and C. P. Kuo, OMVPE Growth of AlGaInP for High Efficiency Visible Light-Emitting Diodes
F. A. Kish and R. M. Fletcher, AlGaInP Light-Emitting Diodes
M. W. Hodapp, Applications for High Brightness Light-Emitting Diodes
I. Akasaki and H. Amano, Organometallic Vapor Epitaxy of GaN for High Brightness Blue Light Emitting Diodes
S. Nakamura, Group III-V Nitride Based Ultraviolet-Blue-Green-Yellow Light-Emitting Diodes and Laser Diodes

Volume 49 Light Emission in Silicon: from Physics to Devices

D. J. Lockwood, Light Emission in Silicon
G. Abstreiter, Band Gaps and Light Emission in Si/SiGe Atomic Layer Structures
T. G. Brown and D. G. Hall, Radiative Isoelectronic Impurities in Silicon and Silicon-Germanium Alloys and Superlattices
J. Michel, L. V. C. Assali, M. T. Morse, and L. C. Kimerling, Erbium in Silicon
Y. Kanemitsu, Silicon and Germanium Nanoparticles
P. M. Fauchet, Porous Silicon: Photoluminescence and Electroluminescent Devices
C. Delerue, G. Allan, and M. Lannoo, Theory of Radiative and Nonradiative Processes in Silicon Nanocrystallites
L. Brus, Silicon Polymers and Nanocrystals

Volume 50 Gallium Nitride (GaN)

J. I. Pankove and T. D. Moustakas, Introduction
S. P. DenBaars and S. Keller, Metalorganic Chemical Vapor Deposition (MOCVD) of Group III Nitrides
W. A. Bryden and T. J. Kistenmacher, Growth of Group III-A Nitrides by Reactive Sputtering
N. Newman, Thermochemistry of III-N Semiconductors
S. J. Pearton and R. J. Shul, Etching of III Nitrides

S. M. Bedair, Indium-based Nitride Compounds
A. Trampert, O. Brandt, and K. H. Ploog, Crystal Structure of Group III Nitrides
H. Morkoc, F. Hamdani, and A. Salvador, Electronic and Optical Properties of III–V Nitride based Quantum Wells and Superlattices
K. Doverspike and J. I. Pankove, Doping in the III-Nitrides
T. Suski and P. Perlin, High Pressure Studies of Defects and Impurities in Gallium Nitride
B. Monemar, Optical Properties of GaN
W. R. L. Lambrecht, Band Structure of the Group III Nitrides
N. E. Christensen and P. Perlin, Phonons and Phase Transitions in GaN
S. Nakamura, Applications of LEDs and LDs
I. Akasaki and H. Amano, Lasers
J. A. Cooper, Jr., Nonvolatile Random Access Memories in Wide Bandgap Semiconductors

Volume 51A Identification of Defects in Semiconductors

G. D. Watkins, EPR and ENDOR Studies of Defects in Semiconductors
J.-M. Spaeth, Magneto-Optical and Electrical Detection of Paramagnetic Resonance in Semiconductors
T. A. Kennedy and E. R. Glaser, Magnetic Resonance of Epitaxial Layers Detected by Photoluminescence
K. H. Chow, B. Hitti, and R. F. Kiefl, μSR on Muonium in Semiconductors and Its Relation to Hydrogen
K. Saarinen, P. Hautojärvi, and C. Corbel, Positron Annihilation Spectroscopy of Defects in Semiconductors
R. Jones and P. R. Briddon, The Ab Initio Cluster Method and the Dynamics of Defects in Semiconductors

Volume 51B Identification of Defects in Semiconductors

G. Davies, Optical Measurements of Point Defects
P. M. Mooney, Defect Identification Using Capacitance Spectroscopy
M. Stavola, Vibrational Spectroscopy of Light Element Impurities in Semiconductors
P. Schwander, W. D. Rau, C. Kisielowski, M. Gribelyuk, and A. Ourmazd, Defect Processes in Semiconductors Studied at the Atomic Level by Transmission Electron Microscopy
N. D. Jager and E. R. Weber, Scanning Tunneling Microscopy of Defects in Semiconductors

Volume 52 SiC Materials and Devices

K. Järrendahl and R. F. Davis, Materials Properties and Characterization of SiC
V. A. Dmitriev and M. G. Spencer, SiC Fabrication Technology: Growth and Doping
V. Saxena and A. J. Steckl, Building Blocks for SiC Devices: Ohmic Contacts, Schottky Contacts, and p-n Junctions
M. S. Shur, SiC Transistors
C. D. Brandt, R. C. Clarke, R. R. Siergiej, J. B. Casady, A. W. Morse, S. Sriram, and A. K. Agarwal, SiC for Applications in High-Power Electronics
R. J. Trew, SiC Microwave Devices

J. Edmond, H. Kong, G. Negley, M. Leonard, K. Doverspike, W. Weeks, A. Suvorov, D. Waltz, and C. Carter, Jr., SiC-Based UV Photodiodes and Light-Emitting Diodes

H. Morkoç, Beyond Silicon Carbide! III–V Nitride-Based Heterostructures and Devices

Volume 53 Cumulative Subject and Author Index Including Tables of Contents for Volume 1–50

Volume 54 High Pressure in Semiconductor Physics I

W. Paul, High Pressure in Semiconductor Physics: A Historical Overview
N. E. Christensen, Electronic Structure Calculations for Semiconductors under Pressure
R. J. Neimes and M. I. McMahon, Structural Transitions in the Group IV, III-V and II-VI Semiconductors Under Pressure
A. R. Goni and K. Syassen, Optical Properties of Semiconductors Under Pressure
P. Trautman, M. Baj, and J. M. Baranowski, Hydrostatic Pressure and Uniaxial Stress in Investigations of the EL2 Defect in GaAs
M. Li and P. Y. Yu, High-Pressure Study of DX Centers Using Capacitance Techniques
T. Suski, Spatial Correlations of Impurity Charges in Doped Semiconductors
N. Kuroda, Pressure Effects on the Electronic Properties of Diluted Magnetic Semiconductors

Volume 55 High Pressure in Semiconductor Physics II

D. K. Maude and J. C. Portal, Parallel Transport in Low-Dimensional Semiconductor Structures
P. C. Klipstein, Tunneling Under Pressure: High-Pressure Studies of Vertical Transport in Semiconductor Heterostructures
E. Anastassakis and M. Cardona, Phonons, Strains, and Pressure in Semiconductors
F. H. Pollak, Effects of External Uniaxial Stress on the Optical Properties of Semiconductors and Semiconductor Microstructures
A. R. Adams, M. Silver, and J. Allam, Semiconductor Optoelectronic Devices
S. Porowski and I. Grzegory, The Application of High Nitrogen Pressure in the Physics and Technology of III-N Compounds
M. Yousuf, Diamond Anvil Cells in High Pressure Studies of Semiconductors

Volume 56 Germanium Silicon: Physics and Materials

J. C. Bean, Growth Techniques and Procedures
D. E. Savage, F. Liu, V. Zielasek, and M. G. Lagally, Fundamental Crystal Growth Mechanisms
R. Hull, Misfit Strain Accommodation in SiGe Heterostructures
M. J. Shaw and M. Jaros, Fundamental Physics of Strained Layer GeSi: Quo Vadis?
F. Cerdeira, Optical Properties
S. A. Ringel and P. N. Grillot, Electronic Properties and Deep Levels in Germanium-Silicon
J. C. Campbell, Optoelectronics in Silicon and Germanium Silicon
K. Eberl, K. Brunner, and O. G. Schmidt, $Si_{1-y}C_y$ and $Si_{1-x-y}Ge_xC_y$ Alloy Layers

Volume 57 Gallium Nitride (GaN) II

R. J. Molnar, Hydride Vapor Phase Epitaxial Growth of III-V Nitrides
T. D. Moustakas, Growth of III-V Nitrides by Molecular Beam Epitaxy
Z. Liliental-Weber, Defects in Bulk GaN and Homoepitaxial Layers
C. G. Van de Walle and N. M. Johnson, Hydrogen in III-V Nitrides
W. Götz and N. M. Johnson, Characterization of Dopants and Deep Level Defects in Gallium Nitride
B. Gil, Stress Effects on Optical Properties
C. Kisielowski, Strain in GaN Thin Films and Heterostructures
J. A. Miragliotta and D. K. Wickenden, Nonlinear Optical Properties of Gallium Nitride
B. K. Meyer, Magnetic Resonance Investigations on Group III-Nitrides
M. S. Shur and M. Asif Khan, GaN and AlGaN Ultraviolet Detectors
C. H. Qiu, J. I. Pankove, and C. Rossington, III-V Nitride-Based X-ray Detectors

Volume 58 Nonlinear Optics in Semiconductors I

A. Kost, Resonant Optical Nonlinearities in Semiconductors
E. Garmire, Optical Nonlinearities in Semiconductors Enhanced by Carrier Transport
D. S. Chemla, Ultrafast Transient Nonlinear Optical Processes in Semiconductors
M. Sheik-Bahae and E. W. Van Stryland, Optical Nonlinearities in the Transparency Region of Bulk Semiconductors
J. E. Millerd, M. Ziari, and A. Partovi, Photorefractivity in Semiconductors

Volume 59 Nonlinear Optics in Semiconductors II

J. B. Khurgin, Second Order Nonlinearities and Optical Rectification
K. L. Hall, E. R. Thoen, and E. P. Ippen, Nonlinearities in Active Media
E. Hanamura, Optical Responses of Quantum Wires/Dots and Microcavities
U. Keller, Semiconductor Nonlinearities for Solid-State Laser Modelocking and Q-Switching
A. Miller, Transient Grating Studies of Carrier Diffusion and Mobility in Semiconductors

Volume 60 Self-Assembled InGaAs/GaAs Quantum Dots

Mitsuru Sugawara, Theoretical Bases of the Optical Properties of Semiconductor Quantum Nano-Structures
Yoshiaki Nakata, Yoshihiro Sugiyama, and Mitsuru Sugawara, Molecular Beam Epitaxial Growth of Self-Assembled InAs/GaAs Quantum Dots
Kohki Mukai, Mitsuru Sugawara, Mitsuru Egawa, and Nobuyuki Ohtsuka, Metalorganic Vapor Phase Epitaxial Growth of Self-Assembled InGaAs/GaAs Quantum Dots Emitting at 1.3 μm
Kohki Mukai and Mitsuru Sugawara, Optical Characterization of Quantum Dots
Kohki Mukai and Mitsuru Sugawara, The Photon Bottleneck Effect in Quantum Dots
Hajime Shoji, Self-Assembled Quantum Dot Lasers
Hiroshi Ishikawa, Applications of Quantum Dot to Optical Devices
Mitsuru Sugawara, Kohki Mukai, Hiroshi Ishikawa, Koji Otsubo, and Yoshiaki Nakata, The Latest News

Volume 61 Hydrogen in Semiconductors II

Norbert H. Nickel, Introduction to Hydrogen in Semiconductors II
Noble M. Johnson and Chris G. Van de Walle, Isolated Monatomic Hydrogen in Silicon
Yurij V. Gorelkinskii, Electron Paramagnetic Resonance Studies of Hydrogen and Hydrogen-Related Defects in Crystalline Silicon
Norbert H. Nickel, Hydrogen in Polycrystalline Silicon
Wolfhard Beyer, Hydrogen Phenomena in Hydrogenated Amorphous Silicon
Chris G. Van de Walle, Hydrogen Interactions with Polycrystalline and Amorphous Silicon—Theory
Karen M. McNamara Rutledge, Hydrogen in Polycrystalline CVD Diamond
Roger L. Lichti, Dynamics of Muonium Diffusion, Site Changes and Charge-State Transitions
Matthew D. McCluskey and Eugene E. Haller, Hydrogen in III-V and II-VI Semiconductors
S. J. Pearton and J. W. Lee, The Properties of Hydrogen in GaN and Related Alloys
Jörg Neugebauer and Chris G. Van de Walle, Theory of Hydrogen in GaN

Volume 62 Intersubband Transitions in Quantum Wells: Physics and Device Applications I

Manfred Helm, The Basic Physics of Intersubband Transitions
Jerome Faist, Carlo Sirtori, Federico Capasso, Loren N. Pfeiffer, Ken W. West, Deborah L. Sivco, and Alfred Y. Cho, Quantum Interference Effects in Intersubband Transitions
H. C. Liu, Quantum Well Infrared Photodetector Physics and Novel Devices
S. D. Gunapala and S. V. Bandara, Quantum Well Infrared Photodetector (QWIP) Focal Plane Arrays

Volume 63 Chemical Mechanical Polishing in Si Processing

Frank B. Kaufman, Introduction
Thomas Bibby and Karey Holland, Equipment
John P. Bare, Facilitization
Duane S. Boning and Okumu Ouma, Modeling and Simulation
Shin Hwa Li, Bruce Tredinnick, and Mel Hoffman, Consumables I: Slurry
Lee M. Cook, CMP Consumables II: Pad
François Tardif, Post-CMP Clean
Shin Hwa Li, Tara Chhatpar, and Frederic Robert, CMP Metrology
Shin Hwa Li, Visun Bucha, and Kyle Wooldridge, Applications and CMP-Related Process Problems

Volume 64 Electroluminescence I

M. G. Craford, S. A. Stockman, M. J. Peanasky, and F. A. Kish, Visible Light-Emitting Diodes
H. Chui, N. F. Gardner, P. N. Grillot, J. W. Huang, M. R. Krames, and S. A. Maranowski, High-Efficiency AlGaInP Light-Emitting Diodes
R. S. Kern, W. Götz, C. H. Chen, H. Liu, R. M. Fletcher, and C. P. Kuo, High-Brightness Nitride-Based Visible-Light-Emitting Diodes
Yoshiharu Sato, Organic LED System Considerations
V. Bulović, P. E. Burrows, and S. R. Forrest, Molecular Organic Light-Emitting Devices

Volume 65 Electroluminescence II

V. Bulović and S. R. Forrest, Polymeric and Molecular Organic Light Emitting Devices: A Comparison
Regina Mueller-Mach and Gerd O. Mueller, Thin Film Electroluminescence
Markku Leskelä, Wei-Min Li, and Mikko Ritala, Materials in Thin Film Electroluminescent Devices
Kristiaan Neyts, Microcavities for Electroluminescent Devices

Volume 66 Intersubband Transitions in Quantum Wells: Physics and Device Applications II

Jerome Faist, Federico Capasso, Carlo Sirtori, Deborah L. Sivco, and Alfred Y. Cho, Quantum Cascade Lasers
Federico Capasso, Carlo Sirtori, D. L. Sivco, and A. Y. Cho, Nonlinear Optics in Coupled-Quantum-Well Quasi-Molecules
Karl Unterrainer, Photon-Assisted Tunneling in Semiconductor Quantum Structures
P. Haring Bolivar, T. Dekorsy, and H. Kurz, Optically Excited Bloch Oscillations—Fundamentals and Application Perspectives

ISBN 0-12-752176-3